UNRAVELLING STARLIGHT

Challenging traditional accounts of the origins of astrophysics, this book presents the first scholarly biography of nineteenth-century English amateur astronomer William Huggins (1824–1910). A pioneer in adapting the spectroscope to new astronomical purposes, William Huggins rose to scientific prominence in London and transformed professional astronomy to become a principal founder of the new science of astrophysics. The author re-examines his life and career, exploring unpublished notebooks, correspondence, and research projects to expose the boldness of this scientific entrepreneur. While Sir William Huggins is the main focus of the book, the involvement of Lady Margaret Lindsay Huggins (1848–1915) in her husband's research is examined, where it may have been previously overlooked or obscured. Written in an engaging style, this book has broad appeal and will be valuable to scientists, students, and to anyone interested in the history of astronomy.

BARBARA BECKER taught history of science at the University of California, Irvine, until her recent retirement. A leading authority on William Huggins, her research interests also include the role of amateurs in the development of nineteenth-century professional astronomy, and the role of controversy in shaping the substance and structure of scientific knowledge.

Sir William Huggins, c. 1905. (*PRS* **86A**)

UNRAVELLING STARLIGHT

William and Margaret Huggins and the Rise of the New Astronomy

BARBARA J. BECKER

University of California, Irvine

CAMBRIDGE
UNIVERSITY PRESS

University Printing House, Cambridge CB2 8BS, United Kingdom

Cambridge University Press is part of the University of Cambridge.

It furthers the University's mission by disseminating knowledge in the pursuit of education, learning and research at the highest international levels of excellence.

www.cambridge.org
Information on this title: www.cambridge.org/9781316644171

First published 2011
First paperback edition 2016

A catalogue record for this publication is available from the British Library

Library of Congress Cataloguing in Publication data
Becker, Barbara J., 1946–
Unravelling starlight : William and Margaret Huggins and the rise of the new astronomy / Barbara J. Becker.
p. cm.
ISBN 978-1-107-00229-6 (hardback)
1. Huggins, William, Sir, 1824–1910. 2. Huggins, Lady, d. 1915. 3. Astronomers – Great Britain – Biography. 4. Astrophysics – History – 19th century. I. Title.
QB35.B43 2011
520.92–dc22
[B]
2010043090

ISBN 978-1-107-00229-6 Hardback
ISBN 978-1-316-64417-1 Paperback

For Madge

Contents

List of illustrations		*page* xi
Acknowledgements		xiii
List of abbreviations		xvii

1 Introduction — 1
1.1 The retrospective narrative — 3
1.2 Chapter summaries — 5
1.3 A note on the unpublished sources — 8

2 'The astronomer ... must come to the chemist' — 11
2.1 Astronomy in nineteenth-century Britain to 1860 — 12
2.2 The spectroscope — 15
2.3 The puzzle of Fraunhofer's lines — 20
2.4 '... something like Qualitative Analysis!' — 21
2.5 '... the astronomer ... must come to the chemist' — 22

3 The young observer — 28
3.1 Early life and education — 28
3.2 Interest in science — 31
3.3 Interest in astronomy — 32
3.4 Tulse Hill — 33
3.5 An observatory notebook — 34
3.6 Developing a research agenda — 38

4 'A sudden impulse ...' — 46
4.1 The Pharmaceutical Society soirée — 47
4.2 William Allen Miller — 47
4.3 Chemical spectrum analysis — 48
4.4 Collaboration — 50
4.5 'Mr. Huggins ... on the "Stellar Spectrum"' — 52
4.6 Spectra of terrestrial metals — 54
4.7 'On the spectra of some of the fixed stars' — 57

5   The riddle of the nebulae   64
  5.1   Astronomical questions: summer 1864   65
  5.2   Variable nebulae   66
  5.3   The 'interminable wilderness of nebulae'   69
  5.4   'No spectra such as I expected!'   72
  5.5   A paper of 'interest & importance'   75
  5.6   Fellowship   76

6   Moving in the inner circle   82
  6.1   Cultivating advantageous alliances   82
  6.2   Opportunism and eclecticism   84
  6.3   The 'willow leaves' controversy   84
  6.4   The nova in Corona Borealis   86
  6.5   The spectra of variable stars   88
  6.6   A new star   88
  6.7   The red flames   91
  6.8   Fireworks and shooting stars   93
  6.9   Crater Linné   95
  6.10   Thermometric research   96
  6.11   Achieving 'a mark of approval and confidence'   98

7   Stellar motion along the line of sight   104
  7.1   The colours of stars   106
  7.2   26 May 1864   109
  7.3   Stellar motion in the line of sight   111
  7.4   Observations   115
  7.5   Publication   120
  7.6   Response   120

8   A new telescope   126
  8.1   '… discussing the size & plumage of the chicken'   127
  8.2   The strains of diversity   134
  8.3   The 'insufficiency of national observatories'   135
  8.4   The Devonshire Commission   136
  8.5   Dissension in the ranks   141
  8.6   The Lockyer factor   143

9   Solar observations   149
  9.1   The 'Great Indian Eclipse'   149
  9.2   Viewing the red flames without an eclipse   154
  9.3   The eclipse expedition to Oran   156
  9.4   Planning the expedition   158
  9.5   A registering spectroscope   163
  9.6   22 December 1870   164

10   An able assistant   170
  10.1   The solitary observer   171
  10.2   An able assistant   174
  10.3   Margaret Lindsay Murray   176
  10.4   Interest in astronomy   176
  10.5   The 'two star-gazers'   178
  10.6   Celestial photography   179
  10.7   Photography at Tulse Hill   182

11   Photographing the solar corona   192
  11.1   The Egyptian eclipse   192
  11.2   Photographing the corona   193
  11.3   The Caroline Island eclipse   197
  11.4   The Riffel expedition   199
  11.5   The Bakerian lecture   204
  11.6   The Cape Observatory   208

12   A scientific lady   221
  12.1   '... zeal and perseverance ...'   221
  12.2   The Henry Draper Memorial   224
  12.3   The 'meteoritic hypothesis'   226
  12.4   The 'chief nebula line'   229
  12.5   'I have added the name of Mrs. Huggins ...'   232
  12.6   A scientific lady   235

13   Foes and allies   240
  13.1   Controversy   240
  13.2   American allies   242
  13.3   Irreconcilable differences   244
  13.4   President of the BAAS   246
  13.5   George Ellery Hale   247
  13.6   The President's address   249
  13.7   Nova Aurigae   251
  13.8   The Yerkes telescope   255
  13.9   Photographing the corona without an eclipse   256
  13.10 The *Astrophysical Journal*   258
  13.11 The Yerkes Observatory   260

14   The new astronomy   267
  14.1   Helium   268
  14.2   Accolades and achievements   271
  14.3   Radium   275

15   'One true mistress'   291
  15.1   Passing the baton   292
  15.2   The Great Grubb telescope   296

15.3   *Scientific Papers*   298
15.4   'Life is work, and work is life'   300
15.5   '... guardian of my Dearest's reputation'   301
15.6   'I now withdraw ...'   309
15.7   The new Huggins Observatory   312
15.8   Wellesley College   313

16   Conclusion   322

Appendix: 'The new astronomy: A personal retrospective'   328
Bibliography   347
Index   375

Illustrations

Frontispiece. Sir William Huggins, c. 1905.   *page* ii
Figure 2.1   William Huggins, c. 1860.   24
Figure 3.1   Title page, Notebook 1.   35
Figure 3.2   Drawings of Mars by Warren De la Rue.   36
Figure 3.3   Observationes Marti by William Huggins.   37
Figure 3.4   Observations of Jupiter by William Huggins.   40
Figure 3.5   Drawings of Jupiter by William Huggins.   41
Figure 4.1   Observations of Saturn by William Huggins.   50
Figure 4.2   Spectroscope of Kirchhoff and Bunsen.   55
Figure 4.3   Improved spectroscope of Kirchhoff and Bunsen.   56
Figure 4.4   William Huggins's chemical spectroscope.   57
Figure 4.5   Interior of the Tulse Hill observatory in the 1860s.   59
Figure 4.6   Spectra of Betelgeuse (α Orionis) and Aldebaran.   60
Figure 5.1   The dark nebula surrounding η Argus [Carinae].   68
Figure 5.2   Plate X. Figs. 1–5, *PTRSL*, **154** (1864).   74
Figure 6.1   Drawings of sunspots, 23 March 1864, by William Huggins.   85
Figure 6.2   Handheld spectroscope.   93
Figure 6.3a Notebook sketch of thermopile, 24 May 1867, by William Huggins.   97
Figure 6.3b Published drawing of William Huggins's thermopile.   97
Figure 7.1   William Huggins's compound spectroscope.   114
Figure 7.2   Line-of-sight observations, drawings from February 1868 by William
             Huggins.   116
Figure 7.2a 11 February 1868.   116
Figure 7.2b 24 February 1868.   116
Figure 7.3   Line-of-sight observations, drawings from March 1868 by William
             Huggins.   118
Figure 7.3a 6 March 1868.   118
Figure 7.3b 10 March 1868.   118
Figure 7.3c 12 March 1868.   118
Figure 7.4   Line-of-sight observations, drawings from March 1868 by William
             Huggins.   119
Figure 7.4a 18 March 1868.   119
Figure 7.4b 30 March 1868.   119
Figure 7.5   Published line-of-sight diagram.   121

Figure 8.1   New Tulse Hill Observatory.   133
Figure 8.2   Joseph Norman Lockyer (1836–1920).   143
Figure 10.1   Grubb's automatic spectroscope.   172
Figure 10.2   William Huggins at the star-spectroscope, c. 1904.   174
Figure 10.3   Margaret Lindsay Huggins (1848–1915).   175
Figure 10.4   'A Photographic Positive'.   183
Figure 10.5   First notebook entries by Margaret Huggins.   184
Figure 10.6   Drawing of a camera by Margaret Huggins.   185
Figure 11.1   William Huggins's prototype coronagraph.   197
Figure 11.2   Sketches of the solar corona by William Huggins.   200
Figure 11.3   Drawings of the solar corona by William Wesley.   201
Figure 11.4   An improved coronagraph built by Howard Grubb.   209
Figure 12.1a Spectra of nebulae compared with spectra of hydrogen, cool
              magnesium and meteorite glow.   227
Figure 12.1b Comparison of visible magnesium spectrum with that of nebula.   227
Figure 12.2   Drawing of the spectrum of Comet *b* 1881 by Margaret Huggins.   228
Figure 12.3   Observations of a 'brilliant aurora', 4 February 1874, by William
              Huggins.   234
Figure 15.1   Twin-equatorial telescope.   297

Acknowledgements

This book grew out of my dissertation, 'Eclecticism, Opportunism, and the Evolution of a New Research Agenda: William and Margaret Huggins and the Origins of Astrophysics' (The Johns Hopkins University, 1993). I still feel the great debt I owe all who helped bring that work to life, particularly to Robert Smith, Peter Hingley and the late Maire Brück, individuals whose encouragement kept me poking around and scribbling over the years.

It was Robert who, back in 1988, first urged me to pursue my interest in investigating William Huggins's role in the origins of astrophysics. After I completed my graduate work, he continued to serve as mentor and friendly nudge. His scholarship, his breadth and depth of knowledge, his clarity of thought and expression, his incredible memory and attention to detail, his wit, his gentility, and above all his candor and honesty have been an endless source of inspiration for me in my teaching, research and writing.

Peter has taken an avid and active interest in my research from the very beginning. Sitting squarely in the midst of a modern information crossroads, he keeps his attentive and discerning ear to the ground and is generous in keeping scholars abreast of what he hears. Without his enthusiastic intercession on behalf of my manuscript, it would still be filed away out of public view.

I deeply regret being unable to complete this work in time for Maire to see it in print. She was so full of ideas and energy that I foolishly imagined she would always be here. We began our wonderful long-distance conversation in May 1991. The wealth of information she had ready at her fingertips (and floating around on the ends of her cranial neurons) never ceased to amaze me. Her warm generosity in sharing all she knew inspired me to do the same at every opportunity. Whenever I became overwhelmed or disheartened by the task that lay ahead, her letters spurred me on. At each step, she led the way by her own example. It was a joy and a privilege to call her my friend. *Ar dheis Dé go raibh sí.*

Now, over two decades after embarking on my journey, I have a multitude of new debts to declare. Much of my research for the present work was made possible thanks to the kind assistance and dedication of the librarians and staff who both maintain the archival material in their collections and facilitate scholars' access to them. Throughout the years, each of my many questions and requests received a cheerful response from this lively corps of hard-working individuals who know and understand the value of the treasures they keep. In particular, I would like to thank Peter Hingley (Royal Astronomical Society Library); Adam Perkins and Zoe Rees (Cambridge University Library, Manuscripts Room); Rupert Baker,

Joanna Corden and Clara Anderson (Royal Society Library); Dan Lewis (Huntington Library); Gregory Shelton (US Naval Observatory Library); Ruth Rogers and Mariana Oller (Wellesley College Library, Special Collections); Mark Hurn (Cambridge University, Institute of Astronomy Library), April Brewer (University of North Carolina, Wilson Library, Rare Book Collection); and Elizabeth Campbell and Melinda McIntosh (University of Massachusetts, Amherst, W. E. B. DuBois Library).

To complement the archival material, my research required direct access to hard-to-locate nineteenth- and early twentieth-century journals, books and other documents. I had already transcribed and photocopied much of what I needed years ago from volumes on the library shelves at Hopkins and the US Naval Observatory, or obtained through interlibrary loan. But the advent of the internet and digitisation of print material brought fully searchable copies of all these resources – and more – to my fingertips whenever I needed them. I (like many other scholars) am extremely grateful for the efforts of everyone responsible for the SAO/NASA Astrophysics Data System (ADS), a Digital Library portal operated by the Smithsonian Astrophysical Observatory (SAO) under a NASA grant and hosted by the High Energy Astrophysics Division of the Harvard-Smithsonian Center for Astrophysics. I also wish to acknowledge my debt to the libraries and other institutions around the globe that have contributed to the growing online availability of books and journals under the aegis of groups like Project Gutenberg, Google Books, Internet Archive and Gallica.

No scholar or her work can thrive in isolation. I am indebted to Helena Pycior, Michael Hoskin, Bernard Lightman, Karl Hufbauer, Michael Crowe, Matt Dowd, David DeVorkin, Peggy Kidwell, Joe Tenn, Wayne Orchiston, Falk Rieß, Tom Williams, the Royal Astronomical Society, the Astronomical Society of the Pacific, the California Academy of Sciences, the Historical Astronomy Division of the American Astronomical Society, the biennial History of Astronomy Workshops and the History of Science Society for the opportunities they have given me to share, discuss and refine my work over the years.

I extend heartfelt thanks to the many people – some friends, others complete strangers – who thoughtfully responded to my queries when my own expertise was inadequate to process the information at hand. John McMahon translated Latin phrases on the title page of William Huggins's first observatory notebook. The page's Hebrew inscription was translated and transliterated by Dan Schroeter. Alan Dehmer's mastery of the art and craft of nineteenth-century photographic techniques and methods enabled me to decipher and interpret barely legible handwritten notes on the subject with greater confidence. Jürgen Teichmann clarified the complex chronology of Joseph von Fraunhofer's employment and responsibilities at the Optical Institute in Benediktbeuern and generously provided me with copies of essential but otherwise inaccessible reference material. Robin Mason and Marilyn Morgan shared valuable information on the casting of Mt Wilson's 100-in glass disc. Art Champagne helped me look at the Hugginses' spectroscopic investigation of the radium glow with modern eyes. David Dewhirst linked me in a very personal way to William Huggins and his Great Grubb telescope through his firsthand account of using the instrument more than a half century ago at Cambridge's Institute of Astronomy.

Space does not permit me to enumerate my many obligations to Jim Bennett, Allan Chapman, Ian Elliott, Owen Gingerich, Ian Glass, the late Owen Hannaway, Peter Harman and Simon Schaffer. Suffice it to say that, without their knowledgeable guidance, kind assistance and encouragement I would still be wandering in Dante's 'dark wood'. Special thanks to Richard G. French for his personal interest in my work over the years. He is currently overseeing a major renovation of Wellesley College's Whitin Observatory which promises to modernise its instructional operations while restoring its historical beauty. Particularly exciting is the planned restoration and reinstallation of three stained glass windows given to the Observatory by Margaret Huggins in 1914. I am deeply grateful to Dick for his kind permission to use an image of the panel 'Sun with Chromosphere, a Nebula, & other Figures' on the cover of this book. I remember the moment I first saw this extraordinary window. It required quite an effort to remove it from the drawer in the observatory where it had been stored for many years. I had seen a black-and-white photograph of the window, but was totally unprepared for the powerful impact of its vivid colour and exquisite design. It took my breath away.

Most of the illustrations in this book were scanned by me from the volumes in which they originally appeared. One of these figures deserves special mention (see Figure 4.6). It is a reproduction of the frontispiece of William Huggins's booklet *On the Results of Spectrum Analysis Applied to the Heavenly Bodies* (1866) from the library of the late Gerald James Whitrow, FRAS (1912–2000). I was honoured to receive this small book as a gift in 2001 from Professor Whitrow's widow, Magda, through Peter Hingley's thoughtful agency. All of the book's illustrations are original photographs pasted neatly in place. They have faded over the years, of course. Readers wishing to see the details of the spectral map depicted in the frontispiece will find it reproduced more clearly elsewhere (see, for example, Huggins and Miller, *PTRSL* **154**, Plate XI; Huggins and Huggins, *Scientific Papers*, opposite p 60) I wished to include a copy of this historic image in my own book both as a tribute to Professor Whitrow memory and to express my sincere gratitude to all responsible for the gift.

Other important historic images are included in this book thanks to the efforts and generosity of the following individuals and institutions. A portrait of William Huggins near the beginning of his long career (see Figure 2.1) appears in this volume with the kind permission of the Royal Astronomical Society and Photo Researchers, Inc. Mariana Oller scanned pages from the Tulse Hill Observatory notebooks, portions of which are reproduced in this volume with the kind permission of Wellesley College Library's Special Collections (see Figures 3.1, 3.3, 3.4, 4.1, 6.1, 6.3a, 7.2, 7.3, 7.4, 10.5, 10.6 and 12.3). Gregory Shelton successfully scanned the delicate – and difficult to capture – image of the nebula surrounding η Argus [Carinae] drawn by John Herschel and reproduced on Plate IX of his *Results of Astronomical Observations made ... at the Cape of Good Hope* (1847). The image is reproduced here (see Figure 5.1) courtesy of the US Naval Observatory Library. Imaging Services of the Cambridge University Library scanned an illustrated page from a letter written by William Huggins to George Gabriel Stokes (Add MS 7656.H1168, 2 January 1882). It is reproduced here (see Figure 11.2) by kind permission of the Syndics of Cambridge University Library. Ian Glass generously sent me an historic photograph which shows the coronagraph built by Howard Grubb and used by David Gill at the Cape Observatory to take routine photographs of the solar corona without an eclipse using William Huggins's prescribed method. It is reproduced here (see Figure 11.4) courtesy of the South African Astronomical Observatory.

I wish to thank my colleagues and students in the History Department of the University of California, Irvine, for their support and encouragement for my research. It was truly a privilege to be part of such an intellectually stimulating community of scholars. I am particularly grateful to Marc Kanda who kept my access to campus resources active after my retirement. Special thanks to Ingrid Wilkerson and Kim Weiss for sharing their infectious curiosity. Such restless itches (I hope) will never be satisfied.

Finally, a few, albeit inadequate, words of appreciation to my friends and family. Despite her distance, Gillian Wu continues to be an example, an anchor and a cheering fount of unbridled enthusiasm. Louis Carlat's youthful blend of humour and sagacity has kept my thoughts focused and my nose to the grindstone. Thanks to my brother, Pat, for his genealogical research assistance and for helping me locate copies of relevant letters from the Henry Draper papers at the New York Public Library. My daughter, Misha, all the while juggling the demands of motherhood and career, eased my transition to the academic world in Chapel Hill by toting books on arcane subjects from and to the university's various libraries for my use. My son, Aaron, now a father himself, interrupted his busy schedule to venture gamely into the bowels of the microfilm room of UMass, Amherst's DuBois Library to copy pages I had missed in the Hale papers. Their delightful spouses, meanwhile, have come to accept (I think) a pixilated mother-in-law whose head appears to be hopelessly trapped in the nineteenth century. The newest stars in my constellation – granddaughters Olivia, Josephine and Lyra – never cease to energise me and light my path. My best friend and loving husband and I have shared more than four decades together, half of which have been spent co-habiting with the Hugginses! In all that time, Hank has never begrudged their presence. Instead, he has provided me with untiring support for all that I set out to do. Whenever computer glitches halted my progress, he dropped whatever he was doing and calmly came to the rescue. He eagerly stepped into his new role of grandfather to allow me precious time to complete my manuscript. When chapters emerged, his cogent comments and suggestions helped me to see what I had written afresh. Now, as I prepare to resume life in the twenty-first century (already in progress), the doors and windows he has opened make it possible for me to hear the sweet Carolina air resound with strains of bluegrass and children's laughter.

Chapel Hill
North Carolina
12 May 2010

Abbreviations

Journals

AA:	*Astronomy and Astrophysics*
AC:	*Annales de Chimie*
ACP:	*Annales de Chimie et de Physique*
ADB:	*Australian Dictionary of Biography*
AJ:	*Astronomical Journal*
AJP:	*American Journal of Physics*
AJS:	*American Journal of Science*
AJSA:	*American Journal of Science and the Arts*
AN.	*Astronomische Nachrichten*
AP:	*Annalen der Physik*
APC:	*Annalen der Physik und Chemie*
APJ:	*Astrophysical Journal*
AR:	*Astronomical Register*
BJHS:	*British Journal for the History of Science*
BJP:	*British Journal of Photography*
CM:	*Cornhill Magazine*
CR:	*Comptes Rendus Hebdomadaires des Séances de l'Académie des Sciences*
DNB:	*Dictionary of National Biography*
DSB:	*Dictionary of Scientific Biography*
EJS:	*Edinburgh Journal of Science*
EM:	*English Mechanic*
EPJ:	*Edinburgh Philosophical Journal*
ER:	*Edinburgh Review*
IAJ:	*Irish Astronomical Journal*
IO:	*Intellectual Observer*
JAD:	*Journal of Astronomical Data*
JAHH:	*Journal of Astronomical History and Heritage*
JBAA:	*Journal of the British Astronomical Association*
JHA:	*Journal for the History of Astronomy*
JRASC:	*Journal of the Royal Astronomical Society of Canada*

MLPSM:	*Memoirs of the Literary and Philosophical Society of Manchester*
MNRAS:	*Monthly Notices of the Royal Astronomical Society*
NRRSL:	*Notes and Records of the Royal Society of London*
OBS:	*Observatory*
PA:	*Popular Astronomy*
PANSP:	*Proceedings of the Academy of Natural Sciences of Philadelphia*
PASA:	*Proceedings of the Astronomical Society of Australia*
PASP:	*Publications of the Astronomical Society of the Pacific*
PJ:	*Pharmaceutical Journal*
PM:	*Philosophical Magazine*
PN:	*Photographic News*
PRI:	*Proceedings of the Royal Institution*
PRS:	*Proceedings of the Royal Society*
PTRSL:	*Philosophical Transactions of the Royal Society of London*
QJRAS:	*Quarterly Journal of the Royal Astronomical Society*
QJS:	*Quarterly Journal of Science*
RBAAS:	*Report of the British Association for the Advancement of Science*
SM:	*Sidereal Messenger*
SP:	*Science Progress*
SS:	*Social Studies of Science*
ST:	*Sky and Telescope*
TC:	*Technology and Culture*
TOS:	*Transactions of the Optical Society*
TRMSL:	*Transactions of the Royal Microscopical Society of London*
TRSE:	*Transactions of the Royal Society of Edinburgh*

Archival sources

CIT:	California Institute of Technology
	GEH: George Ellery Hale papers
CUL:	Cambridge University Library
	ERP: Ernest Rutherford papers
	GGS: George Gabriel Stokes papers
	JCM: James Clerk Maxwell papers
	RGO: Royal Greenwich Observatory
DCL:	Dartmouth College Library
	CAY: Charles Augustus Young papers
HLA:	Huntington Library Archives
HUA:	Harvard University Archives
	ECP: Edward C. Pickering papers
LO:	Lick Observatory
	MLSA: Mary Lea Shane Archives
	ESH: Edward Singleton Holden papers

NYPL: New York Public Library
     HD: Henry Draper papers
RASL: Royal Astronomical Society Library
RSL: Royal Society Library
     AS: Arthur Schuster papers
     JFWH: John F. W. Herschel papers
     JL: Joseph Larmor papers
SI: Smithsonian Institution
     MAH: Museum of American History
SAAOA: South African Astronomical Observatory Archives
     DG: David Gill papers
UEL: University of Exeter Library
     JNL: Joseph Norman Lockyer papers
WCL/SC: Wellesley College Library, Special Collections
YUL: Yale University Library
     DPT: David Peck Todd papers

Organisations/institutions

BAAS: British Association for the Advancement of Science
IUCSR: International Union for Co-operation in Solar Research
JEC: Joint Eclipse Committee
MIT: Massachusetts Institute of Technology
NAS: National Academy of Science
RAS: Royal Astronomical Society
RS: Royal Society

1

Introduction

where in the murky depths of mind
do idea-seeds sleep?
what seasons do they know
that make them wont to wake and grow
in spasms –
madly now, dormant then –
entangling all within their ken:
synaptic vines
to clamber
to the very stars.

My interest in the life and work of English astronomer William Huggins (1824–1910) began over twenty years ago in a graduate seminar on conceptual transfer within and among specialised scientific communities. The exchange of cultural baggage is a subtle dynamic that practitioners, particularly those working in long-established disciplines, usually take care to shield from public view. Our little group spent the semester analysing the far more transparent machinery of newer, still-developing hybrid disciplines like geophysics, biochemistry and astrobiology.

The topic meshed well with my own research interests at the time. I wanted to learn more about how the boundaries of scientific disciplines are established, policed and altered: What are the rules members must follow in investigating the natural world? What questions are deemed appropriate to ask? What do good answers to such questions look like and how can they be recognised? What constitutes an acceptable way of finding those answers? Who is allowed to participate in the search? Who says?

How better to find answers to these questions than to watch a scientific discipline during a period of change? I chose to explore the origins of astrophysics, a mature hybrid science that blends the methods, instruments and theories of chemistry and physics, as well as those of both mathematical and descriptive astronomy. The investigation offered me an opportunity to analyse the dynamics of cross-disciplinary borrowing, conceptual transfer and boundary realignment. It also shed light on the social, cultural and intellectual factors which catalyse and nurture the development of a new scientific discipline.

I was guided in my investigation by the fruitful model for scientific change proposed by Polish physician, Ludwik Fleck (1896–1961) in the 1930s.[1] Later models including those of

Thomas Kuhn[2] and Michel Foucault[3] also describe the complexities of scientific change, but they do so within the context of the systemic inertia that protects scientific endeavour from developing too eclectically. Kuhn's 'paradigm' and Foucault's 'episteme' represent the agreement that frames the linguistic and cognitive work space scientists share with their fellows. Change, in their models, occurs when agreement breaks down. 'Normal science' can only resume once the amplitude of disagreement within the community has been reduced to invisibility.

Rather than focus on scientific communities' drive to achieve conformity and consensus, Fleck drew attention to the creative force of the diversity within them. He recognised that every scientist is, more often than not, a member of more than one group of like-minded fellows who share a particular view, skill or interest. Fleck called these groups 'thought collectives' [*Denkkollectiven*]. Each thought collective is composed of both a small, specialised esoteric core surrounded by a larger, more peripheral, exoteric circle.[4] An individual may be situated within the esoteric core of one thought collective because of expertise or leadership qualities, and yet be accepted as (or simply consider himself or herself to be) a fringe, or exoteric, member of another thought collective of which other individuals form the core. For Fleck, one's place within this dynamic work space is a complex, and richly ambiguous, amalgam of self-perception and group selection that gives rise to what he called the 'intercollective communication of ideas'.[5]

Herein lay the utility of Fleck's model for my studies of change in the boundaries of scientific disciplines. Individuals with differing backgrounds and viewpoints, but who feel they speak a common language, co-exist within areas of overlapping thought collectives. They are exposed routinely to opportunities to apply their specialised theories and methods to the solution of a broad range of problems. Exoteric members of a thought collective thus play important roles in the process of concept change and transfer because they are in a position to act not so much as couriers between two different thought collectives, but as bilingual translators.

Astrophysics is built on a range of questions and methods that were unimaginable to individuals in the first half of the nineteenth century. At that time, positive knowledge of physical and chemical structure of celestial bodies was presumed to be unattainable by proper scientific methods, and hence relegated to the no-man's-land of mere speculation. One might entertain any number of untestable ideas about the origins of stars, or the reasons for their differences in colour, brightness and distribution, but such was not the stuff of science. What *was* considered positively knowable was the location of a celestial body on the sky. Gathering and interpreting this information defined the mission of the practising astronomer and determined the structure of his creative thought and work space. Astrophysics' emergence and efflorescence in the second half of the nineteenth century required the wholesale restructuring of the boundaries surrounding the theory and practice of astronomy. I wanted to know what made it possible to include such an unorthodox line of investigation within the traditional astronomical community. What prompted a segment of that community to move into this new thought space? If science is (to its credit) an inertial system, how do changes like this come about?

My preliminary seminar investigations formed the basis of my dissertation research.[6] I started by poking around in the theoretical and methodological stew of Britain's astronomical community during the last half of the nineteenth century using William Huggins as an historical probe. Huggins struck me as a logical choice to begin what I imagined would be a straightforward investigation into a well-documented episode in the history of science. After all, he was celebrated in his own lifetime as a self-taught pioneer who played a key role in introducing spectrum analysis into astronomical work. He began his career on the periphery of scientific London where he had ample opportunity to interact with astronomers, chemists, physicists, mathematicians and instrument makers of all stripes. He authored numerous articles documenting his ground-breaking use of the spectroscope to analyse the light of celestial bodies. Most notable among them was a retrospective essay, 'The new astronomy', in which he laid out in painstaking detail each step of his discovery-laden career.

To date, historians and scientists who have discussed William Huggins's contributions to astronomical spectroscopy have drawn on this considerable body of evidence.[7] I turned to the unpublished record. Indeed it is from unpublished documents that I have drawn the stuff and substance of the present work. Personal correspondence and observatory notebooks yielded what seemed at first to be an odd assortment of details and anecdotes. You may well wonder, *Gentle Reader*, why historians bother with all these cumbersome and potentially distracting episodes? I hope to convince you that lives are mosaics fashioned out of happenstance and numerous incremental day-to-day decisions rife with clutter and confusion, dead ends and mistakes. Private accounts expose the rough edges of decision making. More important, by putting human flesh on the skeletal prose of the public accounts, they show that even scientists are complex people with annoying personalities, with uncertainties and fears, with charm and pluck and wit.

What follows is less a definitive biography than an expeditionary report based on two decades of digging up and collecting the buried shards of Huggins's long and productive life.[8] I have carefully and, I hope, faithfully reassembled them to enable you to see, as I have, the hand, the instrument, the laboratory; to hear the voice, the scraping of the observatory dome, the zapping of the Geissler tube; to feel the frustration, the joy, the fear; to smell the battery's noxious fumes, the burning magnesium, the sweet night air; to taste the sweat, the ink, the long-forgotten cup of tea.

I also hope you may be motivated to join me in looking for the missing pieces.

1.1 The retrospective narrative

Permit me to illustrate the importance of the unpublished record in documenting the origins of astrophysics with the following story. In February 1893, Huggins received a query from Herbert Hall Turner (1861–1930), Chief Assistant to the Astronomer Royal at Greenwich. Turner was composing an obituary for the Royal Astronomical Society's (RAS) *Monthly Notices*. The subject was Lewis Morris Rutherfurd (1816–92), an American amateur astronomer, photographer and instrument designer, who had attracted the attention of colleagues on both sides of the Atlantic back in 1863 with news of his pioneering programme of stellar spectrum analysis. But ill-health and an aversion to publication had kept Rutherfurd's name out of sight and out of the collective mind of Britain's scientific community for many years.

Turner, a mere infant when Rutherfurd had made his mark, hoped that Huggins, who had twice met the man, would be able to provide some 'local colour' drawn from stellar spectroscopy's early days.[9] Unfortunately, the aging astronomer had no anecdotes to share. But he was happy to oblige Turner's request for clarification on the important, yet somewhat confused, matter of priority in the early efforts to map stellar spectra: who, exactly, among the first mappers of star spectra had done what, and when?

Huggins began with Joseph von Fraunhofer's (1787–1826) first observations of the spectra of a few bright stars in 1823[10] and concluded with the resurrection of research interest in this subject four decades later by a handful of astronomers including himself, Rutherfurd, and Italian astronomer Giovanni Battista Donati (1826–73).[11] He attributed the near-simultaneous burst of activity to Gustav Kirchhoff's (1824–87) 'discovery of the true meaning of the Fraunhofer lines' announced in 1859.[12] But, as for assigning priority for stellar spectroscopy, he argued that the concept had little meaning because he and these few others took their initial steps *independently and unknown to each other* [original emphasis]' – all were 'first' in a sense. In fact – he took pains to note – Rutherfurd's 1863 paper on the subject had completely surprised (and probably annoyed) him, arriving as it did on the very evening he was preparing to publicise the results of his own research undertaken with chemist William Allen Miller (1817–70).

Huggins had been the last of these three men to subject starlight to spectroscopic study. In 1893, he could afford to regard priority with an air of gracious disinterest. As the trio's sole surviving member, he now found the question of who had taken this or that first little step so long ago to be less critical in terms of the long view of history than who had done the most over the intervening years to carry the field forward. He dismissed Donati's early measures as unreliable, and, in a postscript, reminded Turner that Rutherfurd had abandoned this line of investigation soon after conceiving it, thus forfeiting any claim to be considered a 'founder' of stellar spectroscopy. That title, in Huggins's view, clearly belonged to himself. It had been bestowed upon him by Agnes Mary Clerke (1842–1907) in her *Popular History of Astronomy*, and he was not going to cede it willingly.[13]

His comments betray the intensity of his anxiety about the public perception of his contributions to astronomical physics – an anxiety that became more central to his thinking as he grew older. Throughout his lengthy career, his discoveries drew colleagues' acclaim and imitation as well as their criticism and reinterpretation, a consequence of working on the cutting edge of a new scientific specialty. Early on, he learned the hard and, for him, discomfiting lesson that making a discovery guarantees the discoverer neither lifelong credit for it, nor control over its perceived place in the evolving history of the field. To retain control requires vigilance and continual reinsertion of one's own version of events into the public record.

In his published essay on Rutherfurd's life, Turner included a near-verbatim, yet unattributed, recitation of Huggins's personal recollection of events related to spectroscopy's introduction into astronomical investigation.[14] It formed the seed of what eventually became the traditional story of the origins of astrophysics. Huggins diligently nurtured this precious seed over the next few years. It germinated in the text of a stirring essay he wrote titled 'The new astronomy: A personal retrospect' (see Appendix). When the popular magazine *Nineteenth Century* published the piece in 1897, the now-hardy sprout took root in fertile soil.

'The new astronomy' exemplifies what Warren Hagstrom has called the 'scientist's account': an uncomplicated story of a scientific community's origins that both legitimates the work of its past and present researchers and socialises its new recruits.[15] Constructed with tautological clarity, these synthetic narratives reduce the muddle of theoretical and experimental options that really confronted the community's early researchers to a set of clear-cut alternatives.

To take narratives like the 'The new astronomy' at face value is to fall into an alluring trap. Like deftly embroidered curtains, they delight their intended audience while keeping the embarrassing clutter of regrettable missteps and frustrating reversals out of sight, out of mind and ultimately out of the collective's memory. By masking, even deleting from the record, the complexities and uncertainties that mark the first forays into a new realm of scientific investigation, they diminish the role played by calculated risk, negotiation and persuasion in establishing it as a valid branch of scientific inquiry.

1.2 Chapter summaries

In the chapters that follow I will lift the obscuring curtain of 'The new astronomy' to reveal Huggins as less of a single-minded, focused and exhaustive researcher, than a scientific entrepreneur who possessed considerable skill at selecting research projects, designing and manipulating instruments for specific mensurational tasks, and rallying influential colleagues' support for his investigative ventures. The task is threefold: first, to present a new interpretation of the events that marked the development of William Huggins's career based on an in-depth examination of his unpublished notebooks and correspondence; second, to bring to light the research options he perceived were available to him and analyse the actions he ultimately took in the context of mid-nineteenth-century British amateur astronomy; and finally, to present a new account of the synergy of his career and the rise of astronomical physics in light of his private and public accounts.

To lay some essential groundwork, Chapter 2 begins with a look at the community of astronomers in the first half of the nineteenth century.[16] What puzzles piqued their interest and challenged their ingenuity? How did they define the boundaries of acceptable research? What theories, methods and instruments did they consider to be legitimate means for attacking these problems? How did the spectroscope, a laboratory instrument used by physicists and chemists to analyse light, work its way into the core of the astronomical enterprise?

Chapter 3 reviews significant events in the life of William Huggins up to the time he was elected to fellowship in the Royal Astronomical Society in 1854 and constructed his home observatory. What avocational choices did he perceive were available to him as a young and maturing adult? When and why did he become interested in astronomy? How did he acquire and develop sufficient technical and methodological expertise to make the transition from novice to serious amateur? Who influenced his choice of methods, instruments and observational agenda?

The fourth chapter focuses on the reception of Kirchhoff's radiation law in England. How might a novice amateur astronomer like Huggins have become aware of the German physicist's work? How do his later recollections compare with contemporary records of events? Can these accounts be reconciled to restore elided details of his plan to apply Kirchhoff's method to the stars?

In 1864, Huggins shifted his attention from scrutinising the spectra of stars to examining those of nebulae. Chapter 5 explores this bold move, which ultimately propelled him to a position of prestige and authority among fellow astronomers. At the time, there were still questions that remained unanswered concerning the physical nature of these celestial objects. For many astronomers, the 'riddle of the nebulae' was simply irresolvable. How did Huggins convince them to change their minds?

Chapter 6 examines Huggins's observational programme in the period immediately following the announcement of his spectroscopic observations of nebulae.[17] Although his reminiscences later in life give the impression that he spent all, or much, of his time on the spectroscopic study of stellar and nebular objects, ambition and curiosity led him to explore a number of different subjects – the light of a nova, the heat of celestial bodies, the glow of fireworks and meteors – in innovative and often technically challenging ways. How did he maximise his opportunities for new discoveries and avoid being identified as a speculative or impulsive dilettante?

Chapter 7 looks at Huggins's most influential contribution to astronomy: his application in 1868 of the spectroscope to detect – and Doppler's principle to measure – stellar motion in the line of sight. Undertaken entirely by visual means, it was an audacious effort fraught with overwhelming mensurational and interpretive difficulties. How did Huggins overcome these challenges and ultimately persuade his contemporaries that he had, in fact, accomplished what he claimed? What impact did being able to measure the radial velocity of a star have on the research interests of positional astronomers and their attitudes toward celestial spectroscopy?

Chapter 8 focuses on Huggins's career in the wake of his line-of-sight research. He developed a reputation for care in making observations and caution in suggesting explanations for the phenomena he observed. By cultivating important alliances to great personal advantage, he was awarded custodianship of a state-of-the-art telescope paid for with funds appropriated by the Royal Society. How did taking responsibility for the instrument affect Huggins's research agenda? How did he handle the logistics of carrying out a productive observing programme, and fend off criticism from disgruntled colleagues?

Chapter 9 joins William Huggins on an expedition he led to Oran, Algeria, in 1870 to analyse the light of the solar corona during a total eclipse. It is the only eclipse expedition in which he played a role, yet mention of it is conspicuously missing from his retrospective account. What motivated him and his colleagues to embark on such a difficult journey? In organising the expedition, what challenges did he face in terms of leadership and resourcefulness? How did the expedition affect his future research efforts?

Accounts of Huggins's work always mention that he was assisted in his research by his wife, Margaret Lindsay (née Murray) Huggins (1848–1915). Chapter 10 presents evidence from her lengthy and detailed notebook entries to demonstrate that she was more than an able assistant, amanuensis and illustrator whose work conformed to her husband's research interests. What do these unpublished sources reveal about the effect of Margaret's presence and expertise on the research agenda at the Tulse Hill observatory? Why do published accounts trivialise and even obscure the extent of her contributions to the astronomical investigations she performed with her husband despite the fact that she is often the principal source of information about that work?

In 1882, over a decade after his expedition to Oran, Huggins launched an effort to photograph the solar corona without an eclipse. The years-long project is the subject of Chapter 11.[18] His initial perception of success in this project led him to pursue it for many years with great zeal and conviction. Correspondence and notebook entries show that it tested the strength of his persuasive power and encouraged him to build an international network of confirmatory witnesses. The difficulties he faced in achieving his goal, rather than stifling his research efforts, motivated him to improve his research methods and instrumentation. By what means did he hope to convince others of the validity of his coronal photographs? How does a scientific community achieve consensus on what counts as conclusive evidence?

Chapter 12 investigates the circumstances leading up to Margaret Huggins's public debut as her husband's collaborative assistant. Beginning in 1887, the Hugginses photographed and analysed the spectrum of the Orion nebula hoping to identify the physical and chemical cause of the bright lines William had described in his first observations of nebular spectra more than two decades earlier. Their findings embroiled them in a controversy with Joseph Norman Lockyer (1836–1920), founder and editor of *Nature*. What was Margaret's role in this challenging research effort? Why, after fifteen years of collaborative work, did the Hugginses choose to publish their first joint paper on this controversial subject?

Chapter 13 discusses the steps taken by the Royal Society to control the intensifying dispute between Huggins and Lockyer and restore civility within its ranks. What strategies did Huggins employ to maintain his position on the cutting edge of astronomical research? How did his cultivation of alliances with prominent American astronomers influence his professional standing with astronomers both at home and abroad?

Even as he approached eighty years of age, Huggins continued to search for new and innovative ways to apply the spectroscope's analytical power. Chapter 14 brings to light one of his lesser-known research efforts: the spectroscopic study of radium's natural glow. Why have these investigations been forgotten? He also became increasingly nostalgic and wary of encroachment upon his past accomplishments. He penned his stirring retrospective essay, 'The new astronomy' (1897), and, with Margaret's invaluable assistance, he edited a collection of essays on stellar classification titled *The Atlas of Representative Star Spectra* (1899). What role have these works played in the construction of William Huggins's historical legacy?

Chapter 15 begins with a look at the final years of Huggins's life and the steps he took to pass on responsibility for the 'new' astronomy's future development to the next generation of researchers. After relinquishing the telescope that had been on loan to him from the RS for nearly four decades, William worked with Margaret to edit a collection of his scientific papers (1909), which they arranged topically and introduced with excerpts from

'The new astronomy'. How did Margaret establish herself as architect and vigilant guardian of her husband's public image after his death in 1910?

1.3 A note on the unpublished sources

Tracking down Huggins's correspondence has been, and continues to be, a worthy challenge. The archival record is extensive, but it is far from complete. Unpublished documents that would shed new light upon Huggins's innovative work in ultraviolet spectroscopy, for example, have yet to be uncovered. In addition, I have found no collection of his received correspondence, save copies made by his correspondents for their own records, and the handful of letters pasted into Huggins's observatory notebooks. They may have been lost or destroyed after his death. I am sure there are more materials yet to be discovered.

There is no one repository for his outgoing letters. Principal collections can be found in the Scientific Manuscripts Collection, Manuscripts and Archives of the Cambridge University Library (CUL), the Library of the Royal Society of London (RSL), the Library of the Royal Astronomical Society of London (RASL), the Huntington Library in San Marino, California (HL), and the Mary Lea Shane Archives (MLSA) of the Lick Observatory (LO) in Santa Cruz, California.

The papers of George Gabriel Stokes (1819–1903), long-time physical secretary of the Royal Society, are held in the CUL Manuscripts Room. Stokes's papers contain over three decades of letters from William Huggins as well as correspondence from Lockyer, Miller, Warren De la Rue (1815–89), Thomas Romney Robinson (1792–1882), Henry Enfield Roscoe (1833–1915), Alexander Strange (1818–76) and others.

The CUL is also home to the massive collection of documents from the Royal Greenwich Observatory (RGO), including, it would seem, every scrap of paper on which a spot of ink or stray pencil mark was deposited during the lengthy tenure (1835–81) of Astronomer Royal George Biddell Airy (1801–92). Airy's papers are a rich source of information about his personal and administrative concerns as he and his Greenwich staff confronted the need to include spectroscopic measures in their daily routine.

Among the archived RGO records are those from the Royal Observatory at the Cape of Good Hope. Of particular relevance to the story of William Huggins are the numerous letters he exchanged with Her Majesty's Astronomer at the Cape, David Gill (1843–1914), as he guided Gill in implementing his method of photographing the solar corona without an eclipse.[19]

In addition to the correspondence of William Huggins, the RSL holds significant collections of correspondence of such scientific notables as Lockyer, Miller, William Crookes (1832–1919), John Frederick William Herschel (1792–1871), Joseph Larmor (1857–1942) and Arthur Schuster (1851–1934), to name a few. Of special interest are the numerous letters from Margaret Huggins to Larmor, particularly those written after her husband's death. The RSL has also preserved valuable referee reports written in review of colleagues' papers. These documents offer a glimpse of the critical reception of the scientific elite to new ideas being presented in papers before the Royal Society.

The RASL contains many letters written by Huggins, chiefly on matters of Society business. Of particular interest are his written communiqués to the gifted illustrator

William Henry Wesley (1841–1922), who served as the Society's assistant secretary from 1874 until his death. Huggins presented the RAS with some of his own drawings made during observations throughout his career. These items deserve more intensive examination. The library also holds letters from his RAS colleagues including De la Rue, Lockyer and American astronomer George Ellery Hale (1868–1938).

What has survived of Huggins's direct correspondence with Lockyer is held at the University of Exeter (UEL). These letters provide insight into the evolution of the working relationship between these two men. The Library of King's College, London, holds a number of interesting documents and letters related to Miller's tenure there as Professor of Chemistry. Letters Huggins wrote to mathematician Alfred Bray Kempe (1849–1922), dealing principally with administration of the Royal Society and Kempe's tenure as the Society's treasurer, can be found in the collection of the West Sussex Record Office, Chichester. Several letters from Huggins to physician Henry Wentworth Acland (1815–1900) are held in Oxford's Bodleian Library. Letters from Huggins to Lawrence Parsons, 4th Earl of Rosse (1840–1908), can be found in the Rosse papers at Birr Castle in Ireland.

In the United States, the MLSA of the Lick Observatory hold a number of letters of interest to the history of early astrophysics. A well-indexed collection, it contains letters Huggins wrote to Edward Singleton Holden (1846–1914) and William Wallace Campbell (1862–1938), as well as many exchanged among prominent astronomers of the day including Holden, Campbell, Hale, Hugh Frank Newall (1857–1944), Lockyer, James Edward Keeler (1857–1900) and Henry Draper (1837–81). The David Peck Todd (1855–1939) papers at Yale University Library (YUL) and the Charles Augustus Young (1834–1908) papers at Dartmouth College (DCL) contain important letters from both William and Margaret Huggins. Most of George Ellery Hale's papers are held at the California Institute of Technology (CIT) in Pasadena, California. Others remain in the collection of the Huntington Library (HL) in San Marino, California. Most of the papers at the CIT have been placed on microfilm and can be examined at other libraries (the Museum of American History of the Smithsonian Institution in Washington, DC, for example, and the University of Massachusetts, Amherst). The Hale papers include Hale's correspondence with Huggins and Lockyer, as well as other international figures who played a role in the early development of astrophysics.

The Library of Congress Manuscript Collection holds Simon Newcomb's (1835–1909) papers and the correspondence of Thomas Jefferson Jackson See (1866–1962). Worthy of special note in the Newcomb papers is the diary of Mrs Newcomb describing her experiences on the solar eclipse expedition to Gibraltar in December 1870. The Harvard University Library (HUL) holds important correspondence between Huggins and Harvard Observatory Director, Edward Charles Pickering (1846–1919). In addition, the New York City Public Library (NYPL) is the repository for the Henry Draper papers which contain important letters from Huggins, John William Draper (1811–82), Holden, William Lassell (1799–1880) and Richard Anthony Proctor (1837–88), among others.

In 1914, Huggins's widow, Margaret, gave six observatory notebooks to Wellesley College near Boston, Massachusetts, along with a wide range of small astronomical instruments and other items from the Tulse Hill observatory, as well as many personal items and objets d'arts that she and her husband had accumulated during their travels. Currently held in Wellesley College's Special Collections (WCL/SC), the notebooks span forty-five years of his observing career from 1856 to 1901.[20] Because Margaret assumed the task of recording the couple's observations in the notebooks following their marriage in 1875, examination of the notebook accounts from that time forward brings into vivid relief, for the first time, the full extent of her collaborative role in the work done at Tulse Hill.

Notes

1. Fleck, *Genesis and Development of a Scientific Fact* (1979).
2. Kuhn, *The Structure of Scientific Revolutions* (1970).
3. Foucault, *The Order of Things* (1970).
4. Fleck, *Genesis and Development of a Scientific Fact* (1979), p. 104.
5. *Ibid.*, pp. 109–110.
6. B. J. Becker, 'Ecclecticism, opportunism, and the evolution of a new research agenda' (1993).
7. See, for example, Maunder, *Sir William Huggins and Spectroscopic Astronomy* (1913); Dyson, 'Sir William Huggins' (1910); Newall, 'Sir William Huggins' (1910–11) and 'William Huggins' (1911); W. W. Campbell, 'Sir William Huggins' (1910); H. S. Williams, *The Great Astronomers* (1932); Meadows, 'The origins of astrophysics' (1984); Glass, *Revolutionaries of the Cosmos* (2006), ch. 5.
8. For earlier reports, see B. J. Becker, 'Margaret and William Huggins at work in the Tulse Hill Observatory' (1996); 'Priority, persuasion, and the virtue of perseverance' (2000); 'Visionary memories' (2001); 'Celestial spectroscopy' (2003); and 'From dilettante to serious amateur' (2010).
9. The substance of Turner's request, whether oral or written, must be inferred from Huggins's reply. See W. Huggins to Turner, 15 February 1893, Correspondence of the Society, RASL.
10. Fraunhofer had actually observed stellar spectra before 1815: 'With the same set up, I made observations of a few first magnitude fixed stars.' Fraunhofer, 'Bestimmung des Brechungs- und Farbenzerstreuungs-Vermögens …' (1817), p. 220. Huggins's mention of 1823 must refer to the paper's appearance in English translation in the *Edinburgh Philosophical Journal*.
11. For whatever reason, he made no mention of James Carpenter, whose comparisons of stellar spectra were performed at Greenwich under the direction of the Astronomer Royal, George Airy.
12. Kirchhoff, 'Über die Fraunhofer'schen Linien' (1860).
13. Clerke named Vatican astronomer Father Angelo Secchi as another founder of this new science. Secchi began his spectroscopic investigations shortly after Huggins, so was not among the very first stellar spectroscopists, thus technically justifying Huggins's failure to name him. Clerke, *A Popular History of Astronomy* (1885), p. 421. Unless otherwise noted, all citations from *A Popular History of Astronomy* are from this first edition.
14. Turner, 'Lewis Morris Rutherfurd' (1893).
15. Hagstrom, *The Scientific Community* (1965), pp. 211–15.
16. Earlier versions of portions of Chapters 1–3 appeared in B. J. Becker, 'Visionary memories', *JHA* **32** (2001); 'Celestial spectroscopy' (2003); and 'La spettroscopia e la nascita dell'astrofisica' (2003).
17. An earlier version of portions of this chapter appeared in B. J. Becker, 'From dilettante to serious amateur' (2010).
18. An earlier version of portions of this chapter appeared in B. J. Becker, 'Priority, persuasion, and the virtue of perseverance' (2000).
19. The South African Astronomical Observatory also holds a number of letters Huggins sent to David Gill during his tenure as Astronomer Royal at the Cape.
20. S. F. Whiting, in 'Diaries' (1917). For a summary of the contents of the notebooks, see Morgan, 'Huggins archives' (1980).

2

'The astronomer ... must come to the chemist'

This is something like Qualitative Analysis!

— Henry Roscoe[1]

In her 1885 history of nineteenth-century astronomy, Agnes Mary Clerke enumerated the discoveries that marked the recent progress in that science. She drew her readers' attention to the founding of what she called 'astronomical or cosmical physics', a new species of astronomy that was markedly different in goals as well as methods from its older mathematical cousin. 'It is full of the audacities, the inconsistencies, the imperfections, the possibilities of youth', she wrote. 'It promises everything; it has already performed much; it will doubtless perform much more.'[2] Clerke was not alone in her enthusiasm.

Britons who played a role in the development of this hybrid discipline, either as active contributors or sideline boosters, hailed it as a happy consequence of the growing inter-dependency among a heterogeneous subset of scientists and instrument makers. For dec-ades – whether probing the physical properties of light, chemically analysing terrestrial materials or perfecting the production of optical glass – these practitioners had pursued their disparate research agendas using a common, and historically rich, line of investigation: the careful scrutiny of dispersed natural and artificial light. Appropriating elements from the methodological legacy of such notable forebears as Isaac Newton and William Herschel, they examined new types of light sources, tinkered with alternative apparatus arrangements and tested the efficacy of a range of viewing and recording aids. It was a powerful and productive process. As puzzles and puzzle solvers became entwined in their symbiotic give-and-take, increasingly sophisticated methods of spectrum analysis emerged in the first half of the nineteenth century that tamed the individual components of the mechanical menagerie with which investigators worked. The result was a whole greater than the sum of its parts – the spectroscope – a versatile and adaptable device that integrated a prism to disperse light from a targeted source with a modified goniometer consisting of a sighting telescope to view details in the resulting spectral display and a scale to mark and measure small angular displacements.[3]

In the 1860s, a few individuals became attracted by the spectroscope's potential as a tool for analysing the light of celestial bodies. By coupling it with the astronomical telescope, they not only appropriated an instrument of chemistry and physics, they adapted the attendant methods and theories of these disciplines to a new purpose: to uncover the stuff and structure of the stars, a line of inquiry long considered beyond the scope of legitimate science. Their efforts signaled a growing intimacy and necessary interdependency among chemists, physicists and astronomers. Within a relatively short time span – less than two decades – a 'new' astronomy emerged that transformed existing social and professional networks among astronomers, revamped their work space and ultimately realigned the familiar boundaries of acceptable research in the field. By the turn of the twentieth century, celestial spectroscopy boasted its own set of questions, its own instruments and its own journals, as well as its own standards for assessing technical expertise and interpretational competence.[4]

The emergence of astronomical physics in Britain was not a product of the social, economic or intellectual influences that some historians and sociologists of science have suggested play a role in infusing new research agendas into existing disciplines.[5] It did not grow out of epistemological or methodological concerns facing mid-century celestial mechanicians who set their goals and measured their accomplishments against the problems prescribed by the great observatories at Greenwich, Paris, Berlin and Pulkovo.[6] These workers, dedicated to the arduous task of mapping the stars with the precision required to track the motion of the planets, had little time to speculate on the physical and chemical structure of their targets. By the same token, the new astronomy did not arise out of a Malthusian struggle among professional astronomers for a limited supply of new research problems, resources, priority or prestige. Nor did it develop in response to growing dissatisfaction in the ranks with the prevailing professional research agenda.

Instead, astronomical physics got its start thanks to a scant handful of individuals operating outside the boundaries of professional astronomy.[7] To bring these intrepid souls into relief, we shall begin with a description of England's astronomical community in the first half of the nineteenth century. Who were its members? What puzzles did they deem worthy and permissible challenges to their ingenuity? What theories, methods and instruments did they consider legitimate means for attacking these problems? By what standards were observational claims judged to be reliable and confirmable descriptions of Nature? How was authority in such matters assigned?

2.1 Astronomy in nineteenth-century Britain to 1860

In mid-nineteenth-century England practising astronomers were a diverse lot. For a few, it was how they earned their living. But a growing number received no salary for their astronomical contributions. Working for love not money, these individuals were *amateurs* in the true sense of the word.[8] They were merchants, artisans, lawyers, clerics, physicians and landed gentry whose interests, abilities, resources and personal ambition led them to pursue astronomy with varying amounts of rigour and seriousness ranging from dabbler to full-time observer. Largely untrained in mathematics and generally less interested in the quantitative mechanics of the celestial realm than in its more qualitative aspects, amateur astronomers occupied a wide range of investigative niches. Among them were natural historians and collectors of celestial curiosities who captured their observations of sunspots, variable stars and the expanding family of solar satellites in carefully kept notebook records. Some were also dilettante chemists, photographers, microscopists and electricians: avocations that enabled them to see and interpret the astronomical phenomena they observed with compound eyes and brains. The tinkerers among them used celestial bodies as extreme test cases for the sensitivity or accuracy of their instruments. Most sifted patiently and tirelessly through the heavenly haystack hoping to be the first to discover just one of the proverbial needles that lay hidden there. Representatives of all these types could be found among the ranks of the Royal Astronomical Society (RAS).[9]

The RAS had its beginnings in 1820 when a small but dedicated group of astronomers with the broadly defined purpose of encouraging and promoting astronomy founded the Astronomical Society of London.[10] In 1827, one RAS founder, then-President John Frederick William Herschel, described the astronomer's mission as a relentless quest to fix each star in its proper celestial location for the benefit of astronomy, geography, navigation and surveying. Son of renowned astronomer William Herschel (1738–1822), John Herschel was already respected in Britain and abroad for his surveys of double stars as well as continuing his father's projects of star-gaging and cataloguing of nebulae. The eyes and instruments of many working together would make the task that astronomers faced less onerous, he claimed. The Astronomical Society would facilitate the coordination of such efforts.[11]

German positional astronomer *par excellence*, Friedrich Wilhelm Bessel (1784–1846), shared Herschel's sentiments. In 1832, speaking on the state of current astronomy, Bessel invoked the name of Isaac Newton, who, 'by giving his immortal *Principia* to the world, brought unity into Astronomy and banished all arbitrariness from it'. Now that the kindred nature of terrestrial and celestial motion had been revealed, Bessel declared:

the job of Astronomy was clear: to supply the instructions by which Earth-bound observers can compute the movements of the heavenly bodies. Everything else that one might learn about these bodies – the appearance and constitution of their surfaces, for example – may be worthy of attention, but it is of no real concern to Astronomy.[12]

In 1833, John Narrien (1782–1860), a celestial mechanician and Professor of Mathematics at the Royal Military College, London, published an acclaimed account of the history of astronomy from ancient times through the work of Newton.[13] He both prefaced and concluded his work with commentary on the astronomy of his own day. There were, he claimed, two types of astronomical research: practical astronomy devoted to calendrical improvement, and physical astronomy motivated by a desire to further elucidate Newton's mechanical laws.[14] In addition to refining measures of the aberration of starlight, searching for evidence of stellar parallax, and improving achromatic telescopes and chronometers, Narrien emphasised the need for astronomers to continue work on the three-body problem, precession and planetary perturbations.[15] It was, he believed, a waste of time and energy to construct telescopes larger than those currently in use because larger instruments are subject to increased 'derangements'.[16] In terms of astronomical investigation, he concluded that 'human ingenuity will, probably, in future, be able to accomplish little more than an improvement in the means of making observations, or in the analysis by which the rules of computation are investigated'.[17]

In June of that year, polymath William Whewell (1794–1866) inaugurated the third annual meeting of the British Association for the Advancement of Science (BAAS) with a stirring address. Along with the customary survey of 'the recent progress, the present condition, the most pressing requirements of the principal branches of science at the present moment', he offered the assembly some 'general reflexions' that he hoped would 'preside over and influence the aims and exertions of many of us, both during our present discussions and in our future attempts to further the ends of science'.

Astronomy [he asserted] is not only the queen of sciences, but, in a stricter sense of the term, the only perfect science; – the only branch of human knowledge in which particulars are completely subjugated to generals, effects to causes; ... an example of a science in that elevated state of flourishing maturity, in which all that remains is to determine with the extreme of accuracy the consequences of its rules ...[18]

These sentiments were echoed by French positivist philosopher Auguste Comte (1798–1857). Because earthbound astronomers are limited in the direct sensory information they can gather about the stars and other celestial bodies, Comte argued, certain knowledge about the heavens must be based only on what can be seen. All else is untestable conjecture.

We can imagine the possibility of determining the shapes of stars, their distances, their sizes, and their movements; whereas there is no means by which we will ever be able to examine their chemical composition, their mineralogical structure, or especially, the nature of organisms that live on their surfaces ... Our positive knowledge with respect to the stars is necessarily limited to their observed geometrical and mechanical behaviour.[19]

But, in fact, epistemological restrictions like those expressed by Bessel and Comte did not constrain astronomers' investigative imagination. A rigorous programme of positional astronomy could and did reveal new and important knowledge of the heavens. The successful determinations of solar and stellar parallax, the discovery of Neptune and the confirmation of the existence of an unseen companion to the star Sirius – to name a few of the more familiar discoveries of this period – encouraged more intensive and extensive studies of the stars and their mutual interactions in order to measure their masses, distances, intrinsic brightnesses and, perhaps, even determine with certainty the how and the why of their distribution in space.

Browsing the pages of the RAS's *Monthly Notices* and *Memoirs* from the 1850s, we find that astronomers pursued a variety of observational projects. In addition to the usual tasks of timing celestial events and mapping celestial objects, they searched for new asteroids and planetary satellites, located and observed double stars, noted changes on planetary surfaces, counted and mapped sunspots, and documented coincident solar and terrestrial phenomena.

In 1852, Robert Grant (1814–92) described mid-nineteenth-century astronomers as energetic seekers after new knowledge of the heavens in his 'well-conceived, well-executed, and greatly-wanted' *History of Physical Astronomy*.[20] The recent success of Irish nobleman William Parsons, 3rd Earl of Rosse (1800–67), in constructing, operating and making new discoveries with reflecting telescopes of unparalleled size left Grant more optimistic than Narrien had been twenty years earlier concerning the potential for new knowledge to be gained from the use of such instruments.[21] Grant devoted his book's final chapter to an inventory of challenges facing contemporary astronomers, including the need to substantiate claims of having observed changes in the structure and appearance of various celestial bodies, to determine with greater certainty the mass of the newly discovered planet Neptune, to measure stellar parallax, to determine the direction and magnitude of the solar system's motion in space, to uncover the nature of nebulae, and to acquire sufficient information concerning the distribution of stars in space to allow some understanding of the structure of large-scale stellar systems.[22]

In 1860, the annual 'Report of the Council' portrayed the RAS, particularly its amateur contingent, as a vibrant and enthusiastic collection of individuals excited about the numerous reports of the recent solar eclipse, about the use of photography to capture stellar as well as lunar and solar images, and about the discovery of several new minor planets. The Society's sense of purpose had taken on a protean quality that Fellows found especially appealing. In 1862, rather than invoke the value of a concerted effort by all astronomers to effect the completion of one grand project, President John Lee (1783–1866) praised the division of labour among those drawn to study the heavens and encouraged the pursuit of complementary research agendas based on individual skills and inclinations.[23]

January 1863 marked the appearance of a new periodical, the *Astronomical Register*, launched by Sandford Gorton (c. 1823–79), then a relative newcomer to the RAS.[24] By publishing 'stray fragments of information' gleaned at Society meetings and the substance of 'passing conversations' deemed too insignificant to appear in the *Monthly Notices*, Gorton envisioned the *Register* as a 'medium of communication for amateurs and others'.[25] To that end, he announced that each number would contain a summary of the latest RAS meeting, as well as excerpted transcriptions of discussions arising from the presentation of papers for the benefit of those unable to attend. In February of that year, lawyer and mathematician Arthur Cayley (1821–95), who had recently become editor of the *Monthly Notices*, introduced a new section – 'The Progress of Astronomy' – to the Council's annual report.[26]

2.2 The spectroscope

The mingled yarn of mid-nineteenth-century astronomers' diverse research interests created the weft and texture of the tapestry that became astrophysics, but spectrum analysis would be its warp. Its first tenuous threads had already been strung by the time that Irish-born gentleman experimentalist Robert Boyle (1627–91) declared the prism – a novelty of visual wonder and amusement – to be the 'usefullest Instrument' for gaining insight into the fleeting array of colours generated when sunlight passes through it.[27] When, two hundred years later, two German scientists – chemist Robert Bunsen (1811–99) and physicist Gustav Kirchhoff – made the prism a key component in their sun- and lamplight analysing '*Apparat*', the loom was dressed and ready for the weavers.[28]

Britons were quick to recognise the analytical and mensurational power of Bunsen's and Kirchhoff's *Apparat* and to stake their claim on its ancestry. On 6 September 1861, chemist William Allen Miller presented a discourse on spectrum analysis before a well-attended general session at the BAAS's Manchester meeting. 'It must not ... be supposed that the subject is a new one', the veteran spectrum analyst told his audience before tracing 'the principal steps of discovery, from the time of Newton, who first examined the solar spectrum, to the present day'.[29]

Signs on English scientific roads often point back to Newton, but he was hardly the first to take up this line of investigation. Miller can be forgiven for failing to mention the optical studies of English virtuoso Thomas Harriot (1560–1621). Evidence of his fruitful experiments on refraction lay hidden in archives for centuries after his death. But, in quiet collaboration with a few close companions, Harriot employed spheres and 'polished triangle[s] of glass', some of them hollow and filled with transparent fluid, to conduct his successful search for an algorithm that would describe and prescribe light's behaviour when it crosses the boundary separating transparent materials of differing optical density.[30] Thanks to the work of modern scholars, we now know that he was likely the first to discover the so-called 'law of sines', which he used to investigate light's dispersive behaviour.[31]

There were other significant and more widely publicised precedents to Newton's prismatic studies which Miller did not mention. In his 1637 treatise *Les Météores*, for example, René Descartes (1596–1650) – ever the simplifier – pronounced a flat-sided 'prism or triangular crystal' to be equivalent in efficacy to the water-filled glass-sphere he had been using to investigate the reflective and refractive behaviour of rainbow-producing sunlight.[32] A series of methodical experiments had convinced him that a transparent material did not need a curved surface to generate a spectrum.[33] In fact, he concluded, curved boundaries only introduce unnecessary complications to the investigative process.[34]

Over a quarter century later, armed with such a prism, Robert Boyle projected a rainbow-like spectrum onto a piece of white paper and scrutinised the colourful array with a microscope. Although he saw no more detail than he could observe with his naked eye, his experiments on colour, coupled with his enticing speculation that 'perhaps' the prism would prove helpful in resolving questions about colour's nature and cause, encouraged others to employ it in their own optical studies.[35]

Indeed, Isaac Newton (1642–1727) was one such investigator. Motivated in part by enchantment with the spectral appearance of Boyle's 'emphatical' Iris, in part by conviction that Descartes was surely wrong to assert these colours arise in consequence of light's turbulent encounter with the surface of a transparent medium, and in part by a practical desire to rid telescope images of their distorting extraneous colours, young Newton began a series of prism experiments in the 1660s to identify the cause of, and a cure for, the 'celebrated *Phaenomena* of *Colours*' in refracting optical instruments.[36]

Miller's view of the long journey launched by Newton's famed prism experiments was informed by the fact that he had walked part of it himself. He knew all too well that the route had been neither clear nor direct. Nevertheless, the good professor did not burden his audience with a mind-numbing recitation of reality's twists, turns and dead ends. Instead, he constructed a pedagogically useful, albeit illusory, linear progression of pivotal methodological developments each of which, he could argue in retrospect, had made the latest breakthroughs in spectrum analysis possible. After appearing in print, Miller's history of events became an enduring template for future renditions.

2.2.1 William Hyde Wollaston

The first milestone on Miller's list was established at the start of the nineteenth century by William Hyde Wollaston (1766–1828), a retired physician turned entrepreneurial chemist. In 1802 Wollaston announced that he had devised a clever new way to determine a material's index of refraction.[37] He also presented the results from his recent study of variations in the dispersive power of different materials. In the course of these latter investigations, he introduced two modifications to his experimental design. First, rather than rely on the usual circular hole for an aperture, he took an old piece of advice from Newton and admitted the light he wished to examine through an elongated narrow 'crevice'.[38] Second, instead of projecting the spectrum onto a screen, he let the analysed light fall directly on his eye.[39] Observing the solar spectrum in this way, he was surprised to find several 'distinct dark lines' separating the otherwise continuous colourful band into four regions: red, yellow-green, blue and violet.[40]

It is a tribute to Wollaston's wide-ranging curiosity that he elected to scrutinise rather than ignore this unexpected apparition. His sideline venture proved a wonder-filled divertissement. Observing the light of a flame, he saw '5 images, at a distance from each other. The 1st is broad red, terminated by a bright line of yellow; the 2d and 3d are both green; the 4th and 5th are blue.' The spectral display produced by electric light, meanwhile, appeared oddly similar to, yet 'somewhat different' from, both dispersed sunlight and flame. Perplexed, Wollaston was forced to concede, 'It is ... needless to describe minutely, appearances which vary according to the brilliancy of the light, and which I cannot undertake to explain.'

If he wondered privately what had occasioned his serendipitous discovery, he did not hazard a public guess.[41] Intriguing as the lines may have been, they were nevertheless of secondary importance to the real matter at hand. Wollaston remained focused on measuring the refractive power of materials including various types of natural crystals, glass and oils; creating metallic solutions of just the right concentration to cancel the dispersive action of plate, crown and flint glass; and ranking materials of known refractive index according to their dispersive power to illustrate the independent nature of these two optical properties.

For Miller, on the other hand, the lines Wollaston had seen were hardly sideline phenomena. Of course, he was viewing the whole of Wollaston's work through the corrective lens of advances in spectrum analysis over the last half century. Indeed, his own firsthand experience had taught him the investigative value of the dark lines in the solar spectrum and the bright images in flame and electric spectra. He pointed to Wollaston's use of a narrow slit as their cause. No one 'till Wollaston's time', he told his audience (forgetting the example of Newton for the moment), had thought to make use of such an aperture. In his mind, the innovation marked a crucial step forward in the development of spectrum analysis.

Experimentalists who trod more closely on Wollaston's heels had also found the dark lines useful, but more as visual guides to the boundaries separating general regions of colour in the spectrum. David Brewster (1781–1868), for example, used them as fiducial marks for gauging the effectiveness of different filtering materials in his search for a monochromatic light source.[42] But in Brewster's view the lines were not what gave Wollaston's work on refraction and dispersion its lasting significance. Opticians trying to construct achromatic lens systems needed to stop relying on *ad hoc* assessments of individual pieces of glass. It was time to work toward building a scientific understanding of dispersion, its causes and its relationship, if any, to other, better understood physical properties of transparent materials.[43]

For Brewster, Wollaston's research efforts established a solid, albeit qualitative, foundation for his own systematic and quantitative study of materials' optical properties.[44]

2.2.2 Joseph von Fraunhofer

The next of Miller's milestones, and the one upon which all subsequent developments were contingent, was set about a decade after Wollaston by a savvy young Bavarian optician and glassmaker, Joseph von Fraunhofer, who devised a reliable and efficient method of measuring with precision the refrangibility of light, that is, the degree to which individual rays are bent after passing through a prism.[45] Fraunhofer worked at the glassmaking arm of the Munich-based Mathematical-Mechanical Institute (*Mathematisch-mechanische Institut*), a manufactory of achromatic sighting and surveying instruments. He had been brought into the firm as a journeyman lens maker in May 1806 after impressing its directors with his interest in and grasp of textbook optical theory. He advanced quickly by virtue of his uncommon ambition, ingenuity and skill. He wrote a scholarly paper on improving reflecting telescopes. And he developed an innovative method of perfecting a lens's figure that exploited the phenomenon of Newton's rings to expose the smallest of irregularities in the glass.[46]

In February 1809, Fraunhofer became a partner in the Institute and was placed in charge of its optical works at Benediktbeuern. The Institute's premier glassmaker at the time was a talented Swiss artisan named Pierre Louis Guinand (1748–1824).[47] Guinand had developed his own secret method of making homogeneous, unblemished optical glass, including a substitute for English flint glass, the highly prized and scarce commodity required for assembling achromatic lens systems. In August 1809, he began instructing Fraunhofer in the craft.

When Fraunhofer was named director of optical glass production in September 1811, he effectively assumed responsibility for the entire Institute. Although he worked with Guinand over the next two years to improve the quality and increase the quantity of glass they made, Fraunhofer eschewed Guinand's cut-and-try approach to problem solving. Instead, he developed a new, iterative product improvement process that married innovative tinkering with the humdrum routine of methodical testing. Putting the ideals of textbook optical theory into practice in this way left little, if anything, to chance inspiration.

To limit the variables in his investigations, Fraunhofer first looked for a source of monochromatic light. Filtering sunlight through coloured liquids and glass yielded unsatisfactory results. He met with more success examining the dispersed light emitted by coloured flames. Fraunhofer observed the detail in these spectra through the sighting telescope of a surveyor's repeating theodolite.[48] As one of the Institute's principal products, the theodolite was a natural instrument to repurpose in order to conduct his optical experiments. Fraunhofer was disappointed at first to find that each coloured flame produced a spectrum composed of a broad range of colours rather than the monochromatic signature he needed. But his attention was drawn to a bright and sharply defined orange streak that appeared to be common to all flame spectra.

Some time in late 1813, building on experiments he had been developing over the years, he directed narrowly constricted beams of light from six carefully arranged sodium lamps through a prism. The result was an array of six differently coloured spots of light that he was able to observe from a distance of some 225 metres. Again, instead of trying to observe the spots directly with his unaided eye, Fraunhofer looked at them with the telescope of a theodolite. Using the theodolite's alidade, he diligently measured and remeasured the apparent position of each spot with precision. Throughout this challenging process, he relied on the familiar bright orange streak in the lamps' light to calibrate the delicate instrumental arrangement.

When he searched for the bright orange streak in the solar spectrum, he found instead a closely spaced pair of dark lines. Indeed, he saw that the solar spectrum was interrupted by 'innumerable' [*unzählige*] dark lines of varying widths and intensities.[49] Fraunhofer mapped the relative positions of 574 of these lines, labeling selected reference lines with a simple alphabetic system.

In some respects, Fraunhofer's discoveries resemble Wollaston's. Both noted the presence of bright and dark lines in the spectra of light from similar target sources that had passed through narrow slits of nearly identical dimensions. The question naturally arises as to what extent the work of these two investigators was related. In fact, Wollaston's *Philosophical Transactions* paper on refraction and dispersion had been translated into German and republished in two parts in the *Annalen der Physik* about the time Fraunhofer began receiving glassmaking instruction from Guinand in late 1809.[50] The first instalment of Wollaston's translated article appeared in a special number of the *Annalen* devoted entirely to the subject of the behaviour and nature of light.[51]

It is easy for us today to imagine Fraunhofer reading Wollaston's paper with great enthusiasm. But he was first and foremost an optician, not a physicist. At the time he discovered the dark lines in the spectrum, he was living and working a day's journey from Munich, the closest centre of scientific activity. Thus limited in terms of both interest and access to specialised journals like the *Annalen*, it is likely he was unaware of Wollaston's paper.[52] No volumes of the journal were found in his study.[53] During a visit from John Herschel in September 1824, Fraunhofer denied knowing about Wollaston's work.[54]

Miller underscored the importance of Fraunhofer's work to the development of spectrum analysis by drawing attention to the optician's study of celestial light sources other than the Sun. Fraunhofer worried that the 'inflection' (diffraction) of light from an extended source as it passed by the edge of the aperture could influence, perhaps even cause, the appearance of the spectral lines. For this reason, he sought a bright and steady point source of light that he could examine without the need for an intervening aperture. He turned to Venus and several bright stars, not out of any astronomical interest in these bodies, but to have the opportunity to examine diffraction-free spectra. He found it impossible to discern the spectral characteristics of these dispersed point sources of light. Fraunhofer added a cylindrical lens to the optical system to elongate the point of light along an axis perpendicular to the prism's dispersive action. He noted that the spectrum of Venus contained lines identical to those in sunlight, but that stellar spectra exhibited greater variation. In fact, Fraunhofer conjectured that each star's spectrum was unique.[55]

Shortly after French physicist Augustin Jean Fresnel (1788–1827) publicised his pioneering work on diffraction, Fraunhofer undertook quantitative studies of diffraction's dispersive behaviour.[56] He found that the spread of the resulting spectrum was inversely related to the width of the slit used to produce it. His discovery led him to design and make several diffraction gratings, first by winding wire neatly and tautly around a ruled frame, and later by etching parallel lines on a glass plate covered with gold foil to create hundreds of narrow, evenly-spaced slits.[57] Finally, guided by the mathematics of optical interference and armed with diamond-ruled gratings boasting thousands of lines to the inch, Fraunhofer calculated individual wavelengths of spectral lines.[58] Though the normalised distribution of colours in diffraction spectra gave them a clear advantage over those produced by refraction, it would be several decades before diffraction gratings became widely incorporated in spectroscope design and they remained rarities until the late 1860s.[59]

2.3 The puzzle of Fraunhofer's lines

Fraunhofer's lines intrigued practitioners and theorists alike for over forty years after the publication of his spectral maps. It was a wonderful puzzle that encouraged, rather than stifled, their speculation on the mechanism behind the dark lines' appearance. Although they assumed some kind of absorption process was at work, none among them could identify with certainty the absorbing agent or explain the why and how of its narrow selectivity.[60]

In Britain, Brewster and Herschel were joined by other experimentalists like Miller, William Henry Fox Talbot (1800–77) and Charles Wheatstone (1802–75). All were convinced that optical phenomena and physical chemistry are intimately related. Their modes of attack and interpretive frameworks were wide-ranging, unrelated and generally inconclusive.[61] Working independently from the 1820s through the 1840s, they hunted down monochromatic terrestrial light sources, and examined the solar spectrum transmitted through coloured glass and through vapours. They investigated the flame, arc and spark spectra of hundreds of substances. Their aim: to convert the muddle of their empirical results into something resembling a coherent explanation for the appearance and probable cause of Fraunhofer's lines.

Despite their diligence and care, investigators remained stymied by the ubiquity of the prominent D lines, the complexity of spectral signatures and the puzzling observation that some light sources produce both bright and dark lines.[62] The critical insight that helped reduce the confusion came in 1857 as a result of a diversionary inquiry by Scottish professor William Swan (1818–94). Hoping to reveal the mechanism by which light other than that generated by the Sun was produced, Swan observed the spectra of hydrocarbons placed in the colourless flame of a new laboratory burner devised by Robert Bunsen and his student, Henry Enfield Roscoe. Swan became intrigued when he observed a colourful flicker in the outer portion of the lamp's flame whenever the smallest dust speck passed through it. Wondering just how small an amount of a substance would be required to create a colourful spectral display, he dissolved increasingly smaller amounts of common table salt in large quantities of water.[63] To his amazement, he found that less than one-millionth of a grain of salt was capable of colouring the flame with a bright yellow light. Clearly, spectrum analysis was far more sensitive than anyone had imagined. Swan warned spectrum analysts to ensure their test samples contain no foreign material before drawing any conclusions about their chemical constitution.[64]

Taking Swan's caution into account, Bunsen and his collaborator Gustav Kirchhoff studied the flame and spark spectra of light generated by highly purified samples of various salts. They believed their observations confirmed what physical scientists had long suspected, namely, that an individual metal produces its own characteristic pattern of bright spectral lines when it is burned. Conversely, they noted that when white limelight is passed through a cool, vaporous cloud of this same element, the ordinarily continuous rainbow of the white light is broken by dark lines arranged in precisely the same pattern characteristic of the element's flame spectrum. Following up on these experiments, Kirchhoff made additional observations of the solar spectrum formed after sunlight had passed through a flame containing table salt. He concluded that the dark Fraunhofer lines in the Sun's spectrum 'exist in consequence of the presence, in the incandescent atmosphere of the Sun, of those substances which in the spectrum of a flame produce bright lines at the same place'.[65]

2.4 '... something like Qualitative Analysis!'

On 20 October 1859, Kirchhoff presented a paper to the Berlin Academy on the Fraunhofer lines in which he offered both a chemical interpretation and a hint of a physical explanation to account for them. It appeared in English translation six months later.[66] News of his claim spread quickly throughout the scientific world via professional journals, personal correspondence and word-of-mouth. In fact, by October 1860 one would have been hard-pressed to identify a physicist, chemist or optical instrument maker who had not heard something about Kirchhoff's discovery. Bunsen's former student Henry Roscoe, now Professor of Chemistry at Owens College in Manchester, was probably one of the first in England to become aware of it. He quickly recognised the implications of this research. In February 1860 he wrote to George Stokes: 'Have you seen in the last no. of the Annales de Chemie et de Physique a short note about Kirchoff's [sic] discovery of the probable cause of the coincidence of the bands of light ... and dark lines of the spectrum?'[67] Three weeks later, he wrote Stokes again:

I hear from Bunsen that he has detected Lithium in all the Potashes he has examined ... He mixed Mg, Ba, Sr, Ca, Li, Na, K salts together, put some of the mixture on the point of a pin – looked through a telescope & saw at one glance all the substances present! This is something like Qualitative Analysis![68]

But not all of Roscoe's countrymen shared his excitement. As we have seen, it was not as if spectrum analysis was an untried investigative method in Britain. Brewster, Talbot, Wheatstone and Miller were still alive and active in 1860.[69] They saw in Kirchhoff's report something that looked very much like the end they each had had in mind when they designed and performed their own experiments. It may well have been that Kirchhoff received no benefit from their earlier work, but these investigators were understandably disgruntled that Kirchhoff had omitted the usual litany of polite acknowledgement in his paper.[70]

No one disputed that Kirchhoff had observed the behaviour he described, but questions arose concerning his ability to claim he had determined a physical cause for the mysterious dark lines.[71] Some wondered if the lines were caused by absorption of sunlight by Earth's atmosphere. Others complained that insufficient study had been made of the spectra of known terrestrial elements to draw any sensible conclusions from an examination of solar absorption lines.[72]

Despite these reservations, the absorptive and emissive behaviour Kirchhoff had observed in the spectra of luminous gases seemed to many individuals so neat, so law-like, that, in lieu of a theoretical explanation, they were willing to accept empirical evidence strongly suggestive of a physical connection between the spectra of metals and Fraunhofer's lines. Bunsen's discovery of two new elements, caesium and rubidium, in 1860, shortly after beginning work on flame spectra,[73] and the spectroscopic discovery of thallium in 1861 by William Crookes, significantly heightened interest among chemists in the method.[74]

When Roscoe addressed the Chemical Society in June 1861 on ways to extend the utility of spectrum analysis in the laboratory, his lecture generated a spirited discussion.[75] John Herschel asked Roscoe

about the absorption of the sodium ray by the sodium flame. It seems to me that by the explanation given in which it is compared to the action of a sounding-board, that instead of the vibrations being absorbed when they came to it, they ought, on the contrary, to be heightened in intensity, it has always been difficult for me to understand the usual explanation.

Herschel's query was promptly seconded by experimental philosopher Michael Faraday (1791–1867): 'It has always seemed to me a very difficult thing to see how of necessity there was a depression of the light instead of an increase and exaltation of it.'[76] Roscoe agreed, 'Mr. Herschel has hit the nail on the head. It is a very difficult point to answer, and I really do not feel myself to be able to give a satisfactory explanation of it.' From the published record of the ensuing discussion, we are drawn to conclude that none of the distinguished scientists present, including Roscoe, had a firm grasp of Kirchhoff's arguments about the physical process at work, despite the fact that Kirchhoff had, only two months earlier, communicated to the *Philosophical Magazine* a new and more care-fully thought out discussion of what he believed to be the principle behind the dark lines.[77]

2.5 '... the astronomer ... must come to the chemist'

Nevertheless, Roscoe's crusading fervour on behalf of Kirchhoff's method and theory moved discussion moderator, Warren De la Rue, to hail the potential it portended for astronomy: 'The physicist and the chemist', he cheered, 'have brought before us a means of analysis that ... if we were to go to the sun, and to bring away some portions of it and analyze them in our laboratories, we could not examine them more accurately.'[78] Envisioning this chemico-physico-astronomy as an amalgam of specialties united by their method of investigation, he declared:

It is not an uncommon thing for the physicist to tread upon the ground which a chemist thinks belongs to him, and for the chemist to tread upon the ground of the physicist. Now, we have the chemist occupying the ground of the astronomer, and if the astronomer wants to know something of the constituents of the heavenly bodies, he must come to the chemist.

The hue and cry in popular lectures and journal articles about the validity of Kirchhoff's explanation for the Fraunhofer lines, coupled with Roscoe's enthusiastic proselytising, kept it before the public eye long enough to be assimilated into a wider investigative context. In 1860, Florentine comet discoverer Giovanni Battista Donati examined the spectra of fifteen of the brighter stars. Grouping them by their stars' visual colour, he was struck by the 'family likeness' of the spectral patterns in each colour category. Donati drew fire from his colleagues for the small size of his sample, but his study inspired others, most notably Father Pietro Angelo Secchi (1818–78), to launch their own investigations and alerted them to potential observational challenges.[79]

Secchi, Director of the Observatory in the Collegio Romano, had shown interest in spectroscopy years earlier by repeating some of Fraunhofer's observations. But he began a systematic examination of stellar spectra only after he learned of Donati's efforts.[80] Rather than subject a selected list of representative stars to careful examination, Secchi chose to launch a study of as many stars as possible.[81] He assumed that spectral characteristics carried information about the physical properties of the light source and that his survey would, therefore, provide clues to the nature of stellar atmospheres as well as help answer larger questions about the structure of the universe and the motion of stars.

Another pioneer in stellar spectroscopy was American amateur scientist and instrument maker Lewis Morris Rutherfurd. Like Fraunhofer, Rutherfurd was driven less by an interest in the objects he observed than by a passion for the refinement of observing and measuring instruments. Even before he had heard of Kirchhoff's work, Rutherfurd had used spectroscopy to test the achromaticity of lenses so that he could pursue his interest in celestial photography. With no spectroscopes commercially available, he first had to construct his own instrument. Unaware of either Donati's or Secchi's efforts, Rutherfurd observed the spectra of twenty-three stars. Although he was impressed by the variety in the number and positions of the absorption lines they displayed, his concern for instrumental precision led him instead to design and construct improved micrometers, photographic equipment and spectroscopes.[82]

***

Of all the early celestial spectroscopists, the one individual who did the most to shape the emerging disciplinary boundaries of astrophysics was English amateur astronomer William Huggins (see Figure 2.1).[83] He was the first to observe emission lines in the spectra of nebulae, the first to apply Christian Doppler's (1803–53) principle to a star's light in order to determine its motion along the line of sight, the first to suggest a plausible method of observing solar prominences out of eclipse and the first to identify the ultraviolet spectral lines of hydrogen on film. He served as president of the RAS, the BAAS and the RS, and reaped many awards and honours for his scientific contributions, including honorary degrees from Cambridge, Oxford, St Andrews, Edinburgh, Dublin and Leyden. He received the Royal Society's Rumford, Royal and Copley Medals, the Académie des Sciences' Lalande Prize, the Astronomical Society of the Pacific's Bruce Medal, and joined the ranks of those elect few in the history of the RAS to be chosen twice as the recipient of that society's prestigious Gold Medal. In 1871, the enlightened Emperor of Brazil, Pedro II (1825–91), bestowed on him the Order of the Rose. In 1897, he was knighted by Queen Victoria (1819–1901) and, five years later, he was selected to be among the first twelve

Figure 2.1 William Huggins, c. 1860. (RASL/Photo Researchers, Inc.)

individuals awarded the Order of Merit by King Edward VII (1841–1910). As we shall see in the next chapter, he achieved all this with little formal education and no professional or university training in science or mathematics.

Notes

1. Roscoe to Stokes, 19 March 1860, Add MS 7656.R789, GGS, CUL.
2. Clerke, *A Popular History of Astronomy* (1885), pp. 183–4.
3. The name 'spectroscope' first appeared in print in the minutes of a meeting of the American Philosophical Society in Philadelphia on 19 July 1861 when it was recorded that 'Dr. Bache alluded to the subject of spectrum analysis, and the results obtained by the spectroscope.' Its almost casual use here suggests the word had already gained some currency by this time. 'Stated meeting, July 19, 1861', *PAPS* **8** (1861), pp. 277–9; p. 279.
4. Woolf, 'Beginnings of astronomical spectroscopy' (1964); 'Astrophysics in the early nineteenth century' (1968); Meadows, 'The origins of astrophysics' (1984); 'The new astronomy' (1984); and M. E. W. Williams, 'Astronomy in London: 1860–1900' (1987).
5. See, for example, Lemaine *et al.*, *Perspectives on the Emergence of Scientific Disciplines* (1976), pp. 3–23; Kohler, *From Medical Chemistry to Biochemistry* (1982), pp. 1–8; Edge and Mulkay, *Astronomy Transformed* (1976).
6. There was growing interest in London in the work being done by astronomers at observatories in the United States, particularly at the Harvard Observatory which boasted a telescope nearly identical to that at Pulkovo.
7. B. J. Becker, 'Eclecticism, opportunism, and the evolution of a new research agenda' (1993); 'Visionary memories' (2001); Hufbauer, 'Amateurs and the rise of astrophysics 1840–1910', (1986); and Lankford, 'Amateurs and astrophysics' (1981).

8. A. Chapman, *The Victorian Amateur Astronomer* (1998), ch. 1.
9. M. E. W. Williams, 'Astronomy in London: 1860–1900' (1987).
10. Turner, 'The decade 1820–1830' (1987).
11. J. F. W. Herschel, 'Presidential address' (1827), pp. 15–16.
12. Bessel, 'Über den gegenwärtigen Standpunkt der Astronomie' (1848), pp. 5–6.
13. Narrien, *An Historical Account of the Origin and Progress of Astronomy* (1833). The book was praised by the anonymous author of Narrien's obituary as an exact account of ancient astronomy that had surpassed others of its kind in popularity. See 'The late Professor Narrien', *MNRAS* **21** (1861), pp. 102–3.
14. Narrien, *An Historical Account of the Origin and Progress of Astronomy* (1833), p. xi.
15. *Ibid.*, pp. 456 and 503.
16. *Ibid.*, pp. xi–xii.
17. *Ibid.*, p. 520.
18. Whewell, 'Address to the General Assembly' (1833), pp. xii–xiii.
19. Comte, *Cours de Philosophie Positive*, vol. 2 (1833), pp. 8–9.
20. 'Notice of Robert Grant's *History of Physical Astronomy*', *MNRAS* **13** (1853), p. 131.
21. W. Parsons, 3rd Earl of Rosse, 'Observations on some of the nebulae' (1844) and 'Observations on the nebulae' (1850).
22. Grant, *History of Physical Astronomy* (1852), pp. 537–82.
23. Lee, 'Address delivered by the president' (1862), p. 133.
24. Newall, 'The decade 1860–1870' (1987), p. 134. Huggins was listed as a subscriber by June 1863.
25. *AR* **1** (1863), p. 1.
26. Newall, 'The decade 1860–1870' (1987), pp. 133–4; and 'The progress of astronomy', *MNRAS* **23** (1863), pp. 142–5.
27. Boyle, *Experiments and Considerations Touching Colours* (1664), p. 227.
28. Kirchhoff and Bunsen, 'Chemische Analyse durch Spectralbeobachtungen' (1860), p. 162.
29. Miller's address is merely noted in the *RBAAS, Manchester* (1861), p. 1. Fortunately, he repeated his talk before a meeting of the Pharmaceutical Society on 15 January 1862. The text of that address can be found in the *JPS*, 2nd Series **3** (1862), pp. 399–412.
30. Shirley, 'An early experimental determination of Snell's Law' (1951).
31. See Lohne, 'Thomas Harriott (1560–1621)' (1959) and 'The fair fame of Thomas Harriott' (1963).
32. Descartes, 'De l'Arc-en-ciel', *Les Météores* (1637), p. 254.
33. For a discussion of Descartes's investigation of the cause of the rainbow, see Sabra, *Theories of Light from Descartes to Newton* (1981), pp. 60–8.
34. Descartes, 'De l'Arc-en-ciel', *Météores* (1637), p. 255.
35. Boyle, *Experiments* (1664). See Experiments IV and V, pp. 191–3, and Experiments XIV and XV, pp. 227–31.
36. I. Newton, 'A letter of Mr. Isaac Newton ...' (1671/72), p. 3075.
37. Wollaston, 'A method of examining refractive and dispersive powers' (1802).
38. 'Yet instead of the Circular Hole F, 'tis better to substitute an oblong Hole shaped like a long Parallelogram with its Length parallel to the Prism ABC. For if this Hole be an Inch or two long, and but a tenth or twentieth Part of an Inch broad, or narrower; the Light from the Image *pt* will be as simple as before, or simpler, and the Image will become much broader, and therefore more fit to have Experiments try'd in its Light than before.' Newton, *Opticks* (1979), p. 70.
39. There are practical advantages to projecting the spectrum on a screen. More than one observer can witness and examine the coloured array at one time, and notable features of its appearance can be recorded by marking directly on the screen's surface. But such surfaces necessarily absorb and scatter some of the light that falls on them, thus diminishing the spectrum's brightness and obliterating any discernible characteristics of its individual rays.
40. To these Wollaston suggested adding two more classes of light rays, one for the invisible rays beyond the red end of the spectrum, which were discovered (1800) by William Herschel and associated with heat, and one for the rays beyond the violet end, which were discovered (1801) by Johann Wilhelm Ritter (1776–1810) and associated with photochemical action. Wollaston, 'A method of examining refractive and dispersive powers' (1802), pp. 378–80.
41. It may simply be that Wollaston was the first to consider these curious phenomena worth mentioning. Today we recognise a number of factors that, in whole or in part, could have made the dark lines more difficult for Wollaston to ignore. In addition to the fact that he examined the spectra directly with his eye, it is important to note that his prisms were of better optical quality than Newton's. It is also possible that the two men had significant differences in visual acuity. Newton relied on an assistant 'whose Eyes for distinguishing Colours

were more critical than' his own to search for and identify the boundaries of colours in the projected spectra. If the assistant noticed any dark lines separating the colours, Newton did not mention it. Nevertheless, Newton reported, the assistant 'did by Right Lines ... drawn across the Spectrum, note the Confines of the Colours'. These observations were repeated with the same results over several trials. The spacing of these boundaries confirmed Newton's personal suspicion that the colours represented a visual harmonic series. See I. Newton, *Opticks* (1979), Book I, Part II, Proposition, III, Problem 1, pp. 125–32.

42. Brewster, 'Description of a monochromatic lamp' (1823). See especially, Fig. 4, Plate XXVII. A colour reproduction of this illustration can be found in Hentschel, *Mapping the Spectrum* (2002), Plate III.

43. Brewster devised a new instrument to measure dispersive power more handily. It incorporated a telescope to magnify and thus better discern the slight increases and decreases in the spread of aberrant colours seen along the borders of a straight-edged target object. Instead of physically altering the refracting angle of the standard prism, Brewster argued that rotating it about the plane that bisects its refracting angle produced the same effective result. He equipped the instrument with a divided circle [goniometer] to measure incremental angular adjustments in the orientation of the standard prism with unprecedented precision. Brewster, *A Treatise on New Philosophical Instruments* (1813), ch III.

44. Brewster, *A Treatise on New Philosophical Instruments* (1813), pp. 302–3.

45. See Jackson, *Spectrum of Belief* (2000), chs 2–3.

46. Rohr, 'Fraunhofer's work and its present-day significance' (1925–6), p. 284.

47. Guinand's contract with the Institute ended in December 1813. He left Benediktbeuern and returned to Switzerland in May 1814. See Seitz, *Joseph Fraunhofer und sein optisches Institut* (1926), pp. 49–51. I would like to thank Jürgen Teichmann (Deutsches Museum; Ludwig-Maximilians-University) for bringing this information to my attention and for his assistance in clarifying the chronology of Fraunhofer's and Guinand's activities during this period.

48. According to the *Oxford English Dictionary*, the origin of 'theodolite' is unknown. It first appeared in print as 'Theodelitus' in a small book *A Geometrical Practical Treatize Named Pantometria* written by Leonard Digges and published posthumously by his son, Thomas Digges (London: Abell Jeffes, 1592). Digges described the 'theodelitus' as 'a circle divided in 360 grades or degrees' used to 'searche the best proportion or simetrie of many places with a true distance'; see Digges, p. 35. In his *Etymological Dictionary*, philologist Walter W. Skeat amended the original entry for the word noting it appeared to be of English origin, not Greek as some had claimed. Because the precise ruling of a circular disc involved crossing it 'with such numerous slanting strokes as to give it the appearance of being defaced', he suggested the word 'really stands for "The O delitus," i.e. "the circle effaced."' See Skeat, *An Etymological Dictionary of the English Language* (1893), pp. 830–1.

49. Fraunhofer, 'Bestimmung des Brechungs- und Farbenzerstreuungs-Vermögens' (1817), p. 202.

50. Wollaston, 'Neue Methode, die brechenden und zerstreuenden Kräfte der Körper' (1809). Fraunhofer's discovery of the dark lines is always described as having been entirely independent of Wollaston's. I have found no mention of the German translation of Wollaston's paper in any source.

51. Gilbert, [Note to Readers] (1809).

52. Teichmann to the author, 29 March 2010.

53. Jackson to the author, 23 January 2009.

54. 'Article XXXII. – Analysis of Scientific Books and Memoirs', *Edinburgh Journal of Science* 2 (1825), p. 348; Powell, 'Article XIII' (1825), p. 263; Jackson, *Spectrum of Belief* (2000), p. 225, n. 1.

55. Fraunhofer, 'Bestimmung des Brechungs- und Farbenzerstreuungs-Vermögens' (1817), pp. 220–1; Roscoe, *Spectrum Analysis* (1869), p. 181

56. Fresnel, 'Mémoire sur la diffraction de la lumière' (1816); idem, 'Mémoire sur la diffraction de la lumière' (1819).

57. Fraunhofer, 'Neue Modifikation des Lichtes' (1821).

58. Fraunhofer, 'Kurzer Bericht von den Resultaten' (1823).

59. Warner, 'Lewis M. Rutherfurd' (1971), p. 208; Clerke, *A Popular History of Astronomy*, 3rd edn (1893), p. 525. Note: This is the first edition of *A Popular History of Astronomy* in which Clerke discusses diffraction gratings.

60. McGucken, *Nineteenth-Century Spectroscopy* (1969), pp. 14–24.

61. *Ibid.*, chs. 1 and 2.

62. *Ibid.*, pp. 9–14.

63. Swan, 'On the prismatic spectra of the flames of compounds of carbon and hydrogen' (1857).

64. McGucken, *Nineteenth-Century Spectroscopy* (1969), pp. 25–6.

65. Kirchhoff, 'Über die Fraunhofer'schen Linien' (1860).

66. English translation of Kirchhoff, 'Über die Fraunhofer'schen Linien' (1860): Stokes, 'On the Simultaneous Emission and Absorption of Rays' (1860).

67. Roscoe to Stokes, 24 February 1860, Add MS 7656.R788, GGS, CUL. Roscoe is referring to the reprint of Kirchhoff's 'Über die Fraunhofer'schen Linien' (1860) in *AP* **109** (1860), pp. 148–50.
68. Roscoe to Stokes, 19 March 1860, Add MS 7656.R789, GGS, CUL.
69. For a description of the contributions of these individuals and others, see Meadows, 'Origins of astrophysics' (1984) and McGucken, *Nineteenth-Century Spectroscopy* (1969), pp. 1–29.
70. James, 'The creation of a Victorian myth' (1985). See also Sutton, 'Spectroscopy, historiography and myth' (1986), and James, 'Spectro-chemistry and myth' (1986).
71. McGucken, *Nineteenth-Century Spectroscopy* (1969), pp. 24–34.
72. See, for example, Gladstone, 'Notes on the atmospheric lines of the solar spectrum' (1861), p. 141; 'The composition of the solar spectrum', *CN* 4 (1861), p. 293; Morren, 'On spectrum analysis' (1861), p. 302; Giltay, 'On spectrum analysis' (1861).
73. Roscoe, *Spectrum Analysis* (1869), pp. 95–8.
74. Crookes, 'On the existence of a new element probably of the sulphur group' (1861).
75. Roscoe, 'On the application of the induction coil to Steinheil's apparatus for spectrum analysis' (1861).
76. This 'usual explanation' was based on an analogy of light with sound making use of theories of harmonics and resonance. See 'Proceedings of the Chemical Society', *CN* 4 (1861), p. 132.
77. Roscoe was able to provide a clear discussion of Kirchhoff's later explanation for the Fraunhofer lines in his third of three lectures given at the Royal Institution the following spring.
78. 'Proceedings of the Chemical Society', *CN* 4 (1861), p. 130.
79. DeVorkin, *An Astronomical Symbiosis* (1978), p. 77
80. Moigno, *Le Révérend Père Secchi, sa Vie* (1879); McCarthy, 'Fr. Secchi and stellar spectra' (1950); DeVorkin, *Astronomical Symbiosis* (1978), pp. 78–84
81. Secchi, 'Note sur les spectres prismatiques des corps célestes' (1863).
82. Rutherfurd, 'Astronomical observations with the spectroscope' (1863); Warner, 'Lewis M. Rutherfurd' (1971), pp. 194–5; DeVorkin, *Astronomical Symbiosis* (1978), pp. 77–8.
83. Clerke, *A Popular History of Astronomy* (1885), p. 412.

3

The young observer

After a little hesitation ... I decided to give my chief attention to observational astronomy ...

– William Huggins[1]

The introduction of spectrum analysis into astronomical research in the mid-nineteenth century was synchronous with William Huggins's rise to prominence as an amateur astronomer. After his death in May 1910, eulogisers were effusive in their praise of his vision and imagination, which American astronomer George Ellery Hale suggested allowed Huggins 'to divine some of the less obvious applications of the spectroscope'.[2] Appreciation of his willingness to break new ground was tempered by admiration for what the Astronomer Royal for Scotland, Frank Watson Dyson (1868–1939), termed Huggins's 'characteristic thoroughness' and 'care', and what maverick American astronomer Thomas Jefferson Jackson See characterised as his 'judicious habits of weighing evidence', 'wise selection of subjects of research', and 'strict conscientiousness and calm deliberation'.[3] How did so cautious and measured a man come to lead a movement that ultimately revolutionised the theory, technique and practice of astronomy by the turn of the twentieth century?[4]

The question's paradoxical premise, I argue, is founded on a well-crafted and convincing illusion, namely the sturdy façade of Huggins's public persona. Like a precious egg preserved *in situ*, the real stuff of another's life remains undisturbed until we, the curious, penetrate its protective shell. Once inside we may find only dust and musty memories. Not so in the case of William Huggins. As we shall see in the coming pages, he artfully concealed a far more interesting and complex investigator inside his 'disinterested man of science' exterior.

With this chapter I begin the delicate process of bringing the hidden man to light using both published and unpublished accounts to trace the development of his fledgling astronomical career to the end of the 1850s. By that time he had adopted a moderately rigorous, albeit eclectic, research programme based largely on others' interests and questions. Upon what resources did he rely to shape his methods, instruments and observational agenda? How did he acquire his technical and methodological expertise? How did he gain acceptance from his colleagues in the Royal Astronomical Society as a serious amateur?

3.1 Early life and education

What little we know of Huggins's early years is recounted in *A Sketch of the Life of Sir William Huggins*, a small biography penned by Charles E. Mills and C. F. Brooke in accordance with the wishes of Huggins's widow, Margaret Lindsay Huggins. Soon after her husband's death in 1910, she began writing what she envisioned would be the authoritative work on his personal life – a much-needed work in her view. The biography she hoped to write would both complement the published papers that had defined his professional life and correct the blatant inaccuracies she felt abounded in obituaries written by others. Unfortunately, she died before completing her ambitious project. Left in the hands of her old friends, John Montefiore and his sister Julia, it languished for many years. Mills, Julia's executor, finally completed the task with Brooke in 1936. As with many biographical efforts directed by individuals close to their subject, *A Sketch* must be read judiciously particularly when drawing conclusions from uncorroborated anecdotal information.

Born in the City of London on 7 February 1824, William was the only child of William Thomas Huggins of Shoreditch (c. 1780–c. 1855) and his wife Lucy (née Miller) of Peterborough (c. 1786–1868).[5] The senior Huggins was a silk mercer and linen draper – a middleman between the individuals who produce woven fabric and those who put it to use. The Hugginses lived above their shop in Gracechurch Street, a thriving commercial centre with hatters, ironmongers, bootmakers, printsellers, teadealers and wine merchants alongside the coach companies, taverns and life assurance companies that populated its quarter-mile stretch between Cornhill and Eastcheap.[6] It was conveniently located close to the silk-weavers of Spitalfields and the warehouses at St Paul's Churchyard.

The business was a prosperous one with three live-in shop assistants.[7] Resident proprietors of medium-sized operations like this one generally enjoyed a friendly relationship with their employees.[8] A typical work day would run from about 10 o'clock in the morning until as late as 9 or 10 o'clock at night, Mondays through Saturdays, with tea served between 5 and 6 p.m. Living on the premises, the Hugginses could tend their shop in a leisurely way when business was slow and be attentive during peak evening hours.

The bolts and swatches of colourful, luxurious fabrics that filled the shop coupled with the talk of travel to buy cloth elsewhere in Britain and abroad would have stimulated any young child's mind.[9] And William was, according to Mills and Brooke, a precocious boy. At age five, they tell us, he 'ingeniously constructed an electrical machine' – likely a static charge generator – after mastering a primer on electricity.[10] No sooner had he put the machine to work, they report, than he ran through the house shouting, 'I've had a shock, I've had a shock.' Growing up in a silk shop, he would have had access to an abundance of scraps with which to explore the wonders of the fabric's electrical effects.[11]

William received little formal education as a child. Schooling for young children in England in the 1830s was not directed toward practical ends. For the most part the goal was moral, not intellectual, development. Mills and Brooke tell us that William briefly attended a small school on nearby Great Winchester Street. By the age of nine or ten he spent 'an hour in the morning' with the curate of the parish learning 'classics and mathematics'.[12] His appetite for scientific study was whetted, meanwhile, by attending lectures on chemistry and physics at the National Gallery of Practical Science. Established by American emigré Jacob Perkins (1766–1849), this popular science and technology theatre, which Londoners called the Adelaide Gallery, opened its doors in June 1832. There, for a shilling, an inquisitive lad like young Huggins could witness public demonstrations of modern machines and marvellous physical phenomena.[13]

His family's circumstances would have precluded William's attending an elite public school. But by the time he reached school age, increased demand by the upper middle class for comparable educational opportunities had given rise to a new generation of public schools that were selective yet accessible to a broader spectrum of students. Such an institution was the City of London School, established in 1837.[14] In February of that year, at age thirteen, Huggins became one of the school's first pupils. Sources differ on how long he was enrolled there, but the school's records indicate he left in the first term of 1839, perhaps, as Mills and Brooke suggest, in consequence of a bout of smallpox.[15]

Entering students were selected by examination and an interview with the Headmaster. The basic tuition was £8 5s. per year with seven shillings assessed for extra lessons in music, art and foreign languages. Enlightened by the needs of commerce, but not subservient to them, the new school's course offerings extended beyond the confines of the strictly classical curriculum offered at established public schools. William would have received instruction in a broad array of subjects there, including Latin, Greek, French, writing, grammar, bookkeeping, mathematics, natural philosophy, geography, natural history, ancient and modern history, choral singing, chemistry and other branches of experimental philosophy.[16] In February 1838, the school procured experimental apparatus – an air pump, glass tubes, a condensing syringe, a small amount of mercury and a lathe – 'to be used in explaining the principles of Natural Philosophy'. With it, demonstrations – possibly the first of their kind performed in any English school – were immediately introduced by the Revd William Cook as accompaniments to his lectures.[17] Huggins later spoke of his fondness for the eccentric and irascible mathematical master, a Mr Edkins, who, he contended, offered him assistance in mathematical matters even after his formal withdrawal from the school.

Mills and Brooke tell us that Huggins 'studied industriously under private tutors at home' between the time he left City of London School and came of an age to enter university. The wealthy could afford the services of live-in tutors and governesses, of course, but many parents of moderate means were happy to spend a few shillings each week on private lessons for their children taught by peripatetic instructors.[18] Because it was emblematic of the inner drive, ambition and self-motivation that marked a man in control of his destiny, being schooled at home lent a special air of prestige to members of the rising middle and upper middle classes.[19]

William allegedly shared his parents' hope that he would attend Cambridge University, but it is difficult to account for such an ambition. First, as a Nonconformist, he would have been barred from taking a degree at Cambridge at that time.[20] Second, the fees were prohibitive to all but the wealthy. And finally, in the 1840s, a degree from Cambridge offered more prestige than practical scientific expertise. Given his apparent interest in applied rather than mathematical or pure science, it is unclear what end he imagined a Cambridge education would serve.

In the event, just as William reached university age, failing health left his father incapable of continuing to run the family business. Mills and Brooke tell us that, rather than have his parents sell their shop so he could proceed with plans to enter Cambridge, young Huggins dutifully elected to remain at home and assume his father's responsibilities.[21] They assert that, throughout his life, he regretted the 'grievous' but necessary choice that had deprived him of a university education. It was, he felt, an 'irreparable' loss.[22]

Furthermore, they claim, the decision thrust him into a career that he found 'particularly distasteful'. As he began to move more confidently in scientific circles, Huggins may have wished to distance himself from his entrepreneurial past, but his unpublished notebooks and correspondence reveal the extent to which the skills and habits of mind he acquired in the world of business informed his later efforts in scientific investigation. Experience taught him that making profitable choices required navigating wisely between the Scyllas and Charybdises of cost vs. benefit, risk vs. security, change vs. stasis, diversification vs. specialisation. Throughout his life, his actions show he adroitly viewed each new venture – in business or science – as a variation on the knotty dilemma, 'fish? or cut bait?'

Years later, he made the best of his lack of social pedigree, public school education and Cambridge diploma. Having risen above his station despite these handicaps, he bore the stamp of having followed physician-turned-social critic Samuel Smiles's simple recipe for a successful life: self-improvement through a regimen of thrift, perseverance and responsible behaviour.[23]

3.2 Interest in science

Mills and Brooke report that, in his youth, Huggins conducted 'research' in 'chemistry, optics, physics, electricity ... [and] photography' which he pursued independently in 'every spare moment'. Designing and building his own experimental equipment provided opportunities for 'thinking out problems for himself'.[24] His parents supported his investigative passion by giving him a microscope.[25]

In 1852, at the age of twenty-eight, he became a Fellow of the Royal Microscopical Society. It was his first formal attachment to an organisation aimed at the active pursuit of scientific knowledge.[26] The society had been founded only twelve years earlier to disseminate information among individuals engaged in microscopical studies and to encourage scientific instrument makers to further perfect the microscope.[27] Fellows included prominent members of London's medical and physiological community as well as serious amateurs. One such amateur was Warren De la Rue, a stationer, who, like Huggins, ran his father's business in the early 1850s.[28] Microscopy was not De la Rue's only scientific interest. A founding member of the Chemical Society in 1841, he was also an avid amateur astronomer and had joined the RAS in 1851. That same year, after seeing the display of William Cranch Bond's (1789–1859) daguerreotypes of the Moon at London's Great Exhibition, De la Rue was inspired to begin his pioneering efforts in celestial photography using Frederick Scott Archer's (1813–57) newly introduced wet collodion technique.[29] He obtained photographs of the Moon 'possessing great sharpness of definition and accuracy of detail'.[30] As Huggins gradually widened his social network in scientific London, he and De la Rue often crossed paths.

Many years later in his 1897 retrospective essay, 'The new astronomy', Huggins noted that the nineteenth century had witnessed the birth of not one, but two new sciences, which had emerged in natural consequence of the dramatic changes in what traditionally had been considered proper bounds for research in the oldest domains of scientific investigation: astronomy and medicine. Recalling his fascination with microscopy and physiology in the early 1850s, he confessed that, at the time, the exciting developments in those fields made his eventual choice of astronomy an especially difficult one. In fact, his interest in microscopy may have succumbed to a certain squeamishness he experienced while engaged in physiological studies.[31] It may also have fallen victim to the realisation that astronomy offered more opportunities for recognition and advancement to amateurs like himself than the more professionally trained realm of physiology. Nevertheless, Huggins did not immediately abandon his interest in microscopical research.[32]

3.3 Interest in astronomy

Huggins purchased his first telescope for £15 when he was eighteen years old. With it, he attempted to observe the smoke-obscured sky over downtown London by 'lying prone on the floor ... with the telescope pointed out of the window'[33] – surely a challenging position for any celestial observer. Charles Dickens (1812–70) is reputed to have enjoyed watching the coaches come and go from Robert Fagg's busy coach office while sitting across the street at the Spread Eagle Inn just a few doors down from the Hugginses' shop. Perhaps, like Dickens, young William had more terrestrial interests to motivate his early observations.[34]

We can only surmise what prompted him to invest in a telescope at this time, but three celestial events occurred during his childhood and youth that attracted general public attention, and perhaps his as well. In mid-November 1832, when William was only eight years old, Europe was treated to an awe-inspiring, if not frightening, blizzard of falling stars.[35] Nothing like it had been seen since 1799. Although the event seems to have escaped the notice of English observers, they soon received reports of the terror experienced in parts of France where, for two hours, 'innumerable quantities of vivid sparks' filled the otherwise clear and calm night skies.[36]

A few years later, on 15 May 1836, the centre line of an annular solar eclipse passed from Ayr on the west coast of southern Scotland to Alnwick on the northern English coast. On the day of the eclipse, the London sky was clear as residents prepared to witness the obscuration of over 90% of the solar surface. A partial eclipse, even one of such magnitude, lacks the stunning impact of a total eclipse, but the public greeted the spectacle with enthusiasm. Was twelve-year-old William among them? *The Times* of London reported that scheduled church services were cancelled and that hundreds of people gathered at Greenwich to see the event, apparently under the mistaken impression that they would be given an opportunity to observe the Sun with the astronomical instruments there: 'So anxious were many to gain an entrance, that it was necessary to call in the aid of the police to save the gates from being actually borne down by the rush of united hundreds, and a few mathematical ladies were more clamorous than the men.' Fortunately, the crowd was appeased by the presence of some amateur astronomers who had set up their own telescopes in the vicinity.[37]

In 1843, about the time Huggins acquired his first telescope, a bright comet appeared in the sky between February and April. At perihelion, it approached closer to the Sun than any other comet then on record and sported an exceptionally long tail. Observers in Bologna, where weather conditions were favourable, reported seeing the comet in the sky at noon as it passed perihelion on 27 and 28 February. Clouds hid it from view in England until the evening of 17 March, when only its tail appeared above the horizon after sunset. The next night, the whole comet could be seen. A contemporary lithograph shows it stretching across the Parisian sky from the constellation Eridanus to Canis Major on 19 March.[38]

The Great Comet of 1843 remained visible in England until the beginning of April. Did William tote his new telescope to the northern bank of the Thames to watch as the comet shifted its position night after night across the southern constellations?[39] Perhaps. We have found no clues to the what, where, how, or why of his first observational efforts, but there can be no doubt that his interest in astronomy intensified over the years. In 1853, he replaced his first telescope with a 5-in Dollond equatorial that cost £110.[40] And, in 1854, shortly after his thirtieth birthday, he was elected a Fellow of the Royal Astronomical Society.[41] It is within the institutional confines of this group that he found his first opportunities to test his astronomical mettle.[42]

There is no mention of Huggins or his work in the Society's *Monthly Notices* during his first two years as a Fellow. He was still living in the City of London and running the family business. Regular observation would have been easier said than done. Late evening hours and long days in the shop could leave little time or energy to spend gazing at the heavens. When given an opportunity to observe, clouds and/or neighbouring buildings could easily frustrate the best of plans. And even with all other obstacles removed, he still would have to face the practical problem of manoeuvring his cumbersome telescope without assistance.

3.4 Tulse Hill

Mills and Brooke assert that because Huggins 'was astute enough to see that such quiet businesses as his father's were doomed', he decided to 'dispose' of the family shop in the mid-1850s.[43] One might suppose that he was responding to unfavourable fluctuations in the movement of silk or linen brought on by such complex factors as the mechanisation of the textile industry, the elimination of trade protections and the growing popularity of cotton by mid-century. But if there was a sense that the industry was headed for disaster, it was not evident in the number of London's silk mercers. In 1840, Kelly's *London Postal Directory* listed sixty-nine. In 1853, there were eighty-one, including twenty-four from the earlier list.[44] The silk industry, although buffeted by economic tempests throughout the nineteenth century, did not suffer real damage until the discovery of artificial silk (rayon) in 1891, an event which even the most astute market analyst in the 1850s could not have anticipated. Of more immediate concern to mid-century small shop owners like the Hugginses was the introduction of large-sized 'department' stores that sold ready-to-wear clothing, cloaks and prefabricated curtains.[45]

For whatever reason, Huggins and his parents retired from commercial life, sold the lease on their shop and moved in late 1854 or early 1855 to a home in Lambeth, a well-to-do and growing suburb just south of the Thames populated by prosperous businessmen and professionals.[46] Their new residence was a detached house with an extensive garden in the back. Initially the house, like many others in the neighbourhood, lacked a numbered street address. It was simply called 'Alpha Cottage' until enumerated 90 Upper Tulse Hill

Road. Unfortunately, the original structure no longer exists. The neighbourhood suffered extensive damage in World War II and was redeveloped in the 1970s.[47]

The property's elevation offered Huggins a promising vantage point for celestial observation. Prevailing winds blew the smoke of London northward leaving the local skies relatively clear. The garden proved to be an ideal place to set up his telescope on a small wooden stand and keep it ready for use. But he did not remain satisfied with this arrangement for long, and soon contracted a local carpenter to construct a substantial observatory building in his garden in which to keep his telescope, transit-circle and clock.[48] In May 1856, he described the resulting structure in the first of his many communications to the RAS's *Monthly Notices*.[49] Iron columns embedded in concrete lifted the 12×18-ft building 16 feet above the ground in order to clear the neighbouring houses and trees. He supported his instruments on brick columns set firmly in their own concrete foundation free of any connection to the surrounding building. A passageway joined the observatory to the second storey of his home, thus making it 'for all purposes of convenience and access, a part of the house'. Once the observatory was completed and put to use, there could be no doubt among Huggins's friends and colleagues at the RAS as to the seriousness of his commitment to astronomical observation.[50]

3.5 An observatory notebook

In 1856 Huggins acquired a bound notebook to keep a permanent record of his observations. The title page of this first notebook bears the mark of a new enthusiast (see Figure 3.1). Across the top is a phrase, in Hebrew, from Genesis 15:5 that reads 'Look now toward heaven, and count the stars.'[51] He carefully fashioned each letter in a calligraphic style used by scribes to copy the Torah by hand. Below that an exuberant mix of Greek and Latin announces 'PHAENOMENA which Will. Huggins, F.R.A.S., has observed from his observatory'. This is followed by the RAS motto: 'Whatever shines should be observed.' And then, 'Will. Huggins designed and made this observatory for the most commodious exploration of the heavens. 1856. Ed. Leigh, being the builder.' At the bottom of the page is a line in Greek from I Corinthians 15:41 that reads, 'One star differs from another star in glory.'[52] The facing endpaper is inscribed with lines of text in English, Italian and German.[53]

Exactly when Huggins began keeping written records of his observations remains uncertain. Like many neophytes, he probably considered his notebook as a less of a running log or journal than a record of observations he believed were worth preserving. Tersely worded and cleanly illustrated, his first entries seem to be fair copies made at some later time, perhaps from memory, but more likely from notes scribbled on scraps of paper amid the usual confusion of an actual observation. As precious to us as they were to their author, these entries offer a glimpse into what it meant to him to be a serious amateur, what sorts of information he understood to be required in order to participate actively in the astronomical community, and what contributions he was interested in making, or felt qualified to make, to the on-going work of that community.

His very first page of notes is headed 'Observationes Marti. A.D. 1856'. On 2 April 1856, Mars had been at opposition – that is, aligned with, but diametrically opposed to, the Sun as

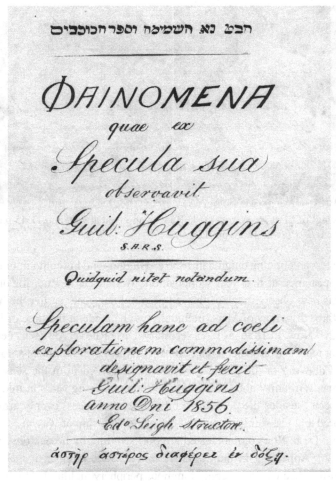

הבט נא השמימה וספר הכוכבים

ΦAINOMENA

quae ex

Specula sua

observavit

Guil: Huggins

S.R.S.

Quidquid nitet notandum.

*Speculam hanc ad coeli
explorationem commodissimam
designavit et fecit
Guil: Huggins
anno Dni 1856.
Ed° Leigh Structor.*

ἀστὴρ ἀστέρος διαφέρει ἐν δόξῃ.

Figure 3.1 Title page, Notebook 1. (WCL/SC)

seen from Earth. Because earthbound observers are at or near their closest approach to Mars at such times, the positional astronomers among them are able to triangulate its distance with greater accuracy, while those curious about the planet's visible features, like Huggins, are treated to their best views of its surface.

The advent of the telescope in the seventeenth century had revealed Mars as a tiny pale disc – a dull and uninspiring sight compared with the drama offered by Jupiter and its satellites, Venus's phases, or Saturn's curious appendages. Over the years, as both instruments and their employers grew in observational capability, Mars became a popular, if somewhat confounding, target. Some of those who tried to make sense of what they were seeing – Giovanni Domenico Cassini (1625–1712) at the Panzano Observatory in Bologna, Dutch physicist and mathematician Christian Huygens (1629–95), and William Herschel, to name a few notable Mars observers – recorded their impressions of the planet's surface.[54] Their drawings testify to the rich variety in visual perception and interpretation among those who trained their telescopes on Earth's intriguing neighbour.

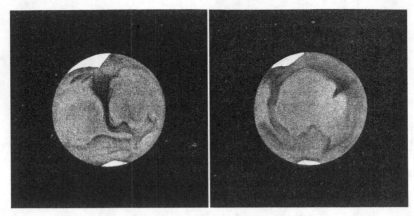

Figure 3.2 Drawings of Mars by Warren De la Rue, as observed at 9:40 and 11:45 p.m. on 20 April 1856. (Flammarion, *La Planète Mars*, pp. 128–9)

Mars was no less an enticing enigma in 1856 as Huggins and his contemporaries prepared to record its appearance at that year's opposition. Warren De la Rue, for one, was busy documenting his new observations of the red planet. Four years earlier, he had presented the RAS with three drawings of Mars, including two he made at the time of its opposition in January 1852.[55] Now, in 1856, as RAS secretary, De la Rue served as the central clearing house for similar reports from other RAS Fellows. In May he communicated to the Society a description and drawing of the planet's disc he had received from the seasoned amateur the Revd Thomas William Webb (1806–85).[56] When observing Mars in mid-April, Webb wrote, he had seen 'besides the pole, three other brighter spaces' evenly arranged around its limb. He noted that Cassini had depicted the planet in a similar way in March 1666.[57]

A month later, De la Rue communicated a report sent him by novice observer Frederick Brodie (1823–96). Appearing in the *Monthly Notices'* June number, it included a sketch showing the martian disc with a generally luminous periphery highlighted by hints of Webb's four bright regions.[58] The dark areas on Brodie's Mars possess an organic character not unlike the flowing patterns De la Rue included in his own drawings from the same opposition.

Although De la Rue's illustrations were not published contemporaneously, two were reproduced decades later by French astronomer Camille Flammarion (1842–1925) in his compendious *La Planète Mars*.[59] Looking at De la Rue's 1856 drawings from his 1892 vantage point, Flammarion considered them to be benchmark examples, indeed 'the best we have had before our eyes from the first pages of this work'.[60] The two form a sequential pair, the second drawn two hours after the first (see Figure 3.2). As such, Flammarion proclaimed, the drawings are 'particularly remarkable'. The shift in position of the features De la Rue depicted in them is consistent with the planet's rotation over the time interval. He rendered surface features in such detail that Flammarion could confidently match them with regions that had since been mapped and named. And the fact that De la Rue aligned Mars's polar caps non-diametrically gave further evidence, if any were needed, of the authority of his representations.[61]

Given this context, what can we learn about Huggins as an observer from his notebook sketches of Mars? First, he planned to make a series of observations of the planet. He drew a

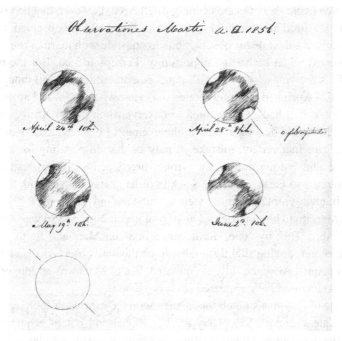

Figure 3.3 Observationes Marti, 1856, by William Huggins, Notebook 1. (WCL/SC)

set of five neat circles on the first page of his observatory notebook, each with a pair of axial guide lines oriented approximately 45° from the vertical. He filled in four of the circles with sketches of the martian surface as he saw it on 24 April, 28 April, 19 May and 2 June (see Figure 3.3). The empty fifth circle hints at unfulfilled plans for at least one additional observation of the red planet before its disc became too small to discern any surface features.[62] Several months later he presented two drawings of Mars (from 28 April and 19 May) to the RAS, although these were never published.[63]

Second, it is likely that he consulted others' reports (past and contemporary) for guidance before making his own observations. In format, and to some extent in execution, his notebook sketches are reminiscent of figures William Herschel included in Table VI of his 1784 paper 'On the remarkable appearances at the polar regions of the planet Mars ...'.[64] Like some of Herschel's illustrations, all of Huggins's drawings show only two, not four, conspicuous bright areas. Unlike De la Rue, Huggins has positioned one at each martian pole, giving the planet an equinoctial appearance. I should note, however, that it was late summer in Mars's northern hemisphere when the planet reached opposition that year, so only its north polar cap would have been visible from Earth.

Third, he came to each observation with an expectation of what he would see. Of special interest are Huggins's representations of the planet's dark surface markings. Those he perceived on 24 April at 10 p.m. seem to have shown up again when he viewed Mars at 8:30 p.m. on 28 April. His sketch from 2 June suggests he witnessed yet another return of the same markings. Did Mars, in fact, present the same face toward Earth on each of these occasions? Mars takes about 39 minutes longer than Earth to complete one full rotation relative to the Sun (solar day). On the evening of 28 April, however, that hemisphere would not rotate into view until after midnight, about four hours after he observed it.

Perhaps he did not schedule his observations to coincide with martian rotations. Perhaps he simply observed when he had an opportunity. Perhaps indeed. But the near-identical appearance of his drawings for 24 and 28 April is extraordinary. Even though the face of Mars visible on 24 April had not fully rotated into view when he looked again on 28 April, he represented it as if it had. If he wanted to observe when the 24 April hemisphere was visible again, why would he look at 8:30 in the evening? I believe the answer could lie in a simple, and perhaps inadvertent, mistake. It may be that, in planning for his observation on 28 April, he adjusted the time in the wrong direction. Perhaps, instead of adding the 39-minute difference to Earth's 24-hour clock in order to stay in sync with Mars's rotation, he subtracted it, giving him an optimal viewing time around 7:30 p.m.

The side of Mars that Huggins viewed at 10 o'clock on 2 June 1856 was almost the same as that on 24 April. But, by June, the distance between Mars and Earth had increased, leaving the red planet looking slightly smaller to earthbound observers than it had appeared at opposition in April. Also, the fully illuminated face of Mars was no longer visible from Earth in June, so it would have appeared slightly gibbous.

The next page of Huggins's notebook is busy with the records of four separate observations made in late summer 1856. They are occasional and out of sequence, enhancing the suspicion that he inscribed them at some later time. A two-month lapse in the record is broken on the next leaf with a note concerning a lunar eclipse on 13 October complete with an artfully coloured illustration of the eclipsed Moon. Over the next two years, Huggins added only seven dated entries. They offer no sign that he was observing to answer questions of his own, or that his observational programme was driven by an unrelenting and urgent desire to serve as an interpreter of Nature. Instead, he seems to have been content to act simply as its witness and interlocutor.

3.6 Developing a research agenda

Huggins needed guidance and encouragement in order to transcend his *ad hoc* observing pattern and learn how to operate within a research agenda. The RAS had served him well up to this time. In September 1858, he attended the Leeds meeting of the BAAS, his first as a member.[65] While there, he came into contact with a wider circle of expert amateurs, most notably the Revd William Rutter Dawes (1799–1868) of Haddenham.

An avid amateur astronomer best known for his careful observations of double stars, Dawes was a quarter century older than Huggins. Like Huggins, he professed a childhood interest in astronomy that he had satisfied with a small telescope. And, like Huggins, he obtained an improved telescope, built an observatory and began more serious observation when he was about thirty years of age.[66] A respected Fellow in the RAS, Dawes received the Society's prestigious Gold Medal in 1855 for his exemplary contributions to the advancement of astronomy including his monumental catalogue of double stars and his simultaneous discovery of the dusky, or so-called crape, ring of Saturn with William Cranch Bond.[67]

An interest in double stars led Dawes to seek objective lenses with superior resolving power. Impressed with those made by the renowned Alvan Clark (1804–87) of Cambridge,

Massachusetts, he purchased a number of them.[68] Perhaps feeling a special kinship with Huggins, Dawes sold the young newcomer, at cost, an 8-in Clark objective for £200, a deal the two men probably negotiated at the Leeds meeting. And Huggins, eager to take advantage of this fine lens's greater resolving and light-gathering power, soon had it mounted in an equatorial, clock-driven telescope built by Thomas Cooke (1807–68) of York, founder of the first telescope factory in England and the nation's premier contemporary telescope maker.

After October 1858, the form and substance of Huggins's observatory record took on an increasingly focused quality. It is possible that simply acquiring a new instrument inspired him to enhance the rigour and sophistication of his observations, but it could not ensure the realisation of such action. Clues to his metamorphosis from a true novice to a confident, self-directed amateur can be found in his cryptic notebook jottings and contributions to the *Monthly Notices*. These point to Dawes's influence on his selection of subjects to observe, his method of observation and his overall sense of purpose in making his astronomical observations.[69]

Huggins's desire to follow his mentor's lead might explain why, despite the fact that a stunning comet discovered by Donati was visible to the naked eye from 19 August to 4 December 1858 and telescopically observable until 4 March 1859,[70] he chose instead to focus on binary stars and changes in Jupiter's surface features. These were projects only observers with sufficiently fine instrumentation and experience could undertake. By comparison, the comet may have seemed like just one more fuzzy little object to him, something even the man on the street could see without instrumental assistance. Subjecting this 'nebulous haze' to telescopic scrutiny would reveal no new information.[71]

On 22 October, Huggins recorded a visit to his observatory by his neighbours, the Clissolds, to see the Moon. He also showed them γ Andromedae (Almach), a binary star consisting of a bright yellow star (A) and its bluish-green companion (B). Because Almach is easily resolved even in small telescopes, it is a popular target for novices. But a closer reading of Huggins's notebook entry reveals that his aim on this particular evening was to observe something far more challenging, a challenge that hints strongly of Dawes's influence. In 1842, Russian astronomer Otto Wilhelm von Struve (1819–1905) had been surprised to find that B, Almach's small blue component, was itself a close binary pair. In 1858, these two stars had an angular separation of 0.55″ – just at the resolution limit of an 8-in telescope.[72] Huggins would have been unable to resolve the close pair with his old 5-in Dollond, and yet he wrote, 'the components of B were beautifully defined & separated. I fully believe', he continued, 'if the atmosphere had not prevented it, that I should have had most excellent definition of the stars, such is the perfection of the image formed by the [8-in Clark] object glass.'[73]

Between October 1858 and July 1860, with few exceptions, Huggins made at least one notebook entry every month, many of them indicative of his new interest in recording subtle changes in a single object over time. From 2 November 1858 to 10 February 1859, a period when Jupiter was favourably placed for viewing, he devoted his attention to observing variations in its surface just as Dawes had done the previous year.[74] Each of Huggins's thirteen recorded observations of Jupiter during this period is accompanied by a drawing, many of which he later excised from his notebook and presented to the RAS (see Figures 3.4 and 3.5).[75]

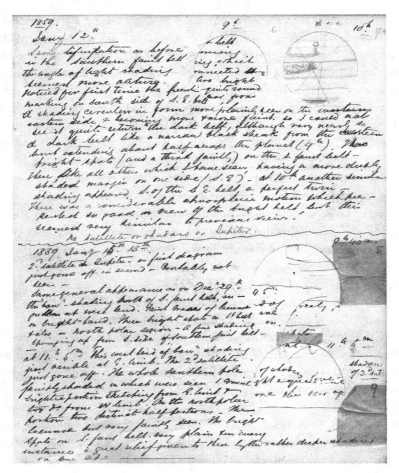

Figure 3.4 Observations of Jupiter by William Huggins. 12 and 15 January 1859, Notebook 1. Note the four neatly cut circles from which drawings of Jupiter (see Figure 3.5) were removed. The original drawings are held in the RASL, Add MS 18. (WCL/SC)

Then, as Saturn moved into view in mid-February 1859, Huggins kept close watch on that planet and its satellites. His observations of Jupiter and, especially, of Saturn continued with great regularity through May 1860, by which time reports out of Berlin were gaining currency in select English scientific circles to the effect that the chemical and physical nature of the Sun could be discerned by analysing its spectrum. But it would be nearly two years before this 'spring of water', as Huggins later described the news, trickled down to the wider audience of which he was a part.

***

At the age of thirty, William Huggins, a businessman with little formal education but an interest in a range of scientific matters, sold his family's business and moved to a suburban location which afforded him both the leisure time and the darkened skies necessary to begin a programme of astronomical observation. His observatory note-book entries and published notices reveal an opportunistic research agenda common

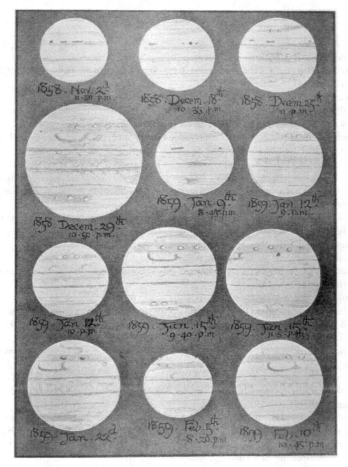

Figure 3.5 Drawings of Jupiter by William Huggins. The drawings of his observations on 12 January (9 and 10 p.m.) and 15 January (9:40 and 11:05 p.m.) were excised from the notebook page shown in Figure 3.4. (W. Huggins and M. L. Huggins, *Scientific Papers*, p. 361)

to most casual amateurs. But it was an agenda that took on an increasingly focused character in consequence of his association with expert amateurs in the RAS and the British Association. By 1860, following his acquisition of a new and more powerful telescope, and adept guidance from the more experienced Dawes, Huggins reached something of an intermediate stage in the development of his identity as a serious observer.

At the same time, his observing choices indicate he appreciated the value of diversification as a way to remain free of onerous constraints on method and problem selection. He wished to remain his own man, a desire born, no doubt, from his experience in the business world. This independence of mind and action served him well throughout his career, particularly in the next phase, when, as we shall see in the next chapter, he embarked upon a new observational programme fraught with substantial risk and promise, namely the application of spectrum analysis to the light of stars.

Notes

1. W. Huggins, 'The new astronomy' (1897), p. 910.
2. Hale, 'The work of Sir William Huggins' (1913), p. 148.
3. Dyson, 'Sir William Huggins' (1910), pp. v and x; T. J. J. See, 'Tribute to the memory of Sir William Huggins' (1910), pp. 391–2.
4. Lankford, 'Amateurs and astrophysics' (1981), p. 277.
5. Christening of William, son of William Thomas and Lucy Huggins (b. 7 February 1824) recorded by Samuel Nichols on 24 August 1824, Register Book, Pinners Hall Meeting, RG4 4226, London 17.
6. For a list of neighbouring businesses, see 'Gracechurch Street', Street Directory, *Post Office London Directory 1853* (London: Frederic Kelly, 1853), pp. 285–6.
7. 1841 Census Return for 97 Gracechurch St, London, St Peter-upon-Cornhill Parish, HO 107/723/f.8.
8. 'Drapers, hosiers, silk mercers, etc.', in Booth, *Life and Labour of the People in London* (1896), pp. 67–87. This information is based on conditions at the turn of the century. Conditions in the Hugginses' shop at mid-century could have been somewhat different.
9. The year William Huggins was born, a law was passed which lifted the prohibition against importing wrought silk creating serious economic hardships for English textile workers, but making a wide range of continental silks available for sale in Britain. See Malmgreen, *Silk Town* (1985), pp. 57–8.
10. Mills and Brooke, *A Sketch of the Life* (1936), p. 7.
11. For some examples of simple electrical machines, see Thompson, *Elementary Lessons in Electricity and Magnetism* (1881), pp. 40–8.
12. Mills and Brooke, *A Sketch of the Life* (1936), p. 8.
13. Inkster and Morrell, *Metropolis and Province* (1983), pp. 98 and 110; Kargon, *Science in Victorian Manchester* (1977), pp. 38–9. On the Adelaide Gallery, see Morus, 'Currents from the underworld' (1993), pp. 53–4, and *Frankenstein's Children* (1998), ch. 3.
14. For news items on the opening of the City of London School see *The Times* of London, 3 February 1837, p. 3f, and 4 February 1837, p. 6e.
15. Mills and Brooke, for example, state that he left the next year after a bout of smallpox, while Dyson reports that he remained there until December 1843. Mills and Brooke, *A Sketch of the Life* (1936), pp. 8–9; Dyson, 'William Huggins' (1910), p. i.
16. I wish to thank William F. Hallett (City of London School) for the valuable information from which I have drawn my description of William Huggins's education there. Following the death of Huggins's widow (Margaret Lindsay Huggins) in 1915, the sum of £2000 was bequeathed to the School for future scholars to pursue the study of astronomy at Cambridge.
17. Douglas-Smith, *The City of London School* (1965), pp. 82–3.
18. Wardle, *English Popular Education: 1780–1975* (1976), pp. 118–19.
19. J. F. C. Harrison, *The Early Victorians 1831–1851* (1971), p. 138.
20. May, *Economic and Social History of Britain 1760–1970* (1987), pp. 84–5. May points out that while Oxford excluded all but Anglicans from entering the university, Cambridge allowed those not affiliated with the Church of England to enter, but prohibited them from taking degrees until they established their membership in the Church. The University Test Act of 1871 opened England's universities to qualified students of all religious faiths.
21. 1851 Census Return for 97 Gracechurch St, London, St Peter-upon-Cornhill Parish, HO 107/1531/1.
22. Mills and Brooke, *A Sketch of the Life* (1936), p. 13.
23. Smiles, *Self-Help* (1884). While Smiles's message struck a particularly responsive chord among the working-class, entrepreneurs also found it appealing. See Briggs, *Victorian People* (1973), pp. 131–3.
24. Mills and Brooke, *A Sketch of the Life* (1936), p. 11.
25. Improvements in microscope design during the 1830s had made them sturdier, easier to use and much more affordable. See Allen, *The Naturalist in Britain* (1976), p. 128.
26. Huggins was certainly not alone in his interest in microscopy. In 1856, prominent physiologist William Benjamin Carpenter wrote a handbook which he hoped would be of 'special interest to the *amateur* Microscopist' on mastering the available apparatus and preparing specimens for observation. By 1875, it was in its fifth edition. See Carpenter, *The Microscope and its Revelations* (1876), p. vii.
27. 'Preface', *TRMSL* **1** (1844), pp. v–vi.
28. Warren De la Rue wrote an article describing his microscopical examination of butterfly markings. De la Rue, 'On the markings on the scales of *Amathusia horsfieldii*' (1844). His father, Thomas De la Rue, had founded a

stationery printing firm on Bunhill Row. See Dreyer and Turner, *History of the Royal Astronomical Society* (1987), p. 154.

29. The daguerreotypes were made by William Cranch Bond of Cambridge, Massachusetts.

30. 'President's address on presenting the Gold Medal to Mr. De la Rue', *MNRAS* 22 (1862), pp. 131–40, p. 135. Knobel, 'Warren De la Rue' (1890). It is interesting to note in this context that there is no reference to the Great Exhibition of 1851 in any account of William Huggins's life.

31. Mills and Brooke, *A Sketch of the Life* (1936), pp. 18–19.

32. W. Huggins, 'Note on the prismatic examination of microscopic objects' (1865).

33. Regarding Huggins's purchase of a telescope, Mills and Brooke state that, 'This was not so momentous an occasion in his life as might at first be thought, since he was prevented by his surroundings from making the best use of the instrument. In fact, it was only by standing upon the roof among the smoky chimneys of the City, or by lying prone on the floor of a room below with the telescope pointed out of the window, that he could make observations with it.' Mills and Brooke, *A Sketch of the Life* (1936), p. 11.

34. 'Gracechurch Street', in Weinreb and Hibbert, *The London Encyclopedia* (1983), p. 319.

35. Every year in mid-November, ten to fifty meteors per hour may be observed at the height of what is known today as the Leonid meteor shower. At approximately thirty-three-year intervals, the number of meteors increases dramatically to an estimated tens of thousands per hour! In Europe, the meteor shower of 1832 was one such event. See R. Burnham, *Celestial Handbook*, vol. 2 (1978), p. 1064; Croswell, 'Will the lion roar again?' (1991).

36. 'Shower of fire', *The Times* of London, 19 December 1832, p. 3c.

37. 'The great annular eclipse', *The Times* of London, 16 May 1836, p. 3b.

38. 'The great comet of 1843 seen in Paris on 19 March 1843', lithograph reproduced in Olson, *Fire and Ice* (1985), p. 90.

39. Almost daily reports on the appearance of this comet were published in *The Times* of London in late March 1843. See the reports of J. F. W. Herschel, James South and others: 21 March, p. 5c; 22 March, p. 5f; 23 March, p. 6b; 24 March, p. 6c; 25 March, p. 7a; 27 March, p. 6e; and 30 March, p. 6e.

40. Mills and Brooke, *A Sketch of the Life* (1936), p. 18.

41. Huggins was elected a Fellow of the RAS on 12 April 1854. *MNRAS* 14 (1854), p. 173.

42. M. E. W. Williams, 'Astronomy in London: 1860–1900' (1987).

43. Mills and Brooke, *A Sketch of the Life* (1936), p. 19.

44. I have selected silk mercers because that was how William Thomas Huggins identified himself in the postal directories, although he referred to himself as a linen draper in the censuses. In practical terms there was little difference in the two professions and how one listed oneself may have had more to do with the clientele one wished to attract than the actual work that was done.

45. A. Adburghan, 'Shops and shopping 1800–1914', in S. Mitchell, *Victorian Britain: An Encyclopedia* (1988), p. 720.

46. The date is uncertain. Mills and Brooke say that they moved in 1854, but there is a letter from William Huggins to the Royal Astronomical Society dated 2 November 1854, which he wrote while still living at 97 Gracechurch Street.

47. Miss Hazel Butterfield, HM Land Registry, to the author, 9 April 1991. The property was purchased c. 1945 by the London County Council.

48. The instruments he possessed in 1856 included his 5-in Dollond telescope, an 18-in transit-circle made by Thomas Jones which Huggins boasted was capable of readings down to 3 arcseconds using the attached verniers, and a clock by T. Arnold. Edward Leigh is listed in Kelly's 1859 Postal Directory as a carpenter, living on New Park Road in Brixton Hill.

49. W. Huggins, 'Description of an observatory erected at Upper Tulse Hill' (1856).

50. A paucity of documentation prevents us from explaining the family's income stream after the sale of their business.

51. The Hebrew phrase reads: *Habet na hashamayma uspor hakokhavim.* The translation and transliteration have been kindly provided by Daniel Schroeter (University of California, Irvine). Schroeter notes that the script is 'written in the calligraphic style that you would find in a Torah, rather than in the way you would learn to write (unless you were a scribe who wrote the Torah by hand)'.

52. I wish to thank John M. McMahon (LeMoyne College) for his valuable assistance in translating the Latin phrases. The translation provided here differs from that given in Sarah Whiting's 'Diaries' (1917) which reads: 'Phenomena which William Huggins [F.R.A.S.] observes with his own telescope'; with the RAS motto: 'Whatever shines is to be noted'; and, 'This glass for exploring the sky was designed and made by him in 1856' along with the name of the builder, Edward Leigh. Note that when presenting her translations in 'Diaries'

(1917), Sarah Whiting inadvertently switched the Italian and German texts. They are given here in their proper order.

53.

Original notebook text	Source	English translation	English translation presented in S. F. Whiting, 'Diaries' (1917), p. 159
Ye are a beauty and a mystery.	Lord Byron, 'Childe Harold's pilgrimage' Canto the Third, LXXXVIII (1816).		
Astri gemelli, pinti in giallo, in azzurro, in verde, in vermiglio, in vivido rosso di porpora! – Rifletti [...] ai gemelli astri, rossi e verdi – o gialli ed azzuri – quale maravigliosa varietà di luce non daranno essi ad ogni pianeta, che loro gira intorno. Oh delizioso contrasto, e giocondo avvicendamento! – Un rosso e verde giorno, alternato con altro bianco, e colle tenebre.	Girolamo Volpe (1824–85) *Il Giglio e l'ape nel Palazzo di cristallo* (Londra: Presso Williams e Norgate, 1852), see pp. 49–50. Volpe's text is an Italian translation of an excerpt from *The Lily and the Bee: an Apologue of the Crystal Palace* (Edinburgh and London: William Blackwood and Sons, 1851), which was originally written in English by Samuel Warren (1807–77), see pp. 52–3.	Double stars – of orange, blue, green, crimson, rich ruddy purple! Think, quoth he, of twin suns, red, and green – or yellow, and blue – what resplendent variety of illumination they may afford to a planet circling about either – charming contrasts and grateful vicissitudes – a red and green day, alternating with a white one, and with darkness.	Star jewels tinted in blue and green, in yellow and vermillion, in the vivid red of porphyry! What marvelous variety of light must they impart to the planets which circle around them.
Sterne werden immer scheinen, / Allgemein auch zum Gemeinen; / Aber gegen Maß und Kunst / Richten sie die schönste Gunst.	The second of two quatrains in 'Leuchtender Stern' (1826) by Johann Wolfgang von Goethe (1749–1832).	Stars will always shed their light / Upon the world all through the night; / But to measurement and art / Their greatest favour they'll impart.	Stars always shine everywhere, giving their light to all, but to those who keep vigils with them they impart their secrets.

54. French astronomer and author Camille Flammarion (1842–1925) compiled the earliest descriptions and sketches of Mars by telescopic observers in 'La première période: 1636–1830', *La Planète Mars* (1892), pp. 3–98.

55. De la Rue had already presented a drawing he made of Mars in October 1851. His illustrations were publicly displayed, but not reproduced in the *Monthly Notices*. His detailed written description of these observations was published in the April number. De la Rue, 'Mars in opposition in 1852' (1852).

56. Webb, 'Note on the telescopic appearance of the planet Mars' (1856).

57. Cassini's drawing was based on his observation of Mars from Bologna on 3 March 1666, just before its opposition on 19 March. The telescope he used was fitted with lenses by renowned Roman optician, Giuseppe Campani (1635–1715), who made his own observations of Mars at that time. A report, along with drawings based on these observations, appeared in 'Observations made in Italy' (1666).

58. Brodie, 'Note on the telescopic appearance of Mars' (1856).

59. De la Rue's drawings were based on his observations made at 9:40 and 11:45 p.m. on 20 April 1856. Flammarion, *Planète Mars* (1892), pp. 128–9.

60. Flammarion, *Planète Mars* (1892), p, 128.

61. In his summary of the first period of Mars observations, Flammarion remarked that 'the snowy polar caps are not situated at the extremities of the same diameter, and they do not mark absolutely the planet's geographic poles. The poles are generally covered. But when they are at their minimum, they reduce to a circular white point located some distance from the polar axis. Herschel found in 1781, a 13°–14° separation between the polar

axis and the center of the northern cap (then very small after its summer) while the southern cap was very extensive with its center near the pole.' Flammarion, *Planète Mars* (1892), p. 97.

62. The observations were made on 24 April (10h), 28 April (8 1/2h), 19 May (10h) and 2 June (10h) 1856. See Notebook 1, WCL/SC.

63. Huggins did not excise these, or the other drawings he presented to the RAS, from his notebooks. His presents may have been his original observation sketches on which he based his notebook illustrations, or vice versa. A search for the illustrations in the RASL was unsuccessful. W. Huggins, 'Note accompanying drawings of Jupiter, Mars, &c.' (1856).

64. W. Herschel, 'On the remarkable appearances at the polar regions of the planet Mars' (1784).

65. 'List of members', *RBAAS, Leeds* (1858).

66. 'The Rev. William Rutter Dawes', *MNRAS* **29** (1869), pp. 116–20.

67. Airy, 'Address of the president on presenting the Gold Medal' (1855). The name of the dark ring of Saturn is spelled either 'crape' or 'crepe'. The former spelling was used by Huggins's contemporaries and so is used here.

68. For a brief discussion of Dawes's efforts in comparing the quality of objective lenses, see 'The president, referring to the excellent opportunities. . .', *MNRAS* **25** (1865), pp. 230–1.

69. Apparently Huggins's close association with Dawes on astronomical matters was noted by Dawes's survivors. In a letter to the Revd Thomas Romney Robinson, Huggins remarked, 'Mr. Dawes's nephew & executor has most kindly presented me with Mr. Dawes' observatory journals.' See W. Huggins to Robinson, 23 January 1869, Add MS 7656.TR75, GGS, CUL. Dawes's observatory notebooks are now held in the RASL.

70. See the numerous individual reports of this comet, *MNRAS* **19** (1858), pp. 12–28.

71. See W. Huggins, 15 October 1858, Notebook 1, WCL/SC. For artists' renderings of this impressive comet, see Olson, *Fire and Ice* (1985), pp. 99–100.

72. I have determined this using 450 nm to approximate the wavelength of incoming light from these stars and then applying the formula for determining the resolving power of a telescope (separation of objects in arcseconds = 3.72 ÷ diameter of objective lens in inches). It should be noted that it is only because of the short wavelength of the light emitted by these stars that an 8-in telescope was able to resolve them.

73. W. Huggins, 22 October 1858, Notebook 1, WCL/SC.

74. Dawes, 'On the appearance of round bright spots on one of the belts of Jupiter' (1857).

75. Huggins's original drawings of Jupiter are held at the RAS. See MSS Add. 18, RASL. They were presented to the Society at the meeting held on 14 April 1871. See *AR* **9** (1871), pp. 107–9.

4

'A sudden impulse . . .'

> . . . news reached me of Kirchhoff's great discovery. . .
>
> — *William Huggins*[1]

It has long been accepted by historians of astronomy that William Huggins underwent a dramatic change in his research interests and methods in the early 1860s. And no wonder, for he has told us so himself in 'The new astronomy':

I soon became a little dissatisfied with the routine character of ordinary astronomical work, and in a vague way sought about in my mind for the possibility of research upon the heavens in a new direction or by new methods. It was just at this time . . . that the news reached me of Kirchhoff's great discovery of the true nature and the chemical constitution of the sun from his interpretation of the Fraunhofer lines.

This news was to me like the coming upon a spring of water in a dry and thirsty land. Here at last presented itself the very order of work for which in an indefinite way I was looking – namely, to extend his novel methods of research upon the sun to the other heavenly bodies. A feeling as of inspiration seized me: I felt as if I had it now in my power to lift a veil which had never before been lifted; as if a key had been put into my hands which would unlock a door which had been regarded as for ever closed to man – the veil and the door behind which lay the unknown mystery of the true nature of the heavenly bodies.[2]

The passage is a favourite among chroniclers of the origins of astrophysical research.[3] Not only does it possess all the authority of an eyewitness account, but its wry understatement of the deadly tedium faced by serious observers pursuing a nightly regimen of traditional astronomical mapmaking and clockwatching struck a chord with turn-of-the-century readers who had spurned such drudgery themselves and opted for a career in astrophysics. It was invigorating to trace one's professional roots to a pioneer of such vision and daring – someone obviously on the inside track who correctly assessed the state of affairs and acted to rectify things.

But Huggins was not the prescient genius his statement may suggest. He could describe the events that influenced his early career choices in such clear-cut terms because he was writing about them some thirty-five years after the fact. The image of a sudden leap forward with no turning back called up by this passage does not mesh with the patchwork of observations that he recorded between 1860 and 1865. His transformation was in reality a complex and gradual process, which, as with many other retrospective accounts, became simplified and foreshortened when viewed with the long focus of his own mnemonic lens.

4.1 The Pharmaceutical Society soirée

It all started, Huggins tells us, in January 1862, when 'I happened to meet at a soirée of the Pharmaceutical Society, where spectroscopes were shown, my friend and neighbour, Dr. W. Allen Miller, Professor of Chemistry at King's College, who had already worked much on chemical spectroscopy.'[4] Miller was at the soirée to deliver, once again, the well-received lecture he had given months before at the BAAS meeting in Manchester on using spectroscopy for qualitative analysis.[5] He began his talk by introducing Fraunhofer's early examination of the spectra of Venus, the Moon, Sirius, Castor, Procyon, Capella and Betelgeuse.[6] His allusions to celestial spectroscopy were merely a prologue to his real purpose, namely to bring his audience up to date on what was currently known about the spectra observed in coloured flames, the power of the method for chemical analysis, and Kirchhoff's recently completed map of the solar spectrum.

Miller acknowledged it was conceivable that Kirchhoff's interpretation of spectral patterns would provide a 'further glimpse into the machinery of the universe', but he cautioned his audience that, 'it appears by some to have been hastily assumed, in a spirit of self-confidence, that we already have the key to everything upon this subject'. He was sure additional research would ultimately teach investigators a 'lesson of reverent humility'.[7]

After the soirée, Huggins recalled, '[a] sudden impulse seized me to suggest to [Miller] that we should return home together. On our way home I told him of what was in my mind, and asked him to join me in the attempt I was about to make, to apply Kirchhoff's methods to the stars.'[8] Did Miller's talk inspire Huggins to consider including the veteran spectroscopist in his evolving astronomical research plans? Possibly. It is curious that Huggins makes no mention of Miller's talk in his retrospective essay, recounting only that there had been a display of spectroscopes and a chance meeting with his 'friend and neighbour'.[9] They lived across the street from one another, so it is quite easy to imagine that their subsequent collaboration grew out of a conversation the two men shared during a carriage ride home that evening.[10] But, if this was indeed Huggins's first introduction to spectrum analysis, he had somehow remained ignorant of a subject that had achieved, by then, a degree of popular currency. Of course, in early 1862, Huggins was still operating on the periphery of London's professional scientific community. Thus it is possible that he was unaware of Miller's BAAS address the previous year, the recent discussions on spectrum analysis at the Chemical Society, and the papers that had appeared in the *Philosophical Magazine* and in the publications of the Royal Society.

4.2 William Allen Miller

Miller was Chair of Chemistry and a dean at King's College, London. He was a founding member of the Chemical Society and had recently been elected RS treasurer and vice-president, positions he held until his untimely death in 1870. An experienced spectroscopist and photographer, he began his first experiments in prismatic analysis in the 1840s purportedly in a lumber-room located below the chemical lecture theatre at King's that he had converted into a make-shift laboratory.[11] Unlike his colleagues in physics, who wished to reconcile selective absorption of light by gases with then-current views on the nature of light, Miller aimed to uncover the relationship between the chemical properties of the substances he examined and the pattern of colourful lines observed in their flame spectra. He wanted to use these spectral signatures to expedite the positive identification of chemical elements. At the annual BAAS meeting in 1845, he reported on his examination of the dark lines observed when light passed through coloured vapours. It was a study he hoped would aid in determining the cause of the dark lines in the solar spectrum, a problem of 'much difficulty and obscurity'.[12]

In his teaching, he emphasised methods of qualitative analysis to prepare his students in the theory and practice of modern chemistry. One of his students kept a diary of his medical school years (1860–64) and often referred to his classes with Miller. Although the study of chemistry did not seem to be this young man's strong suit, he commented favourably on his professor's presentation of the polarisation of light, marvelled at his electrical demonstrations, and – just months after a translation of Bunsen's and Kirchhoff's paper on chemical spectrum analysis appeared in the *Philosophical Magazine*[13] – he wrote: 'Professor Miller regaled us with some beautiful experiments, showing the different colours of the various metals during combustion.'[14] The demonstration indicates nothing about Miller's knowledge of, or reaction to, Bunsen's and Kirchhoff's spectral work, but it does show that, as an empirical method of chemical analysis, flame testing had become a matter of routine at the introductory level in the universities by 1860, even with no physical theory on which to base it.

Miller had written an influential textbook, *Elements of Chemistry*, which appeared in a second edition in 1860. In it he addressed the problem of the fixed lines in the solar spectrum, noting that they 'are independent of the nature of the refracting medium, and occur always in the same colour, and at corresponding points of the spectrum'. While admitting their utility for opticians in determining accurately a substance's index of refraction, he added, 'No satisfactory explanation has yet been found for the cause of this phenomenon.'[15]

4.3 Chemical spectrum analysis

On 1 March 1861, Miller's young colleague Henry Roscoe gave an invited lecture at the Royal Institution on Bunsen's and Kirchhoff's observations of chemical spectra, including their recent historic discovery of the first new element to be found using the method of spectrum analysis. The two collaborators named this new element 'caesium' for the two blue lines that distinguish its spectrum.[16]

Meanwhile, William Crookes was preparing to announce his own spectroscopic discovery of a new element he called 'thallium' for the green line in its spectral signature.[17] This may explain, in part, why the noisy attention being paid to the German spectroscopists failed to impress Crookes, and why, as editor of *Chemical News*, he felt the need to remind his readers of the method's English ancestry by reprinting the text of Miller's 1845 BAAS report.[18] 'Professor Miller', he declared, 'has anticipated, by nearly sixteen years, the remarkable discovery, ascribed to Kirchhoff, of the opacity of certain coloured flames to light of their own colour.'[19] In fact, he claimed, the pioneer spectroscopist's earlier work surpassed that being done more recently, a clear reference to the work of Bunsen and Kirchhoff. And he pronounced Miller's old spectral maps of the elements to be more accurate than Kirchhoff's.[20]

Crookes's words set off a minor priority dispute that kept Miller and the subject of spectrum analysis in the public eye. Roscoe complained in a letter to Stokes:

I see that Crookes wishes to claim for W. Allen Miller the discovery of the power of luminous gases to absorb light of the same kind as they emit, but I cannot believe . . . that Miller could think of claiming this point. His sentence on this subject is not only vague, but as I read it, positively incorrect.[21]

Anxious to set the record straight, Kirchhoff prepared his own chronicle of the history of spectrum analysis in which he questioned the value of Miller's early spectroscopic work, and angrily chastised Crookes for wrongly stating that it had anticipated his own:

[B]y way of experiment, I have laid Professor Miller's diagrams before several persons conversant with the special spectra requesting them to point out the drawing intended to represent the spectrum of strontium, barium, and calcium respectively, and . . . in no instance have the right ones been selected.[22]

Later that year the BAAS met in Manchester. As president of the Chemistry Section, Miller delivered the annual report on progress in that science. The year had seen great strides in practical chemistry from the development of new methods for generating gases for industrial use to the discovery of new processes for creating artificial crystallised minerals. But, in Miller's view, nothing had captivated the public imagination quite like Bunsen's and Kirchhoff's spectroscopic discovery of caesium, and, a few months later, a second new element, which they named 'rubidium' for the dark red line in its spectrum.[23]

Recall that Miller also offered an evening lecture at the meeting that was, as Agnes Clerke later noted, of 'considerable historical value'.[24] The richly illustrated talk on 'The new method of spectrum analysis' was crowded with individuals who had come to hear about Kirchhoff's method and theory. Instead, Clerke recalled, 'they were not a little surprised to find Kirchhoff occupying the end of a long series of illustrious names, from Newton in 1701 to Wollaston in 1802 and Fraunhofer in 1815'.[25] It was this lecture that Miller repeated at the Pharmaceutical Society soirée in January 1862, the very soirée Huggins had recalled in 'The new astronomy'.

Roscoe, for his part, gave a series of three lectures on spectrum analysis at the Royal Institution in March and April 1862 that would have held great interest for Huggins and Miller as they launched their own research programme.[26] Roscoe focused the first two on Fraunhofer's early work on stellar spectra and described how the examination of spectra could be employed in chemical analysis. But in the third talk, he turned to the application of spectrum analysis to the luminous bodies in the heavens, likening the possibilities for extending human understanding in this area of research with the marvels that could be imagined to follow the discovery of the philosopher's stone. He exhorted his colleagues 'to prove that chemical analysis has now stretched forth her hand beyond the limits of our earth, and that she now affords us certain information respecting the chemical constitution of the sun and the far distant fixed stars'. In conclusion, Roscoe raised the prospects of founding a 'new stellar chemistry'.

Miller had emphasised the *similarities* mentioned by Fraunhofer, in particular that Fraunhofer had seen the sodium D lines in Procyon's spectrum, and the magnesium *b* group as well in the spectra of Capella and Betelgeuse. Roscoe, on the other hand, stressed the *differences* noted by Fraunhofer between the Sun and the individual stars he had observed. 'All we know' about the chemical composition of the fixed stars, Roscoe opined, 'is that they contain substances not found in the solar atmosphere'. In fact, he declared, 'it appears that each fixed star possesses a spectrum peculiar to itself'.[27] Lest his audience misconstrue such a difficulty as insurmountable, he reminded them, 'We are ... only on the threshold of these subjects; the possibility of obtaining an insight into the composition of stellar matter has indeed been shown, but it remains for the astronomer and the physicist to obtain for us that knowledge which we so much desire.'[28]

4.4 Collaboration

When did Miller join Huggins in his effort to 'apply Kirchhoff's methods to the stars'? According to Huggins, Miller agreed – after some initial hesitation – to come to work in his observatory 'on the first fine evening' following the soirée.[29] But his notebook records do not corroborate this claim.

Huggins recorded nothing in his notebook between 18 July 1860 and 31 March 1862, not even a note or comment on the brilliant daytime comet of 1861 (Tebbutt's),[30] or the transit of Mercury on 11 November 1861,[31] both of which attracted the attention of English observers. It is a perplexing – and disappointing – gap.

When he resumed his notebook entries in March 1862, it was not to document what he and Miller might have accomplished on their 'first fine evening' of collaborative celestial spectroscopy, but rather to record his return to observing Saturn (see Figure 4.1). His fascination with Saturn at this time is easy to explain. For one thing, the planet was at opposition which meant it was favourably placed for viewing that spring. But its real appeal to amateurs like Huggins stemmed from the fact that, between November 1861 and August

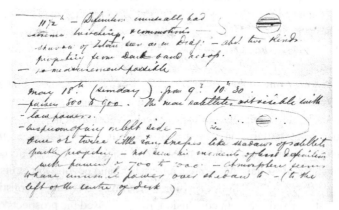

Figure 4.1 Observations of Saturn by William Huggins. 17 and 18 May 1862, Notebook 1. Huggins refers to 17 May as the 'night of Transit of Titan's shadow. Calculated by Mr. Dawes to enter on disk 9 h 41 m'. His sketch shows the shadow of Titan observed at 11:30 p.m. about 'two thirds projecting from dark band across'. On 18 May, Huggins noted a 'suspicion of [Saturn's] ring on left side' and 'little roughnesses like shadows of satellites partly projecting' on the disk when the atmosphere permitted. His sketch shows the positions of (from left to right) Titan, Iapetus, Dione, Tethys, Saturn, Enceladus [?] and Rhea. Enceladus was just emerging from behind Saturn at the time. (WCL/SC)

1862, earthbound observers could see Saturn's ring system edge-on, an auspicious orientation that recurs roughly every fourteen years in consequence of the planet's axial tilt. The rings are so thin that they virtually vanish from sight at such times.

In addition, all but one (Iapetus) of the eight then-known saturnian moons orbit in, or very close to, the ring plane. When that plane is edge-on, a variety of satellite transits, eclipses and occultations naturally occur. Observing and timing such events – notably the passage of the shadow of Saturn's largest moon Titan across the planet's surface – were alluring challenges for amateur observers, including Huggins. From his notebook entry for 3 June 1862, we learn that 'Dr. Miller & family' visited his observatory hoping to catch a glimpse of a 'shadow', possibly the thin dark line cast on Saturn's ball by its rings.[32] This is, incidentally, Huggins's only reference to Miller in his 1862 notebook records.

It is worth noting that amateur astronomers who were watching Saturn so closely in 1862 had more than transits of Titan's shadow on their minds. The temporary absence of the highly reflective rings made it easier to see Saturn's more elusive satellites. In fact, during Saturn's previous edge-on apparition in the autumn of 1848, William Lassell (1799–1880) in England and George Phillips Bond (1825–65) in the United States independently announced the presence of what appeared to be a new moon (Hyperion) orbiting Saturn.[33] Lassell, a brewer by trade, was a veteran satellite discoverer. In September 1846, he had announced the probable existence of a satellite (Triton) orbiting the newly discovered planet, Neptune, a claim that was verified the following year.[34] And, in 1851, he found two small moons of Uranus (Ariel and Umbriel).[35] Thus, within five years, one planet and four moons had been added to the Sun's family, and the moons had all been found by amateurs. In 1862, the *Monthly Notices* carried reports from observers who had seen bright points in the ring plane.[36] Huggins's notebook entries and correspondence show that he shared this interest.[37] Clearly all hoped that, with Saturn's rings once again reduced to near invisibility, yet another one among them would be rewarded for his keen eyes and diligent observation with the discovery of a new moon.

Remarkably, all of the observations Huggins recorded from 29 August 1862 until 6 July 1865 occupy just two sides of a single leaf. There are a few pages of notes in Notebook 4 that were originally thought to have been records of experiments performed in 1871.[38] But the data they contain make it clear that Huggins recorded them in the autumn of 1863 while working with Miller on the chemical spectra of terrestrial elements. Are there other, as yet uncovered, notes from this particular research effort? Did Miller, the experienced spectroscopist, write up the rest of their observations? That Huggins kept these particular records separate from his log of visual and telescopic observations suggests he viewed his spectroscopic work at that time as part of a different enterprise.

The 1862–5 leaf's recto contains two entries made in 1863. The first, in April, includes drawings of Saturn and Jupiter. Six months later, Huggins recorded the barometric fluctuations generated by a violent thunderstorm. The newsworthy weather prompted him to submit a brief report of his observations in a letter to *The Times* of London.[39] The leaf's verso contains a single entry for 1864 (sketches of sunspots; see Figure 6.1), and two entries from July 1865 (micrometer calibrations).[40]

Despite the lack of notebook records during this time period, we can try to establish a rough timeline of their collaborative observations by noting when the particular stars and planets they targeted were visible in the London sky between the date of the Pharmaceutical Society soirée (15 January 1862) and the date they submitted their preliminary report to the Royal Society (19 February 1863).[41] Huggins and Miller provide a few specific dates in their full report to the Royal Society, read 28 April 1864, that we can use to scaffold this timeline. Jupiter, for example, would have been well-placed for viewing from late February of 1862 and throughout the spring and summer.[42] They mention observing the spectrum of Jupiter, as well as those of the Moon and Saturn, on 12 April 1862, the earliest date to which they refer in the paper. During the summer, the constellation Cygnus, and the stars in the handle of Ursa Major would have been seen easily after sunset.[43] Late autumn would have been a good time to examine Mars and the constellations Lyra, Andromeda, Pegasus, Cassiopeia and Triangulum. Indeed, Huggins and Miller tell us that they observed the spectrum of Mars on 6 November, and that of the star β Pegasi (Scheat) on 10 November 1862. The brighter stars like Betelgeuse (α Orionis), Aldebaran (α Tauri) and Sirius (α Canis Majoris), whose spectra Huggins and Miller observed well enough to sketch schematically, as well as Procyon (α Canis Minoris), Capella (α Aurigae) and Castor (α Geminorum), whose spectra they measured only approximately, are all well-placed in the northern hemisphere during January and February. Thus, they could have been the first and/or the last objects observed.[44]

Regardless of what the heavens had to offer, Miller was preoccupied at the start of 1862 with his own research problems. In his continuing efforts to photograph metallic spark spectra, he had found, much to his dismay, that although the sensitivity of photographic plates was greater in the more refrangible, or blue-violet end of the spectrum, the prisms he was using to disperse the arclight absorbed those rays. He was busy searching for materials with sufficient dispersive power that were also transparent to shorter wavelengths of light.[45] It was a daunting task complicated by the discovery that he was in danger of being partially forestalled, albeit inadvertently, by Lucasian Professor of Mathematics, George Gabriel Stokes.[46] The Irish polymath had developed a method of observing the spectrum of an electric spark using a fluorescent material in lieu of photography, a project he pursued in complete ignorance of Miller's investigation.[47] Thus, feeling pressed early in 1862 to ready his results for publication, Miller would have been freer to collaborate with Huggins in a spectroscopic examination of these stars at the conclusion of the period in question.

4.5 'Mr. Huggins ... on the "Stellar Spectrum"'

The earliest print reference to Huggins's spectral work appeared in the first number of the *Astronomical Register*. There we learn that 'Mr. Huggins read a paper on the "Stellar Spectrum"' at the RAS meeting on 14 November 1862. It was the Society's first gathering of the new season, and Fellows were eager to catch up on all the latest news. The *Register* observed 'The Meeting was very numerously attended, the rooms in fact being inconveniently crowded.'[48] The crowding may have exposed more people to the news of Huggins's pioneering study. On the other hand, it may simply have made it harder for everyone to hear what he said. Unfortunately, no details of his paper are provided, nor is any discussion recorded.

The January 1863 number of the *Monthly Notices* included a translation of a paper by Donati in which he described and illustrated the Fraunhofer lines he had observed in the spectra of fifteen of the brighter stars in 1860.[49] Pressed now by priority concerns, Huggins and Miller composed a brief note announcing to the RS that they had observed multitudes of absorption lines in the spectra of a number of selected fixed stars. They submitted their preliminary report, along with schematic drawings comparing the spectra of Sirius, Betelgeuse, Aldebaran and the Sun, on 19 February. That very day, word reached the Society that Lewis Rutherfurd had observed and recorded the spectra of twenty-three stars.[50] When Huggins and Miller read their paper a week later, they let it be known that their work was not confined to the study of three stars alone, but that they had, in fact, spent the past twelve months examining the spectra of thirty to forty other stars as well as those of Mars, Jupiter and the Moon.[51]

Their brief paper soon appeared in the Royal Society's *Proceedings*.[52] Joseph Norman Lockyer, a newly elected Fellow of the RAS, wrote a brief but enthusiastic review of it for the *Monthly Notices*.[53] At the time, Lockyer was an up-and-coming clerk in the War Office. In part to supplement his income, but mostly to satisfy his own desire to remain abreast of and involved in all the very latest scientific discoveries and pronouncements, he accepted an invitation to serve as science writer for *The Reader*, a weekly review of books and contemporary issues in science and the arts launched in January 1863.[54] In that magazine's first number, Lockyer drew attention to Donati's observations of stellar spectra.[55] In the second, he noted Secchi's spectroscopic work.[56] And, on 7 March, Lockyer reported at length on Huggins's and Miller's preliminary report to the Royal Society.[57]

In April, Astronomer Royal George Airy, then Society President, presented an illustrated report on observations of stellar spectra at Greenwich to the RAS.[58] An abbreviated transcript of the comments and questions that followed Airy's talk appeared in the *Register*'s May number.[59] Admiral Russell Henry Manners (1800–70), presiding in Airy's stead, expressed the Society's thanks for the Astronomer Royal's report. 'Although many private observers were giving attention to the subject', Manners proclaimed, 'it was to Greenwich Observatory that all looked as a directing star.'[60]

In contrast to the chair's decorous remarks, Huggins offered some critical comments of his own. At this early stage of the *Register*'s existence, reports from the meetings tended to concentrate on what Hugh Frank Newall later characterised as 'pithy remarks'.[61] In this case, the *Register*'s report does not disappoint and provides a glimpse of an assertive side of Huggins normally hidden in published accounts:

Mr. Huggins ... wished to ask whether the lines shewn in the diagram were the whole of the lines seen. He considered his method of observation was calculated to give more perfect results than that employed at Greenwich. The Spectrum of Capella, for instance, which in the Astronomer Royal's drawing has only two bars, he had observed to be full of fine lines, so fine and so many as apparently to defy accurate delineation. This Star, as well as several others, seems to exhibit quite as many lines as would be seen in the Solar Spectrum under similar circumstances of diminished light and of dispersive power. It, as well as Aldebaran and Arcturus, possesses the "D" line and altogether resembles the Sun very closely. Sirius in addition to the three bars, has lines corresponding with B and C, and indications of other lines between. Mr. Huggins remarked on the extreme importance of obtaining the *exact position* of the lines relatively to those in the Solar Spectrum, and stated that his measures were taken by a direct comparison with a Sodium line thrown into the apparatus, and viewed with the Stellar Spectrum. Mr. Huggins had also compared directly the Spectra of other metals with the Star Spectra.[62]

Never one to miss an opportunity to find out more about a promising new method of astronomical investigation, the Revd Charles Pritchard (1808–93) spoke up first, not to respond to Huggins's concerns about the Greenwich data, but to press Huggins on the reliability of his own observations. How was it 'possible that the exact line could be known', asked Pritchard, since 'the least movement of the instrument would cause an alteration of place'? Huggins confidently replied that 'the apparatus was so constructed that the least movement of the telescope would cause the Spectrum to flash out of view; but not alter its position relatively to the spectrum of comparison'.

As for Airy, rather than respond directly to Huggins's bold remarks, he delegated his assistant, James Carpenter (1840–99) 'the actual observer of the Spectra at Greenwich', to discuss the results of his spectral research project and their visual representation. Although many lines had been observed, Carpenter tersely explained, 'only those measured were introduced into the diagram'.

The last word went to Huggins. Clearly still unsatisfied with the quality of the spectro-scopic work at Greenwich implied by the report, he let it be known that 'the shaded bars in α Orionis [Betelgeuse], as represented in Greenwich's diagram, were not truly bars or bands, but groups of fine lines, several of which he had measured'. A 'pithy' exchange indeed! It is easy to understand why, despite Airy's initial enthusiasm in announcing the results of Greenwich's first spectroscopic investigations, they were discontinued for nearly a decade.[63]

4.6 Spectra of terrestrial metals

Huggins had every right to boast of having observed and measured 'groups of fine lines' in Betelgeuse's spectrum. He and Miller were the first to discern this level of detail in stellar spectra. Their technical achievement alone was worthy of applause. But what did the lines signify? Miller's years as a pioneer in the emerging science and art of qualitative spectrum analysis exposed him to similar features in terrestrial spectra. Mapping them was a daunting work in progress. Until spectrum analysts made more headway in this research effort, no one could say with any certainty if, or how, terrestrial and celestial spectra might be related.

In 1863, working alone, Huggins undertook the task of remedying the lamentable lack of precise spectral maps of terrestrial metals. It is the records of these experiments that he entered in Notebook 4 – the only ones from the early phase of his spectroscopic work that have been uncovered. He presented his results to the RS in November 1863. Because Huggins was not yet a Fellow of the Society, his paper was communicated by Miller, whose assistance in obtaining pure samples of the various metals Huggins gratefully acknowledged. Nineteen of the metals he examined were identical to those whose ultraviolet spectra Miller had recently photographed.[64]

The paper was refereed by the Revd Thomas Romney Robinson.[65] The septuagenarian director of Ireland's Armagh Observatory (since 1835) had been elected to Fellowship in the RAS when Huggins was just a boy of six. Throughout his long life, Robinson maintained a youthful curiosity concerning new phenomena and methods of investigation. Indeed, his own recent study of electric spark spectra left him no stranger to the practical challenges that faced the spectroscopist.[66]

Declaring Huggins's paper 'a very good one', Robinson announced that he had 'no hesitation in recommending [it] for publication in the Transactions'. He applauded the young amateur for employing a train of six prisms to produce a highly dispersed spectrum. Furthermore, the fine telescope with which Huggins had viewed the spectra introduced a remarkable increase in precision.

Robinson grumbled that Kirchhoff's measures were 'so far arbitrary' that it was 'not in every case an easy task to identify their lines'. Perhaps 'arbitrary' was not the right word here, but it was the adjective of choice at the time among Kirchhoff's English-speaking critics. It is true that Kirchhoff's measures were peculiar to his own apparatus. After all, dispersion is a non-linear phenomenon and every material has its own dispersive properties. Others had to adjust their instruments and scales to match Kirchhoff's in order to make any reasonable comparison of spectra. But, even with the same apparatus, local fluctuations in temperature and humidity were known to introduce discrepancies in the apparent location of spectral lines from one trial to the next. Finding agreement was not a simple matter.[67]

In their historic first paper on 'chemical analysis', Kirchhoff and Bunsen had included an illustration of the apparatus with which they examined the flame spectra of several pure salts (see Figure 4.2).[68] It was a simple construction consisting of collimating and viewing telescopes fixed into opposite sides of a trapezoidal box. Inside the box was a prism mounted on a turntable. The observer slowly rotated the prism and watched as the entire spectrum passed across the viewing telescope's field. Once a particular line of interest appeared at its sharpest and brightest – its minimum angle of deviation – the prism's precise orientation could be noted by reading the fiducial marks of a horizontal scale. The scale was not an integral part of the spectroscope, however. It was 'fixed at some distance' from a small mirror secured to the prism's rotational axis. To read the scale it was necessary to view its reflected image through another 'small telescope placed some way off'.

This 'detached scale' design troubled Robinson. It must have bothered Bunsen and Kirchhoff as well. In a 'Second memoir', they described an 'improved form of apparatus'

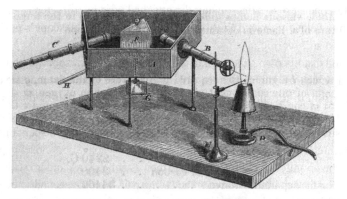

Figure 4.2 Spectroscope of Kirchhoff and Bunsen. (Kirchhoff and Bunsen, *PM* 4th Series, **20**, p. 91)

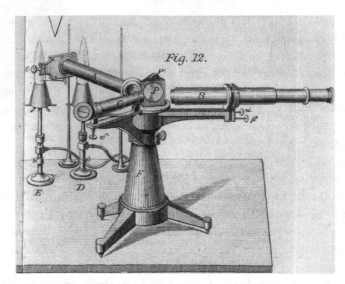

Figure 4.3 Improved spectroscope of Kirchhoff and Bunsen. (Plate VI, Fig. 12, Kirchhoff and Bunsen, *PM* 4th Series, **22**, opposite p. 420)

which was 'in every respect ... to be preferred to that described in our first memoir' (see Figure 4.3). It was, they proclaimed 'more manageable' and produced 'more distinct and clearer images'. In addition to these advantages, its more sophisticated optical arrangement allowed the observer to view two different spectra at once along with a superposed image of the horizontal scale, thus making it possible to both examine and measure the spectra simultaneously.[69]

Robinson was glad to see that Huggins had abandoned the Germans' 'detached scale' comparison scheme in favour of a fixed brass arc divided into 15″ intervals.[70] Both the graduated arc and the radial motion of the viewing telescope were centred about a point near the middle of the last prism face, enabling Huggins to scan the spectrum emerging from the stationary prism train by moving the telescope slowly along the brass arc (see Figure 4.4). In this way, he could map individual spectral lines easily and precisely by first aligning each one on the telescope's cross-hairs – using the attached wire micrometer when necessary to check positions of closely spaced lines – and then reading the telescope's position directly off the graduated arc's vernier.

Huggins noted that his own spectral line measures did not correspond with those reported by Bunsen and Kirchhoff. He blamed the differences on the Germans' practice of constantly adjusting their prism to the angle of minimum deviation for the part of the spectrum being observed. To create a more standardised spectral map, one that any observer working with a prism-based instrument could use as a reference, Huggins fixed his six dispersing prisms in the position of minimum deviation for Fraunhofer's D line – actually a closely spaced pair of lines – and left this arrangement undisturbed. He then established a standard scale for the entire spectrum based on the separation between the D line pair, which he found to be equivalent to five intervals on his graduated arc, or sixty parts of one revolution of his micrometer screw. He

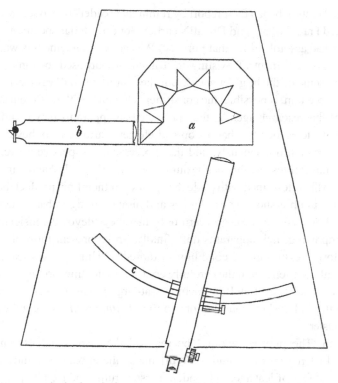

Figure 4.4 William Huggins's chemical spectroscope. (W. Huggins, *PTRSL* **154**, p. 140)

designated the position of line D_1 as '1000', and that of line D_2 as '1005'. Accordingly, he assigned numerical values for the positions of all other spectral lines.

Huggins introduced another significant methodological change. Because he was primarily interested in studying the spectra of night sky objects, he needed to find an alternative to the solar spectrum to serve as his standard reference. He chose the spectrum of air, which he could produce at will in his observatory by generating a spark between two platinum electrodes. By freeing himself, and others, of the need to track the Sun or await its appearance, he moved the mensurational focus of celestial spectroscopy away from the light source associated with Bunsen and Kirchhoff to one that was new, more neutral, and laboratory-based. It was a welcome shift for British investigators.

4.7 'On the spectra of some of the fixed stars'

On 28 April 1864 Huggins and Miller submitted a full report of their work 'On the spectra of some of the fixed stars' for publication in the *Philosophical Transactions*.[71] The paper's referee, George Stokes, acclaimed it as a 'model of patient and accurate investigation'.[72] He was especially impressed that the authors considered only two stars in the paper despite the large number they had observed. By wisely rejecting observations that presented interpretive problems, Stokes noted, they ensured the accuracy of their statements.

The two collaborators began their report by reminding readers that, over two years earlier, they had verified Fraunhofer's and Donati's earlier work on stellar spectra using a 'form of apparatus' they had assembled for that purpose. We can well imagine this was a shoestring and sealing-wax arrangement of existing instruments repurposed to bring into view 'the stronger lines in some of the brighter stars'. But their ultimate goal was far more ambitious. They aimed 'to ascertain, if possible, the constituent elements of the different stars'. To meet the demands of this research agenda, they perfected an apparatus they could attach to the observing end of a telescope that had its optical elements arranged so they could view and compare two spectra simultaneously, and that possessed a dispersive power sufficient to resolve the sodium D lines and the magnesium *b* band (see Figs. 1, 2 and 3 in Figure 5.2).[73]

Huggins and Miller described with pride the features of their final product, but masked the nitty-gritty of the trouble-shooting challenges and disappointing setbacks they faced in the process. Instead of messy details, we learn only that they 'devoted considerable time and attention' to improving their apparatus until finally, 'in its present form of construction it fulfils satisfactorily several of the conditions required'.[74] Huggins was learning the art of converting actual experience into the carefully contrived scientific jargon which the elites of the community would accept and the members-at-large would expect. He had had some practice doing this in his short contributions to the *Monthly Notices*, but now he was being tutored by a master.

Years later in 'The new astronomy' he recalled the resourcefulness and hard work required to find alternatives to liquid-filled prisms so the apparatus could move freely in any plane without fear of leakage, to broaden the spectrum produced by point-like stars so the dark absorption lines could be examined more easily, to carefully align both telescopic and comparison spectra, and to alter drastically the traditional organisation of observatory space in response to new demands made by the intrusion of chemical laboratory instruments.[75] The problems mentioned are ones whose solutions really mattered in the end. Huggins could recognise those problems in 1897. By then, the dead ends he and Miller may have pursued in the early 1860s were forgettable wastes of time. And as he looked back to the 'moment' when everything was finally in place, he could proclaim:

Then it was that an astronomical observatory began, for the first time, to take on the appearance of a laboratory [see Figure 4.5]. Primary batteries, giving forth noxious gases, were arranged outside one of the windows; a large induction coil stood mounted on a stand on wheels so as to follow the positions of the eye-end of the telescope, together with a battery of several Leyden jars; shelves with Bunsen burners, vacuum tubes, and bottles of chemicals, especially of specimens of pure metals, lined its walls. The observatory became a meeting place where terrestrial chemistry was brought into direct touch with celestial chemistry.[76]

The new observatory *cum* laboratory was ready just 'before the end of the year 1862'. At the time, Tulse Hill was home to the only work space of its kind in the world. In it, Huggins and Miller conducted their 'more than usually toilsome' work to extend their survey of stellar spectra to include some fifty stars.[77] Although they found Sirius too difficult to observe because of its proximity to the horizon, they mapped details in the spectra of Aldebaran and Betelgeuse to a level of completeness that satisfied them (see Figure 4.6). To discover what could be learned of planetary surfaces and atmospheres, they examined the spectra of light

INTERIOR OF THE OBSERVATORY, 1860—1869.

Figure 4.5 Interior of the Tulse Hill observatory in the 1860s, by J. Swain, possibly noted engraver Joseph Swain (1820–1909). The illustration – probably based on William Huggins's recollection rather than drawn from life – shows the mounting and tube built by Thomas Cooke for the 8-in Alvan Clark objective Huggins purchased from William Rutter Dawes. The star-spectroscope, depicted in Figs. 1 and 2 of Plate X, *PTRSL* **154** (see Figure 5.2), is attached to the viewing end of the telescope. Note the elaborate spark apparatus positioned prominently on the floor in the foreground. (W. Huggins and M. L. Huggins, *Atlas of Representative Spectra*, p. 4).

from the Moon, Venus, Mars, Jupiter and Saturn. They identified spectral signatures of many known terrestrial elements in the spectra of fixed stars. Having found that the spectra of orange stars differ from those that are bluer, they suggested that stars are a chemically heterogeneous population. They recorded the spectra of at least two stars (Sirius and Capella) on wet collodion plates. Unfortunately, the resulting images displayed no lines.

'[S]tars', they concluded, 'while differing the one from the other in the kinds of matter of which they consist, are all constructed upon the same plan as our sun, and are composed of matter identical, at least in part, with the materials of our system.' Because stellar spectra

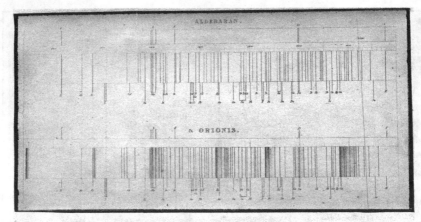

Figure 4.6 Spectra of Betelgeuse (α Orionis) and Aldebaran compared with the solar spectrum, and the spectra of chemical elements. (Frontispiece, W. Huggins, *On the Results of Spectrum Analysis Applied to the Heavenly Bodies*)

display the signatures of elements associated with life such as hydrogen, sodium, magnesium and iron, they suggested that 'at least the brighter stars are, like our sun, upholding and energising centres of systems of worlds adapted to be the abode of living beings'.

Huggins and Miller argued cogently, on the strength of carefully considered evidence, that an earthbound observer in possession of the right instruments and attendant skills could discover the true chemical and physical nature of celestial bodies. Thus, they changed the boundaries of acceptable research by which astronomers, physicists and chemists defined their communities, and exploded the Comteian notion that 'Our positive knowledge with respect to the stars is necessarily limited to their observed geometrical and mechanical behaviour.'[78]

***

To attain the recognition of his fellows, Huggins had to demonstrate both a commitment and a willingness to conform to the RAS's pre-existing performance standards. He began to move in this direction by constructing an observatory building attached to his home, purchasing a high-quality refracting telescope with which to make his observations, and contributing reports from his observatory for publication in the RAS's *Monthly Notices*.

But his entrepreneurial bent drove him to diversify as well. Always on the lookout for an unoccupied research niche, he eagerly acted on the opportunity to try something entirely new: applying the chemical method of spectrum analysis to the light of celestial bodies. It was a risky line of investigation to pursue for a tyro hoping to move from the periphery into the inner circle of serious amateur astronomy. To do both successfully, he had to convince those within that circle of spectroscopy's value and importance to the increase of astronomical knowledge – and of his own technical competence in this area.

As Huggins subjected more terrestrial and celestial objects to spectroscopic scrutiny, he worked to sustain his visibility among his fellow amateur astronomers. He confidently and rightly advertised his instruments, methods and resulting observations as clearly superior to any others engaged in similar tasks. He also strove to build on the potentially fleeting notoriety he achieved in the wider scientific community stemming from his collaborative work with William Allen Miller, the well-respected chemist, recognised pioneer spectroscopist and high-ranking official in the Royal Society. By wisely nurturing his relationship with Miller, Huggins gained a leg up into that elite scientific circle, and increased opportunities for personal recognition and advancement.

Notes

1. W. Huggins, 'The new astronomy' (1897), p. 911.
2. *Ibid.*
3. See, for example, Maunder, *Sir William Huggins and Spectroscopic Astronomy* (1913), p. 9; Dyson, 'Sir William Huggins' (1910), p. ii; Newall, 'Sir William Huggins' (1910–11), p. 177; Newall, 'William Huggins' (1911), p. 261; W. W. Campbell, 'Sir William Huggins' (1910), p. 308; H. S. Williams, *The Great Astronomers* (1932), pp. 345–6; Meadows, 'The origins of astrophysics' (1984), p. 13.
4. Although the implication in the quote as taken from context is that this is coincident with his first hearing of Bunsen and Kirchhoff's work at some time shortly after its original announcement, the soirée to which he refers took place on 15 January 1862.
5. 'Report on presidential address by William Allen Miller to the Chemical Section of the BAAS', *CN* 4 (1861), pp. 159–61; p. 159.
6. The text of Miller's speech includes a footnote in which these celestial objects are identified. We cannot know whether he named these particular objects in the actual presentation. Miller, 'On spectrum analysis' (1862), p. 400. The text of Miller's address was reprinted in *CN* 5 (1862), pp. 201–3 and pp. 214–18.
7. Miller, 'On spectrum analysis' (1862), p. 412.
8. W. Huggins, 'The new astronomy' (1897), pp. 911–12.
9. *Ibid.*, p. 911.
10. Miller lived at 103 Tulse Hill Road.
11. Although this is reported in Miller's RS obituary [Trotter, 'William Allen Miller' (1870)] and repeated in the 1894 edition of the *DNB* [Clerke, 'William Allen Miller' (1894)], there is no record at King's College as to where this room might have been located or how it might have been altered to make such research possible. I was given a tour of the depths of the college's old building by Professor D. I. Davies (King's College), who pointed out rooms that he conjectures may have been appropriated by Miller for his spectroscopic experiments.
12. Miller, 'On the action of gases on the prismatic spectrum' (1845), pp. 28–9.
13. Kirchhoff and Bunsen, 'Chemical analysis by spectrum-observations' (1860).
14. See entries for 25 October 1860; 22, 23 and 28 November 1860; 1 December 1860 in Taylor, *The Diary of a Medical Student* (1927). I am indebted to Professor Davies for bringing this book to my attention.
15. Miller, *Elements of Chemistry* (1860), p. 146.
16. Roscoe, 'On Bunsen and Kirchhoff's spectrum observations' (1861).
17. Crookes, 'On the existence of a new element' (1861).
18. Miller, 'Experiments and observations on some cases of lines in the prismatic spectrum produced by the passage of light through coloured vapours and gases, and from certain coloured flames' (1861). [Reprinted from *PM*, 3rd Series **27** (1845), pp. 81–91.]
19. *Ibid.*
20. Crookes republished papers written decades earlier by other British researchers who had investigated spectral phenomena in the *Chemical News* in three instalments: 'Early researches on the spectra of artificial light from different sources' (1861). See pp. 303–4 for his comparison of Miller's and Kirchhoff's spectral maps.
21. Roscoe to Stokes, 6 February 1862, Add MS 7656.R794, GGS, CUL.
22. Kirchhoff, 'Contributions towards the history of spectrum analysis' (1863), p. 255.
23. Miller, 'Address to the Chemistry Section' (1861).
24. Clerke, 'William Allen Miller' (1894), p. 429.
25. Trotter, 'William Allen Miller' (1870), p. xxii.
26. Roscoe, 'A course of three lectures on spectrum analysis' (1862).
27. The difference between Miller's and Roscoe's representations of Fraunhofer's observations is subtle, yet interesting in terms of what the two men seem to have believed further investigation of stellar spectra might reveal. See Fraunhofer, 'Kurzer Bericht von den Resultaten' (1823).

28. Roscoe, Lecture III in 'A course of three lectures on spectrum analysis' (1862), p. 292.

29. W. Huggins, 'The new astronomy' (1897), p. 912.

30. See, for example, Main, 'Observations of Comet II, 1861' (1862). The comet appeared in the summer of 1861 and sported a coma as large and bright as the Moon and a tail between 100° and 120° in length. It was reportedly so bright between 29 June and 1 July that it cast shadows at night, making it possibly one of the brightest comets of the century. See Olson, *Fire and Ice* (1985), p. 97. For drawings of Comet 1861 II by Secchi and De la Rue, see Sagan and Druyan, *Comet* (1985), pp. 175, 179, and 180.

31. Observations on the transit were noted by others in England. See, for example, the reports of J. W. Jeans and J. Baxendell in *MNRAS* 22 (1862), pp. 42–3.

32. W. Huggins, 3 June 1862, Notebook 1, WCL/SC. The full entry states 'Very fine. Dr. Miller & family. The shadow not visible.' It is unclear to what shadow he is referring. Dawes had published in the *Monthly Notices* a list of dates and times when the transit of Titan could be observed. One scheduled for 17 May was observed and recorded by Huggins. A second one, scheduled for 2 June, was unfortunately clouded out. Titan's shadow would not have been visible on 3 June.

33. W. Lassell, 'Discovery of a new satellite of Saturn' (1848). W. C. Bond, 'Discovery of a new satellite of Saturn (1848). Bond actually saw the moon first, on 16 September 1848, while Lassell, the amateur, did not spot it until the 18th. However, both confirmed their first observation on 19 September, making them co-discoverers. [Note: the actual sighting was by George Phillips Bond, but as his father, William Cranch Bond (1789–1859), was a Fellow of the RAS, the note was communicated by him. This has occasionally led to some confusion.]

34. Lassell's original announcement concerning the alleged satellite of Neptune was couched in tentative terms. The observation on which this claim was based was made when Neptune was positioned too near the Sun to permit a second search. Lassell stated 'One, or perhaps two, luminous points have been seen, which *may be* satellites; but this will require further scrutiny.' ['Observations of Leverrier's planet', *MNRAS* 7 (1846), pp. 154–7, p. 157.] In this same announcement he claimed to see what appeared to be a ring crossing Neptune's disc, a feature which he later reported others had seen as well. The following summer, when Neptune had returned to a more favourable position, Lassell confirmed his sighting of the satellite, as did Otto Struve and G. P. Bond. With such a questionable first observation, however, it is not clear that these subsequent sightings can reasonably be called confirmations. See also Lassell, 'On a satellite of Neptune' (1850); R. W. Smith, 'William Lassell and the discovery of Neptune' (1983); R. W. Smith and Baum, 'William Lassell and the ring of Neptune' (1984).

35. Lassell, 'In a letter dated . . .' (1851).

36. Birt, 'On the appearance of Saturn's ring, 1862' (1862); Dawes, 'Saturnian phenomena' (1862).

37. In a lengthy and detailed note to the RAS, Huggins reported his own observations of Saturn including its colour, relative brightness and other salient features. W. Huggins, 'On some phenomena attending the disappearance of Saturn's ring, May 19th, 1862' (1862).

38. For a discussion of the contents of the individual notebooks, which is essentially correct except on this one point, see Morgan, 'Huggins archives' (1980).

39. W. Huggins, 13 April and 30 October 1863, Notebook 1, WCL/SC. This is the only such observation he recorded in the notebooks. W. Huggins, 'To the editor of the Times', *The Times* of London, 2 November 1863, p. 10d.

40. W. Huggins, 23 March 1864, Notebook 1, WCL/SC. A miscellany of entries, most recorded in 1865, fill the next five leaves. Turning the last of them reveals a bold new hand – that of Margaret Lindsay Huggins who, in 1875, became William Huggins's wife and collaborator. The page is clearly headed '1889'. For whatever reason, Huggins abandoned this first notebook in 1865/6 to start afresh in Notebook 2, WCL/SC.

41. J. N. Lockyer, 'Note on communication . . .' (1863); W. Huggins and Miller, 'Note on the lines in the spectra of some of the fixed stars' (1863).

42. W. Huggins and Miller, 'On the spectra of some of the fixed stars', *PTRSL* 154 (1864).

43. The paucity of summertime objects is undoubtedly due to the late hour at which the sky becomes dark at London's latitude.

44. Huggins and Miller noted an attempt to capture the spectra of Sirius and Capella on a photographic plate during the course of these difficult observations. It is likely that they did their visual work during the first apparition of these stars in order to gain some experience before tackling the photographic work.

45. See Miller, 'On the photographic transparency of various bodies' (1862).

46. Stokes, 'On the long spectrum of electric light' (1862).

47. Stokes to Roscoe, 7 February 1862, Add MS.7656.R795, GGS, CUL. After discovering their overlapping efforts, Miller and Stokes exchanged letters in which they discussed the nature of their work. See Miller to Stokes, 25 April 1862, Add MS.7656.M533 and 29 April 1862, Add MS.7656.M535, GGS, CUL. It is interesting that, in these letters, Miller makes no reference to any concurrent work on stellar spectra.

48. *AR* 1 (1863), p. 2. The paper is not on the annual list of 'Papers read before the Society . . .', *MNRAS* 23 (1863), pp. 149–52.
49. Donati, 'Memorie Astronomische' (1863); originally published in the *Annals of the Museum at Florence*, 1862.
50. Rutherfurd, 'Astronomical observations with the spectroscope' (1863); Warner, 'Lewis M. Rutherfurd' (1971). Huggins later claimed to have heard of Rutherfurd's work just as he arrived at the Royal Society to deliver his own paper on the same subject. See W. Huggins to Turner, 15 February 1893, Correspondence of the Society, RASL.
51. Certainly the length of study and the extent of Miller's and Huggins's work impressed Lockyer in his report on the paper in the *Monthly Notices*.
52. W. Huggins and Miller, 'Note on the lines in the spectra of some of the fixed stars' (1863).
53. Lockyer, 'Note on communication. . .' (1863).
54. *The Reader* aimed at encouraging what Meadows has characterised as 'free and uninhibited discussion of contemporary controversial topics in science, religion and the arts'. Meadows, *Science and Controversy* (1972), p. 16.
55. J. N. Lockyer, 'Science' (1863), p. 20.
56. *Ibid.*, p. 47.
57. J. N. Lockyer, 'The paper on stellar spectra' (1864).
58. Airy, 'Apparatus for the observation of the spectra of stars' (1863). A brief announcement concerning the completion of an apparatus necessary for the prosecution of this work was included in the 'Progress of astronomy' section of the annual 'Report of the Council', *MNRAS* 23 (1863), pp. 143–4.
59. *AR* 1 (1863), pp. 68–70.
60. *Ibid.*, p. 70.
61. Newall, 'The decade 1860–1870' (1987), p. 134.
62. *AR* 1 (1863), p. 70.
63. Airy was publicly and loudly criticised in 1872 for not having included stellar spectroscopy in Greenwich's research agenda. As we shall see, Airy argued with some conviction that this was less a case of neglect than that spectroscopic study of celestial bodies neither conformed to his understanding of Greenwich's purpose and mission, nor did it fit into the daily routine there. No mention was made at that time of Carpenter's 1862–3 spectroscopic efforts or of Huggins's blunt criticism of them.
64. Miller had not examined the spectrum of lithium, osmium, iridium, thallium, strontium, barium or calcium. Huggins did not examine the spectrum of tungsten, molybdenum, nickel, copper, aluminum, magnesium, or graphite/gas-coke. W. Huggins, 'On the spectra of some of the chemical elements' (1863).
65. Robinson, 'On the spectra of some of the chemical elements by William Huggins', RR.5.122, RSL.
66. Robinson, 'On spectra of electric light . . .' (1862).
67. The first map of the so-called normal solar spectrum – a spectrum generated using diffraction rather than dispersion and, hence, free of dispersive variation – was not published until 1869 by A. J. Ångström. See Ångström, *Recherches sur le Spectre Solaire* (1868); *Spectre Normal du Soleil* (1869).
68. Kirchhoff and Bunsen, 'Chemical analysis by spectrum-observations' (1860), p. 91.
69. Kirchhoff and Bunsen, 'Chemical analysis by spectrum-observations – Second memoir' (1861), pp. 505–7 and Plate VI.
70. Huggins used apparatus designed by John Gassiot. For a description of Gassiot's design, see J. P. Gassiot, 'On spectrum analysis; with a description of a large spectroscope having nine prisms, and achromatic telescopes of two-feet focal power', *PRS* 12 (1863), pp. 536–8.
71. W. Huggins and Miller, 'On the spectra of some of the fixed stars', *PTRSL* 154 (1864).
72. Stokes, 'On spectra of fixed stars by Huggins & Miller', RR.6.147, RSL.
73. Miller's influence can be inferred in this if we recall his allusion during his Pharmaceutical Society lecture to Fraunhofer's having detected these very lines in the spectra of a few bright stars. W. Huggins and Miller, 'On the spectra of some of the fixed stars', *PTRSL* 154 (1864), pp. 414–19.
74. W. Huggins and Miller, 'Note on the lines in the spectra of some of the fixed stars' (1863), p. 444; 'On the spectra of some of the fixed stars', *PTRSL* 154 (1864), p. 414.
75. W. Huggins, 'The new astronomy' (1897), pp. 912–13.
76. *Ibid.*, p. 913.
77. W. Huggins and Miller, 'On the spectra of some of the fixed stars' (1864), p. 414.
78. Comte, *Cours de Philosophie Positive* (1833), p. 9.

5

The riddle of the nebulae

Was I not about to look into a secret place of creation?

– William Huggins[1]

In 'The new astronomy' Huggins tells us: 'I was fortunate in the early autumn of the . . . year, 1864, to begin some observations in a region hitherto unexplored' namely, the nebulae.[2] He could indeed feel 'fortunate' to have turned his spectroscope on this class of celestial objects. They are among the faintest on the sky. It had been over thirty years since he caught his first glimpse of a nebula's spectrum. Even so, Huggins recounted the experience in riveting detail as if it were only yesterday. Readers could only imagine the diligent care with which he must have recorded the event in real time.

In 1914, Huggins's widow, Margaret, shipped six bound notebooks containing the records of nearly a half century of work done at their Tulse Hill observatory off to Wellesley College near Boston, Massachusetts. When the notebooks arrived at the private school for women, Professor of Astronomy Sarah Frances Whiting (1846–1927) was the first to examine them. She found what she believed to be the notes from his early observations of nebular spectra. 'In 1864', she wrote, '[Huggins] records his observations of the green lines in the nebulae, and scores of nights were spent trying to match these lines with magnesium, lead, iron, what-not.'[3] Entries fitting Whiting's description can be found in Notebook 1. However, they are clearly dated 1889 and 1890, and they are written in Margaret Huggins's hand.[4]

In fact, there is nothing in Huggins's notebooks, correspondence or published papers to bridge his abrupt shift in attention from common terrestrial elements and stellar spectra to the unresolved nebulae. Whatever encouraged him to subject their light to spectroscopic scrutiny, it was a bold stroke which ultimately propelled him to a position of prestige and authority within the wider science community. The results of his investigation captured his colleagues' imagination and heightened their awareness of the potential of prismatic analysis to generate new knowledge of the heavens.

Over the years, as the new path he opened up became a well-worn thoroughfare, the public – probably even Huggins himself – lost sight and memory of the research options he and his colleagues had available to them in August 1864. Until now, that has not been a problem for historians. They had 'The new astronomy' upon which to rely. Its endearing simplicity and first-hand authority established his decision to aim his star spectroscope at the faint light of a nebula as less a matter of choice than an inevitable next step in the advance of scientific knowledge. But this account has been allowed to stifle curiosity for too long. It is time to turn back the clock, albeit imperfectly, to the summer of 1864, to situate Huggins more authentically within the prevailing climate of interest that prompted him to tackle what he later called the 'riddle of the nebulae'.[5]

5.1 Astronomical questions: summer 1864

Without unfiltered first-person documentation, we lack the palette necessary to paint a faithful portrait of William Huggins as he stood on the verge of his landmark discovery in the summer of 1864. Fortunately there is ample ink available in the public record to bring the man and his motives into relief against the busy background of the astronomical community at large. To fill our pen, we turn first to 'Progress of astronomy' in the February 1864 number of the *Monthly Notices* for a review of the past year's notable developments.[6] Important advances had been made in key, albeit workaday, problem areas. New and improved methods of determining solar parallax, for example, promised to reduce the uncertainty still muddling measures of this crucial celestial yardstick, and invaluable contributions had been made toward unravelling the enigma of sunspots and other solar surface phenomena. But there were no 'salient' astronomical discoveries to report and no mention of nebulae.

In truth, the quest for greater understanding of the nature and structure of nebulae had reached something of an impasse. There really had been little new to say on the subject since the landmark observations of William Parsons, 3rd Earl of Rosse. In 1850, after an exhaustive examination of numerous nebulae with his unparalleled 6-ft reflector, Lord Rosse despaired that, 'as observations have accumulated the subject has become, to my mind at least, more mysterious and more inapproachable'. He concluded that 'when certain phenomena can only be seen with great difficulty, the eye may imperceptibly be in some degree influenced by the mind' causing reports to conform to the observer's own predispositions.[7] Given the real physical limits of instrumentation and atmospheric distortion, it seemed to Lord Rosse that the debate over nebulae would remain, for the time being at least, unanswerable in any positive, testable way.[8]

Even the text of the chapter on 'Clusters of stars and nebulae' in the latest edition of John Herschel's authoritative *Outlines of Astronomy* remained exactly as it appeared when it was first published in 1849.[9] For, despite their increased light grasp and optical refinement, when it came to nebulae, the best telescopes of the day were still less instruments of measurement and analysis than optical cabinets of curiosities through which astronomers, professional and amateur alike, armed with catalogues of nebulous objects could hunt down, view, describe and attempt to categorise the myriad species of luminous smudges that populate the night sky.[10]

This is not to say that astronomers in February 1864 had lost hope of ever knowing the true nature of nebulae with certainty. On the contrary, they remained cautiously optimistic. Like their predecessors, they believed that if and when the great breakthrough were to come, it would be thanks to expert use of a fine telescope. It was an unquestioned conviction that both guided and constrained all their thinking about the celestial challenges they faced. Having lived so long inside this deceptively comfortable box, those actively working on the nebular riddle expected the answer would arise out of imaginable improvements to methods and instruments with which they were familiar. Thus, until August of that year, they continued to row through the fog in a boat firmly tied to the pier.

5.2 Variable nebulae

There were two papers on nebulae among the technical presentations on instruments old and new, exposés on the challenges of precision measurement, and reports from widely viewed celestial events read before the Society between February 1863 and January 1864.[11] They revisited a quiescent thread of discussion on the question of nebular variability that had erupted in February 1862, when *The Times* of London published a letter from seasoned observer John Hind (1823–95) announcing a bewildering astronomical 'undiscovery'.[12] According to Hind, it seemed certain that a small nebula he found barely a decade earlier had 'totally vanished from its place in the heavens'! Ludwig d'Arrest (1822–75) in Copenhagen reported the nebula missing from its place in the constellation Taurus the previous December. Now its absence had just been confirmed by Urbain Jean Joseph Leverrier (1811–77) in Paris.

Hind was perplexed. He had noted nothing unusual about the nebula when it first caught his eye in 1852. The irresolvable spot appeared to be just another distant globular cluster. But the news from Copenhagen and Paris caused him to question that designation. How could a vast system of individual stars lose all its brilliance in such a short period of time? In search of a more plausible explanation, his thoughts turned to the nebula's unusual neighbour on the sky, a star so close it 'almost touched' the nebula. In fact, it was the sudden appearance of this star back in 1852 that had attracted his attention to the nebula in the first place. Then, as quickly as it appeared, the star dimmed to near invisibility, confirming Hind's suspicion that it was a variable. Were the fading star and disappearing nebula connected in some way? Had some dark matter intervened to block the light of both bodies? Did the nebula's apparent brightness depend on the star's own light production? Or could the nebula itself, like the star, be a variable celestial phenomenon?

Hind's announcement sparked immediate discussion in the *Monthly Notices*. Arthur von Auwers (1838–1915), then at the observatory at Königsberg, informed his colleagues that he had been watching the nebula for many years. Indeed, he had noticed it was fainter than expected back in 1858 and had been unable to find it since January 1861, despite repeated attempts.[13] Though absorbed in his own investigation of suspected variability in the Orion nebula, Pulkovo's Director Otto Wilhelm von Struve immediately followed up on d'Arrest's missing nebula report.[14] In December 1861, he detected only 'some traces' of the object. But by March he thought it had brightened again. Was the nebula's apparent increase in brightness due to intrinsic variability, Struve wondered, or simply to differences in the sky's transparency on the nights he had observed it? Uncertain of the answer, he urged more observers to investigate the situation.

All the talk of missing and variable nebulae attracted the attention of John Herschel, whose considerable experience observing nebulae over the decades had convinced him that accounts of short-term variations in a nebula's appearance were more likely to be produced by differences in perception and local observing conditions than any real physical change in the object itself. But Hind's report and the authority with which it was supported clearly gave even this old sceptic pause. In April 1862, he apprised the RAS of the apparent disappearance of yet another previously 'well-authenticated Nebula'.[15] It seems that three nebulae listed in his father's second catalogue (H-II.114, 115 and 116) were missing from one of d'Arrest's reports. Instead, d'Arrest listed two 'new' nebulae in precisely the locations William Herschel had given for 114 and 115. Could these really be new, Herschel wondered, or had d'Arrest inadvertently overlooked their earlier discovery?

This was, of course, a rhetorical question aimed at preserving his illustrious father's priority. His real anxiety stemmed from d'Arrest's failure to mention any nebula in 116's location. Granted, it was a smaller, fainter object, but his father had characterised it as 'resolvable' and 'pretty bright'. Could it now be missing entirely? Happily for Herschel, the answer in this case was 'no'. Paris-based astronomer Jean Chacornac (1823–73) soon reported all three nebulae to be safely situated in their proper Herschelian places.[16]

The sudden swell of interest in nebular variability receded gently into the background where it remained until January 1863, when a brief unattributed essay by Herschel on the 'Variability of nebulae' appeared in the popular *Cornhill Magazine*.[17] After reviewing the particulars of three 'unequivocal' exemplars, including Hind's, Herschel announced the discovery of another 'still more extraordinary' case by Madras astronomer Eyre Burton Powell (1819–1904), namely a recent change in the nebulosity surrounding the southern star η Argus (now η Carinae).

The star itself had quite a history of variability, as Herschel could personally attest. It attracted little attention for over a century and a half after Edmond Halley (1656–1742) first recorded it in 1677 while surveying the southern sky from the island of St Helena.[18] But thanks to notes kept by botanist William John Burchell (1781–1863) during a research trip to Brazil, we know η Argus suddenly increased in brightness in February 1827.[19] A year later, the sky-savvy Burchell recorded that, although the star had dimmed somewhat, it was still far brighter than usual.[20]

In 1837, while Herschel was at the Cape of Good Hope (1833–8) classifying the southern hemisphere's stars, another dramatic surge in brilliance made η Argus a rival to the brightest stars on the sky. He was unaware of its flare-up a decade earlier. Indeed, he later confessed, he never 'had reason to suppose its magnitude variable'.[21] Nevertheless, he had observed η Argus often since his arrival there in 1833 because of his interest in the nebula that enveloped the star.

'Judge of my surprise', he wrote to Johann Mädler (1794–1874) in January 1838, 'to find on the 16.17. December that it had suddenly become a Star of the first magnitude and almost equal to Rigel [β Orionis]'.[22] By 1843, η Argus was nearly as bright as Sirius (α Canis Majoris). Then, beginning in 1850, it slowly began to fade. A handful of southern hemisphere observers tracked the star's decline including Powell in Madras, and Francis Abbott (1799–1883), an Australian watchmaker and amateur astronomer, who observed it from Hobart, Tasmania.[23] After 1858, η Argus dropped precipitously in brightness.[24]

In his *Cornhill Magazine* essay, Herschel described the nebula around the star as 'the largest and finest ... in the southern hemisphere'. During his sojourn at the Cape, he had devoted several months to the painstaking task of preparing an 'accurate representation' of it

Figure 5.1 The dark nebula surrounding η Argus [Carinae]. (Detail from Plate IX, J. F. W. Herschel, *Results of Astronomical Observations*, US Naval Observatory Library)

(see Figure 5.1).[25] A principal subject of his drawing was the striking dark nebula, later nicknamed the Keyhole, which possessed 'a shape somewhat resembling the figure 8, only with its two compartments communicating'.[26] For the benefit of readers who lacked access to a copy of the illustration, Herschel vividly described this 'singular oval vacuity' as 'completely closed, the southern [end] especially being bounded by a strongly-marked and definite outline, as if cut out of paper'. Now, however, a mere quarter century later, Powell had seen the oval '*decidedly open* at the south end'! If Powell's claim could be confirmed, Herschel declared, it would represent 'perhaps the most startling thing which has yet occurred in sidereal astronomy'. Indeed, this exceptional circumstance underscored for him the need to expedite the plans that had, by then, been under consideration for over a decade to erect a 'great reflecting telescope' at Melbourne.[27]

Powell had long since published his record of variations in η Argus's brightness in the *Monthly Notices*, but Herschel's note in a London-based literary magazine was the first, and for well over a year the only, public announcement of the changes Powell had observed in its nebula.[28] Living in the antipodes, well out of earshot of the *Cornhill Magazine* and its news of this late development, Abbott submitted his own lengthy report to the RAS on the 'very remarkable' variable star η Argus and the nebula surrounding it.[29]

While documenting the star's decline over the years, Abbott, like Powell, familiarised himself with others' descriptions of its appearance, paying particular attention to Herschel's esteemed Cape monograph and its carefully executed illustration of η Argus nestled in a luminous nebula. Like Powell, he compared Herschel's drawing with what he had seen recently in his own telescope. And, like Powell, he was surprised to find the configuration of the bright and dark nebulae completely changed. According to Abbott, Herschel's dark closed oval had opened up at *both* its extremes. Even more astonishing, η Argus was no longer embedded in a glowing nebula. Instead, it now appeared as one among a crowd of colourful stars glimmering against a black and vacant backdrop.

5.3 The 'interminable wilderness of nebulae'

In November 1863, around the same time Abbott's paper appeared in the *Monthly Notices*, a report on yet another possible variable nebula appeared in the weekly periodical *The Reader*. In a letter to the editor, the Revd Thomas William Webb let it be known that he had seen the 'curious nebula' that was said to surround Merope, a prominent star in the Pleiades. This nebula had been discovered in 1859 by avid comet hunter Wilhelm Tempel (1821–89) who likened its appearance to that of a 'beautiful comet'.[30] Webb's announcement was news-worthy because Tempel's 'large bright nebula' had proved an elusive target for others, especially observers with large telescopes. D'Arrest, for example, had had no success in seeing the nebula for himself and suggested it might be variable. But now even some of those who had seen the nebula easily were reporting that it was no longer visible.

Webb's note not only kept variable nebulae on the front burner of astronomers' minds, but, just as Herschel's *Cornhill* essay had done months before, it brought the subject to the attention of a wider audience.[31] And it arrived early in the Pleiades-viewing season, prompting a promise from *The Reader*'s editor, presumably Norman Lockyer, to give the matter of nebulosity in the familiar asterism his immediate attention.

Webb was the respected author of the practical observing guide *Celestial Objects for Common Telescopes*[32] and a regular contributor of informative essays on astronomical topics to the monthly science magazine, *Intellectual Observer*. Just a few months earlier, acting in the latter capacity, he had launched a series of essays devoted to star clusters and nebulae.[33]

In the series' first instalment, he acquainted his readers with the history of nebula observation from Ptolemy to the present, and laid out for them the basic dilemma of the nagging nebular riddle: does the 'milky' appearance of 'irresolvable' nebulae offer direct evidence that at least some regions of the heavens are filled with luminous fluid? Or is their haziness an illusion caused when large aggregates of stars are situated at 'unapproachable' distances? The trouble was, as Webb explained, no means had yet been developed to enable earthbound observers to measure the distance to any of these morphologically complex and varied bodies. He lamented that 'after so many years we find ourselves without a guide in the interminable wilderness of nebulae'.[34]

To illuminate the puzzle's challenges facet by facet, Webb built each instalment of his series around an annotated list of selected clusters and nebulae. In November, he suggested his readers observe and compare for themselves the resolvable double cluster in Perseus (h-χ) and irresolvable regions in the Great Nebula in Orion (M42).[35] In light of recent reports of alleged changes in nebulae like Hind's and those surrounding the stars Merope and η Argus, he speculated on the fruitful avenues of investigation that variable nebulae could open up for astronomers. 'We may never be able to unveil the whole mystery,' he opined, 'but it is reasonable to hope somewhat from the future. Larger telescopes are called into existence every year . . .'[36]

In December, Webb focused on three instructive targets easily found in small telescopes at that time of year. The Great Nebula in Andromeda (M31), for example, cannot be resolved in even the largest telescopes, but its hazy neighbour (M33) can. By contrast, the naked-eye speck in nearby Perseus (M34) can be resolved easily into a beautiful stellar group by the merest of optical aids, a telescope finder. In Webb's view, these examples justified his cautious optimism that 'possibly, results as yet unlooked for may . . . be in store for instruments such as the $18\frac{1}{2}$-inch achromatic of Clark, the 20-inch of Buckingham, or the 25-inch on which Cooke is said to be at present engaged'.[37]

In January 1864, Webb brought his readers thoroughly up to date on what noted astronomers had to say about the nebula surrounding the Pleiad Merope, 'one of the most curious objects in the heavens'.[38] Webb was always on the lookout for interesting celestial targets that his readers could access with their small-aperture telescopes. The Merope nebula would give a wide population of observers the opportunity to make important contributions to the nebular puzzle. 'Obscure as this object is,' he emphasised, 'it may possibly prove a most important witness as to the existence of luminous matter in an unconcentrated form.'[39]

Soon after this article appeared, *The Reader*'s editor (again, presumably Lockyer) fulfilled the promise he had made a few months earlier by devoting an entire column to the nebulosity in the Pleiades.[40] 'What food for thought', he exulted in conclusion, 'and what a field of observation undreamt of a year or two ago, have we here!' Paraphrasing solar observer Richard Christopher Carrington (1826–75), who had recently expressed the hope that the years of sunspot data he had collected would help to answer the simple, yet profoundly difficult question, 'What is a Sun?',[41] *The Reader*'s editor now asked, 'And what *are* nebulae, since they certainly are not what our text-books describe them? . . . [A] thousand other questions are suggested by the scant observations we at present possess relating to the unanticipated phenomena of the variability of nebulae.'[42]

In February, Webb returned to a favourite telescopic object, the Great Nebula in Orion. He had recently noticed 'obvious' differences between his own assessment of the nebula's telescopic appearance and those reported previously by others. He concluded it deserved additional scrutiny and, as always, encouraged amateurs to get involved. 'Where the highest authorities differ', he emphasised, 'there is room for every man's independent work.'[43] In particular, he drew attention to the 'trapezium', a region within M42 that seemed to be dotted with resolvable stars swimming in a sea of irresolvable luminous haze. Careful study of this area with improved telescopes would finally break this 'most remarkable of all astronomical enigmas', namely, 'what is the true character of many of the nebulae? Are they merely remote collections of stars? or are they luminous mists, of a nature wholly unknown . . .?'[44]

The Orion Nebula was also the topic of a lively discussion at the RAS's June meeting, the last before the summer recess, when a letter was read from the Director of Harvard Observatory, George Phillips Bond. Like Struve at Pulkovo, Bond had studied the object with great care in recent years.[45] Back in 1861, he had made the startling and highly controversial claim that the nebula was actually spiral in structure, like 'the celebrated nebula 51 *Messier*' described a decade earlier by Lord Rosse.[46] Now, Bond wanted to make a case for the proofs of engravings he had sent to Greenwich months ago that mapped the entire region in great detail. Unfortunately, he complained in his letter, these engravings had been misinterpreted by the Greenwich observers, who in January compared them unfavourably against both their own direct telescopic observation of the nebula and John Herschel's drawings of the nebula made in 1847.[47] The repartee that followed in the wake of reading Bond's letter centred on the challenge of documenting the true appearance of such celestial objects so that subtle changes in their delicate features could be distinguished with certainty.

After Charles Pritchard graciously acknowledged the difficulty of 'engraving drawings of such objects', septuagenarian engineer Charles Blacker Vignoles (1793–1875) offered his own ready solution to the problem: 'Photograph them.' Richard Hodgson (1804–72) doubted that photography would be able to capture the detail necessary. His past experience with daguerrotypes and his successful reproduction of illustrations from his own microscopical observations gave authority to his criticism, but Mr Vignoles would not be dissuaded. 'Excuse me,' he persisted, 'but I am in the habit of having maps and drawings constantly photographed with the best result, and the Government do so extensively, reproducing plans most perfectly.' Vignoles was seconded in his opinion by optician Joseph Beck (1829–91), who added, 'Photographs can be taken so perfectly from pencil drawings as not to be distinguished from the original.'[48]

No one was suggesting photographs be taken of the nebula directly through a telescope. Astronomical photography was in its infancy at the time.[49] Bond's meticulously executed drawings were assumed to be accurate representations of the nebula as he had observed it. The new idea Vignoles and Beck were promoting simply involved photographing those drawings in order to disseminate the images widely for the purpose of comparison. Participants in this particular discussion were still comfortable with the notion of the telescope as principal conduit of information, and the observer as principal receiver and recorder of that information. Nevertheless, we see evidence that the methodological box that nebula observers had built around themselves was beginning to feel a bit too tight. For some of them, at least, a new technology – one tested and found to be successful in solving similar problems in other professional endeavours – had begun to erode their old ways of thinking about the process of reproducing and disseminating visual information.

All the best puzzles, like earthquake-prone landscapes, nurture a stealthy yet steady buildup of fertile creative tension beneath their seemingly tranquil surfaces. In the case of the nebular puzzle, the dynamic of that tension began to change in the early 1860s as reports on variable nebulae trickled in from observers around the globe. Each contributed something new to the puzzle enterprise. Some named new objects or classes of objects to study; others made improvements to existing instruments. The most provocative suggested new questions, or asked old questions in new ways. But none triggered the seismic shift in thinking about the nebular problem that was required to make a real difference in the terrain. That would come with William Huggins's ground-breaking decision to find out what a nebula's light looked like when viewed through his star-spectroscope.

5.4 'No spectra such as I expected!'

What really started wheels turning in Huggins's mind as he cast about for new investigative targets for his star-spectroscope and new opportunities for discovery in August 1864? As he explained it then, his interest in attempting the difficult task of subjecting the extremely faint light of nebulae to prismatic analysis stemmed from the growing suspicion that celestial objects shared spectral similarities that made it possible to draw conclusions concerning their chemical and physical makeup. He and Miller had already made a strong case for a 'unity of plan' throughout the universe encompassing the substance, structure and dynamics of organic and inorganic matter.[50] 'It became ... an object of great importance', he wrote, 'to ascertain whether this similarity of plan observable among the stars and uniting them with our sun into one great group, extended to the distinct and remarkable class of bodies known as nebulae.'

What he wanted to determine, and what he hoped the spectroscope would reveal, was the 'essential *physical* distinction' that 'separates the nebulae from the stars'.[51] In other words, he expected nebulae to differ from stars not so much in terms of the stuff from which they were made as in their temperature or density. If nebulae were distant, irresolvable clusters of stars, their spectra should be star-like, that is, a faint continuous background of colour interrupted by recognisable patterns of absorption lines. If, on the other hand, they were indeed gaseous stellar embryos, an observer would still see the familiar and characteristic spectral signatures of the many elements known to exist in the Sun and other stars. But, instead of dark absorption lines common to ordinary star spectra, bright-line emission spectra, like those produced by flames or electric sparks, would be observed.

Though he would have vehemently denied it, Huggins, like all observers, was prone to suggestion. What he was looking for informed his observational interpretations. Some observers, like William Herschel, saw the symbiosis of expectation and observation as a training for the eye that, when properly balanced, could make one a truly great observer.[52] Lord Rosse worried that a predisposed eye could lead one to erroneous conclusions especially when the object under scrutiny is just at the edge of perception.[53] Huggins, by contrast, believed his observations were bias-free *because* he selectively rejected what he considered substandard measurements or qualitative descriptions. He would not have agreed that such behaviour was predicated upon an internalised set of expectations. To interpret his reported observations, it is essential to know what he expected to see.

When Huggins began his investigation on 29 August 1864, he could not focus on Orion's trapezium or any of the variable nebulae that Webb and others had advertised as likely keys to solving the nebular puzzle. These objects were not in London's night sky at that time of year. Even if they had been, their faintness would have made them less than ideal targets for his spectroscope. Instead, he 'directed the telescope for the first time to a planetary nebula in Draco', popularly known today as the 'Cat's Eye'.[54]

It was a good choice. One of the brighter of these distinctive, disc-like nebulae, the Cat's Eye is perched high in the sky in late August thus clear of the horizon's obscuring haze. More important, it is one of the more colourful of that class, described by some as bluish, by others as bluish-green.[55] Indeed, I would argue that it was the colourful nature of the planetary nebulae that led Huggins to select one as his first nebula to examine spectroscopically.

In 'The new astronomy', he wrote:

The reader may now be able to picture to himself to some extent the feeling of excited suspense, mingled with a degree of awe, with which, after a few moments of hesitation, I put my eye to the spectroscope. Was I not about to look into a secret place of creation?[56]

Carrying the wide-eyed reader along in this suspenseful and well-crafted narrative, Huggins continued:

I looked into the spectroscope. No spectrum such as I expected! A single bright line only! At first I suspected some displacement of the prism, and that I was looking at a reflection of the illuminated slit from one of its faces. This thought was scarcely more than momentary; then the true interpretation flashed upon me. The light of the nebula was monochromatic, and so, unlike any other light I had as yet subjected to prismatic examination, could not be extended out to form a complete spectrum ...[57]

There can be no question that what he observed in his spectroscope was not what he expected to see. But he was less surprised to see an emission line spectrum than by its near-monochromatic character. Closer examination revealed not one, but three, individual lines, each with its own colour and separated by some distance from the others. The spectra of seven of the eight other planetaries he examined shared this remarkable appearance. The one outlier – a pale blue planetary in the southern constellation Eridanus (NGC 1535) that would have appeared low on the horizon in the wee hours of the morning at that time of year – showed no lines.

In 'The new astronomy', Huggins minimised his initial concern that this unexpected result was in some way erroneous, perhaps stemming from instrumental difficulties. He did not treat it quite so lightly in his *Philosophical Transactions* paper. It was prudent to be wary when unexpected lines appeared. Bands of light flashing into view by virtue of prismatic internal reflection could easily mislead the observer. Just a few years earlier, chemists Friedrich Wilhelm Dupré (c. 1835–1908) and his brother August (1835–1907) had become victims of just such a rush to judgement when they observed what they believed to be a previously unobserved blue line during their spectroscopic examination of London water. In their enthusiasm, they announced the probable discovery of a new element in the calcium group of metals.[58] But, even before the Duprés's article appeared in print, further investigation showed the blue line was, in fact, 'formed by internal reflection from the different surfaces of the prism and lenses'.[59]

Whether driven by simple curiosity or an awareness of the evidentiary authority that would be needed to convince others to accept this new and surprising result, Huggins did not limit his observations to planetary nebulae. In the *Transactions*, he drew attention to the very different looking spectra of another group of nebulous objects that included two globular clusters in Hercules (M92 and NGC 6229), two prominent irresolvable nebulae in Andromeda (M31 and M32), and a so-called 'nebulous star' (Andromeda 55). Like the planetary in Eridanus, none of these spectra exhibited any bright lines. Some, he reported, looked decidedly 'stellar' – their spectra were mainly continuous.

What sort of physical entities were the enigmatic emission nebulae? How, exactly, could they be fitted into the great scheme of celestial things? Huggins was not entirely sure. Whatever they were, these nebulae were not even 'a special modification ... of our own type of suns'. Instead, he proclaimed, we 'find ourselves in the presence of objects possessing a distinct and peculiar plan of structure'. He cautiously concluded that 'we must probably regard these objects, or at least their photo-surfaces, as enormous masses of luminous gas or vapour'. Only 'matter in the gaseous state' emits 'light consisting of certain definite refrangibilities only'.[60] To bolster his argument, he reminded his readers that no less of an authority than John Herschel had suggested a similar explanation for the uniform luminosity seen in the discs of planetary nebulae.[61]

Huggins was unsure how to interpret the chemical meaning of the three lines found in nearly all the planetary nebulae's spectra (see Fig. 5 in Figure 5.2). He admitted that it was

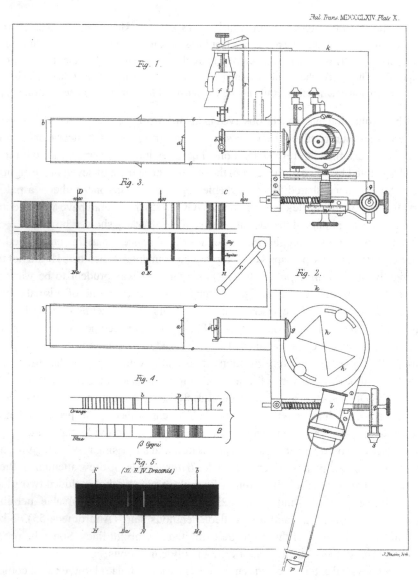

Figure 5.2 Plate X. Figs. 1 and 2: schematic drawings of Huggins's star-spectroscope. Fig. 3: comparison of solar spectrum with those of Jupiter and the sky. Fig. 4: comparison of spectra of Albireo A and B. Fig. 5: spectrum of the planetary nebula, H-IV 37 [NGC 6543] in Draco. (W. Huggins, *PTRSL* **154**)

possible to imagine, although very unlikely, that the light was produced by a cluster of peculiar stars totally unlike our Sun or any of the other individual stars that had yet been examined, each possessing a spectral signature comprised solely of those three lines. And what chemical elements did these lines represent? A mixture of hydrogen and nitrogen? A substance as yet unknown on Earth? 'May it not indicate a physical difference in the atoms', he conjectured, 'in connexion with the vibrations of which the lines are probably produced?'[62]

How the historian longs for a notebook record, or some other contemporaneous clue to how the mystery unfolded before his eyes. In what order did Huggins view each of these objects? On how many occasions did he return to each of them for just one more look? How quickly did he come to separate them all into two distinct groups based on the appearance of their spectra? What hypotheses, if any, did he entertain to explain the lack of visible bright lines in the spectrum of the planetary nebula in Eridanus? How long did it take him to conclude:

The riddle of the nebulae was solved. The answer, which had come to us in the light itself, read: Not an aggregation of stars, but a luminous gas.[63]

Less than a year earlier, in his paper on stellar spectra, Huggins had hailed the similarity among stellar, solar and terrestrial spectra as evidence for a fundamental uniformity of nature. Now he felt forced to conclude that some nebulae were 'systems possessing a structure, and a purpose in relation to the universe, altogether distinct and of another order from the great group of cosmical bodies to which our sun and the fixed stars belong'.[64] It took some time for him to become comfortable with this new conclusion.

Still, it was fortuitous that Huggins selected this class of objects to begin his work. Planetary nebulae present emission spectra which are strikingly different from stellar spectra or the continuous spectra of galaxies. The lack of ambiguity in Huggins's initial observations of planetary nebulae lent an air of confidence to subsequent work on a broader range of nebular types many of which presented observers with less clear-cut spectra to interpret.

5.5 A paper of 'interest & importance'

Huggins submitted his paper to the Royal Society on Thursday 8 September, a little over a week after he first observed the bright-line spectrum of the Cat's Eye. It was the third paper he had submitted to the Society in ten months, and his second as sole author. All were communicated on his behalf by his neighbour and collaborator, William Allen Miller. Conforming to the accepted prescription for scientific papers at the time, his paper on nebulae is an artificial but logical reconstruction of the sequence of challenges imposed on the would-be discoverer by Nature's unwillingness to reveal her secrets to any but the worthy. Huggins placed these challenges within an ordered pattern of unfolding discovery. Rather than survey all the difficulties and dead ends he had encountered, he limited his discussion to the problems he had solved.

Huggins was delighted to learn that the RS had accepted his paper for publication. He sent word to Norman Lockyer that, because of its 'interest & importance', the paper was to be printed 'at once' and, he boasted, '*in my name only*'. 'Perhaps', he hinted to Lockyer, 'you may be inclined to write an article [in *The Reader*] on the nebulae when you see my paper, which contains some interesting observations & deductions.'[65]

At the time, Lockyer was more interested in observing the Moon and the planet Mars than in celestial spectroscopy.[66] But he was also in the business of writing and editing articles on scientific subjects for *The Reader*. Huggins clearly perceived him as an ally with valuable connections to the popular press. The two men came to have a long and complex relationship ranging from open and friendly camaraderie to spiteful and rancorous contention. Huggins composed these letters during the early amicable stage of their acquaintance when their astronomical interests did not overlap. His letters served a dual purpose: ostensibly he intended them to convey pertinent information to Lockyer as a man of science, but they were also intended to cajole and impress him as a potential proselytiser.[67] He was writing to Lockyer, the science editor, not to a fellow spectroscopist. In doing so, he hoped to advertise his work's importance, enhance his visibility, and extend his renown outside the astronomical community.

In the event, Lockyer waited until November before commenting on the paper in *The Reader*.[68] He drew attention to Huggins's claim that the simplicity of the nebular spectra indicated an elemental constitution, perhaps hydrogen, nitrogen, or some other as yet unknown primitive matter that may be peculiar to the celestial material from which nebulae are composed. What might be ascertained in terrestrial laboratories about the physical structure of nebulae, given what is known from chemists' studies of spectral patterns?

Months later, in June 1865, inspired by a lecture Huggins had delivered before the Royal Institution, Lockyer wrote another column bearing a title that reprised the question he had raised back when Huggins turned his spectroscope on the Cat's Eye for the first time: 'What is a Nebula?'[69] Now Lockyer cheered the 'vast progress' that had been made 'towards answering this and other equally momentous questions! Ten years ago', he reminded his readers:

the light of the sun . . . travelled no farther than our retinas, and all was mystery and mental darkness; but at the present time, although the riddle of the sun is not quite solved, the coming day has more than dawned upon our minds, and stars and nebulae are also rising on our inner horizons . . .

5.6 Fellowship

Huggins was well aware that exposure attained by publishing in the RS journals accorded him no small degree of acclaim in a group whose recognition he valued and of which he yearned to be a part. By the early 1860s, the Royal Society had regained much of the prestige it had gradually lost during the eighteenth and the first half of the nineteenth centuries when Fellowship was accorded to increasingly larger numbers of non-scientific men at the expense of the aims its originators had intended.[70] Rules imposed in 1846 limited the election of new Fellows to fifteen each year.[71] The new competitive nature of its election process made Fellowship in the Royal Society a goal that was just beyond the reach of many aspirants and hence all the more tantalising a prize.

Huggins's paper on nebular spectra represented something very new. Edward Sabine (1788–1883), President of the Royal Society, exulted, '[H]ere we have a totally different view opened.'[72] It was welcome news indeed for those like Sabine, who saw the question of the nature of nebulae as a crisis for observational astronomy. In fact, 'crisis' is the very word he used to describe the impasse that had been reached on the matter due to the fact that even the largest telescopes then built were unable to resolve many of the nebulae on the sky. In Sabine's view, Huggins's ability to breach this impasse was due to his 'referring the spectra lines to no mere instrumental scale ... but to standard spectra of known elements which are formed in juxtaposition with that to be examined, and to which its lines can be compared with extreme precision'.[73]

Huggins successfully communicated the confidence he felt that his particular spectro-scopic apparatus and method constituted an entirely new yet reliable means for attacking the problem of the nebulae. It was an ingenious design probably developed in response to difficulties he experienced in the course of his work on terrestrial spectra.[74] He introduced a sequence of mirrors that directed the light from the comparison source into the collimating tube so that an observer could see and compare simultaneously the spectrum of the celestial body under scrutiny and that of a known terrestrial substance (see Figs. 1 and 2 in Figure 5.2).

His colleagues in the RAS commended not only his instrumental design, but also his shrewd application of the new method of spectrum analysis to an old and difficult problem. Much of their enthusiasm can be traced to Warren De la Rue, whose presidential address to the RAS in November 1864 gave voice to the subtle spirit of change afoot within the Society's ranks. Decrying those who would retain old methods for tradition's sake, De la Rue exhorted his Fellows:

Let us not indulge in vain regrets for superseded methods; but, on the contrary, let us reflect how rich in promise is its future, if, in rejecting all idea of finality in any of its methods or procedures, we gladly adopt the best aids which the progress of physical inquiry places within our reach; and in this, as in all our contests for new scientific territory, admit no rallying-cry but – Forward![75]

De la Rue was well-acquainted with the potential power of the method of spectrum analysis for resolving questions concerning the physical and chemical constitution of celestial bodies. In his presidential role, he helped shape the positive response of the RAS to Huggins's work. The *Astronomical Register* praised Huggins's new research on nebulae 'the fruits of which are of the most interesting character, and tend to prove conclusively that the constitution of *some at least* is truly gaseous or vaporous, and that the generalisation of considering all these bodies as clusters of stars must be given up'.[76] Similarly, the 'Progress of astronomy' section of the annual 'Report of the Council' for 1865 announced:

the relation between terrestrial physics and the physics of the sidereal heavens is rapidly becoming more intimate, the boundary line once supposed to divide them is gradually disappearing, while new and unexpected fields open for the application of results of experimental philosophy, in distant regions of space, where formerly they were supposed to have little or no concern.[77]

Expanding on De la Rue's remarks, the Report lauded the recent publication of both John Herschel's monumental 'Catalogue of nebulae and clusters of stars' and Huggins's work on nebular spectra, which provided the 'first glimpse of our knowledge of the constitution of the mysterious systems of matter which have formed the [nebulae]'.[78]

It marked the first significant reference to Huggins's spectroscopic research in the *Monthly Notices*. Two full pages were allotted to discussion of it in a subsection of the Report titled 'Analysis of light from the nebulae'. The unusual length of this section and its placement as first among a number of other, briefer, special reports provide some indication of the importance that Fellows in the Society accorded to Huggins's contribution.

At the same time, Huggins received considerable acclaim from within the ranks of the Royal Society. On 1 June 1865, he was elected to Fellowship. It was a major turning point in his career.[79] Less than a year later, he began recording spectroscopic observations of celestial bodies in a new notebook: Notebook 2 in the Wellesley collection.

***

Beginning with his study of nebulae, William Huggins's investigative successes convinced many of his contemporaries of the spectroscope's broad utility in generating new knowledge about old and familiar telescopic subjects. They raised important methodological and theoretical questions for the wider community of practising astronomers, both amateur and professional. And they contributed to the growing fuzziness of disciplinary boundaries that had defined the traditional limits of acceptable astronomical research in the mid-nineteenth century.

Notes

1. W. Huggins, 'The new astronomy' (1897), p. 916.
2. *Ibid.*, p. 915.
3. S. F. Whiting, 'Diaries' (1917), p. 160.
4. The entries made by Margaret Huggins in the back of Notebook 1 are a direct continuation of those found in Notebook 2, WCL/SC. It seems that the Hugginses were not ones to let good paper go to waste.
5. W. Huggins, 'The new astronomy' (1897), pp. 916–17.
6. 'The progress of astronomy during the past year', *MNRAS* **24** (1864), pp. 102–7.
7. W. Parsons, 'Observations on the nebulae' (1850), pp. 503–4.
8. See Grant, *History of Physical Astronomy* (1852), pp. 569–71. Jonathan Crary has drawn attention to the interaction between the human visual response and the prepared mind by reminding us of the fact that the Latin root for the word 'observer', *observare*, 'means "to conform one's action to comply with," as in observing rules, codes, regulations, and practices.' In Crary's view, 'There never was or will be a self-present beholder to whom a world is transparently evident. Instead there are more or less powerful arrangements of forces out of which the capacities of an observer are possible.' Crary, *Techniques of the Observer* (1990), p. 6. For examples from the history of astronomy, see Hetherington, *Science and Objectivity* (1988).
9. Compare 'Chapter XVII. Of clusters of stars and nebulae' in the first edition of J. F. W. Herschel, *Outlines of Astronomy* (1849), with the same chapter in the 5th edition (1858) and 7th edition (1864). Herschel finally revised the chapter to include Huggins's spectroscopic discoveries in the 10th edition (1869). The so-called 'new edition' published in Philadelphia in 1869 was basically a reprint of the 5th edition (1858).
10. Lacaille, 'Sur les étoiles nébuleuses du ciel Austral' (1756). Messier, 'Catalogue des Nébuleuses et des amas d'étoiles' (1781). William Herschel's catalogues were published in the *Philosophical Transactions*. See his 'Catalogue of one thousand new nebulae and clusters of stars' (1786); 'Catalogue of a second thousand of new nebulae and clusters of stars' (1789); and 'Catalogue of 500 new nebulae' (1802). See Hoskin, *William Herschel and the Construction of the Heavens* (1963). John Herschel's catalogues were also published in the *Philosophical Transactions*. See 'Observations of nebulae and clusters of stars, made at Slough, with a twenty-feet reflector, between the years 1825 and 1833' (1833) and 'Catalogue of nebulae and clusters of stars' (1864).
11. Abbott, 'Observations on η *Argus*' (1864) was read on 15 November 1863, and Hind, 'Note on the variable nebula in *Taurus*' (1864) was read on 8 January 1864.
12. Hind, 'Changes among the stars', *The Times* of London, 4 February 1862, p. 8b. Hind's variable nebula continued to attract the attention of observers for decades. See, Hind, 'Auszug aus einem Schreiben des Herrn

Hind an die Redaction' (1853). The nebula was found again in 1890 by American astronomer Sherburne Wesley Burnham (1838–1921) at the new Lick Observatory in California. Its neighbouring star, T *Tauri*, is the eponym of a special class of young unstable stars. See S. W. Burnham, 'Note on Hind's variable nebula in Taurus' (1890).

13. 'Extract of a letter from Mr. A. Auwers ...', *MNRAS* **22** (1862), pp. 148–50.

14. Struve, 'On the missing nebula in Taurus ...' (1862).

15. J. F. W. Herschel, 'Letter ... to Mr. Hind on the disappearance of a nebula in Coma Berenices', (1862). The three nebulae in question comprise one end of a long string of fourteen galaxies known as Markarian's Chain which spans the boundary between Coma Berenices and Virgo. These three nebula are known today as NGC4473, 4477 and 4479.

16. Chacornac, 'On the missing nebula in Coma Berenices' (1862).

17. J. F. W. Herschel, 'Variability of nebulae' (1863).

18. Halley designated η Argus as a fourth magnitude star in the branches of 'Charles's Oak' (*Robur Carolinum*), a new, but short-lived, constellation he carved out of stars in Argo to honour King Charles II. Halley, *Catalogus Stellarum Australium sive Supplementum Catalogi Tychonici* (1679). For an exhaustive history of observations of η Argus, see Frew, 'The historical record of η Carinae ' (2004).

19. J. F. W. Herschel, *Results of Astronomical Observations* (1847), pp. 35–6.

20. Lynn, 'William John Burchell and η Argus' (1907). Burchell did not publicise his 1827–8 observations until 1845 when he wrote a letter to Manuel Johnson at Oxford.

21. J. F. W. Herschel, 'Extract of a letter from Sir John Herschel to the president...' (1838).

22. J. F. W. Herschel, 'Auszug eines Schreibens von Sir John Herschel...' (1838).

23. The English-born watchmaker had made a new life for himself in Hobart, rebuilding his personal and artisanal reputation after being transported to the colony in 1844 for 'obtaining two watches under false pretences'. See Rimmer, 'Francis Abbott' (1969).

24. By 1868, η Argus had all but disappeared from view.

25. J. F. W. Herschel, *Results of Astronomical Observations* (1847), p. 37 and Plate IX. Clearly satisfied with the results of his effort, Herschel reproduced the portion of Plate IX that focused on η Argus and the nebula immediately surrounding it in Figure 2 Plate IV of each edition of his acclaimed *Outlines of Astronomy*. See, for example, 4th edn (Philadelphia, PA: Blanchard & Lea, 1861).

26. [J. F. W. Herschel], 'Variability of nebulae' (1863), p. 144.

27. Royal Society, 'Extract from Council minutes of the Royal Society (November 25, 1852)', *Correspondence Concerning the Great Melbourne Telescope* (1871), Part 1, p. ii.

28. E. B. Powell, 'Variations in the light of η Argus' (1861). Powell's own report on variations in the nebula appeared in May 1864, when he wrote to confirm Abbott's observation. See E. B. Powell, 'Notes on α Centauri and other southern binaries, and on the nebula about η Argus' (1864).

29. Abbott, 'Notes on η Argus' (1863).

30. Tempel, 'Schreiben des Herrn Wilh. Tempel' (1861).

31. T. W. Webb, 'Nebula in the Pleiades' (1863).

32. T. W. Webb, *Celestial Objects for Common Telescopes* (1859).

33. T. W. Webb, 'Clusters of stars and nebulae ...' (August, 1863). Huggins contributed an article to the magazine on the findings from his stellar spectroscopic investigations. See W. Huggins, 'Spectrum analysis applied to the stars; or, the stars, what are they?' (1863).

34. T. W. Webb, 'Clusters of stars and nebulae ...' (August, 1863), p. 59.

35. T. W. Webb, 'Clusters and nebulae ...' (November, 1863).

36. *Ibid.*, p. 263.

37. T. W. Webb, 'Clusters, nebulae, and occultations' (December, 1863), p. 351.

38. T. W. Webb, 'Clusters and nebulae ...' (January, 1864), p. 448.

39. *Ibid.*, p. 450.

40. 'Variable nebulae', *The Reader* **3** (1864), p. 109.

41. Carrington, *Observations of Spots on the Sun* (1863), p. 17.

42. 'Variable nebulae', *The Reader* **3** (1864), p. 109.

43. T. W. Webb, 'Clusters and nebulae ...' (February, 1864), p. 58.

44. *Ibid.*, p. 60.

45. See G. P. Bond, *Observations upon the Great Nebula of Orion* (1867).

46. G. P. Bond, 'On the spiral structure of the Great Nebula of Orion' (1861), p. 206.

47. G. P. Bond, 'Remarks upon the statements of Messrs. Stone and Carpenter ...' (1864). For the text of the original report from Greenwich, see 'Proceedings of various observatories: Royal Observatory, Greenwich', *MNRAS* **24** (1864), pp. 91–3.

48. 'Eighth meeting, June 10th, 1864', *AR* **2** (1864), pp. 152–3.

49. In fact, M42 was the subject of the first successful photograph taken of a nebula, but the achievement was still over a decade in the future. American physician Henry Draper took this photograph on 30 September 1880. Ranyard, 'Note on Dr. Henry Draper's photograph of the nebula in Orion' (1881). Gingerich, 'The first photograph of a nebula' (1980).

50. W. Huggins and Miller, 'On the spectra of some of the fixed stars', *PTRSL* **154** (1864), pp. 433–4.

51. W. Huggins, 'On the spectra of some of the nebulae', *PTRSL* **154** (1864), p. 437. Emphasis added.

52. Schaffer, 'Herschel in Bedlam' (1980), p. 216.

53. W. Parsons, 'Observations on the nebulae' (1850), pp. 503–5.

54. The class of nebulae to which it belongs was dubbed 'planetary' by William Herschel because of their characteristic disc-like appearance in the telescope. Herschel designated this particular nebula as H-IV.37, the thirty-seventh object in his fourth catalogue. Today it is catalogued as NGC 6543.

55. R. Burnham, *Celestial Handbook*, vol. 2 (1978), pp. 870–2, with photograph.

56. W. Huggins, 'The new astronomy' (1897), pp. 915–16.

57. *Ibid.*, pp. 916–17.

58. F. W. and A. Dupré, 'On the existence of a fourth member of the calcium group of metals' (1861).

59. Crookes to G. Williams, 11 February 1861, cited in James, *The Early Development of Spectroscopy and Astrophysics* (1981), p. 203.

60. W. Huggins, 'On the spectra of some of the nebulae' (1864), p. 442.

61. *Ibid.*, pp. 442–3. J. Herschel, *Outlines of Astronomy*, 7th edn, p. 646.

62. W. Huggins, 'On the spectra of some of the nebulae' (1864), p. 444.

63. W. Huggins, 'The new astronomy' (1897), pp. 916–17.

64. W. Huggins, 'On the spectrum of the Great Nebula in the sword-handle of Orion' (1865), p. 42. That Huggins was uncomfortable with this conclusion can be inferred from his emphasis on the uncertainty of current knowledge about the structure and composition of the nebulae in an address before the BAAS in 1866. See W. Huggins, *On the Results of Spectrum Analysis Applied to the Heavenly Bodies* (1866), pp. 29–43. As we shall see in Chapter 13, in 1891, with years of examining nebular spectra behind him and recently having had the opportunity to examine photographs of nebulae taken by Isaac Roberts (1829–1904), Huggins was emboldened to distance himself publicly from his 1864 position that nebulae were of a totally different form and nature from all other celestial bodies. He had embraced that earlier view, he claimed, while under the influence of religious dogma. Now that he had shed these fetters, he claimed he was once again able to recognise his observations of nebular spectra as providing evidence that celestial bodies are 'obviously not a haphazard aggregation of bodies, but a system resting upon a multitude of relations pointing to a common physical cause.' Specifically, he argued, these observations support the nebular hypothesis and a universal evolutionary scheme for all celestial bodies. See W. Huggins, 'Address of the president' (1891), p. 20.

65. W. Huggins to J. N. Lockyer, 20 September 1864, JNL, UEL. Although Huggins was recognised as sole author of the paper on nebular spectra, as a non-Fellow, he was required to communicate it to the Royal Society through someone who was. Not surprisingly, he chose W. A. Miller.

66. For a discussion of Lockyer's early astronomical work, see Meadows, *Science and Controversy* (1972), pp. 41–6.

67. In the first letter which Huggins closed 'In haste', he took the time to add a note following his signature suggesting that Lockyer, who was considering relocating closer to London, 'take a *house* here [Tulse Hill] with a *good garden.*' Huggins even offered to help Lockyer find a suitable place. W. Huggins to J. N. Lockyer, 20 June 1864, JNL, UEL. Meadows draws attention to the fact Huggins may have been more successful than he planned in persuading Lockyer to the benefits of spectroscopic research – shortly after receiving this letter from Huggins, Lockyer began to make arrangements to obtain a spectroscope for his own telescope. See Meadows, *Science and Controversy* (1972), p. 46.

68. J. N. Lockyer, 'Celestial analysis' (1864).

69. J. N. Lockyer, 'What is a nebula?' (1865). See W. Huggins, 'On the physical and chemical constitution of the fixed stars and nebulae' (1865).

70. For a discussion of the changes in Royal Society Fellows, see Lyons, *The Royal Society 1660–1940* (1968), pp. 125–6; pp. 165–6; pp. 204–5; pp. 232–4; and pp. 275–8. An overview of the changes undergone by the Society between 1830 and 1872 can be found in M. B. Hall, *All Scientists Now* (1984), pp. 62–112.

71. Lyons, *The Royal Society 1660–1940* (1968), pp. 259–62.

72. Sabine, 'President's address' (1864), p. 502.

73. *Ibid.*, p. 501.
74. Huggins does not tell us this himself. In his review of Huggins's recent work on the spectra of stars and nebulae for the *AR*, T. W. Burr states that 'the labours of Mr. Huggins on the metals taught him the risk of relying on small measured differences, which might be due to variations of temperature and other derangements of the apparatus, and he therefore conceived the idea of exhibiting the spectra of the metals in the same field of view with the stellar spectra'. Burr, 'Spectrum analysis of the stars and nebulae' (1864), p. 254.
75. De la Rue, 'Presidential address' (1864), p. 9. See also 'Analysis of light from the nebulae', in 'Report of the Council', *MNRAS* **25** (1864), pp. 112–14.
76. Burr, 'Spectrum analysis of the stars and nebulae' (1864), p. 256.
77. 'Progress of astronomy during the past year', *MNRAS* **25** (1865), pp. 109–19; pp. 109–10.
78. *Ibid.*, p. 112.
79. 'Annual meeting for the election of Fellows ...', *PRS* **14** (1865), p. 299.

6

Moving in the inner circle

> I saw ... the creation of a great method of astronomical observation which
> could not fail in future to have a powerful influence on the progress of
> astronomy ...
>
> — *William Huggins*[1]

Huggins could have set himself the arduous task of examining the spectrum of every known nebular object, or systematically cataloguing the spectra of northern hemisphere stars. Instead, he pursued a varied and opportunistic research programme like many other amateur astronomers of his day, devoting considerable time and serious attention to research problems generated by others, and to the exotic rather than the mundane.[2] As an independent observer he was free of the obligations and commitments that restricted his institution-bound contemporaries. Driven by broad interests and an insatiable curiosity, he explored a number of different subjects in innovative and often technically demanding ways. His challenge was to maximise his exposure to opportunities for new discoveries without becoming identified as a speculative or impulsive dilettante.

It was a challenge his years as an entrepreneur had prepared him well to meet. He developed a reputation for care in making observations and caution in suggesting explanations for the phenomena he observed. His successes led to more opportunities for success, and he became recognised as one upon whom colleagues could rely for advice on spectroscopic matters.[3]

6.1 Cultivating advantageous alliances

In June 1865, the Royal Society called on Huggins to verify a discovery recently announced by Father Angelo Secchi, director of the Vatican observatory. It was not the first time Huggins faced such a task, nor would it be the last. A few months earlier, Huggins had loudly criticised the methods and results of Secchi's spectroscopic observations of the Orion nebula.[4] He had particularly harsh words for Secchi's suggestion that the spectrum of Betelgeuse appeared to show that the object may not be a star at all, but rather a transitional form between a nebula and a star.

Now Secchi claimed to have found a red star which, despite appearing quite unremarkable in the telescope, displayed a peculiar spectrum when examined with the aid of a spectroscope.[5] Instead of the usual continuous background of colour interrupted by discrete dark lines, this star's spectrum was composed exclusively of broad bands of red, yellow and green, each separated by a dark interval. Indeed, he reported, the star's unusual spectrum was surprisingly similar in appearance to that produced by a Geissler tube: a cylindrical, electrode-capped glass tube developed in 1857 by German instrument maker Heinrich Geissler (1814–79), inventor of the mercury vacuum pump. When connected to a battery, such a tube glows with a light characteristic of the pure but rarefied gas within it. By 1865, Geissler tubes had become widely used as standards of comparison in qualitative analysis.

Secchi admitted he did not know what gas was in the Geissler tube he was using as his comparison light source. But he was not convinced that chemical composition really mattered in this case. It was of greater significance, in his view, that the tube's light was sparked by electricity. Perhaps, he speculated, the light of this particular star (and possibly others) is produced electrically. The dark gaps in its spectrum might indicate, not absorption of light by a cooler gaseous atmosphere, but a true absence of selected coloured rays in its light.[6]

Huggins also received a request to examine Secchi's red star from Thomas Romney Robinson at Armagh. The men were members of two largely non-intersecting circles of astronomers, but Robinson had recently visited Huggins's Tulse Hill observatory. Separated by age, geography and observational interests, they had little in common to bring them together until now.[7]

Unfortunately, the star in question was a denizen of the winter sky, and Huggins was unable to observe it until March of the following year.[8] By the time he responded to Robinson's request, the old astronomer's interest had shifted from Secchi's star to a more pressing concern, namely the construction of a 4-ft Cassegrain to survey the southern sky from Melbourne, Australia. A contract had just been signed with Dublin-based instrument maker Thomas Grubb (1800–78).[9] Robinson was excited about the plans.[10] He had been advocating for just such an instrument since 1849.[11] Now that his dream was finally going to be realised, he wanted the telescope to have every modern astronomical convenience. Not only would the Great Melbourne boast resolution comparable to that of Lord Rosse's 6-ft Leviathan, but its special attachments for photography and spectroscopy would rival the best known at the time.

Happily, Huggins's belated response on Secchi's red star caught Robinson as he contemplated the design of this new state-of-the-art instrument. The letter's arrival seems to have presented the old astronomer with a timely opportunity to tap his younger colleague's spectroscopic expertise, for Huggins wrote again on 13 March, this time with a list of suggestions for fitting up the 'great Southern reflector' for spectroscopic work.[12] He provided Robinson with information that was both rudimentary and practical – designed to communicate the sort of tacit knowledge required to reduce frustrating, yet often unavoidable, trial and error in equipment arrangement and thus facilitate observation. Unable to demonstrate in person how best to arrange the apparatus, he described it in plain language, bolstering his instructions with the kind of rationale a man of Robinson's experience could appreciate: 'With a little practice', he counselled the seasoned astronomer, 'the irregularities of the clock may be compensated for in part by varying the pressure of a finger placed against the eye-tube.'[13]

In time, Huggins became identified as a valued resource by a small but influential circle of experienced observers who actively sought his advice on how to make their instrumentation capable of collecting spectroscopic information on celestial bodies. In the autumn of 1866, for example, the Royal Society formed a committee to establish an astronomical and meteorological observing station in India. With £200 at their disposal, committee members pondered how best to spend the grant. 'Many of those most competent to judge', Vice-president and Treasurer William Allen Miller wrote to President Edward Sabine, 'are agreed that a powerful telescope provided with a suitable spectroscope would be a most valuable thing under a competent and permanent observer at the present time.'[14] Huggins was consulted as were John Herschel and George Stokes. And after the spectroscopic instruments had been constructed, Lieutenant John Herschel (1837–1921), the elder Herschel's son, who was to oversee their operation in India, visited the Tulse Hill observatory for instruction in their proper use.[15]

6.2 Opportunism and eclecticism

As mentioned at the end of the previous chapter, it was about this same time that Huggins began a new notebook. His entries in it have a different character from those of the first notebook. For one thing, they are more complete, including more background information in each entry as to observing conditions and instrumentation employed, even occasional interpretive remarks. Still, there are many gaps.

Its title page is not as elaborate as that of his first one. On it he has written simply 'Observatory-Book, 1866' in large script letters above the RAS motto.[16] The first thirteen pages are filled with lists and descriptions of instruments, as well as a lengthy roster of visitors to the observatory from as far back as 1859.[17]

Huggins's entry for 14 March 1866 is the first related to celestial spectroscopy to appear in the notebooks. Only six of his fifteen entries made during the next two months involved examination of nebulae. The rest concerned observations of the Moon, the Sun and various stars as well as an investigation into the effect of aperture on the size of the ring visible around the telescopic image of a star.

Judging from the form and variety of the observations Huggins recorded in his new notebook, he still eschewed a programme devoted to a single type of object or methodological approach in favour of one that left him free to explore whatever interested him at any given time. He monitored a number of different celestial objects on an irregular basis: the surface of the Sun, the spectra of various nebulae and the spectra of stars brought to his attention by others. He seems to have had no regular observing schedule. Whenever he was notified of something new or unusual in the sky, he immediately subjected it to scrutiny. He was also tolerant of visitors to his observatory, including neighbours or friends who simply wished to see the Moon through a telescope.

6.3 The 'willow leaves' controversy

By the spring of 1866, Huggins was no stranger to solar observation, but it was not his forte. Recall that his only notebook entry for 1864 consisted of a cluster of small sketches of sunspots (see Figure 6.1).[18] In the even sketchier notes that surround those illustrations he

Figure 6.1 Drawings of sunspots by William Huggins. 23 March 1864, Notebook 1. Huggins wrote: 'luminous thatch very bright – luminous spots over the whole disk of sun – not of willow shape'. (WCL/SC)

explained that patterns could be seen in them representing 'luminous thatch very bright – luminous spots over the whole disk of sun – not of willow shape – tolerably uniform in size – of different forms' – a reference to the so-called 'willow leaves' controversy, a heated dispute then brewing in the RAS over how best to describe and interpret the mottling of the Sun's surface.

In March 1861, James Nasmyth (1808–90), whose invention of the steam hammer had brought him fame and fortune enough to retire and pursue his love of astronomy, reported that the solar disc had a filamentary structure that he likened to strewn willow leaves in appearance.[19] According to Nasmyth, these long and slender shapes, though fairly uniform in size, were layered in a helter-skelter fashion over the entire solar surface, but more clearly organised near the edges of sunspots. Nasmyth's claims were enthusiastically confirmed by distinguished solar observers including Pritchard, De la Rue and Carrington. There the matter rested until the autumn of 1863.

William Rutter Dawes also had considerable experience studying the Sun. To date, the eagle-eyed binary star resolver, whom we met in Chapter 3 and upon whom Huggins had relied while taking his first steps as a serious amateur, had remained silent on the subject of Nasmyth's 'willow leaves'. But recent events compelled Dawes to speak out. On 26 August, in his presidential address to the BAAS, industrialist William George Armstrong (1810–1900) hailed fellow-engineer Nasmyth's 'willow leaves' as a 'remarkable discovery'. Indeed, Armstrong could 'imagine nothing more deserving of the scrutiny of observers than these extraordinary forms'.[20] Three months later, at the November RAS meeting, Charles Pritchard announced that he and De la Rue – Nasmyth's mentor – had been in correspondence with John Herschel concerning the 'willow leaves'. While preparing the seventh edition of *Outlines of Astronomy*, Herschel saw a need to amend article §387 on the nature of the Sun's visible surface to reflect the promise of Nasmyth's 'remarkable discovery'. The possibility that Herschel would do such a thing, thus giving the 'willow leaves' his coveted seal of approval, roused Dawes from complacency.[21]

What was wrong with the original wording of §387, Dawes wondered? Who could argue with the truth and accuracy of Herschel's long-standing statement that the Sun's surface 'is finely mottled with an appearance of minute, dark dots, or *pores*'? With over a decade of experience as a solar observer using superior instruments, Dawes had seen these dynamic features, too. But he would not call them 'willow leaves'. Nor would he willingly grant Nasmyth, a relative novice at this sort of thing, credit for a new discovery. How an individual perceives and interprets the Sun's appearance in any given observation, Dawes argued, depends on the size of the telescope and the degree of magnification employed. Care must be taken to avoid the error of 'discovering' what has already been seen more clearly (and, hence, described differently) by others. Indeed, the mottling on the Sun's disc had been likened to 'brain coral', 'soapsuds in hard water' and 'rice-grains' by some; others regarded the whole debate as a war of words. In Dawes's view, it would be better to liken the striated borders of sunspots to 'small bits of straw or thatching' and the bright solar surface to 'minute *fragments of porcelain*'.[22]

For his part, Huggins cast his lot with his friend Dawes. His own observations, he asserted confidently, led him to conclude that, whatever they might be, the 'bright particles' lacked the uniformity of size and shape necessary to be classified as willow leaves.[23] But he could hardly present himself as an expert on the matter, and the controversy continued to simmer without resolution for some time.

In April 1866, the subject was raised again by 'willow leaves' advocate Edward James Stone (1831–97), first assistant at Greenwich.[24] Two weeks later, on the morning of 26 April, Huggins spent two hours scrutinising the solar surface. He took detailed notes, dividing his remarks into categories headed 'distribution', 'form', 'size' and 'brightness'. Before preparing his report for delivery at the next RAS meeting (10 May), he examined the solar disc at least three more times to confirm these observations.[25]

He bravely advertised his stance on the 'willow leaves' question by titling his paper 'Results of some observations on the bright granules of the solar surface'. Aware of the need to tread with great care through the field of egos that would hear and/or read his words, he couched his statements in neutral terms, offering constructive criticism and well-considered commentary to all who had voiced opinions on the issue. The name 'granule', he argued, is purely descriptive and free of any hypothesis as to the nature of the phenomenon.

Stone expressed his great pleasure 'that observers were getting so close together on the subject of the solar photosphere'. He personally preferred the term 'willow leaves', but he acknowledged the aptness of 'granule' to describe the elongation common to these features on the Sun's surface. De la Rue expressed his satisfaction that 'all observers were agreed on there being elongated forms' regardless of what one called them. He congratulated Huggins and others on their efforts.

The controversy did not end so much as it faded away, thanks in large part to Huggins's astute presentation of his case. Rather than review the fractious past and reinforce the personal antagonisms that were blocking productive exchange between the pro and con 'willow leaves' camps, Huggins treated all views, including his own, as worthy but in need of improvement – improvement that could only come through working together. He used the generic term 'granule' to point the way toward a common middle ground where it was likely no one would be completely satisfied, but all would find enough agreement to move forward.

6.4 The nova in Corona Borealis

In mid-May 1866, an unexpected celestial event temporarily grabbed Huggins's attention. Alerted to the appearance of a nova in the constellation Corona Borealis by Irish amateur astronomer John Birmingham (1816–84), he acted immediately to analyse its spectrum. The identification of this nova as a long-period variable star drew the attention of variable star observers to the results of his spectroscopic examination of this one celestial body.

On 18 May 1866, he wrote to John Herschel.[26] Then 74 years old, the eminent astronomer was one of the few surviving charter members of the RAS. Herschel, it will be recalled, had a long-standing interest in the variability of celestial objects, dating back to the sudden outburst of η Argus in the southern sky in December 1837.[27] Huggins moved quickly not merely to notify the elder astronomer of the new star's appearance, but more importantly to announce his own unprecedented spectroscopic study of it.

Stellar outbursts bright enough to be seen with the naked eye have rarely been noted. There are practical reasons for this. Those classified today as true supernovae are indeed rare: by modern estimates, only about three or four such events occur, on average, in any given galaxy during a millennium. These dazzling stars appear with remarkable suddenness, often rival the brighter planets in brilliance and occasionally remain visible during daylight hours. Their luminosity fades slowly but steadily over a period of twelve to eighteen months.

By contrast, novae occur with relative frequency – as many as thirty or forty may burst forth in our galaxy each year, but their small increase in brightness makes them less likely to reach naked-eye visibility. And their increase in luminosity is of such short duration that the entire event can be obscured by a week or two of overcast sky, or washed out by sun- or moonlight. Even under ideal conditions, registering such an event requires that someone is watching, and is watching with an eye willing and able to recognise what is seen.[28]

The nineteenth century witnessed an increase in the number of amateur astronomers, an increase in telescopic power available to them, and a rise in interest in noting and measuring variability in celestial objects. The likelihood of someone observing an event of this type was greater than ever before. Observers like Birmingham, Joseph Baxendell (1815–87), John Russell Hind and Norman Pogson (1829–91) were particularly intrigued by stars like Mira (o Ceti) and Algol (β Persei) that fluctuate in brightness with some degree of predictability.

Some variable stars, like R Coronae, attracted attention because of their idiosyncratic behaviour. First noted by Edward Pigott (1753–1825) in 1795, this star usually maintains a brightness just at the limit of human vision. Pigott and subsequent observers found that, unlike most other variable stars which flare up briefly, R Coronae occasionally dims quite suddenly for periods of a few months, and does so at highly irregular intervals.[29] In late 1863, the star dimmed dramatically as it had before, but, by 1866, it had failed to regain its usual level of luminosity despite a brief and slight brightening in mid-1865. Variable star observers were watching it with interest in 1866. It may not have been entirely coincidental, then, that the appearance of a nova in the same constellation and located within a few degrees of R Coronae should be observed independently by several individuals in May 1866.[30]

However, there is no indication that Birmingham was paying special attention to that region of the sky. He was simply 'struck with the appearance of a new star in *Corona Borealis*' as he walked 'home from a friend's house'.[31] On 12 May, he sent a note to *The Times* of London announcing his discovery of the new star. Fearing (correctly) that the note might not be published, and believing the new star's spectrum would make an interesting subject to analyse, he sent another directly to Huggins, who received it about 5 o'clock on the evening of 16 May 1866.[32] Just two hours later, a similar message arrived at Huggins's

Tulse Hill home from the experienced variable star observer Baxendell in Manchester.[33] These unexpected announcements would have caught Huggins far less prepared had it not been for a fortuitous sequence of events that brought the problem of variable stars to his attention shortly before the nova made its appearance.

6.5 The spectra of variable stars

Stellar variability had come up for discussion at the RAS's March meeting after two communiqués on the subject were read. In one, German-born astronomer Herman Goldschmidt (1802–66) claimed that a new star seen in 1827 had, in fact, been recorded on several occasions in the past. His assertion that the nova was not so much a 'new' star as a variable of exceptionally long period prompted one attendee to comment that several other novae could turn out to be long-period variables as well.[34] Indeed, there was growing suspicion at the time that novae were examples of variables with greater or shorter periods.

The other, a letter from Angelo Secchi, was ostensibly addressed to the Society as a whole, but contained remarks intended for Huggins alone.[35] This time, Secchi wished to draw the Society's (read, Huggins's) attention to a discrepancy between the positions of lines he had observed recently in the spectrum of Betelgeuse and those Huggins had ascribed to the same lines some two years earlier.[36] Secchi suggested that the difference might be due to a physical change within the star. After all, Betelgeuse already had a reputation as an irregular variable star. And, although astronomers had noted only subtle changes in the star's visual appearance of late, a dramatic increase in its luminosity had been reported as recently as 1852. Secchi reasoned it was likely that variations in a star's luminosity would be accompanied by observable changes in its spectrum.

Society President Charles Pritchard invited Huggins to respond.[37] Thus put on the spot, he offered only a few general comments preferring to hold his opinion until he could study the matter fully. To that end, he soon got together with Miller to re-examine Betelgeuse's spectrum for signs of change. He wrote to Baxendell to ascertain the current phase of the star's variation and then examined the star himself. After weighing all the evidence, Huggins (with Miller) submitted a detailed critical evaluation of Secchi's methods and conclusions to the *Monthly Notices*.[38]

Contacting Baxendell initiated a timely exchange that led Huggins to undertake a brief investigation of the spectrum of another red variable star, μ Cephei (William Herschel's 'Garnet Star').[39] Because Baxendell knew Huggins was interested in the spectra of variable stars, he alerted Huggins as soon as the new star in Corona Borealis burst on the scene. Certainly Huggins's investigations into the spectral signatures of variable stars in the weeks immediately preceding the nova's appearance readied his mind and his method to make the most of this unexpected observational opportunity.

6.6 A new star

The sky was clear on 16 May when the letters from Birmingham and Baxendell arrived. Huggins took advantage of the time remaining before darkness fell to invite Miller to join him in observing the new star spectroscopically. Together, they prepared the spectroscope and arranged a sodium sample for comparison. In short order they were ready to go. After sighting the new star, Huggins noted, and Miller confirmed, that the nova's spectrum was compound, that is, comprised of a series of bright lines superposed on a nearly continuous background. Huggins attributed the continuous spectrum broken by absorption lines to the body of the star and likened the general appearance of the spectrum to that of the Garnet Star.[40] He believed the bright lines were produced by glowing hydrogen gas and drew attention to what appeared to him to be a nebulous region immediately surrounding the star.

The Royal Society was scheduled to meet the very next day. Huggins and Miller wasted no time preparing a brief paper describing their preliminary observations of the nova's spectrum.[41] Miller's official duties required his presence at the meeting, but Huggins chose to stay at home and make additional observations.[42] He invited another friend, Sidney Bolton Kincaid (1849–98), to work with him. Already, the star's luminosity had dropped to near the limit of unaided human vision. Huggins induced a spark between platinum wires covered with wet cotton to produce a comparison hydrogen spectrum. What he saw confirmed his suspicion that the bright lines emitted by the star coincided with those produced by hydrogen. He was less certain about the nebulosity he had seen around the star the night before.

On Friday 18 May, Huggins attended Pritchard's talk at the Royal Institution on the construction of telescopes, and so was only able to observe the nova by eye on his way home.[43] Over the next week, he observed its spectrum every evening, even though the waxing Moon coupled with the star's diminishing brightness made it difficult to see. Henry Acland of Oxford, Baxendell and even chemist Henry Roscoe visited his observatory to catch a glimpse of the new star's spectrum. Kincaid and Miller made return visits, but, as the star grew fainter, and the evening hours grew shorter, Huggins returned to objects of routine interest: the solar surface and various nebular spectra. Between 28 May and 13 September 1866, he recorded only four more observations of the nova in his notebook.

He began to develop his own theory to explain the unusual appearance of the nova spectrum. His interpretive framework was guided by, indeed limited to, the match of the nova's spectral line patterns with their artificially produced terrestrial counterparts. The new star was dubbed T Coronae by the Astronomer Royal, with the roman letter 'T' identifying it as a previously unrecorded star and the third variable found in Corona Borealis.[44]

At the first RAS meeting after the new star appeared, Huggins obliged the president's request for a report on his novel spectroscopic observations of T Coronae.[45] He announced to his curious colleagues that the nova's unusual spectrum included both dark absorption lines (interpreted by many as associated with a celestial body sharing many of the Sun's physical characteristics) as well as bright emission lines (accepted as indicative of luminous gas). Huggins speculated that the nova was, in fact, a 'star on fire', which, by virtue of some cataclysmic event, had let loose a large quantity of hydrogen gas into its immediate surroundings. In his view, the intense heat of the star had ignited and consumed the gas in a short period of time – hence explaining the sudden rise and rapid decline in the nova's luminosity.

In July, an article he had written on the new star appeared in the *Quarterly Journal*.[46] In it, he emphasised that he was only able to provide an explanation for the nova's change in brightness because of his careful spectroscopic examination of the star's light. And he posed the provocative question of what the star's spectrum might have been like just before the outburst occurred. He wondered about the bright lines seen in other stars. Could such a feature portend a similar cataclysm in these stars sometime in the near future? Huggins believed that proper interpretation of the differences in stellar spectral signatures would lead to an understanding of the physical causes of variation in stellar luminosity. If a non-varying star with bright lines in its spectrum could be observed methodically over time, perhaps a longer chain of events could be formed linking nebulae, novae and stars together in some progressive scheme.[47]

On 24 August 1866, Huggins delivered an evening address at the BAAS meeting in Nottingham titled 'On the results of spectrum analysis applied to the heavenly bodies'.[48] In it, he devoted considerable attention to his spectroscopic examination of the new star in Corona Borealis, presenting essentially the same information he had given in the *Quarterly Journal*.

Even though Huggins no longer referred to the nova as an example of a 'star on fire', the phrase must have captured Charles Pritchard's imagination. In an article for the popular magazine *Good Words* titled 'A true story of the atmosphere of a world on fire', the RAS president credited Huggins and Miller with the observation that founded the emerging hypotheses on the cause of stellar outbursts.[49] Thus, Pritchard introduced both the two investigators and the fruitfulness of stellar spectroscopy to a wide audience.[50]

Huggins's interest in the nova picked up briefly again in September after Baxendell reported another flare-up, but by early October, Huggins recorded, 'Not brighter certainly – appeared about the same as on Sept 28 ... I saw nothing remarkable.'[51] Even if there had been something to note, the star was moving quickly into the western twilight and would soon be unobservable. He turned his attention to more accessible variables – Algol and Mira – bright stars with reasonably well-established periods that were high in the autumn sky.

He also examined the spectrum of γ Cassiopeiae (Tsih, the middle star of the familiar W-shaped constellation) after learning of Secchi's claim that he had observed bright lines in its spectrum.[52] The star has achieved some notoriety since 1910 when it first began displaying signs of the erratic variability that has characterised it to the present day.[53] But, in 1866, it was just another unremarkable stable star. For this reason, Secchi's claim presented a special personal challenge to Huggins. If, on the one hand, bright lines really did exist in the spectrum of a non-variable star like γ Cassiopeiae, then confirming their presence would have far-reaching theoretical implications. On the other hand, if the bright lines were only an illusion generated by the contrast between a narrow isolated region of the continuous spectrum that happened to be surrounded by dark absorption bands, confirming them could brand him as a careless and uncritical observer. Huggins's previous experience with Secchi's work led him to strongly suspect the latter case was true. His notebook entries reveal his uncertainty over how best to interpret his own observations.

Following his first observation of γ Cassiopeiae on 8 October, Huggins noted 'A bright line sharply defined in the red. There is a faint shading on both sides. I am almost certain, that the brightness does not arise from contrast [*sic*], I think it may be considered a bright line ... Mr. Kincaid present & saw the star.'[54] The next time he observed, he looked at Algol and the unusual variable star β Lyrae. According to Secchi, both β Lyrae and γ Cassiopeiae exhibited similar spectra, a claim Huggins was unable to confirm. Another look at γ Cassiopeiae's bright red line with his direct vision spectroscope left him wary of Secchi's claim. The red line, he recorded, 'falls exactly in a space between absorption lines. Still doubtful if truly a bright line ... These bright lines are less marked with *increased dispersion*. I was much impressed with necessity for caution.'[55]

Although Huggins was generally disposed to distrust Secchi's claims, it is interesting to note that he was equally uncertain about similar findings in other stars reported by Charles Wolf (1827–1919) and Georges Rayet (1839–1906) of the Paris Observatory. Nevertheless, he publicly supported their observations and left his concerns hidden from public view.[56]

6.7 The red flames

In autumn, Cassiopeia is high in London's sky in the early evening, and Huggins returned to γ Cassiopeiae from time to time, to check on the contrast and character of its alleged bright lines. But it was not a project he felt compelled to pursue and he was soon distracted from this effort by a renewed interest in solar phenomena.

While observing an annular eclipse in May 1836, Francis Baily (1774–1844) saw a 'row of lucid points, like a string of beads' shine through the nooks and crannies of the trailing limb of the lunar disc at second contact.[57] He presumed the beads to be a momentary divertissement and expected an annulus to form once all the lunar mountains cleared the solar perimeter. Instead, much to his surprise, the beads not only persisted, they became elongated strands of liquid sunshine separated by pronounced parallel black lines. Seconds passed before the black lines dissolved and the familiar shimmering circle of sunlight appeared around the Moon. Was this something like the infamous 'black drop' effect that muddled the timing of transit events? If so, how could the attentive observer mark the true beginning and end of each phase in an eclipse? Baily urged colleagues to look carefully for signs of these 'remarkable' phenomena and for clues to their cause during future eclipses.

A perfect opportunity arose six years later when a total eclipse crossed Europe in July 1842. George Airy, for whom accurate timing of celestial events was everything, failed to see the beads from his viewing station near Turin, Italy.[58] Baily observed the eclipse from Pavia. He did report seeing them again, but this eclipse introduced two other spectacles that absorbed his attention during totality: a 'corona, or kind of bright *glory*' surrounding the black lunar disc, and '*three large protuberances*' resembling 'Alpine mountains ... coloured by the rising or setting sun'.[59] His vivid description and beautiful illustration of the latter in his report on the 1842 eclipse inspired new questions: Were the rose-coloured protuberances illusions brought on by eye fatigue or an over-active imagination? Were they some sort of dazzling atmospheric effect? Or were they true solar phenomena?

Being limited to momentary glimpses of these and other eclipse phenomena by the brevity of totality and the capriciousness of the attending weather made it difficult to obtain confirmatory observations.[60] In the 1840s, astronomer Temple Chevallier (1794–1873) experimented with placing a small metal disc in the focus of his telescope's eyepiece to produce an artificial eclipse that would make the protuberances visible on any clear day.[61]

A major breakthrough came in 1860 when Warren De la Rue claimed success in photographing the near-solar atmosphere during totality. He interpreted the images he had obtained as showing the limb of the Moon sequentially occulting the flame-like protrusions, and thus convinced his fellow astronomers that the prominences were solar in origin rather than transient features in the terrestrial atmosphere or simply illusions brought on by the sharp contrast of dark and light.[62]

His photographs confirmed once and for all the flames' reality and solar origin.[63] They also conjured multiple new mysteries that left solar specialists on tenterhooks until they could view the flames again. The total phase of the next total eclipse, in December 1861, was expected to be barely two minutes long. In April 1865, another promised over five minutes of totality, but to observe it required travel to South America or Portuguese West Africa. A third eclipse, with almost three minutes of predicted totality, was expected in August 1867, but it was even less inviting as an expedition prospect. Its centre line was due to cross Argentina and then plunge southeast over the Atlantic before terminating near the Antarctic circle.

By the latter part of that decade, as we shall see in Chapter 9, interest in the nature of solar prominences was again on the rise. Norman Lockyer, for example, whose earlier astronomical interests had centred on the Moon and the planet Mars, turned his attention to the Sun in the mid-1860s in part because of the excitement generated by the 'willow leaves' controversy.[64] In March 1866, he began a spectroscopic study of sunspots using a clever method of his own design. He projected the Sun's image onto a screen that had a small slit. The screen could be moved to position the slit across a sunspot. In this way a linear segment of the sunspot as well as a portion of the adjoining photosphere were thrown into the attached spectroscope allowing the spectra of both sunspot and photosphere to be compared simultaneously.

Lockyer's analysis of these observations formed the basis of his first paper to appear in the *Proceedings*.[65] Titled 'Spectroscopic observations of the Sun', the paper was communicated on his behalf by the RS secretary, physiologist William Sharpey (1802–80), on 10 October 1866, and read at the 15 November meeting. The dates are significant in terms of the dynamics of Lockyer's increasingly competitive relationship with Huggins. Indeed, the ensuing commotion over who first conceived, developed and executed a successful plan to observe solar prominences without an eclipse diverted the attention of the paper's readers from Lockyer's sunspot findings to his suggestions for possible future applications of the spectroscope to solar research. In particular, his query, 'and may not the spectroscope afford us evidence of the existence of the "red flames" which total eclipses have revealed to us in the sun's atmosphere; although they escape all other methods of observation at other times?' became a central point of contention.[66] At the time he submitted the paper, Lockyer noted that his spectroscope possessed insufficient dispersing power to render the prominence spectral lines visible without an eclipse.

Meanwhile, on 10 November 1866, one day after the regular monthly meeting of the RAS, and not quite a week before the RS meeting at which Lockyer's paper on solar spectroscopy was to be read, Huggins wrote in his observatory notebook, 'I tried a new method of endeavouring to see the red-flames' by a method that 'had appeared to me probable (for some weeks)'.[67] The parenthetical phrase here is telling. But to whom was it addressed? Did he anticipate competition in the quest for viewing the elusive solar prominences without an eclipse? Did he intend this entry to stand as evidence of his priority?

The method Huggins believed would render the prominences wholly visible to an observer – and which he was then claiming to have been piecing together in his mind for some time – was not spectroscopically based. If, as reported, the prominences were red in colour, he reasoned it should be possible to filter out most other regions of the solar spectrum using a stack of differently coloured pieces of glass held together with Canada balsam. Did he fail to recognise the potential of prismatic analysis as a practical means by which the solar prominences might be rendered visible? If he had been contemplating viewing them by filtering the Sun's light 'for some weeks' already, what motivated him to test his method at this particular time? Did Lockyer reveal something of his own intentions in informal conversation at the RAS meeting the night before?[68]

For the moment, at least, it did not matter. Lockyer's inadequate apparatus prevented him from executing his clever plan, while Huggins's could not be coaxed to perform as he had hoped. Besides, searching for prominences was just one of many irons Huggins had in the fire at the time: he was, as we shall see in the following sections, busy measuring the heat of celestial bodies, observing changes in a lunar crater, and preparing his assault on the problem of measuring stellar motion in the line of sight.[69] Aside from a few notes on sunspots, he recorded very little related to solar investigation during this period. A view of the red flames without an eclipse would have to wait.[70]

6.8 Fireworks and shooting stars

At the RAS meeting in November 1866, president Charles Pritchard called Fellows' attention to a much-anticipated meteor shower expected in the middle of the month. It was the first such event to have been predicted in advance. Professor Hubert Anson Newton (1830–96), of Yale College, had examined historic accounts of notable so-called 'star-showers' and announced the night of 13–14 November 1866 as the probable time for a repeat of 1833's spectacular display.[71] Pritchard warned the assembly, 'If any man went to bed on either of those nights, he was not worthy to be called an astronomer.'[72]

Huggins had already been preparing himself by viewing the spectra of sudden flashes of flaming metallic substances produced by fireworks displays in September and October at the Crystal Palace, not far from his home. He used an instrument he called a meteor-spectroscope.[73] The handheld instrument was a small direct-vision spectroscope with three contiguous prisms, one of flint glass inverted and sandwiched between two of crown glass (see Figure 6.2). The prism train was held in place in a tube attached to a small viewing telescope with a field of view of about 7 degrees. The records of Huggins's fireworks observations indicate that he had no difficulty spotting transient events and felt confident that he could detect spectral characteristics in the light produced.[74]

Figure 6.2 Handheld spectroscope. Schematic drawing of William Huggins's direct-vision spectroscope. (W. Huggins, *PRS* **16**, p. 241)

Huggins spent the evening of 13 November re-examining the star γ Cassiopeiae with Miller, while in the early morning hours of 14 November he made an effort to view the meteors between 1:45 and 3:15 a.m. He reported seeing many small meteors during the first hour of his vigil, but very few afterwards. Only one bright meteor appeared, but it was behind a cloud. 'Saw one or two faint ones through prism, but nothing satisfactory. The display at this time, a very poor one.'[75]

Meanwhile, other observers reported the display as being especially fine. It came just as Professor Newton had predicted and did not disappoint most of those who reported their observations. Accounts of observations made under excellent weather conditions filled almost half a page in *The Times* of London the next day. They variously described the 'fiery shower' as surpassing 'anything that the present generation has witnessed', 'like sparks flying from an incandescent mass of iron under the blows of a Titanic hammer', 'bursting globes of fire' and a 'magnificent spectacle'.[76] Years later, astronomer Robert Stawell Ball (1840–1913) recalled being encouraged to look for the meteor shower during his two-year tenure with Lord Rosse at Parsonstown:

I saw a spectacle which, even after an interval of forty years, was the grandest I ever remember having seen ... Each of them was bright enough and sufficiently conspicuous to arrest attention. But when they came in dozens, in scores, in uncounted hundreds, and finally in myriads, the scene was unspeakably sublime.[77]

According to Agnes Clerke:

The display, although ... inferior to that of 1833, was of extraordinary impressiveness. Dense crowds of meteors, equal in lustre to the brightest stars, and some rivalling Venus at her best, darted from east to west across the sky with enormous apparent velocities ... Nearly all left behind them trains of emerald green or clear blue light, which occasionally lasted many minutes, before they shrivelled, and curled up out of sight. The maximum rush occurred a little after one o'clock on the morning of November 14, when attempts to count were overpowered by frequency. But during a previous interval of seven minutes five seconds, four observers at Mr. Bishop's observatory at [the London suburb of] Twickenham reckoned 514, and during an hour 1120.[78]

The shower generated considerable discussion at the RAS meeting of 11 January 1867. If Huggins participated, his comments were not reported in the *Register*.[79] In fact, it appears probable he did not attend the meeting, for he recorded in his notebook that on that very evening a Mr Leaf and his sons called to have a look at Mars through the telescope. He also noted that his mother was very ill. Perhaps he was unable to leave her on her own.[80]

Huggins's only published comment on his meteor observations did not appear until some time later in a brief paper on the handheld spectroscope. In it he succinctly and unapologetically stated 'Unfortunately, I was prevented from making the use of the instrument which I had intended at the display of meteors in November 1866.'[81] What pressed him to make this unfounded claim? Did his need to guard his reputation as a careful and capable observer intensify his ever-present worry that colleagues would respond unfavourably to news that, despite his advanced preparation and expertise, he had, in fact, failed to observe many meteors, or their spectra?

6.9 Crater Linné

In the 1820s, cartographer Wilhelm Gotthelf Lohrmann (1796–1840) and astronomer Johann Heinrich Mädler described the lunar feature 'Linné' – named in honour of Swedish taxonomist Carl von Linné (1707–78) – as a deep crater with a diameter of some five to six miles, a size that made it the third largest crater in an otherwise smooth and barren plain. Located near the western edge of Mare Serenitatis, it had been noted simply as a round white spot with no mention of any crater-like features by German astronomer Johann Hieronymous Schröter (1745–1816) as early as 1788.[82] When, in 1830, Mädler teamed up with Berlin banker Wilhelm Wolff Beer (1797–1850) to produce their renowned lunar map, crater Linné was clearly depicted.

In October 1866, however, the German-born director of the Athens Observatory, Johann Friedrich Julius Schmidt (1824–84), announced that the crater had suddenly and inexplicably vanished. He had seen Linné in the early 1840s looking as it had been mapped by Mädler and Beer.[83] But now, observing it again nearly a quarter century later, Schmidt concluded that a real and significant change had recently taken place on the lunar surface. He communicated his observation by letter to the avid English lunar observer, William Radcliff Birt (1804–81), who immediately set to the task of corroborating the finding. Birt alerted fellow observers through the *Astronomical Register* and the *Monthly Notices*.[84]

The news broke at a time when interest in the study of lunar features was increasing among British astronomers, and it stimulated a great deal of speculation.[85] Some saw the crater's alleged disappearance as evidence of recent volcanic activity on the Moon, while others thought the crater may have been erased by a disturbance in the lunar atmosphere. 'A change', Agnes Clerke wrote, 'always seems to the inquisitive intellect of man like a breach in the defences of Nature's secrets, through which it may hope to make its way to the citadel.'[86]

Huggins first examined Linné in December 1866 and monitored it sporadically until December 1873. Although he had shown no interest in lunar surface features before 1866, he had searched for evidence of an atmosphere on the Moon two years earlier by observing, through a spectroscope, the extinction of the light from a star during a lunar occultation. He interpreted the negative results of this effort as probable, though not conclusive, evidence against a lunar atmosphere.[87] He was drawn to examine the crater by reports from regular observers of the Moon that changes in lunar features might be caused by the weathering effects of an atmosphere.

In his notebook entries on Linné, Huggins referred to the region ascribed to the crater as a 'white hazy patch' and 'less defined' than other areas on the lunar surface.[88] On 8 May 1867, he suggested that the crater Hercules also presented what he called a 'twilight' appearance. He claimed this twilight effect was absent in other more sharply defined craters, but did not view this as evidence of a lunar atmosphere. Instead he attributed Linné's 'cloudy appearance' to a 'peculiar, partly reflective property of the material of which *Linné* consists'.[89]

In January 1874, he submitted a summary of his six years of observations of Linné[90] including selected extracts from his notebook records of the appearance of the crater under different degrees of illumination[91] from which he concluded that changes in the crater were, in fact, illusions caused by variations in the direction of the light hitting the Moon's surface in that region.[92]

6.10 Thermometric research

Contemporaneous with his observations of the crater Linné, a new and completely different type of observation captured Huggins's attention, namely measurement of heat reaching the Earth from the Moon and brighter stars. Between 14 February and 3 June 1867, he recorded observations on eighteen separate occasions, thirteen of which included some kind of thermometric work.[93] He made no public announcement of these efforts, however, until February 1869 when he described what he had done both in his yearly observatory report for the *Monthly Notices* and in a brief paper submitted to the *Proceedings*.[94] In the *PRS* paper, he explained that he undertook this difficult work because he believed reliable quantitative measures of the heat of stars could be used to complement spectral data in the important task of determining the 'condition of matter from which the light was emitted in different stars'.[95]

Huggins's thermometric research has been ignored by his biographers and by historians of astronomy. Lawrence Parsons, 4th Earl of Rosse (1840–1908), and Edward Stone are the individuals normally associated with thermometric observations of celestial bodies during this period. Both of these men, however, began their work ignorant of Huggins's earlier efforts and long after he had given it up.[96]

In the decades preceding Huggins's stellar heat measures, a number of individuals developed ingenious methods of adapting the thermopile to the telescope to measure the quantity of radiant heat that reaches the Earth from celestial bodies.[97] Thus, there were precedents for his project and published reports from which he could have drawn motivation and technical assistance. But if his previous performance is any clue, Huggins did not derive his research questions from the existing literature. His venture into celestial thermometry at this particular time, a task which involved the acquisition and mastery of an entirely new kind of instrumentation and investigative method, presents something of a puzzle.

One clue may be found in the minutes of the RAS meeting on 10 January 1867, just one month before Huggins recorded his first thermometric observation. At that meeting, James Park Harrison read a paper on the radiation of heat from the Moon.[98] Harrison, a member of the Royal Meteorological Society, sought to show, through analysis of long-term records of terrestrial temperatures kept at the world's major observatories, that these temperatures were directly related to lunar phase. Sunlight reflected from the Moon's surface, he claimed, had the capacity to evaporate cloud cover on the Earth. What Harrison was arguing was not new. Nearly twenty years earlier, John Herschel had presented nearly identical views in the first edition of his classic *Outlines of Astronomy*.[99]

De la Rue expressed doubt that Harrison's work fell within the purview of the RAS suggesting that perhaps the Meteorological Society might be better suited to handle such discussion. Fortunately for Harrison, the more indulgent Charles Pritchard was presiding. He declared, 'I think we have the right to keep the Moon here.' Undaunted, De la Rue pressed the speaker on his research methods: 'Could not Mr. Harrison measure the heat from the Moon by a thermo-electric apparatus?' Harrison was convinced that the heat was 'used up in the atmosphere' leaving little or nothing to measure. The subsequent discussion was lively if inconclusive, drawing on the authority of Greenwich and the absent John Herschel for confirmation of Harrison's claims.[100] It was left to Baxendell to refute them in the next month's number of the *Astronomical Register*. Although Harrison continued to investigate the phenomenon, he never raised the subject again at the RAS.[101]

Huggins had no interest in accounting for terrestrial temperature fluctuations, but it is intriguing to speculate on the influence Harrison's presentation may have had on him at that time. Because no human sense can directly receive the information being examined and measured, thermometric work required a greater degree of instrumental intervention than was commonly used by astronomers. Ideally, if a thermopile were placed at the focus of a telescope aimed at a particular celestial body, any heat emitted by that body should trigger a differential expansion in the two metals comprising the receiving end of the thermopile. That slight temperature difference would then be converted into an electrical impulse, the strength of which could be inferred by the degree of deflection observed in the attached galvanometer's needle.

Huggins may have been encouraged to try to measure the heat of the Moon and stars from an interest in the instrumentation and the gadgetry rather than any theoretical concerns. His thermopile was constructed by a Mr Becker of Messrs Elliott Brothers (see Figures 6.3a and 6.3b).

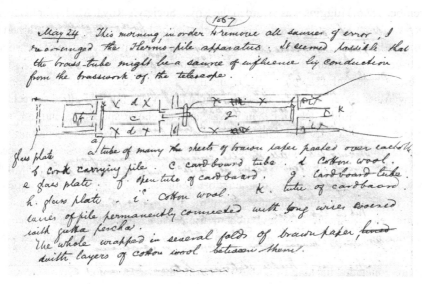

Figure 6.3a Notebook sketch of thermopile by William Huggins. 24 May 1867, Notebook 2. To insulate the thermopile from external heat sources, Huggins used cork to secure the pile inside a cardboard tube surrounded by cotton wool. (WCL/SC)

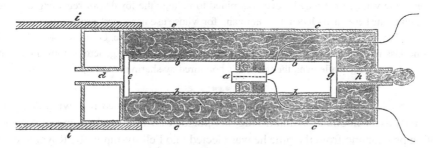

Figure 6.3b Published drawing of William Huggins's thermopile. (W. Huggins, *PRS* 17, p. 310)

John Peter Gassiot (1797–1877), an eminent electrical researcher and maker of thermopiles who lived near Huggins in Clapham, may have served as his expert link to past investigative efforts in this unusual area.[102]

Huggins worked hard to cajole consistent results from his apparatus. He drew diagrams of his equipment, gauged the accuracy of his measures on the basis of the consistency of the data he collected, suggested possible sources of error and described modifications which he felt would reduce those errors. He even provided advice to Thomas Romney Robinson on techniques of carrying out such research in early 1869.[103] In the end, however, his disappointment over the unreliability of his results, coupled with his difficulty in converting deflections of the galvanometer's needle into an equivalent quantity of heat, persuaded him to abandon thermometrics in favour of other projects.

6.11 Achieving 'a mark of approval and confidence'

In November 1866, Huggins was awarded the Royal Society's Royal Medal for his work on the spectra of both terrestrial chemicals and celestial bodies. Much of this work had been done with Miller, who, as an officer in the Society, was prevented by virtue of the Society's own 'self-denying ordinance' from being named co-recipient.[104]

In February 1867, after a vote by the Council of the RAS to waive the usual prohibition against awarding a shared medal, Huggins and Miller were jointly named to receive the Society's coveted Gold Medal for their researches on nebular spectra.[105] In the Council's view, these investigations had laid the foundation for the eventual resolution of the decades-old problem of the nature of nebulae. In his presidential address on the occasion of the Medal's award, Pritchard nested his tribute to Huggins's and Miller's nebular spectra work in the midst of his congratulatory remarks on Huggins's observations with Miller on T Coronae and his innovative use of the air spectrum as a standard against which to compare celestial spectra. These, Pritchard argued, marked yet 'another instance of that good fortune and success which so commonly attends, not so much the bold, as the well prepared'.[106]

Pritchard concluded his lengthy address by noting the 'vast amount of scientific resource' required for the successful prosecution of astronomical physics:

The most delicate appliances of optical science, the most refined in chemistry, the most profound arrangements of the electric force, have been called into requisition . . . For to-day . . . we may perhaps be pardoned the vanity, or it may be even permitted to indulge the loyalty, of regarding the science which we cultivate here, as a Queen, to whom and for whose use her sisters present the tribute of the fruits which they have gathered in the varied fields of human knowledge. Of herself and of her sister sciences no less may be affirmed than this: the lowliest of them finds a generous home in the ample threshold of her palace, while even the noblest is honoured by sharing a seat upon her throne.[107]

***

Contrary to what might be expected given the acclaim he received following his spectroscopic analysis of nebulae, William Huggins pursued an independent and often eclectic observing programme from the time he was elected into Fellowship in the Royal Society in June 1865 until he was officially awarded responsibility for the Grubb telescope paid for by the Royal Society in November 1869. At times, as we have seen in the case of the nova in Corona Borealis or the lunar crater Linné, the objects of his study were opportunistic responses to reports of others' findings. In other instances, as in the case of his thermometric studies, he was completely original, albeit unsuccessful, in developing a new method of acquiring useful information about the physical nature of celestial bodies. Although he was not the first to observe solar prominences without an eclipse, Huggins noisily claimed priority for suggesting that they could be observed in the first place. He thus intruded upon the claims of his chief competitor, Norman Lockyer. In the process, he gained a healthy respect for the constructive potential of establishing priority. In all of these efforts, Huggins betrayed his skill, energy, ambition and enterprise as he continually sought new ways to make contributions to astronomy worthy of recognition and prestige.

The next chapter will focus on one other of his eclectic research activities during this period, namely his efforts to measure the relative motion of celestial objects along an observer's line of sight. In conceiving a method by which Doppler's principle could be applied to astronomical research, contriving instrumentation to meet the exacting specifications that the measurements required, and introducing simple adjustments to improve the precision of those instruments, William Huggins gave astronomy an elegant and reliable research tool of broad utility.

Notes

1. W. Huggins, 'The new astronomy' (1897), p. 920.
2. This is not to say that Huggins's contributions to large-scale projects being conducted by others was insignificant. David DeVorkin has discussed Huggins's participation in the development of theories of stellar evolution based on competing systems of stellar classification. See DeVorkin, *An Astronomical Symbiosis* (1978), especially Chapter 2.
3. In fact, it is because of the eminence of his correspondents that a number of his letters from this phase of his career survive.
4. 'On the spectrum of the nebula of Orion', *MNRAS* **25** (1865), pp. 153–5; *AR* **3** (1865), pp. 105–6.
5. W. Huggins, 13 June 1865, MM 16.55, RSL.
6. Secchi, 'Étoile rouge singulière' (1865).
7. A date of 3 June or July can be inferred since Robinson's visit is noted in between a visit by the Earl of Rosse [William Parsons] and his son, Lord Oxmantown [Lawrence Parsons] on 26 May and the 29 July visit of Otto von Struve. 'Visitors to the Observatory, 1865', Notebook 2, WCL/SC.
8. W. Huggins to Robinson, 7 March 1866, Add MS 7656.TR66, GGS, CUL. Huggins found the star's spectrum was not limited to less refrangible rays, although he noted its more refrangible rays were quite faint.
9. The signing of this contract in late February 1866 concluded nearly four years of argument over the design of the telescope. See Robinson and T. Grubb, 'Description of the Great Melbourne Telescope' (1869), p. 135; Bennett, *Church, State and Astronomy in Ireland* (1990), pp. 130–2.
10. Not all astronomers shared Robinson's enthusiasm. John Herschel, for example, did not approve of mounting the telescope equatorially, that is, aligning its principal axis parallel to that of the Earth's rotation. He feared the telescope's clock drive would generate image-distorting vibrations and argued for a more traditional horizon-based altazimuth mount claiming 'there must be no *daring experiment* incurring *risk of failure*'. Letter from J. Herschel to Sabine, 10 December 1862, published in Royal Society, *Correspondence Concerning the Great Melbourne Telescope* (1871), Part 2, pp. 19–21; cited in Bennett, *Church, State and Astronomy in Ireland* (1990), p. 131. Bennett provides a summary of the history of the Melbourne telescope (pp. 130–4), which is a welcome elaboration on the meagre discussion in King, *The History of the Telescope* (1955), pp. 264–7. For Robinson's views on the equatorial mount, see Robinson to the Earl of Rosse, 28 December 1852, cited in Royal Society, *Correspondence Concerning the Great Melbourne Telescope* (1871), Part 1, pp. 13–15.
11. Bennett, *Church, State and Astronomy in Ireland* (1990), pp. 117–18.

12. This is surmise, however, as only Huggins's letters to Robinson survive in the Stokes collection at the University of Cambridge. See W. Huggins to Robinson, 13 March 1866, Add MS 7656.TR67, GGS, CUL.

13. *Ibid.*

14. Miller to Sabine, 29 August 1866, Add MS 7656.RS535, GGS, CUL.

15. W. Huggins, 'Visitors to the Observatory, 1867', Notebook 2, WCL/SC; Sabine, 'President's address' (1868), p. 138. For examples of Huggins's advisory relationship with Greenwich, see 'Suggestions for the new spectroscope', W. H. M. Christie to Airy, 24 October 1872, RGO 6.174/3/8/96–103, CUL; 'Spectroscopic apparatus &c recommended by Mr. Huggins', W. Huggins to Airy, 23 January 1874, RGO 6.174/3/8/83, CUL; Christie to W. Huggins, 4 February 1874, RGO 6.174/3/8/84, CUL; W. Huggins to Christie, 5 February 1874, RGO 6.174/3/8/85, CUL; W. Huggins to Christie, 20 February 1874, RGO 6.174/3/8/89, CUL; W. Huggins to Christie, 19 March 1874, RGO 6.174/3/8/91, CUL. Finally, on 20 March 1874, Christie wrote to Huggins, 'Your advice has been of so much service to us that we shall be very glad indeed to have the advantage of your examining the spectroscope and making any practical suggestions for which I needn't say, we shall be most grateful.' RGO 6.174/3/8/94, CUL.

16. Title page, Notebook 2, WCL/SC.

17. 'Visitors to the Observatory', Notebook 2, WCL/SC.

18. W. Huggins, 23 March 1864, Notebook 1, WCL/SC.

19. Nasmyth, 'On the structure of the luminous envelope of the Sun' (1862).

20. Armstrong, 'Address' (1863), pp. lviii–lix.

21. J. F. W. Herschel, *Outlines of Astronomy*, 7th edn (1864), pp. 695–6.

22. For a record of the discussion at the RAS on solar surface features, see *AR* **2** (1864), pp. 4–6 and pp. 98–101. For an historical analysis of the debate, see Bartholomew, 'The discovery of the solar granulation' (1976).

23. 'Discussion on the willow-leaved structure of the Sun's photosphere', *MNRAS* **24** (1864), pp. 140–2; p. 141.

24. Bartholomew, 'Discovery of the solar granulation' (1976), pp. 276–9. 'Sixth meeting, April 13, 1866', *AR* **4** (1866), pp. 139–40.

25. W. Huggins, 'Results of some observations on the bright granules of the solar surface' (1866).

26. W. Huggins to J. F. W. Herschel, 18 May 1866, HS.10.40, JFWH, RSL.

27. J. F. W. Herschel, 'Extract of a Letter from Sir John Herschel to the President ...' (1838). Herschel also submitted papers on the variability of other stars as well, including α Cassiopeiae, α Orionis, η Cygni and others.

28. For a discussion of variable stars from a mid-nineteenth-century viewpoint, see Grant, *History of Physical Astronomy* (1852), pp. 540–1. Agnes Clerke also discusses stellar variability in *A Popular History of Astronomy* (1885), pp. 426–34. For a modern survey of the history of variable star observation, see Hogg, 'Variable stars' (1984).

29. Pigott, 'On the periodical changes of brightness of two fixed stars' (1797); R. Burnham, 'R Corona Borealis', *Celestial Handbook*, vol. 2 (1978), pp. 702–8, p. 702.

30. These included a Mr Barker in London, Ontario, who claimed to have sighted it on 4 May. Barker's priority claim produced a brief but energetic exchange of letters in *AR* **4** (1866), p. 268 and pp. 310–11.

31. From a letter to E. J. Stone, dated 7 July 1866, in Birmingham, 'The new variable near η Coronae' (1866).

32. Huggins cited the contents of Birmingham's note in his *Nineteenth Century* retrospective article written in 1897. In this later account, Huggins mistakenly reports that he received the letter on 18 May when it is clear from his notebook entry and his letter to Herschel cited earlier that Birmingham's letter arrived on the 16th. This is a relatively insignificant error, and may say nothing about the accuracy with which the rest of the excerpt has been transcribed, but, as I shall show in a few later instances, Huggins took liberties in transcribing other things. Thus, we must look with some suspicion on this alleged quote from Birmingham.

33. By 1866, Baxendell had been observing variable stars for nearly twenty years and had recently married the sister of veteran variable star observer Norman Pogson. See 'Joseph Baxendell', *MNRAS* **48** (1888), pp. 157–60.

34. See *AR* **4** (1866), p. 97.

35. An extract of the letter, dated 10 February 1866, was published in *MNRAS* **26** (1866), p. 214.

36. W. Huggins, 17 April 1866, Notebook 2. W. Huggins and Miller, 'On the spectra of some of the fixed stars', *PTRSL* **154** (1864).

37. For a summary of the discussion which ensued surrounding the reading of Secchi's letter see *AR* **4** (1866), pp. 97–8.

38. W. Huggins and Miller, 'Note on the spectrum of the variable star α Orionis, with some remarks on the letter of the Rev. Father Secchi' (1866).

39. W. Huggins, 4 and 7 May 1866, Notebook 2, WCL/SC.

40. W. Huggins, 16 May 1866, Notebook 2, WCL/SC.

41. W. Huggins and Miller, 'On the spectrum of a new star in Corona Borealis' (1866). The paper they submitted on 17 May dealt primarily with observations made on the previous evening. They added notes to the text before it was published containing the results of subsequent observations.

42. W. Huggins, 17 May 1866, Notebook 2, WCL/SC.

43. C. Pritchard, 'On the telescope, its modern form, and the difficulties of its construction' (1866); W. Huggins, 18 May 1866, Notebook 2, WCL/SC.

44. Variable stars not already assigned a letter of the Greek alphabet designating its brightness relative to other stars in the constellation [a system developed by German astronomer Johann Bayer (1572–1625) in 1603], were named according to a scheme developed by Friedrich Wilhelm August Argelander (1799–1875) in 1844 which uses a roman capital letter in the order in which their variability was discovered beginning with the letter 'R' and going through 'Z'. See Hogg, 'Variable stars' (1984). See also W. Huggins and Miller, 'On the spectrum of a new star in Corona Borealis' (1866), p. 146. W. Huggins, 'On a new star' (1866), p. 277.

45. See 'Mr. Huggins, at the president's request . . .', *AR* **4** (1866), p. 181.

46. W. Huggins, 'On a temporary outburst . . .' (1866).

47. W. Huggins and Miller, 'On the spectrum of a new star in Corona Borealis' (1866); W. Huggins, 'On a new star' (1866), and 'On a temporary outburst . . .' (1866).

48. This address was soon published as a small book titled *On the Results of Spectrum Analysis Applied to the Heavenly Bodies* (1866), a book which gained wide circulation and was translated into French and German.

49. See C. Pritchard to J. F. W. Herschel, 27 March 1867, reprinted in A. Pritchard, *Charles Pritchard: Memoirs of his Life* (1897), pp. 257–8. 'In *Good Words* of next April I give some account of Frauenhofer's [*sic*] Lines and T. Coronae, and at last just venture on the speculation that it may be an old worn-out sun waxed dim – Faye says with a crust on it; and then a planet and a satellite went crash or scrape into it. If so, the ocean of the planet *might* afford the hydrogen and the *blue* that (as when the wind blows over gas), and the crash of the satellite (oceanless) *might* have caused the second outburst!'

50. This may be the article that first attracted the attention of young Margaret Lindsay Murray – William Huggins's future wife – to the subject of spectrum analysis of the light of celestial bodies.

51. W. Huggins, 8 October 1866, Notebook 2, WCL/SC.

52. Secchi, 'Stellar spectrometry' (1866)', and 'Spectrometric studies' (1866). See also 'Schreiben des Herrn Prof. Secchi . . .' (1866); 'Communications relative à l'analyse spectrale de la lumière de quelques étoiles' (1866); and 'γ Cassiopeiae', *MNRAS* **27** (1867), p. 135.

53. R. Burnham, 'Gamma Cassiopeiae', *Celestial Handbook* (1978), vol. 1, pp. 489–92.

54. W. Huggins, 8 October 1866, Notebook 2, WCL/SC.

55. W. Huggins, 12 October 1866, Notebook 2, WCL/SC.

56. In his notebook account of his observation of Wolf and Rayet's stars on 18 November 1866 (Notebook 2, WCL/SC), Huggins states, 'Looked at three stars with bright lines seen by M. M. Wolf & Rayet. Saw with great distinctness one very strong bright line in the first two. Uncertain about other lines. Spectrum appeared continuous.' In his published account of work done at his observatory for the year, he simply reported, 'Mr. Huggins has confirmed the observations of M. M. Wolf and Rayet so far as to the presence of bright lines in the three small stars described by them' [*MNRAS* **28** (1868), p. 87].

57. Baily, 'On a remarkable phenomenon that occurs in total and annular eclipses of the Sun' (1836), pp. 15–19.

58. Airy, 'Observations of the total solar eclipse of 1842, July 7' (1842).

59. Baily, 'Some remarks on the total eclipse of the Sun, on July 8th, 1842', *MNRAS* **5** (1839–1843), pp. 212–13. The coloured drawing of the Sun at totality (located opposite page 212) depicted what Baily referred to as 'protuberances'. In the text, he noted their 'colour was red, tinged with lilac or purple; perhaps the colour of the peach blossom would more nearly represent it'. To view a colour reproduction of this remarkable illustration, see B. J. Becker, 'From dilettante to serious amateur' (2010), Figure 2, p. 115.

60. The desire to view the prominences of the Sun without having to wait for an eclipse led astronomers to try to find some way to obscure their view of the central disc of the Sun in order to reveal the fainter features at the solar limb. Airy recounted cutting a hole in a screen to match precisely the size of the image of the solar disc. That way the brilliance of the disc's image would not interfere with the prominences that were expected to appear arrayed around the hole's circumference. Airy had no success with this method. Charles Piazzi Smyth also failed in a similar trial during his expedition to Tenerife in the late 1850s. See discussion following Huggins's presentation of his paper, 'On a possible method of viewing the red flames without an eclipse', *AR* **7** (1868), pp. 263–5; p. 264.

61. 'A letter from Professor Chevallier', *MNRAS* **5** (1842), pp. 186–7. About the same time, the Revd Baden Powell also devised a method for imitating a solar eclipse. See 'The president announced . . .', *MNRAS* **5** (1843), p. 264.

62. De la Rue's 1860 eclipse photographs were highly regarded by his fellow astronomers as a significant achievement in both celestial photography and solar astronomy. See R. W. Smith, 'The heavens recorded' (1981).
63. De la Rue, 'The Bakerian lecture: On the total solar eclipse of July 18th, 1860' (1862); R. W. Smith, 'The heavens recorded' (1981); and Rothermel, 'Images of the Sun' (1993).
64. Meadows, *Science and Controversy* (1972), pp. 44–7.
65. J. N. Lockyer, 'Spectroscopic observations of the Sun' (1866). Lockyer's paper was communicated by his friend, William Sharpey, because Lockyer was not yet a Fellow of the Royal Society. He was elected in 1869.
66. J. N. Lockyer, 'Spectroscopic observations of the Sun' (1866), p. 258.
67. W. Huggins, 10 November 1866, Notebook 2, WCL/SC.
68. The informal minutes of the meeting published in the *Astronomical Register* shows most attention was directed to the upcoming meteor shower and reports of spectroscopic observation of the nova in Corona Borealis. Given that Lockyer's paper on observing solar prominences was due to be read before the Royal Society in less than a week, Lockyer would have wanted to assure that it was not prematurely announced at the RAS. Even though he would not have discussed the matter formally, he may have mentioned it privately.
69. Huggins recorded solar observations on 31 May, 2, 8 and 27 November, and 5 December 1867; and 6 February, 15 April and 19 December (at which time he recorded his own first sighting of a prominence out of eclipse) 1868. See Notebook 2, WCL/SC.
70. See Chapter 9.
71. H. A. Newton, 'The original accounts of the displays in former times of the November star-shower' (1864).
72. C. Pritchard, 'Remarks of the president' (1866).
73. W. Huggins, 'Description of a hand spectrum-telescope' (1868).
74. W. Huggins, 13 September 1866 and 29 October 1866, Notebook 2, WCL/SC.
75. W. Huggins, 14 November 1866, Notebook 2, WCL/SC.
76. See 'The November meteoric shower over the metropolis', followed by six letters to the editor and a brief paragraph report on the meteor shower from Glocester [*sic*], *Times* of London, 15 November 1866, 10b–d.
77. W. V. Ball, *Reminiscences and Letters of Sir Robert Ball* (1915), pp. 70–1.
78. Clerke, *A Popular History of Astronomy* (1885), p. 379.
79. 'Third meeting, January 11, 1867', *AR* 5 (1867), pp. 25–8.
80. W. Huggins, 11 January 1867, Notebook 2, WCL/SC.
81. W. Huggins, 'Description of a hand spectrum-telescope' (1868) p. 242.
82. Clerke, *A Popular History of Astronomy* (1885), p. 315.
83. Schmidt, 'Über den Mondcrater Linné, (1867); Clerke, *A Popular History of Astronomy* (1885), pp. 315–16.
84. Birt responded to questions about Schmidt's announcement at the RAS meeting on 8 February 1867. He submitted a note to the *MNRAS* [27 (1867), pp. 93–4] which spawned a flurry of observations by a number of individuals, including William Huggins, who submitted reports in later numbers. See the indicated volumes of the *MNRAS* for reports on crater Linné: W. R. Birt, 28 (1868), p. 220 and 29 (1869), p. 63; W. Huggins, 27 (1867), p. 291 and 34 (1874), p. 197; Mädler, 27 (1867), p. 303; W. Noble, 28 (1868), p. 48 and p. 187; J. Joynson, 28 (1868), p. 49 and p. 126; Webb, 28 (1868), p. 185 and p. 218; E. Crossley, 28 (1868), p. 187; C. L. Prince, 28 (1868), p. 219; C. E. Burton, 34 (1874), p. 107.
85. In 1864, the BAAS formed a Lunar Committee and, at about the same time, Birt founded the Selenographic Society. For a discussion of this upsurge in interest in the Moon among British astronomers, see Clerke, *A Popular History of Astronomy* (1885), pp. 313–14.
86. *Ibid.*, p. 314.
87. W. Huggins, 'On the disappearance of the spectrum of ε Piscium' (1865). At the end of this paper, Huggins urged caution in using this single observation as a basis for concluding the Moon had no atmosphere. He felt that the difficulties attending spectroscopic examination meant that many more such observations were required before any conclusions could be drawn. However, it was to this particular observation that Agnes Clerke referred when stating that lunar observers had demonstrated the lack of a lunar atmosphere. See Clerke, *A Popular History of Astronomy* (1885), p. 311.
88. W. Huggins, 14 December 1866 and 14 February 1867, Notebook 2, WCL/SC.
89. W. Huggins, 'Note on the lunar crater Linné' (1867), p. 296.
90. In the paper, Huggins mistakenly refers to the 'observations I have made of this object during *nine* years', [emphasis added] an error which could have been readily corrected by the alert reader from the data given in the paper, but which may have easily gone unnoticed by most.
91. W. Huggins, 'Note on the lunar crater Linné' (1874).
92. Contrast Huggins's interpretation of his observations described in his 1874 paper on Linné (cited above) with that proposed in a note of C. E. Burton which immediately preceded Huggins's in that number of the *Monthly*

Notices. Burton compared 1873 measures of the size of the crater, including several made by himself, with those made by Huggins and Tacchini in 1867. Burton found them to be significantly different and concluded the crater had undergone physical change in that short time. See Burton, 'On the present dimensions of the white spot Linné' (1874).

93. W. Huggins, 14 and 23 February; 5 and 29 March; 11 April; 4, 17, 18, 23, 24, 26 and 30 May; 1 and 3 June 1867, Notebook 2, WCL/SC. He recorded one final thermometric observation in this notebook on 6 February 1868.

94. W. Huggins, 'Mr. Huggins' observatory' (1869); 'Note on the heat of the stars' (1869). In his observatory report, Huggins states, 'Time has not been found during the past year to continue some experiments which were conducted in the Observatory in the early part of 1867, for the purpose of detecting the heat received from the stars, and also, if possible, from the Moon.' In the *Proceedings* paper, he claims, 'In the summer of 1866 it occurred to me that the heat received on the earth from the stars might possibly be more easily detected than the solar heat reflected from the moon ... Towards the close of that year, and during the early part of 1867, I made numerous observations on the moon, and on three or four fixed stars.' Of the two accounts, the former seems the more likely. According to his notebook record, his first celestial thermometric observation took place on 14 February 1867 on a first-quarter Moon and the star Sirius with no consistent effect.

95. W. Huggins, 'Note on the heat of stars' (1869), p. 312.

96. See L. Parsons, 'On the radiation of heat from the Moon' (1869); 'Note on the construction of thermopiles' (1870); 'On the radiation of heat from the Moon. II' (1870); 'On the radiation of heat from the Moon ... The Bakerian lecture' (1873). See Stone, 'Approximate determinations of the heating-powers of Arcturus and α Lyrae' (1870). See also 'On the radiation of heat from the Moon', *AR* **6** (1868), pp. 92–3; 'Heating power of the stars', *MNRAS* **30** (1870), pp. 107–9.

97. For a contemporary review of early measures of celestial sources of heat, see Anonymous, 'Lunar warmth and stellar heat' (1870). See also R. S. Brashear, 'Radiant heat in astronomy, 1830–1880', unpublished manuscript, pp. 1–12. These early efforts involved measures of the heat of the Moon. Notable among these investigators were Leopoldo Nobili and Macedonio Melloni (1830), James Forbes (1835), Charles Piazzi Smyth (1856) and John Tyndall (1861). Brashear draws attention to the isolated, relatively primitive and unconnected nature of these different research projects. Thus, the independent and highly individualised work of Huggins, Parsons and Stone in the late 1860s is characteristic of this earlier pattern. See also McRae, *The Origins of the Conception of the Continuous Spectrum of Heat and Light* (1969), pp. 333–68

98. J. P. Harrison, 'Inductive proof of the Moon's insolation' (1868). For a brief account of the discussion which ensued following this paper's presentation, see 'On the radiation of heat from the Moon', *AR* **6** (1868), pp. 39–41.

99. J. F. W. Herschel, *Outlines of Astronomy* (1849), pp. 253–4.

100. 'On the radiation of heat from the Moon', *AR* **6** (1868), pp. 39–41.

101. 'On the radiation of heat from the Moon', *AR* **6** (1868), pp. 92–3.

102. I am indebted to Maire Brück for reminding me of Gassiot's proximity to Huggins's home in Tulse Hill. M. Brück to the author, 8 May 1992. Gassiot had loaned a thermopile to Charles Piazzi Smyth for his thermometric studies in the Canary Islands some ten years earlier. See R. S. Brashear, 'Radiant heat in astronomy', p. 4; H. Brück and M. Brück, *The Peripatetic Astronomer* (1988), pp. 59–60.

103. W. Huggins to Robinson, 17 February 1869, Add MS 7656.TR79; 27 February 1869, Add MS 7656.TR77; and 20 March 1869, Add MS 7656.TR81, GGS, CUL.

104. Sabine, 'President's address' (1866), pp. 280–2. Sabine referred to Miller's contributions several times during his conferral speech.

105. C. Pritchard, 'President's address on presenting the Gold Medal to Mr. Huggins and Prof. Miller' (1867), p. 164.

106. *Ibid.*, p. 160.

107. *Ibid.*, pp. 164–5.

7

Stellar motion along the line of sight

I am almost certain. . . .

– William Huggins[1]

In 1718, after comparing contemporary records of stellar positions with those of ancient times, Edmond Halley determined that the bright stars Palilicium (Aldebaran), Sirius and Arcturus had undergone a greater displacement on the two-dimensional sky than could be accounted for by precession alone. He postulated that these stars possess a 'particular Motion of their own . . . which in so long a time as 1800 Years may shew it self by the alteration of their places, though it be utterly imperceptible in the space of a single Century of Years'.[2] Halley's 'particular' stellar motion is what today's astronomers call 'proper' motion. It constitutes one component of a star's 'space velocity', or motion in space relative to the Sun. The other is its 'radial velocity', or motion in the line of sight.

Since Halley's day, astronomers have measured the proper motion of many stars. But even the nearest of our Sun's stellar neighbours is too distant to exhibit any of the visual cues (e.g. changes in apparent brightness or size) we normally rely on as evidence of motion in the line of sight. Indeed, the ability to detect, let alone measure, a star's radial velocity eluded earthbound observers until the late 1860s when William Huggins brought the new instruments and methods of celestial spectroscopy to bear on the matter. It proved to be the most influential of his contributions to modern day astronomical practice.

To be sure, his earlier discoveries had attracted widespread attention and acclaim by that time. Nevertheless, positional astronomers – the core of the professional astronomical community – still considered the fruits of this harvest as exotic delicacies rather than any real food they might want to cultivate and consume themselves. All this changed after they accepted Huggins's claim that spectrum analysis offered a direct and reliable means to determine the true motion of stars and other celestial bodies in three-dimensional space.

Thanks to automated precision instruments, measuring stellar radial velocity is a straightforward and routine procedure today. For Huggins, by contrast, who had conducted his pioneering efforts entirely and exclusively by visual means, it was an audacious project fraught with overwhelming mensurational and interpretive difficulties. As we shall see in this chapter, his observatory notebook records reveal the steps he took to overcome these challenges and they expose the rhetorical skill with which he attempted to persuade his contemporaries that he had, in fact, accomplished what he claimed.

Not all were persuaded – at least not immediately. But by the time Huggins's retrospective essay appeared in print in 1897, astronomers at Greenwich, Potsdam, Lick and Pulkovo had made measuring stellar motion in the line of sight an integral part of their observing programmes.[3] Scaling the steep learning curve generated by the rapid evolution of this new and very different research agenda required nerve and sure-footed alacrity. The vanguard did their best to remain in the forefront, while colleagues in other observatories around the globe busily adapted their own instruments and skills to keep apace.

In 'The new astronomy', Huggins turned readers' attention to the subject of his own role in the method's conception and early development. He spoke like the proud father of a successful child. Who could blame him? He declared that from 'the beginning of [his and Miller's] work upon the spectra of stars' he had always envisioned spectrum analysis as more than just a means to identify their chemical and physical nature. He was sure of its potential to become a tool of wide utility that 'could not fail in future to have a powerful influence on the progress of astronomy'.[4]

Thirty years earlier, in the introduction to his paper on stellar motion in the line of sight, he was more explicit:

We were at the time [1862–3] fully aware that these direct comparisons were not only of value for the more immediate purpose for which they had been undertaken ... but that they might also possibly serve to tell us something of the motions of the stars relatively to our system. If the stars were moving towards or from the earth, their motion, compounded with the earth's motion, would alter to an observer on the earth the refrangibility of the light emitted by them, and consequently the lines of terrestrial substances would no longer coincide in position in the spectrum with the dark lines produced by the absorption of the vapours of the same substances existing in the stars.[5]

Huggins may have been thinking about measuring stellar motion in the line of sight for a long time, but he left no contemporaneous record to support such a claim in either his observatory notebook, his correspondence, or any published account of his work. Indeed, in their paper 'On the spectra of some of the fixed stars', he and Miller said nothing to indicate they had looked for, or even suspected, a relationship between a star's spectral pattern and its motion.

Their silence notwithstanding, the subject did come up, as we shall see, in the discussion that followed their reading of that paper at the RS meeting on 26 May 1864.[6] Indeed, I would argue it was the discussion that followed the paper, not the work that preceded it, that really spawned Huggins's first thoughts on stellar motion in the line of sight.

Huggins was attracted to this line of investigation during an exceptionally eclectic period in his career. It was just one of many tempting puzzles he could have chosen to unravel. Tracing the ancestry of his interest in measuring the radial velocity of stars is like opening a *matryoshka*. Each layer conceals another beneath it that is at once enchanting and provocative. Delight and curiosity drive the historian inward toward its core. So, we begin this chapter with an examination of that core: the enigma of star colour.

7.1 The colours of stars

As they concluded their paper on 26 May 1864, Huggins and Miller drew heavily from the latest report on star colour by Admiral William Henry Smyth (1788–1865).[7] Smyth was a passionate amateur astronomer whose acclaimed two-volume *Cycle of Celestial Objects* had won him the RAS's prestigious Gold Medal in 1845.[8] After decades of studying the colours of stars, he still maintained the same circumspect view he held back in 1844 that the 'tints of stars require still closer observations before any correct deductions can be drawn'.[9]

At that time, no one knew if star colour was an extrinsic or intrinsic quality. It was difficult to account for recorded differences in the apparent colour of some stars. Smyth pointed to the example of Sirius. Though the ancients classified it as a red star, it 'is now decidedly white, and brilliantly white too'.[10] There were recent examples as well. William Herschel had described all the components of double stars γ Leonis (Al Geiba) and γ Delphini as white. Now, Smyth pointed out, the two stars in γ Leonis appear 'golden yellow and reddish green' while those in γ Delphini are considered 'bright yellow and bluish green'. Were such disparities the result of atmospheric or instrumental distortion? Were they produced by natural frailties of human visual perception and the ease with which even the most careful observer can fall prey to suggestion or illusion? Were they simply emblematic of language's limits to communicate personal sensations? Or do stars really change their colour over time?

Despite, or perhaps because of, these knotty and unresolved questions, Smyth believed it was just as important to record each star's apparent colour as it was to map its position on the sky or gauge its brightness if astronomers were going to come to any rational understanding of the heavens and its occupants. Early on, he had recommended several fruitful avenues of study to frame the work that needed to be done: ascertain the true frequency with which double stars exhibit complementary or contrasting colours; search for patterns in the relationship between brightness and colour in these pairs; test individual perceptual differences by controlling observational variables; uncover evidence of long-term change by comparing contemporary against historic observations; and deduce 'important conclusions respecting the velocity of different coloured rays' by scrutinising the light of variable stars for signs of colour.[11]

For his part, Smyth devoted considerable attention to double stars. They are 'often of very dissimilar colours', he explained, 'and no one who has ever directed a telescope to the heavens, can have failed to be struck with the brilliant hues they present'.[12] His work inspired Father Benedetto Sestini (1816–90) at the Collegio Romano to pursue this challenging research agenda. Sestini had already created a catalogue of coloured stars based on his observations in Rome.[13] In 1849, he informed Smyth of his plan to use the same telescope to re-examine these stars from his new location in Washington, DC, in order to compare the effect of local observing conditions on apparent star colour.

Soon afterward, in the *Aedes Hartwellianae*, Smyth expanded the *Cycle*'s few pages of comments on star colour into a lengthy and detailed account of current theories, unresolved questions and advice on how best to look for answers.[14] He included a list of one hundred pairs of double stars to encourage others to join in the effort. He displayed the data for these stars in tabular form so that readers could easily compare Sestini's descriptions of their colours alongside those from his own original survey and his recent independent reassessment.

Smyth still viewed the cause of star colour as an open question. Until more evidence had been gathered and carefully analysed, he advocated waiting to choose among the many interesting explanations that had been proposed. David Brewster, for example, had found significant dark gaps in the spectra of some colourful stars. Perhaps the resulting imbalance in their spectral array accounts for their tinged appearance.[15] Smyth reminded his readers of the Newtonian conjecture that different colours of light move at slightly different speeds through space, a view supported by the currently favoured undulatory theory of light. Given stars' enormous distances, the smallest of disparities in time required by each colour to make the journey to Earth would be multiplied to easily discerned intervals. He repeated his call for careful observation of new stars and variable stars to test the validity of this explanation for star colour.[16]

Smyth was intrigued, though largely unconvinced, by another proposal. 'In the present incertitude', he noted, 'it is suggested that variations in colour may be owing to variations in stellar velocity.'[17] He dubbed this idea 'Sestini's theory' apparently unaware it was originally proposed nearly a decade earlier by Austrian mathematics professor Christian Andreas Doppler. In fact, few astronomers, or any other men of science, were then familiar with Doppler's name or his work. His lack of notoriety was hardly surprising or unusual. When he presented his theory before the Royal Bohemian Society of Sciences in May 1842, only a scant handful of members were present.[18] Among them was mathematician and logician Father Bernard Bolzano (1781–1848), who became one of the young professor's early advocates. Bolzano penned a warm and generally enthusiastic article on Doppler's new theory which was published the following year in the widely read *Annalen der Physik*.[19]

The journal's audience must not have included Smyth, however. Sestini seems to have been his sole source for information on this new and unusual explanation for the cause of star colour. It is unclear whether Smyth realised that the Italian ex-patriot's devotion to the study of colour in double stars stemmed from his support for Doppler's provocative idea. But, in fact, as Sestini explained in 1850, he intended his catalogue to illustrate 'the hypothesis proposed by Professor Doppler to explain the variation of colors observed in some' double stars.[20]

Doppler considered his theory as a logical extension to – and conceptual generalisation of – stellar aberration, a small annual periodic shift in a star's apparent position on the sky. Aberration was first noted by James Bradley (1693–1762) and Samuel Molyneux (1689–1728) in 1725 during their unsuccessful search for evidence of stellar parallax. Bradley correctly attributed the phenomenon to both the finite speed of light and Earth's orbital motion in the plane perpendicular to that of the star's incoming light.[21]

Doppler believed he had discovered something analogous that occurs as a natural consequence of the relative motion of a star and its observer in a direction along the observer's line of sight. Adhering to the views of Huygens and Leonhard Euler (1707–83), Doppler assumed that light propagates as a wave, like ripples in water or sound in air. A train of waves emitted by a stationary source will travel at a speed that is characteristic of the medium which carries it. Arriving one after another, the individual waves within the train will break upon a stationary observer with a frequency determined by their length and their speed. However, the observer's perception of the waves' frequency will change if the wave source and/or the observer begins to move toward or away from each other. The waves will break more frequently and with greater intensity if the two approach each other, less so if they move apart. Doppler was convinced the difference between the waves' intrinsic frequency and that perceived by the observer would offer a way to calculate the source's relative speed of approach or recession.

Doppler was a mathematician, not an experimentalist. He deduced the elements of his theory from formulas he derived from fundamental principles of wave behaviour.[22] Applying the formulas first to sound and then to light waves, he calculated what an observer would hear or see under a range of set conditions.[23] The results convinced him that his theory could account for many previously unexplained astronomical phenomena such as the complementary colours so often observed in binary star pairs as well as changes in the colour and brightness of periodic variable stars, novae and extinguishing stars.[24]

In June 1845, Dutch meteorologist Christoph Hendrik Diedrik Buijs-Ballot (1817–90) tested Doppler's theory in an elaborate experiment involving musicians and musically trained observers who alternated serving as passengers on, and bystanders alongside, a rapidly moving railway train.[25] Although the musicians struggled to sustain a single tone and observers strained to hear over the noise of the locomotive, Buijs-Ballot felt certain the change in pitch observed in each trial conformed to Doppler's predictions.

Nevertheless, he questioned the propriety of generalising the theory to stars. He doubted stars could be found travelling with sufficient speed to exhibit the effect Doppler proposed. Besides, he argued, even if some were to be found, Doppler was wrong to expect an observer to note any hint of colour due to their motion. The recently discovered invisible rays beyond the red and violet extremes of the visible spectrum cannot be neglected. As the shortest wavelengths of an approaching star's visible light are shifted beyond the violet end of the visible spectrum, invisible rays in what is now called the infrared region of the spectrum will be shifted up into the visible range. The star's overall light would be composed of the same full array of colours regardless of its motion.

In August 1848, Scottish engineer John Scott Russell (1808–82), who was unaware of either Doppler's or Buijs-Ballot's work, reported on the unusual auditory sensations experienced by passengers and bystanders alike when they are exposed to the sounds emitted by a rapidly moving railway train.[26] Bystanders heard a rise and fall in pitch as the train passed. Passengers, meanwhile, heard no change in the train's own sound as it moved. But they did detect a difference in the pitch of echoes heard when the train approached and moved through a tunnel. And they complained of a jarring dissonance when they heard these echoes in combination with the train's own sound.

In a lecture he gave in December of that year, Armand-Hippolyte-Louis Fizeau (1819–96) used Russell's report to support the results of his own laboratory studies on variations in the pitch of sounds produced by a source moving towards or away from a stationary observer.[27] Like Russell, Fizeau was unfamiliar with Doppler's and Buijs-Ballot's previous work. It is interesting to compare his thoughts with theirs as he considered the practical challenges of extending what he had observed in his acoustic experiments to the case of light waves.

First, like Doppler and Buijs-Ballot, Fizeau assumed that the light source and/or the observer would have to be moving at a considerable speed in the line of sight to produce any noticeable shift. But, unlike Buijs-Ballot, Fizeau was sure there were stars that moved at such speeds. Second, in his acoustic experiments, he had designed his sound source to emit a narrow range of tones in order to make a shift in pitch easy to detect. Knowing most natural light sources emit a wide range of wavelengths including rays beyond the visible spectrum, Fizeau, like Buijs-Ballot, concluded that any apparent shift in the lengths of those waves due to the source's or observer's motion would result in no perceptible change in the light's overall colour. Nevertheless, Fizeau happily pointed out, this did not mean the star's motion could not be detected. Because Fraunhofer's dark lines mark the absence of single wave-lengths in a light source's spectrum, they could serve as visual guides in optical experiments just as single-frequency tones had served Fizeau in his acoustic experiments. Indeed, he suggested, any shift detected in the positioning of these lines in a star's spectrum should be interpreted as evidence of the star's relative motion toward or away from the observer. He was genuinely optimistic that astronomers could and would develop instruments capable of measuring such small displacements with precision.

The full text of Fizeau's 1848 lecture was not published until 1870.[28] And his idea to use Fraunhofer's lines to search for evidence of spectral shift due to motion was not proposed independently by anyone else in the interim. However, in 1850, the prolific nineteenth-century intelligencer Abbé François-Napoléon Marie Moigno (1804–84) published a detailed summary of Fizeau's lecture as part of a critical review and analysis of Doppler's theory in the third of his four-volume *Répertoire d'Optique Moderne*, an exhaustive compilation of current optical research.[29] The fact that Doppler and other investigators who had become involved in the debate over his theory remained unaware of Fizeau's confirmatory acoustic experiments as well as his suggestion to look for shifts of the dark lines in stellar spectra illustrates the compartmentalisation of scientific information by language and specialisation. Publishing one's work did not necessarily mean it would reach all with an interest in the subject.

One individual who knew of, and frequently referred to, Moigno's *Répertoire* was the brilliant young mathematician James Clerk Maxwell (1831–79).[30] In 1857, while Chair of Natural Philosophy at Marischal College in Aberdeen, Scotland, Maxwell discovered Moigno's presentation of Fizeau's work on the measurement of the speed of light in different media.[31] It appears likely, therefore, that Maxwell was familiar with Fizeau's 1848 lecture. Coincidently, Moigno's essay on Fizeau's work in volume 3 of the *Répetoire* immediately precedes his review of Doppler's work.

7.2 26 May 1864

It is important to know if Maxwell knew about Fizeau's lecture because he was among the Fellows present at the 26 May 1864 Royal Society meeting when Huggins and Miller read their paper 'On the spectra of some of the fixed stars'. Maxwell was then living in London and teaching natural philosophy and astronomy at King's College, where Miller taught chemistry. Although it had been several years since Maxwell first studied Fizeau's compar-ison of the speed of light in water and air, it is likely these experiments were very much in his mind that evening. Just one month earlier, he had submitted his own paper to the Royal

Society in which he offered a plan for detecting Earth's motion through the luminiferous ether by looking for evidence of small changes in the index of refraction in prisms.[32] In it, he credited Fizeau's experiment as a stimulus and guide. Maxwell reported a null result and promised further trials. Unfortunately, upon reviewing the submission, Stokes discovered a fatal error in the foundation of Maxwell's experimental premise and encouraged him to withdraw the paper.[33]

Maxwell had a long-standing interest in the physics and perception of colour. He would have listened with interest as Huggins and Miller attributed star colour to the chemical composition and physical structure of the atmosphere that envelops each star's incandescent core.[34] Their study confirmed that star spectra are interrupted by an assortment of the dark gaps that Wollaston, Fraunhofer and Brewster had already observed. They concluded that a star's colour is a result of the combination of all the individual colours remaining in its spectral array.

In presenting their evidence, Huggins and Miller made frequent reference to Smyth's new edition of the *Cycle*, which included the latest iteration of his developing treatise on star colour.[35] They noted, for example, that the spectrum of 'orange tinged' Betelgeuse is dark in its green and blue parts. That of white Sirius, by contrast, displays few notable dark lines. Perhaps, they suggested, the star's atmosphere is very thin, or even incandescent itself. They admitted they had not examined the spectrum of a green, blue or violet star. These are often dim, they explained, and their spectra even fainter. Besides, nearly all are located too close to a bright neighbour making them difficult to isolate and examine separately.

Back in 1863, when Huggins and Miller were examining the spectra of stars, Smyth was preparing to publish a monograph on the science of star colour.[36] He asked several seasoned observers, including Huggins, to examine binary pairs of contrasting colour and give their expert opinion on them. Huggins offered his comments on Albireo (β Cygni), describing the colours of its components A and B as 'very dilute bichromate potassa yellow' and 'dilute ammonia-sulphate of copper blue' respectively (see Fig. 4 in Figure 5.2).[37] He informed Smyth that he had augmented his telescopic examination with observations of the individual stars through his 'new spectrum apparatus' and promised to observe the pair again under better viewing conditions so that he could offer a detailed assessment of their spectral appearances.

Following up on this promise, Huggins and Miller reported that they had re-examined the individual spectra of the yellow and blue pair.[38] Although they did not measure the dark lines they found, they noted that the pattern of dark lines in each spectrum matched what would be expected given that star's overall colour. Yellow-orange A showed dark gaps in the violet and blue with only narrow dark lines in the orange and red and no strong lines in the yellow. The spectrum of the bluer B star was faint and difficult to observe. Its orange and yellow regions were covered by 'several groups of closely set fine lines'. Huggins and Miller observed few dark lines in the blue and violet part of the spectrum.

The discussion that followed the reading of their paper was reported by Lockyer in *The Reader* a few weeks later. 'We have pretty certain proof of the idea which has long been floating in many minds as to the cause of the colours of the stars', he announced, 'though their variability in colour, which has lately been so sternly insisted upon, is still yet to be explained.'[39] To quote Huggins and Miller, *'the differences of colour* [depend] *upon the differences of constitution of the investing atmosphere*, and these again intimately [are] connected with the chemical constitution of the stars'.[40]

Details reported in Lockyer's account of the RS meeting hint at the wide-ranging topics that were covered. We can infer, for example, that recent work on star colour by American natural historian Jacob Ennis (1807–90) was mentioned. Ennis was convinced that a star's intrinsic colour is due largely to its chemical composition.[41] Furthermore he believed that these intrinsic colours can be modified by Earth's atmosphere.[42] The investigations of alleged colour anomalies in binary pair 95 Herculis by astronomer Charles Piazzi Smyth (1819–1900), son of Admiral Smyth, seems to have been part of the discussion as well. The younger Smyth and his assistant had witnessed a dramatic change in the colour of one star in that pair over just a few years while its companion appeared to remain constant.[43] The observation called into question any notion of a connection between the orbital motion of binary stars and their colours.

It is perhaps in this context that Doppler's theory was raised, if only to refute it. Lockyer asserted 'it is evident that the colours of the stars must be better watched than they have been; and, should the observation of Mr. Ennis and Professor Smyth be confirmed, some other theory other than Doppler's must be found to account for their variability'.

Then, according to Lockyer, Maxwell added his voice to those who questioned the validity of Doppler's theory. He remarked – in words that seem to have come directly from Fizeau's 1848 lecture – that 'if the colours were really tinged in consequence of the motion either of the star or our earth, the lines in the spectrum of the star would not be coincident with the bands of the metal observed on the earth, which gives rise to them'. Clearly Maxwell took Huggins's and Miller's observations as strong evidence that, contrary to Doppler's views, star colour could not be explained solely as a consequence of motion in the line of sight.

Lockyer concluded his article with confidence:

Messrs. Huggins and Miller, doubtless will not let the matter rest, for they promise to examine the remarkable class of blue, green, and violet stars, which are found alone in close contiguity with usually brighter orange or red stars. Possibly it may be found that changes in our own atmosphere may have more to do with the changing colours of the stars than is imagined.[44]

Still, one year later, in a lecture at the Royal Institution, Huggins failed to mention motion in the line of sight as a possible cause of colour in stars. Instead, he simply reiterated the view that 'the colours of the stars have their origin in the chemical constitution of their atmospheres'.[45]

7.3 Stellar motion in the line of sight

Huggins submitted his paper on the relative motion of the Earth and stars along the line of sight four years after the May 1864 RS meeting. In his introductory remarks, he reminded his readers of the basic principles of Doppler's theory, that it was first enunciated nearly a quarter century earlier, and had been confirmed soon afterward for the case of sound by Buijs-Ballot.[46] He noted others' attempts to measure stellar motion in the line of sight, including recent papers by Ernst Friedrich Wilhelm Klinkerfues (1827–84)[47] and Angelo

Secchi.[48] But he criticised their observational methods, and, in the case of Secchi, the faulty theoretical foundation on which he interpreted his observations.[49] In contrast, Huggins boasted a superior array of instrumentation which, he argued, permitted him to make the fine distinctions required in the positions of the spectral lines found in the light of celestial bodies compared with stationary terrestrial sources. He had acquired a new micrometer and a new and more dispersive train of prisms. He also created an improved method for calibrating the celestial spectrum with that of the comparison terrestrial spectrum. Such instrumental changes, in Huggins's view, made the detection of any slight difference in position easier and the measurement of the difference less prone to error.

In describing these improvements, Huggins displayed a sophisticated appreciation for the causes of observer and/or instrumental error. He also demonstrated a creative knack for developing an experimental design to eliminate, or at least limit, what he perceived as potential sources of those errors. In this way Huggins was able to convince other astronomers of the research potential of this innovative spectroscopic application.

To put some British authoritative weight behind the theoretical foundation of his research, Huggins included an excerpt from a lengthy and technical letter he had received from Maxwell in reply to his request for information on the effect of Earth's motion on observers' perception of light from celestial bodies.[50] It had been three years since the RS meeting when Maxwell had commented on using spectra to detect motion in light sources. Why Huggins chose to write to the young physicist at this particular time is unclear. Maxwell had been a more convenient resource when he taught at King's and lived in London. But, in spring 1865, he had freed himself of teaching responsibilities and returned to Glenlair, his estate near Dumfries, Scotland, to write his treatise on electromagnetic theory. Perhaps Huggins's recent acquisition of improved instruments encouraged him to think he could finally attempt the difficult research. Perhaps investigating stellar radial velocity really had been a back-burner project all along and he was just waiting for time and opportunity. After all, as we have seen in the previous chapter, he had been very busy applying the spectroscope to all sorts of celestial phenomena including nebulae, solar prominences, sunspots, novae, comets and variable stars.

There is one other thing worth mentioning which might have prompted Huggins to sit down and write to Maxwell when he did. At the May 1867 RAS meeting, Sidney Bolton Kincaid described an apparatus which he called the 'Metrochrome'. A staunch adherent to Smyth's plan to observe and record the colours of stars, Kincaid designed this instrument 'for the purpose of determining star colours, by which the tints of fixed stars may be exactly recorded relatively to standards easily reproducible by any observer, with any kind of telescope, any number of years hence'.[51] Advertised by its creator as easy to use and read, the Metrochrome created an artificial star by means of an incandescent platinum wire, an array of filters made of specially prepared chemical solutions and a rotating drum to blend the coloured light they produce. When the coloured filters were properly adjusted, Kincaid boasted, 'it will be possible to produce the exact colour of a particular star; and then the record of the solutions employed, and the dimensions of the several apertures, will enable the exact reproduction of such colour at any future period'.[52] An image of the artificial star could then be directed into a telescope for comparison alongside a real star.

Kincaid's presentation would have been of great interest to Huggins. For one thing, Kincaid was one of his occasional collaborators in Miller's absence. For another, Huggins was fascinated by gadgets. The idea of the Metrochrome would have catalysed his creative energies. It is easy to imagine him mentally tinkering with it in his mind, modifying it to produce a different kind of star-spectroscope. The presentation would also have focused his attention in new ways on the old star colour question.

All these experiences combined to give Huggins a broader sense of celestial spectroscopy's analytical and mensurational capabilities. When he did reconsider the problem of star colour in the spring of 1867, he clearly had a new perspective on Doppler's theory. If he was planning to investigate Doppler's theory, he was wise to seek advice from the one man who seemed to know how to go about designing a method to put it to the test.

Huggins received Maxwell's now-famous reply to his query on 12 June 1867. In it, Maxwell described two experiments based on the application of Doppler's principle to luminous bodies. The second experiment, which Maxwell had in fact performed in his failed attempt to determine Earth's motion through the luminiferous ether, does not seem to have interested Huggins. It involved measuring the small changes in the index of refraction in a prism as light was made to travel through it in different directions relative to that of the ether flow.

But the first experiment, which Maxwell had not attempted, did intrigue Huggins and formed the basis of his research on the motion of stars in the line of sight. Maxwell – again seeming to take his words directly from Fizeau's 1848 lecture – suggested that if an individual line in a star's spectrum could be positively identified with one produced by a known terrestrial element, then observing the two spectra simultaneously should reveal any lack of coincidence in the lines' positions. This difference, analogous to Doppler's change in pitch in the case of sound, could then be used to ascertain the relative radial motion of the star and Earth. Using a simple hypothetical example of a stationary star being observed from Earth as it orbited the Sun, Maxwell calculated that it would be necessary to employ instrumentation capable of detecting a shift on the order of one-tenth of the distance separating the sodium D lines, or about 0.06 nm in modern terms.[53] Each division of Huggins's old wire micrometer only allowed him to register a displacement equivalent to about 0.2 nm.[54] This may have been adequate for measuring distances between binary stars, but it was insufficient to meet the stringent demands of Maxwell's suggested test.

Huggins had purchased a new double-image micrometer from Dawes on 10 May. A few weeks later, he put it to the test. He measured the angular distance between the close binary pair γ Leonis.[55] Then he determined the value of the vernier divisions on this new micrometer.[56] Huggins claimed that an interval measured by two-hundredths of a division of the micrometer screw head was equivalent to 0.046 millionths of a millimetre (or a little less than 0.05 nm in modern terms), indicating that the instrument could be relied upon to measure the minute shifts expected due to the relative motion of stars and Earth.[57]

Armed with his new micrometer and with information about the specific position of emission lines in stationary terrestrial gases, Huggins attempted to observe the shift of stellar spectral lines. On 25 June 1867, less than two weeks after receiving Maxwell's letter, he aimed his telescope and its new arrangement of 'two dense 60° prisms . . . Hoffman's [*sic*] & Ross's dense prism on α Bootis [Arcturus]'.[58]

There are a number of reasons why Huggins may have selected this particular star: Arcturus possesses a visible reddish tinge, it is one of the brightest stars in the sky at that time of year, it is situated high above the horizon in the evening, and, as Halley noted, it exhibits a large proper motion. Perhaps Huggins believed Arcturus was likely to possess a large motion in the line of sight as well. He recorded in his notebook, 'Atmosphere unfavourable. The lines seen, but not with the steady distinction necessary for determining whether they have any motion towards or away from the earth.'[59]

It is unclear what his observational objective was the following night. He selected the beautiful binary pair, ε Bootis (Mirak, or Izar), and took 'one measure on each side of zero . . . with *wire* micrometer'. The striking blue/orange contrast of its components suggests a test of Doppler's contention that a star's colour might be associated with its radial velocity, but Huggins only recorded data useful for determining the angular separation of the two stars.[60]

In late November 1867, he obtained a new and more highly dispersive spectroscope. It was equipped with two dense compound prisms made by Hofmann, two 60° prisms of Guinand's glass made by Simms, and one 45° prism of Chance's flint glass made by Browning (see Figure 7.1). The compound prisms were similar in construction to handheld direct-vision spectroscopes. Each consisted of a train of five prisms, including two very dense flint prisms placed alternately among the other three and cemented together with Canada balsam to reduce light loss at surface boundaries. He boasted that this improved instrumental design enhanced the observer's convenience even more by making it possible to completely remove one of the two compound prisms in order to observe faint objects.[61]

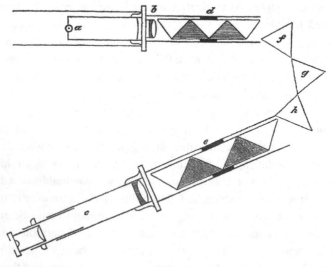

Figure 7.1 William Huggins's compound spectroscope. Huggins created this high-dispersion spectroscope using two direct-vision spectroscopes (*d* and *e*) in combination with three simple prisms (*f, g* and *h*). Spectroscope *e* can be removed when lower dispersion is desired. (W. Huggins, *PTRSL* **158**, p. 536)

If Huggins began immediately to make observations of stars for the purpose of measuring their radial velocities, he did not record any such work in any surviving notebook. Nevertheless, he did record observations of a number of familiar objects over the next four months to gauge the new spectroscope's performance against earlier observations with his old instrument. He seems to have needed this time to become used to manipulating his new instrument.[62]

7.4 Observations

Finally, on 11 February 1868, Huggins directed the new spectroscope with both compound prisms at the star Sirius. He compared the Fraunhofer F line in Sirius to that of a hydrogen spark.[63] Huggins wrote (see Figure 7.2a):

Appeared to me very slightly more refrangible than line of H . . . Afterwards compared the red line. It is defined in H. With a narrow slit could not see line in Sirius distinctly enough, but suspected that it was also very slightly more refrangible.[64]

There is nothing here to indicate that Huggins interpreted this observation as evidence that Sirius and Earth were moving toward one another or that he ascribed any physical interpretation to the observation.

Nearly two weeks later, on 24 February, Huggins made a second comparison; this time his words betrayed the influence of Maxwell's June 1867 letter (see Figure 7.2b):

The bright line [of] H was seen thin and sharp. It was not coincident with the centre of the dark line [in Sirius] but just grazed the more refrangible edge. It was therefore more refrangible than the dark line by half the thickness of the dark line. The thickness of dark line I estimated to be about equal to interval of sodium lines. This would make the line of H more refrangible by rather less than half of width of sod. [sodium] say .4.[65]

The fact that he gave a numerical value to this otherwise rough-sounding visual comparison suggests he intended to calculate the relative velocity of Sirius in the line of sight. The proportional difference he claimed to have observed in the positions of the two lines was four times larger than the shift Maxwell suggested would be observed due simply to Earth's orbital motion. It is important to note that Huggins seems unaware of, or unconcerned by, this observation's direct conflict with the one he made on 11 February, in which he observed the F line of Sirius to be *more* refrangible than that of the comparison hydrogen spectrum. The next night, he examined other stars in a similar fashion, but could not see their spectral lines steadily enough to make comparisons.[66]

At this point, Huggins was still comparing celestial and terrestrial spectra by the same means that he and Miller had used in their early work on stellar spectra. Their goal then had been to identify and match characteristic patterns associated with a particular element. Displacement of a whole group of spectral lines relative to its counterpart in the comparison terrestrial spark was not of investigative interest. But now Huggins made that difference the focus of his attention. Careful alignment of the telescopic spectrum with the comparison terrestrial spectrum was crucial.

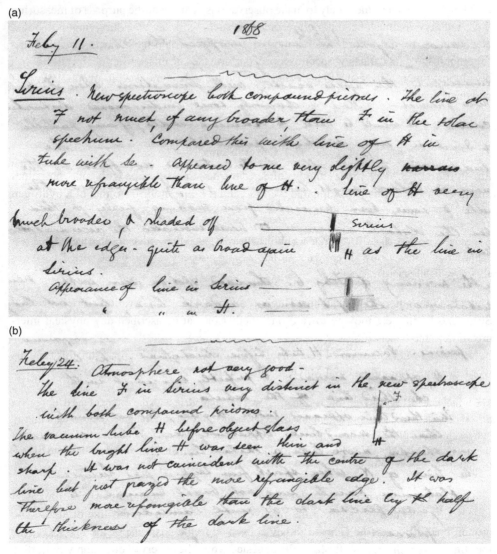

Figure 7.2 Line-of-sight observations. Excerpts from notebook entries made by William Huggins in February 1868 including sketches comparing the observed position of the F line in the spectrum of Sirius with that of a terrestrial hydrogen sample. (WCL/SC). (a) 11 February 1868. (b) 24 February 1868.

Huggins's observations convinced him to develop a new arrangement for throwing the comparison spark into the telescope. In his paper, he stated that the unreliability of his earlier measures using the old method had forced him to reject 'observations of many nights'.[67] The notebook record contains only a few measures, however, raising serious questions about his method of selection.[68] He designed a new comparison arrangement which included two small silvered pieces of glass positioned to produce two comparison spectra, one appearing above and the other below the celestial spectrum being observed.[69] This new apparatus was complete on 5 March.

Huggins returned to comparing the F line in Sirius with that of hydrogen. Now, he complained, he could not be sure of the proper placement of the hydrogen spectrum (see Figure 7.3a): 'Made the observation at least 20 times but without absolute certainty. Upon the whole, my impression was that of coincidence, but if there was any difference line of H on more refrangible side.'[70] A few days later, he observed again using the new apparatus, but this time with only one of the compound prisms. He recorded (see Figure 7.3b), 'I am almost certain after a great number of trials that the line of H is a very little more refrangible than line in Sirius. Certainly not more than $\frac{1}{2}$ its breadth.'[71]

His confidence in these observations was soon shaken, however. On 12 March he wrote (see Figure 7.3c):

Certainly the line of H appeared on the less refr. side of the line in Sirius. The opposite of the very satisfactory observation with tube before object glass. Arranged tube before obj. glass, but clouds prevented any further observation.[72]

By 18 March, Huggins had become convinced that it was difficult to be sure of the exact calibration of his celestial and terrestrial spectra. In the past, he had contented himself that the two instruments were properly aligned by placing a luminous object directly in front of the object glass of the telescope to serve as a substitute star. He could then adjust the apparatus until the terrestrial spectrum appeared to coincide with that produced by the surrogate star. He soon discovered that slight changes in position of the light source serving as his stellar stand-in resulted in dramatic differences in the location of its spectrum, thus throwing into question the accuracy of his earlier calibrations. He wrote (see Figure 7.4a):

Vacuum tube before the Obj. glass. Atmosphere not very good. When best seen the line of vacuum tube always more refrangible than the line in Sirius. Once or twice as the star was going out [of the field of view] I was not quite so sure of this, & I had almost the impression that they were more nearly coincident. I am not sure of this, but certainly when line best seen, then ... the line of H appeared, as above, more refrangible. Still it will be well, while the slit remains as it is, to see with some other stars, whether or not the motion of the star sensibly alters position of line.

Afterwards with same app. both compound prisms being inserted, compared line of Na in induction spark between pt. [platinum] wire & Mg with lines in alcohol lamp placed before the end of the adapter ... I found it was possible by moving the lamp to alter position of lines from lamp relatively to those of spark by a quantity about equal to abt. half the distance of the separation of lines of Na.

I then placed a diaphragm of paper in the end of the adapter [holding the spectroscopic instruments in place] with a small hole about $\frac{1}{4}$ dia. in centre. I then held the lamp before the hole, it was no longer possible to alter position of lines of Na. I found however that these lines *did not coincide* with lines of spark, but were about $\frac{1}{3}$ or $\frac{1}{4}$ of distance of lines of Na more refrangible. This state of things accounts perfectly for my recorded observations of Sirius.

(a)

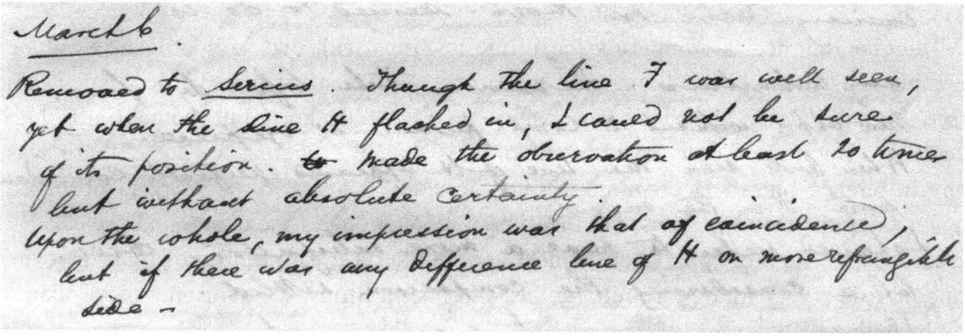

(b)

(c)

Figure 7.3 Line-of-sight observations. Excerpts from notebook entries made by William Huggins in early March 1868 including a sketch comparing the observed position of the F line in the spectrum of Sirius with that of a terrestrial hydrogen sample. (WCL/SC). (a) 6 March 1868. (b) 10 March 1868. (c) 12 March 1868.

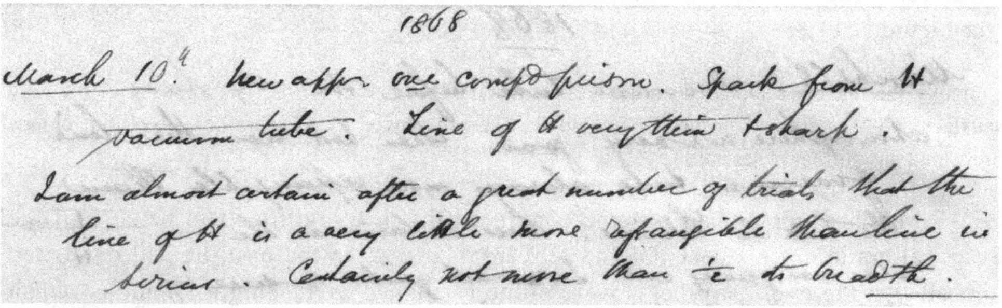

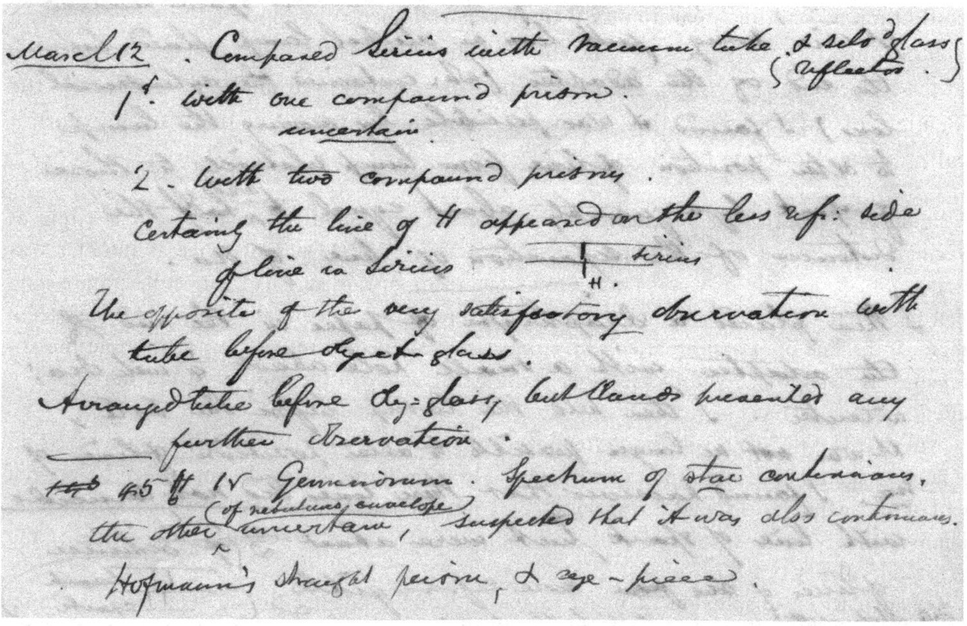

(a)

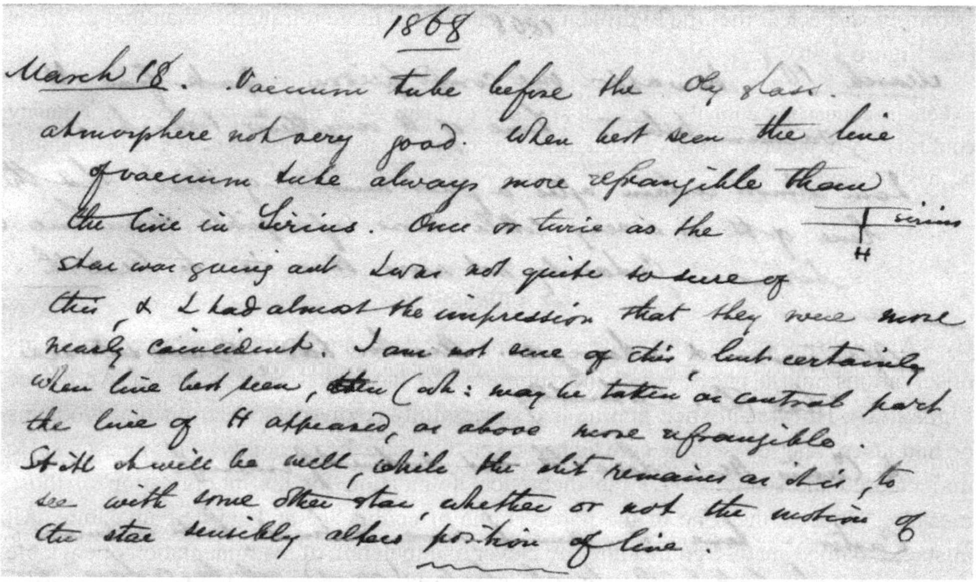

(b)

Figure 7.4 Line-of-sight observations. Excerpts from notebook entries made by William Huggins in late March 1868 including sketches comparing the observed position of the F line in the spectrum of Sirius with that of a terrestrial hydrogen sample. (WCL/SC). (a) 18 March 1868. (b) 30 March 1868.

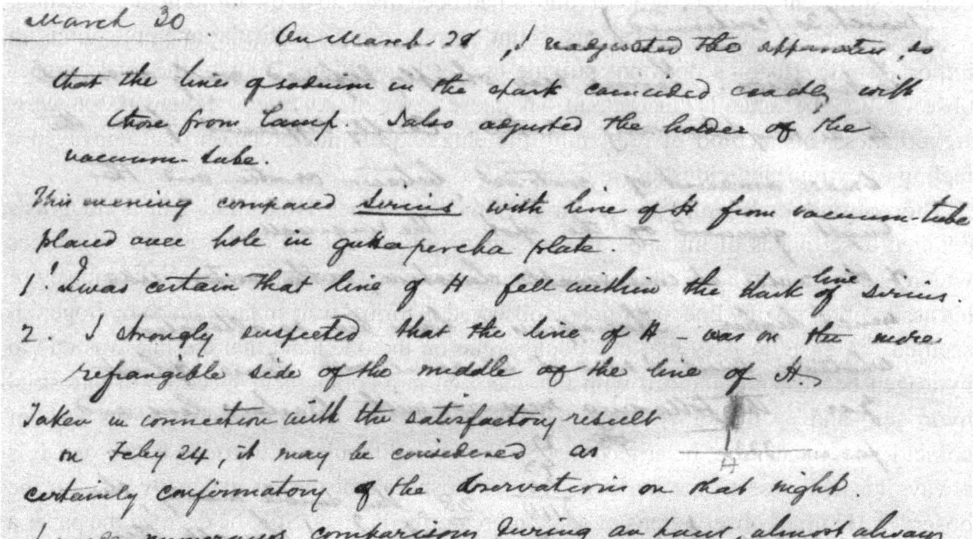

On 30 March, Huggins returned to the comparison of Sirius with hydrogen. This time he 'strongly suspected' that the hydrogen line was indeed more refrangible than that of Sirius (see Figure 7.4b):

Taken in connection with the satisfactory result of Feby 24, it may be considered as certainly confirmatory of the observations on that night. I made numerous comparisons during an hour, almost always the same result.[73]

7.5 Publication

By 4 April, Sirius was too low in the sky to continue the comparisons. Huggins compiled his observations into a paper which he submitted to the Royal Society on 23 April (see Figure 7.5). He was satisfied that he had successfully resolved the instrumental problems he had faced, and stressed the care he had taken to ensure the reliability of his measures. He underscored the soundness of the theoretical foundation for his interpretation of those measures. Indeed, the tone of the paper is one of confidence and spirited adventure. All obstacles named are overcome either by clever manipulation of instrumentation or enviable patience. He believed he was justified in disregarding observations he deemed unworthy because sufficient numbers of observations had been made to allow for judicious weeding.

The question of which observations 'count' as supporting a particular interpretation is an important one. Huggins does not provide the reader with any clues to the total number observations he actually made and, of those, how many were discarded, or why. Nevertheless, his method of presenting the data and his interpretation of it conveyed a feeling of extreme judiciousness and care.

The Astronomer Royal, George Airy, refereed the paper. He complimented Huggins's detailed descriptions of his apparatus and his exposition of the observational difficulties which had led him to adjust this arrangement. However, Airy questioned the legitimacy of his assumption that the line 'at or near F observed in Sirius' was, in fact, due to hydrogen. It seemed 'illogical' to Airy that one could argue on the one hand that the line was due to hydrogen because it coincided with the line seen in a comparison spectrum of terrestrial hydrogen, and on the other hand that the star was in motion because of its lack of coincidence. In addition, he argued that Sirius was a difficult star to use as a test case. It is always located close to the horizon where atmospheric distortion inevitably plagues the observer. He urged observations of some more northerly stars. Still, he deemed the paper a 'very important one', and recommended it be published in the *Transactions*.[74]

7.6 Response

The physical theory supporting Huggins's line-of-sight measures was a challenge to the best of contemporary astronomers. The RAS President, Charles Pritchard, called upon Huggins to elucidate further the basic premise of Maxwell's mathematical method of determining a star's radial velocity. In a rather hastily written letter of 6 July 1868, Huggins recounted the major points of Maxwell's 12 June 1867 letter,

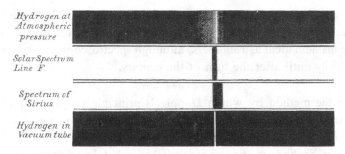

Figure 7.5 Published line-of-sight diagram from Plate XXXIII, W. Huggins, *PTRSL* **158**, comparing the position of the F line in the spectrum of Sirius with that of a terrestrial hydrogen sample. (Roscoe, *Spectrum Analysis*, p. 256)

including the mathematical formulae for making the necessary computations. He also included a brief discussion of Doppler's principle.[75]

A few years later, Airy himself wrote to George Stokes:

In some of the German pamphlets now afloat, on optical subjects, there is repeated allusion to das Doppelsche Princip, or some such term, conveying the idea that a man named Doppel has introduced some optical principle. It has something to do with change in the velocity of light but I see no clear description of it – can you help me?[76]

Stokes replied by citing Huggins's 1868 paper, the very paper Airy had himself refereed!

Recall that Airy had overseen a brief foray into the spectroscopic study of stars in 1863, about the same time Huggins launched his own research efforts in that area. But such observations did not fit into Airy's programme of systematic and repetitive observations of celestial bodies to record changes in their positions. Celestial mechanics was not concerned with the chemical composition of stars. Nevertheless, spectroscopic study of the Sun found its way into the Greenwich routine beginning in 1872.[77] Airy eventually began to enquire about the utility of spectroscopy as a way of corroborating other kinds of observations being planned, such as the transit of Venus.[78]

Airy's newfound interest in astronomical physics was not simply a response to the success of Huggins's spectroscopic work. It was, as we shall see in the next chapter, a visceral reaction to attacks by a vocal and energetic group of British scientists on the perceived inability of Greenwich to do what was necessary to maintain and even enhance Britain's stature in physical science.[79] Led by Lieutenant-Colonel Alexander Strange, an expert designer of scientific instruments and advocate of State support for science, this group wished to establish a number of national physical laboratories which would assume responsibility for prosecuting a research programme which included the new astronomical problems and methods. Airy viewed their move as a moral trespass – a usurpation of his job as Astronomer Royal.[80] He worked hard and quickly to demonstrate that Greenwich was capable of incorporating spectroscopy into its daily routine.[81] As a result of Airy's efforts, Greenwich published for the first time in 1876 results of line-of-sight measurements conducted by Edward Walter Maunder

(1851–1928) and William Henry Mahoney Christie (1845–1922).[82] Thus it was that routine measurement of stellar radial velocities became an element in the research agenda of the mathematical astronomers, although spectroscopic study of any sort was given low priority until after the turn of the century.[83]

***

In conceiving the method by which Doppler's principle could be applied to astronomical research, William Huggins introduced astronomers to an elegant and reliable research tool of broad utility. More importantly, he successfully persuaded his contemporaries that he had, in fact, accomplished what he claimed, in spite of the overwhelming mensurational and interpretive difficulties the method entailed.

Huggins repeated his observations and revised his measures for Sirius's velocity in the line of sight in 1872.[84] But, the truth was, visual observations could never satisfy positional astronomers' need for precision. As Hermann Carl Vogel (1841–1907) pointed out years later, the slight displacement in a target star's spectrum was difficult to view due to both the faintness of the spectrum and the instability of the apparatus that produced them.[85] Although the methods and practice of celestial photography had been improving over the years, particularly after the introduction of the dry gelatine plate in the 1870s, it was not until 1887 that Vogel, then working at Potsdam, captured the first photograph showing displacement of stellar spectral lines.[86] Observatories began constructing apparatus solely for the purpose of detecting and measuring small displacements in star spectra to determine their radial velocity. They commissioned and purchased instruments designed to meet the exacting specifications that the measurements required, and they introduced adjustments to improve the precision of those instruments.

Such undertakings were beyond the capacity of private individuals. As we shall see in the chapters that follow, Huggins had long since moved on to other projects by the time observatories had improved the movement in their clock drives, increased the stability of their telescopes, and introduced adequate equipment to produce photographs of star spectra reliable enough to make visual comparisons from them. As the world's major observatories became immersed in measuring the radial motion of many different types of celestial bodies, it became increasingly clear that Huggins had unwittingly introduced a method the ready implementation of which was beyond the ken, resources and ability of many of his fellow amateurs, thus effectively closing doors to them in favour of the mathematical astronomers.

Notes

1. W. Huggins, 10 March 1868, Notebook 2, WCL/SC.
2. Halley, 'Considerations on the change of latitudes of some of the principal fixt stars' (1718). Arcturus shifts across the sky a little more than 2 arcseconds each year. By comparison, the diameter of the full Moon is roughly 1800 arcseconds.
3. For a contemporaneous overview of the state of radial velocity measures in the world's major observatories, see H. C. Vogel, 'On the progress made in the last decade in the determination of stellar motions in the line of sight' (1900).
4. W. Huggins, 'The new astronomy' (1897), p. 920.

5. W. Huggins, 'Further observations on the spectra of some of the stars and nebulae' (1868), p. 529.
6. J. N. Lockyer, 'Stellar physics' (1864).
7. W. H. Smyth, *The Cycle of Celestial Objects continued* ... (1860), pp. 306–34.
8. W. H. Smyth, *A Cycle of Celestial Objects* ... (1844).
9. W. H. Smyth, *The Cycle of Celestial Objects continued* ... (1860), p. 303.
10. *Ibid.*
11. *Ibid.*, p. 304.
12. *Ibid.*, pp. 300–1.
13. Political unrest forced Sestini to flee Rome before his catalogue of coloured stars could be published. The catalogue was communicated by his colleague, Father Francescoe de Vico (1805–48). Vico, 'Schrieben des Herrn Professors De-Vico ...' (1849).
14. W. H. Smyth, *Aedes Hartwellianae* (1851).
15. *Ibid.*, p. 299. Brewster's observation was brought to Smyth's attention by a reviewer of *A Cycle of Celestial Objects* (1844) writing in *The North British Review* **6** (1847), pp. 236–7. The reviewer recalled that Brewster had announced his finding 'about fifteen years ago' at a meeting of the BAAS. If so, it was announced in unpublished discussion of Brewster's 'Report on the recent progress of optics' (1832).
16. W. H. Smyth, *Aedes Hartwelliana* (1851), pp. 302–4.
17. *Ibid.*, pp. 299–302.
18. Doppler, 'Über das farbige Licht der Doppelsterne ...' (1843).
19. Bolzano, 'Ein Paar Bemerkungen über die neue Theorie in Herrn Professor Chr. Doppler's Schrif' (1843).
20. Sestini, 'On the colors of stars' (1850).
21. J. Bradley, 'A letter ... giving an account of a new discovered motion of the fix'd stars' (1727). Stellar aberration offered evidence of Earth's motion in the plane perpendicular to that of a star's incoming light and an indirect means to measure the speed of light.
22. Doppler, 'Über das Farbige Licht der Doppelsterne ...' (1843), pp. 470–1.
23. *Ibid.*, pp. 471–3.
24. *Ibid.*, pp. 473–82.
25. Buijs-Ballot, 'Akustische Versuche auf der Niederländischen Eisenbahn ...' (1845).
26. Russell, 'On certain effects produced on sound by the rapid motion of the observer' (1849).
27. Fizeau, 'Acoustique et optique' (1848).
28. Fizeau, 'Des effets du mouvement sur le ton des vibrations sonores ...' (1870).
29. Moigno, *Répertoire d'Optique Moderne* (1850), vol. 3, pp. 1165–1203.
30. 'James Clerk Maxwell', *PRS* **33** (1881), pp. i–xvi; pp. ii–iii. Huggins's and Maxwell's relationship was friendly, rather than strictly business. Maxwell, for example, wrote Huggins a lengthy personal letter to express his sympathy on the occasion of the death of Huggins's mother. See Maxwell to W. Huggins, 13 October 1868, in L. Campbell and Garnett, *The Life of James Clerk Maxwell* (1969), pp. xxiii–xxiv. Also, the two men shared a love for dogs which provided them with informal topics of discussion in the remaining few letters that have survived. Maxwell to W. Huggins, 2 May 1872, *ibid.*, pp. xxiv–xv; see also W. Huggins to Maxwell, 1 October 1874, Add MS 7655/II/87, JCM, CUL.
31. See Moigno, *Répertoire d'Optique Moderne* (1850), vol. 3, pp. 1159–65. Maxwell to Stokes, 8 May 1857, Add. MSS. 7656.M408, GGS, CUL, published in Harman, *The Scientific Letters and Papers of James Clerk Maxwell* (1990), vol. 1, pp. 502–4. I wish to thank Peter Harman (Lancaster University) for drawing my attention to this letter.
32. Harman, *The Scientific Letters and Papers of James Clerk Maxwell*, vol. 2 (1990), pp. 9–10 and pp. 148–56; *The Natural Philosophy of James Clerk Maxwell* (1998), pp. 164–5.
33. Harman, *The Scientific Letters and Papers of James Clerk Maxwell*, vol. 1 (1990), pp. 148–56.
34. W. Huggins and Miller, 'On the spectra of some of the fixed stars', *PTRSL* **154** (1864), p. 429.
35. W. H. Smyth, *The Cycle of Celestial Objects continued* (1860), pp. 306–34.
36. W. H. Smyth, *Sidereal Chromatics* (1864).
37. *Ibid.*, pp. 67–8.
38. W. Huggins and Miller, 'On the spectra of some of the fixed stars', *PTRSL* **154** (1864), p. 431; Plate X, Fig. 4.
39. J. N. Lockyer, 'Stellar physics' (1864), p. 15.
40. *Ibid.* The emphasis is in *The Reader* article, not in the original text as written in the *PTRSL*.
41. Ennis, *The Origin of the Stars, and the Causes of their Motions and their Light* (1867), pp. 102–63.
42. Ennis, 'The influence of the Earth's atmosphere on the color of the STARS' (1864).
43. W. H. Smyth, 'Appendix I', *Sidereal Chromatics* (1864), pp. 77–81. The reliability of these observations has been called into question by modern astronomers. See R. Burnham, *Celestial Handbook*, vol. 2 (1978), pp. 971–2.

44. J. N. Lockyer, 'Stellar physics' (1864), p. 16.
45. W. Huggins, 'On the physical and chemical constitution of the fixed stars and nebulae' (1865), p. 446.
46. W. Huggins, 'Further observations on the spectra of some of the stars and nebulae' (1868), p. 530.
47. Klinkerfues, 'Fernere Mittheilungen über den Einfluss der Bewegung der Lichtquelle auf die Brechbarkeit eines Strahls' (1866), pp. 33–60; cited in W. Huggins, 'Further observations on the spectra of some of the stars and nebulae' (1868), p. 531.
48. Secchi, 'Sur les spectres stellaires' (1868); cited in W. Huggins, 'Further observations on the spectra of some of the stars and nebulae' (1868), p. 531.
49. W. Huggins, 'Further observations on the spectra of some of the stars and nebulae' (1868), pp. 531–2.
50. Maxwell's well-earned renown for his mathematical contributions to understanding such physical processes as the kinetic theory of gases and the propagation of electromagnetic disturbances through space was greatly magnified upon his death in 1879 at age 48. As we shall see in Chapters 14 and 15, Huggins was much troubled in later years by claims that Maxwell had assisted him in devising his spectroscopic method of measuring stellar radial velocities. It grieved Huggins to share the credit for this, his most original contribution to the new astronomy.
51. Kincaid, 'On the estimation of star colours' (1867), p. 265.
52. *Ibid.*, p. 266.
53. See Maxwell to W. Huggins, 10 June 1867, cited in W. Huggins, 'Further observations on the spectra of some of the stars and nebulae' (1868), p. 533. In 1894, it was proposed that the unit of linear measure equivalent to 10^{-10} m [or 10 nm, in modern terms] be named the 'angstrom' to honour the Swedish physicist, Anders J. Ångström (1814–74) for his important contributions to solar spectroscopy. Previously, British spectroscopists had become accustomed to referring to these units as 'tenth-metres'. For Huggins's reaction to this proposed change, see W. Huggins to Stokes, 6 December 1894, Add MS 7656.H1275, GGS, CUL; W. Huggins to Hale, 11 December and 19 December 1894, Box 22 [Reel 19], GEH, CIT.
54. W. Huggins and Miller, 'On the spectra of some of the fixed stars', *PTRSL* **154** (1864), pp. 416–17. Here Huggins states that he is able to measure approximately 1800 parts between lines A and H of Fraunhofer's solar spectrum. Line A is now given a value of 759.4 nm while H is 396.8 nm. Dividing the difference between these values by 1800 yields approximately 0.2 nm.
55. At that time the pair of stars were separated by approximately 3 arcseconds.
56. W. Huggins, 31 May 1867, Notebook 2, WCL/SC. Huggins obtained values of 0″.11901 and 0″.12134, with a later determination of 0″.12201. See the opening pages of Notebook 2 under instrumental specifications for the double-image micrometer.
57. W. Huggins, 'Further observations on the spectra of some of the stars and nebulae' (1868), p. 542.
58. W. Huggins, 25 June 1867, Notebook 2, WCL/SC.
59. *Ibid.*
60. W. Huggins, 26 June 1867, Notebook 2, WCL/SC.
61. W. Huggins, 'Further observations on the spectra of some of the stars and nebulae' (1868), pp. 536–7.
62. W. Huggins, 27 November 1867, Notebook 2, WCL/SC: 'Applied for first time new spectroscope furnished with two Hoffmann's [*sic*] prisms, two prisms of flint (Simms) & one flint of 45° (Browning).'
63. The F line is the bright blue-green line in the hydrogen spectrum at 486.1 nm.
64. W. Huggins, 11 February 1868, Notebook 2, WCL/SC.
65. W. Huggins, 24 February 1868, Notebook 2, WCL/SC.
66. Huggins examined Aldebaran, Betelgeuse, α Canis Minoris (Procyon), Castor and Capella. W. Huggins, 25 February 1868, Notebook 2, WCL/SC.
67. See W. Huggins, 'Further observations on the spectra of some of the stars and nebulae' (1868), p. 538.
68. See Notebook 2, WCL/SC. Entries for 11, 24, 25 February 1868 contain comparisons of star and nebula spectra with those of terrestrial elements. On 27 February, Huggins worked in his laboratory on terrestrial spectra because the weather prevented celestial observation. A few days later, on 3 March, he noted 'it will be necessary to invent & construct a new arrangement for the comparison of the spark'.
69. W. Huggins, 3 and 5 March 1868, Notebook 2, WCL/SC. The new apparatus was described in Huggins, 'Further observations on the spectra of some of the stars and nebulae' (1868), p. 529.
70. W. Huggins, 6 March 1868, Notebook 2, WCL/SC.
71. W. Huggins, 10 March 1868, Notebook 2, WCL/SC.
72. W. Huggins, 12 March 1868, Notebook 2, WCL/SC.
73. W. Huggins, 30 March 1868, Notebook 2, WCL/SC.
74. Airy, 'Referee Report on William Huggins's paper "Further observations on the spectra of the Sun and of some of the stars and nebulae, with an attempt to determine therefrom whether these bodies are moving towards or

from the Earth"', RR.6.150, RSL. Judging from the title by which Airy refers to Huggins's paper in his report, at least one of two new sections had been added to Huggins's original paper on stars and nebulae before it reached Airy for comment, namely, the one on the solar spectrum which Huggins submitted 30 April. Huggins submitted a second addition to the paper on 2 July, which included his observations of the spectrum of Comet II, 1868. Airy reported on that section separately. See Airy, Referee Report, RR.6.151, RSL.

75. W. Huggins to Pritchard, 6 July 1868, Correspondence of the Society, RASL.

76. Clearly the aged Airy's memory had failed him here. Airy to Stokes, 15 March 1872, RGO 6.483/12/13c/296, CUL.

77. On 28 February 1872, Huggins wrote to Airy, 'I am glad there is a prospect of the sun being observed spectroscopically at Greenwich. I will think for a few days as to the getting of a suitable spectroscope.' W. Huggins to Airy, 28 February 1872, RGO 6.271/7/2/327, CUL.

78. Letters between Airy and Huggins on this topic began with a note from Airy to W. Huggins on 29 November 1871 requesting Huggins's opinion of Angelo Secchi's article in the *Comptes Rendus* on the usefulness of the spectroscope during eclipses and the forthcoming transit of Venus. This brief series of notes concluded on 8 March 1872 with a message from Huggins to Airy encouraging him to utilise spectroscopy during the transit of Venus in December 1874. See RGO 6.483/7/2, CUL.

79. R. W. Smith, 'A national observatory transformed' (1991), pp. 5–6. See also R. W. Smith 'The Cambridge network in action' (1989); and Meadows, *Greenwich Observatory, Volume 2* (1975), p. 121.

80. M. E. W. Williams, 'Astronomy in London: 1860–1900' (1987), p. 14.

81. For Strange's views on the establishment of national laboratories, see his testimony before the Devonshire Commission, given 24 April 1872 and recorded in Royal Commission on Scientific Instruction and the Advancement of Science, *Minutes of Evidence, Appendices, and Analyses of Evidence*, vol. 2 (1874), pp. 75–6. Airy's testimony given on 26 April 1872 before the same Commission giving his response to Strange's views can be found in the same volume, pp. 93–100. As we shall see in the next chapter, Strange's efforts to arouse the Fellows of the RAS to action on this issue resulted in much turmoil within the Society which was recorded in the *Astronomical Register*.

82. 'Spectroscopic results for the motions of stars in the line of sight, obtained at the Royal Observatory, Greenwich', *MNRAS* **36** (1876), pp. 318–26. The publication of these radial velocity measures in the *Monthly Notices* became fairly routine after this first report.

83. Meadows, *Greenwich Observatory, Volume 2* (1975), pp. 83–105.

84. W. Huggins, 'On the spectrum of the Great Nebula in Orion, and on the motions of some stars towards or from the Earth' (1872), pp. 386–94.

85. H. C. Vogel, 'On the progress made in the last decade in the determination of stellar motions in the line of sight' (1900), pp. 374–5; Hearnshaw, 'Doppler and Vogel – Two notable anniversaries in stellar astronomy' (1992).

86. H. C. Vogel, 'Über die Bestimmung der Bewegung von Sternen im Visionradius durch Spectrographische Beobachtung' (1888).

8

A new telescope

The Royal Society some three or four years ago, when I was myself on the
council, voted a sum of money for the construction of a large telescope to
be placed in the hands of Dr. Huggins . . . . [T]hough I voted with the rest
of the council for this grant of money. . ., I objected then, as I object now,
to an arrangement which I did not think a satisfactory one.
— Lieutenant-Colonel Alexander Strange[1]

On 4 April 1868, at about the same time Huggins was completing his paper on stellar motion
in the line of sight, he received a visit from Howard Grubb (1844–1931). Son – and now
partner – of engineer and telescope maker Thomas Grubb, young Howard had spent the last
two years supervising the construction of the Great Melbourne telescope, a state-of-the-art
instrument for studying nebulae in the southern sky. Its construction had been commissioned
by the Royal Society on behalf of, and with funds provided by, the state of Victoria,
Australia. Recently completed, the monumental 4-ft Cassegrain would soon be bound for
Melbourne.

Accompanying Grubb on his visit was astronomer Albert Adolphus Adalbert LeSueur
(1849–1906), who had observed the figuring of the mirrors and would be assisting in the
instrument's operation at the University of Melbourne's observatory.[2] Huggins took great
pride in showing his observatory to guests like these, who knew and appreciated not only the
fine quality of his instruments, but the skill required to put them to good use.

Grubb had declared the Great Melbourne ready for inspection back in December. But
weeks of inclement weather prevented members of the Royal Society's Telescope
Committee from putting it through the paces until 17 February. Robinson, De la Rue and
Lord Rosse pronounced it 'a masterpiece of engineering' and 'perfectly fit for the work for
which it was destined'.[3] William Lassell, writing separately, affirmed that 'the entire instrument is a great triumph of mechanical engineering and optical skill'.[4]

There was still work to be done on the spectroscope and other instruments being built to
accompany the telescope. The visit from Grubb and LeSueur gave Huggins an opportunity
to see the compound prism Grubb had assembled for the spectroscope. 'It is very good', he
soon wrote to Robinson, 'but it would be better still if they could get the glass which the
Paris opticians use'.[5] Indeed, when plans were being drawn up for the spectroscope two
years earlier, Huggins had advised Robinson 'The best glass for the prisms of the Great
Reflector would be, I think, some Swiss glass which was made some years since', a

reference to the fabled optical glass of one-time Fraunhofer associate, Louis Pierre Guinand. 'The dispersion is greater than that produced by some of Chance's glass', that is, glass made by Chance Brothers, the Birmingham glassworks renowned for its window and optical glass. Chance's glass was manufactured under the expert guidance of French glassmaker Georges Bontemps (1801–82), who, in 1827, had purchased and, soon after, perfected Guinand's secret method for producing optical glass. 'I have had for some years two small prisms of this Swiss glass', Huggins advised Robinson in 1866.[6] 'It does not tarnish, it is, too, exquisitely white, & homogeneous.' Now, in 1868, he noted 'I have some prisms made for me by [J. G.] Hofmann of Paris, which give much greater dispersion relatively to refraction than the glass of Chance.'[7]

For Huggins, the meeting with Grubb and LeSueur must have been a bittersweet experience. Surely he was gratified to know his expertise was so highly valued by all concerned with the Great Melbourne's design and construction. Nevertheless, he would have been keenly aware that, once the new telescope saw first light, his own observatory's spectroscopic capabilities would be likely to pale in comparison. In April 1868, that long-awaited moment was but a gleaming speck on the event-horizon. The nine-plus tons of finely crafted metal that stood assembled at Grubb's works in Dublin had yet to be packed, shipped and erected – a process that could (and did) take months. Huggins still had time to secure his place on the cutting edge.[8]

8.1 '... discussing the size & plumage of the chicken'

Two weeks later, Huggins wrote to Robinson in search of sage advice on purchasing a new telescope 'for the purpose of carrying on more successfully my spectrum observations'. There were so many factors to consider and he was unsure how to proceed. As he explained to Robinson, he had special physical requirements for his future instrument. He wished to be able to collect more light in order to improve his ability to observe faint objects. For these reasons, he longed for a large-aperture telescope, perhaps one measuring 12 inches in diameter. Unfortunately, such a large refractor, if built to the same aperture-to-focal length ratio as his 8-in Alvan Clark (1 : 15), would not fit in his existing observatory building. Huggins feared he was limited to a telescope with a maximum focal length of 10 feet. If he wanted a larger aperture under these restrictions, he would have to accept the risks inherent in a highly curved lens, which, unless ground to perfection, could cause troublesome distortion in the images it produced. He confided that he had already been in contact with Thomas Cooke, who was, in Huggins's words, 'anxious' to make him a new telescope of '10 inches & 10 feet focal length'.[9]

Recalling that Thomas Grubb had made Robinson a 7-in telescope with a 5-ft focal length, Huggins inquired as to its performance. Would an instrument 'of similar construction, say 12 inch aperture and 10 feet focal length' be worthy of consideration?[10] We do not have Robinson's reply, but from Huggins's next letter, dated 6 May, we can infer that Robinson had tried to persuade him to consider a reflector. But, at least for now, Huggins was set on a refractor. 'If I could manage to afford it', he replied, 'I should like at some future time to have a large reflector *as well* as an achromatic, but I think it would not be prudent to depend wholly on a reflector.'[11]

Grubb, he told Robinson, had given him an estimate of £800. Though he did not specify what that price included, he made it clear the amount was out of his range. 'This settles the question', he declared. 'I cannot afford it, & scarcely like to go to the R.S. in forma pauperis.'[12] Even the £560 he hoped to get from the sale of his Alvan Clark would not help cover such a cost.[13] Cooke, he wrote, had agreed to make him a 'dumpy' – a refractor with an 11-in aperture and 10-ft focal length – 'as a great favour'. Robinson and Grubb were very close. Perhaps Huggins hoped that Robinson would convey this information to Grubb and encourage him to lower his asking price.[14]

All these plans became moot, however, upon Cooke's sudden and unexpected death six months later.[15] The bad news reached Huggins a few weeks after the death of his own mother, a loss which left him temporarily paralysed with grief and overwhelmed by indecision.[16] 'We had been all in all to each other for years', he wrote in sorrow to Robinson.[17] He embarked on a brief trip with a friend to Brighton in October, which helped him recover somewhat, but the 'prolonged anxiety' caused by the disarray into which his fledgling plans for a new telescope had fallen took its toll.[18]

In the meantime, Robinson had begun working quietly behind the scenes on his bereaved colleague's behalf. He composed a lengthy letter proposing that the Royal Society commission a telescope to be built by an experienced instrument maker specifically to advance the cause of nebular and stellar spectroscopic research. When Huggins returned from Brighton, he found a draft of the missive awaiting his comments and suggestions. Though he did 'not see anything that requires any alteration', he urged Robinson to underscore the need for a '*clock* of great excellence for keeping the slit upon the linear image of a star'. Robinson sent the proposal off to George Stokes on 31 October, recommending William Huggins as the one individual who could most effectively carry out such a research programme:

Mr. Huggins has done wonders with the means at his disposal; but any one who is familiar with this kind of work must know that his eight-inch object-glass cannot go much beyond what it has already revealed to him, and must regret that one so highly gifted for these investigations should not be enabled to pursue them to the greatest possible extent ... I think [the new telescope] should be intrusted to Mr. Huggins, who, in addition to his devotion to a pursuit where he has already obtained such grand results, would be doubly stimulated to exertion by such a mark of approval and confidence from the Heads of British Science.[19]

It is unclear when Huggins first became aware that he was to be the beneficiary of Robinson's well-wrought scheme. 'Many thanks for your great kindness', he wrote to Robinson on 6 November. Whether he was expressing gratitude for his friend's petition to the RS on his behalf, or for comfort and hospitality offered in a time of abject grief, Huggins was already feeling agitated by the idea that the RS might procure an instrument made to his specifications for his own exclusive use. He confessed, 'I am suffering from my nervous system having been shaken so that I am rather nervous about having a large instrument. I fear I shall not be able to do all that the Society might reasonably expect.'[20]

On 17 December, the RS appointed a committee to consider Robinson's proposal, which included a detailed list of the apparatus he believed necessary to prosecute the research programme he envisioned. Robinson estimated the cost to the Society at around £2,000, a considerable sum, but one he believed would prove a worthy investment. Besides, he reminded his Fellows, concern over cost was:

not the way in which persons who represent the Intellect and Knowledge of a Nation like ours, look at such matters. The power of the Royal Society over men's minds does not rest on the amount of its annual income or its balance in the bank. Its real wealth is whatever its animating influence or its helping hand has added to the treasures of science; its real power consists in the conviction of our countrymen that it is a mighty instrument of the highest and brightest progress, that its motives are as generous as its acts are beneficent and noble.[21]

Two thousand pounds was indeed an extraordinary amount of money for the Society to invest in a single research project. The RS annually dispensed small grants to individuals from a £1,000 apportionment from Parliament called the Government Grant. But a survey of the list of recipients from 1855–70 shows that, while they ranged from as much as £300 for experiments on steamboiler explosions to as little as £7 7s. 9d. to help defray the expense of determining the depression of the Dead Sea, the grants were distributed widely and averaged well under £100 each. In a number of years, the full £1,000 went undistributed. To apportion £2,000 would require special measures. Fortunately, just at the time Robinson was submitting his proposal, the Society was negotiating the receipt of a large bequest from the estate of one Benjamin Oliveira, FRS.

Oliveira had been elected to Fellowship in June 1835. He never contributed any papers to the Society's journals, but he did present a gift of £50 for the general promotion of science in 1854. Upon his death in 1865, which went unmarked by a Society obituary notice, Oliveira left the Society the munificent sum of £4,000. *The Record* notes that this was reduced to £1,506 17s. 1d. following a chancery suit.[22] Further claims on the money left only £1,350 at the Society's disposal.[23]

By January 1869, Huggins was still in a quandary over which instrument maker to employ – Grubb or Thomas Cooke's sons. At least, by then, he had accepted the fact that he would need a larger observatory in order to accommodate a larger telescope. But that decision only presented him with additional worries to share with Robinson:

I feel deeply the great interest you take in the matter of the large telescope. It appeared to me that a dome of 18 feet diam., was the largest that I could hope to move without assistance. It *would cripple me much to need an assistant*. I should lose so many *opportunities* of making observations ...

Mr. Cooke had agreed to construct a very *light* dome for me, with arrangements so that I could work it alone without great fatigue. I am not strong, & it would not do for me to be fatigued before observing.

I fear another foot added to the dome would almost put it beyond my powers.

Could Grubb suggest some very easy way of working the dome? Would his form of equatorial be such that 16 feet could be used in a dome of 18 or 19 feet? ...

P.S. There is also Alvan Clark.[24]

Five days later, sounding more anxious than ever, he wrote again to Robinson:

I am not quite decided (if the telescope should be granted by the R.S.) whether to remain *here* or to move to a greater distance from town.

I have purchased the lease of my present house, so I would have secure possession. (for 50 years & more!!)

It is conveniently near for the scientific societies in London. As I am south of London, I have not to observe over London, & the country round being well-drained I think the air is probably as still as it would be anywhere. Besides this, I find I can, without difficulty, enlarge my present observatory to the required size, with, of course, an entirely new dome. You saw how very convenient it is, as I go from the house into it through a passage. I have an uninterrupted view except for a small space due south which is of little importance. Another inducement is the proximity of my friend Dr. Miller.[25]

So many decisions, and so much at stake! Should the tube be ventilated or closed? Huggins leaned toward the latter 'to keep the internal air snug & quiet'. Besides, a closed tube could be insulated to mediate the effects of external temperature changes. How best to insulate it? Cooke had argued for a tube-within-a-tube assembly, but Huggins thought a thick layer of papier-maché, like that on his present telescope, should suffice. A more important concern was determining the optimal aperture for a 15-ft focal length. Could Chance produce a 16½-in flint disc? How much would Grubb charge to build such an instrument? 'I ought not to trouble you with these details', Huggins wrote in concluding his letter. 'I am very anxious however to have the telescope *as perfect* as possible, as I should hope to make other observations, as well [as] those with the spectroscope.' Following his signature, he quipped: 'We are discussing the size & plumage of the chicken, but it is *not hatched yet*! The Council is still sitting on the egg you laid.'

At least, that was how Huggins believed the situation stood. In fact, Robinson, like a good mother hen, had never let the egg out of his nurturing sight. Irish-born Stokes, the sage RS secretary, was his son-in-law, after all, and Dublin-based Thomas Grubb his good friend and instrument maker of choice for over three decades. While Huggins muddled over the chicken's 'size & plumage', Robinson already had a very good idea how things were going to turn out. But it was a worry to Robinson that Huggins continued to write favourably of Cooke & Sons. 'Mr. Cooke called upon me yesterday', Huggins had written in the postscript to his long letter of 23 January. 'He & his brother are *very anxious* to make a large telescope for me, & would do any reasonable thing to get an order.' What Huggins needed, in Robinson's view, was a personal meeting with Howard Grubb to dispel his lingering sense of customer loyalty to the recently departed Cooke.

Robinson quickly wrote to propose just such a meeting. Although we do not have this note, we can infer from the wording of Huggins's equally prompt reply that Robinson must have made the situation clear: although the size of the chicken was still open for discussion, the question of its plumage had been settled. On 31 January Huggins wrote 'If Grubb is likely to have the order, I think there would be a *great advantage in my having* a conversation on several little matters with Mr. H. Grubb.'[26] In subsequent correspondence with Robinson, he never mentioned Cooke & Sons again.

Howard Grubb travelled to England in February 1869, carrying with him a 12-in doublet to be examined by Stokes at Cambridge.[27] Huggins also went up to Cambridge to observe the trials on the doublet, which he described in a newsy letter to Robinson.[28] He had met and 'discussed several small matters' with Grubb. Of greater importance was the previous day's meeting of the RS committee that was considering Robinson's proposal. Huggins informed Robinson 'It seems pretty well decided to let Grubb have the construction of the instrument.' But the design was still a matter of discussion. De la Rue proposed building a reflector that could be converted quickly and easily into a Cassegrain. A large mirror (22-in) would be positioned permanently in the tube, but a secondary mirror and objective glass would always be standing at the ready for insertion when desired. Of course there were technical details that would have to be worked out . . .

On 27 February Robinson and Huggins each put pen to paper to update the other. Huggins laid out his thoughts on Cassegrains, Gregorians and Newtonians, weighing relative ease of use against suitability to his own investigative interests like the study of non-visible spectral rays. No one design really fitted his needs. 'Can *you* not invent a *new form of reflector* and let your name be associated with the instrument', he asked his benefactor.[29] Robinson, meanwhile, elaborated on Grubb's proposed adaptation of De la Rue's convertible design. He was well aware of the delicate manoeuvring required each time the telescope was transformed. 'I well remember how nervous I used to be when pulling Souths [James South (1785–1867)] 12 inch into its tube', he wrote. But 'as to the [proposed] 15 inch, I don't think one person will be able to manage it however perfect the mechanical contrivances which Grubb will certainly provide'.[30] He continued:

You will I think on such an occasion want a pair of hands for the 'brute force' of hoisting and lower[ing] and another pair for the delicate part, guided by a good head. Now though I know you dont wish the bothers of a formal assistant, yet you must have a Gardener or other Labourer; and whenever you have to change the telescope is there any reason why you could not have him at this work for (~ say) an hour?

By 6 March, Huggins had heard enough about the De la Rue plan. He wrote to Robinson, hoping the thoughts expressed 'will agree with your own wishes':

After mature consideration the whole thing appears to me thus: –

1. The efficiency & ease of manipulation of the achromatic must not be interfered with.
2. The double telescope in one tube appears to me very unwieldy & undesirable on several grounds.
3. The achromatic must be a telescope complete in itself.
4. If a separate tube be used for the reflector then perhaps there would not be necessity for increasing the massiveness of stand in first estimate. Also dome need not be 1 foot larger (I have written to Grubb on these points).
5. The reflector to be a Cassegrain 24 inches diam. with a second *small* speculum of same curvature as the great one. This might be quite small.

The separate tube will not matter if it can be put up with the assistance of two or three carpenters. These could be had & the change made leisurely in the day time when it is intended to work for a time with the reflector.[31]

'I hope', he added with a hint of worry, 'we can depend upon Grubb for a *good* object glass.'

Two weeks later, the RS committee made its decision to entrust the task of building the telescope to Grubb & Son. Huggins expressed his great appreciation to Robinson 'to whose influence and kindness I shall always consider myself indebted for the use of so powerful an instrument'.[32] The instrument would be versatile as well, with two separate telescopes, just as Huggins had insisted: an 18-in Cassegrain reflector (in lieu of the 24-in he had suggested) and a 15-in achromatic refractor. These two tubes were to be capable of being mounted interchangeably on one of Grubb's fine equatorial stands. The work was to be finished by December. Indeed Grubb claimed it was already under way, having begun the preparation of two 15-in Chance glass discs for the achromatic.[33] The intervening months would afford Huggins sufficient time to sell his Alvan Clark, and enlarge his observatory to receive the new instruments.

In August, Huggins sold his telescope, some micrometers and his observing chair to architect and amateur astronomer Charles Joseph Corbett (1823–82).[34] Sincerely hoping the new telescope would soon be completed, he wrote to John Herschel and George Airy with the humorous request that, as he was 'now without a large telescope . . . will you have the kindness to keep all large comets away until I am equiped [*sic*] with the necessary apparatus to attack them'.[35]

News of the Royal Society's generous award of the telescope was to be made public at its upcoming Anniversary Meeting. In early November, Huggins coyly wrote to Robinson: 'I believe now the "chicken" is alive, & just ready to come out of its shell. I should think there is little doubt but that "Grubb" will get the order.'[36] On 30 November, Society President, Edward Sabine, announced:

Celestial spectroscopy has indeed attained such importance, that it requires for its successful prosecution the undivided attention of the astronomer who devotes himself to it, as well as an observatory specially designed for it. Our great national observatories cannot supply this want, for they have their own specific destination; and the high optical power which is required, if we wish to make further progress, is scarcely within the reach of amateurs.

These considerations have induced your Council to believe that an attempt to encourage and aid this most interesting class of researches is an object in full unison with the highest purpose of the Royal Society's existence; and they have therefore, after most careful deliberation, resolved to act on this conviction by providing a telescope of the highest power that is conveniently available for spectroscopy and its kindred inquiries. The instrument will, of course, be the property of the Society, and will be intrusted to such persons as, in their opinion, are the most likely to use it to the best advantage for the extension of this branch of science; and, in the first instance, there can be but one opinion that the person so selected should be Mr. Huggins.

The execution of this project was much facilitated by the receipt of £1350 from a bequest made to the Society by the late Mr. Oliveira; and in the beginning of the year proposals were received from the chief opticians of the time, of which that of Mr. Grubb was accepted last April.[37]

In December, as the original deadline arrived, Huggins received good news from Robinson. 'I am delighted to hear so good an account of the performance of the glass', he replied. 'Especially am I pleased to hear that the chromatic aberration is not unsatisfactory. I have feared that Grubb would not succeed so well on this point as much [as] the spherical aberration.'[38] Grubb was also making progress on the barrel-shaped roof that Huggins had ordered. Unfortunately, he told Robinson, the alterations to his house and observatory had been slowed by bad weather. 'On this ground', he noted, 'I have been less impatient for the completion of the instrument. Now I shall be able to prepare my observatory.'

The weather continued to conspire against the construction work at Tulse Hill leaving it unfinished until February 1870 (see Figure 8.1). And unexpected difficulties in figuring the 15-in achromatic lens meant it took Grubb the better part of a year to complete.[39] Huggins had a small, 2-in portable telescope, but there is no evidence that he used it to make any observations during this hiatus. Meanwhile, other projects kept him busy. He helped William Lassell's daughters, Jane and Caroline, to edit a translation of Heinrich Schellen's book on spectrum analysis.[40] And there was the all-important (and frustrating) planning required to organise an expedition to Algeria to view the solar eclipse coming up in

Figure 8.1 New Tulse Hill Observatory. The photograph shows Margaret Lindsay Huggins standing at the base of the observatory building. (W. Huggins and M. L. Huggins, *Atlas of Representative Spectra*, p. 23)

December – a major undertaking that is the subject of the next chapter.[41] Then, in early August 1870, just as plans for the expedition seemed to be falling apart for lack of government support, the long-awaited shipment from Grubb arrived! It was still in pieces, of course, but Grubb was coming to assemble it.[42]

In November, nearly one year later than originally promised, the instrument Huggins called the Great Grubb Equatorial was ready for use. To test for any positional error in its axes, he turned it on the Ring Nebula in Lyra (M57), Neptune, Jupiter and γ Andromedae, but was hindered in his observations by atmospheric turbulence.[43] On Monday 14 November, Sabine and De la Rue visited the observatory. It was, Huggins declared, 'a fine night' to give his guests a celestial tour with his new instrument.[44] The next evening, he returned alone for another look through the telescope. But further observations had to await his return from the Algerian eclipse expedition.[45] After some additional adjustments by Howard Grubb between 9 and 20 February 1871, the telescope was ready for use, although the spectroscopes were not yet complete.[46]

Thus began a new phase in Huggins's observing career. Accepting the loan of these first-class instruments – costing more than his personal means would allow – carried with it an obligation to the Royal Society, and to its Fellows. From that time on, he would be directly answerable to them for his choice of observational problems, his methods, even his diligence in the application of these coveted instruments. Many agreed with Sabine that 'there can be but one opinion that the person so selected should be Mr. Huggins'.[47] In their minds, he had become identified as *the* British authority on astronomical spectroscopy.

But serious questions were soon raised over the efficacy, and even the propriety, of placing such costly instruments in the hands of a single individual not beholden to an existing institution. As Huggins grappled with the logistics of carrying out a productive observing programme in his private Tulse Hill observatory, he increasingly found himself a target of anxiety-provoking criticism from disgruntled colleagues. Their censure was, in part, a by-product of growing dissension in the ranks of the Royal Astronomical Society over the complex and contentious issue of government support of scientific endeavour. It brings to light the synchrony and interactive dynamics of his personal career trajectory as it intersected the wider sweep of change afoot in London's astronomical community.

8.2 The strains of diversity

Between 1860 and 1870, the number of Fellows in the RAS rose by 34%, the largest increase by far in any decade since its founding.[48] The expansion broadened the spectrum of research interests among the Fellows which, in turn, gave rise to a range of factions within its ranks and keen competition for limited resources and rewards.[49]

In the 1860s, Huggins played a key role in laying the groundwork for the successful transplantation of the spectroscope from the laboratory of the analytic chemist into the observatory of the astronomer. By the early 1870s, increasing attention to spectroscopic research in the pages of the *Monthly Notices*, *Astronomical Register* and the Royal Society's *Proceedings*, made it clear to amateur and professional astronomers alike that new discoveries awaited those willing to incorporate the new methods and instrumentation into their research agenda.

But in Britain, at least, to prosecute a fruitful research agenda in this youthful and still evolving specialty required independent financial means. Popular liberal notions of *laissez-faire* and individual responsibility stood in opposition to State support of private endeavour. The kudos heaped on British participants in the Great Exhibition of 1851 for their technical ingenuity was tangible evidence to many at the time of the wisdom of the liberal position. As the acclaim faded over the years, it was replaced by the wails of jeremiahs who tried, albeit with little effect, to rouse the nation from its complacency regarding the need to attract talented young men into technical fields and prepare them to carry the baton of Britain's past industrial success into the next century.[50]

The growth in public concern over the Government's stance on State support for scientific institutions and science instruction coincided with the rise of astronomical physics and the consequent increase in choices of research opportunities available to Fellows of the RAS. These large-scale developments took place during a critical phase of Huggins's career as he made the uneasy transition from total independence to beholden obligation.

As custodian of the Royal Society's telescope, Huggins became embroiled in a vigorous debate between two hard-nosed factions. One, led by Lieutenant-Colonel Alexander Strange, believed that the new research agenda opened up by astronomical physics, and the skills required to pursue it with diligence and efficacy, warranted the establishment of a network of Government-sponsored astrophysical laboratories. The other, led by Astronomer Royal George Airy, argued that such research was best left to the combined efforts of Greenwich and the contributions arising out of individual initiative. By publicly placing himself in Airy's camp, Huggins incurred the wrath and public criticism of Alexander Strange.

8.3 The 'insufficiency of national observatories'

In his classic *Organisation of Science in England*, English historian of technology Donald Stephen Lowell Cardwell used the words 'alarm' and 'near-panic' to describe the reaction at home to the disappointing showing by British participants in the 1867 International Exhibition in Paris.[51] The great wave of self-congratulation, satisfaction and confidence set in motion by the accolades tendered toward Britain in the spacious galleries of the Crystal Palace sixteen years earlier had now subsided, leaving in its wake the fear that Britain had been soundly beaten at her own game.[52] Chemist Lyon Playfair (1818–98) served as a juror at the Paris Exhibition. He returned to London disheartened by the realisation that foreigners viewed Britons as having passed their peak.[53] Playfair was sure it was the fault of the woeful lack of technical education in Britain, and he was moved to recommend a Government inquiry into what he viewed as a national emergency.[54] Others agreed that an inquiry was necessary even if they did not all agree on where to place the blame. These individuals hoped an inquiry would resolve that question and present the nation with a list of recommendations to help put Britain back on track.

The Society of Arts (originally the Society for the Encouragement of Arts, Manufactures and Commerce) was the first institution to respond to this concern. The Society of Arts was esteemed for having served the congruent needs of British science, technology and society for over a hundred years.[55] It had not only the motivation to construct a forum for this inquiry soon after Playfair and others called for one, but the institutional history and structure to fashion a workable and formalised mechanism to keep the issue of scientific progress alive and before the public eye so that it would not become a mere flash in the pan. Sponsorship of these activities by the Society of Arts assured them the extended exposure needed to garner additional support for the cause of broad-based State sponsorship of science and education from those social visionaries who questioned the basic tenets of the liberal political theory of individualism.

Alexander Strange was one of the more vocal of these crusaders. His efforts on behalf of direct financial assistance from the Government to ensure the growth and development of astronomical physics split the astronomical community along a number of non-orthogonal axes. Debates ensued over where – properly or even rationally – to locate the boundaries for acceptable astronomical research; over what to consider as a legitimate question to ask of celestial phenomena being observed; over how to collect data to help answer these questions; and, finally, over who should be deemed a sufficiently authoritative observer such that their evidence not only counted, but determined the standard against which other evidence was to be weighed. These sometimes cantankerous exchanges both informed the direction and accelerated the pace of future astronomical research.[56]

Strange was a retired officer of the Royal Engineers and Royal Artillery returned from years of service in India. The son of a British jurist in India, he was educated in England, but returned to India at age sixteen when he entered the 7th Madras Light Infantry as a cadet. Given his keen interest in science and his able tinkering with telescopes and other instruments, Strange's superiors deemed him the obvious candidate to assist in the challenging project of triangulating the longitude of India. Following a bout of malaria, he returned to England in 1861, where he won an appointment as Inspector of Scientific Instruments in the India Department of the Government. His excellent design of scientific instruments and careful superintendence of their construction gained him a reputation as a 'mechanician of the highest order' in London's scientific community.[57]

He was a member of a number of scientific societies including the Royal Society, the Royal Astronomical Society, the British Association and the Meteorological Society. Shortly after his return to England from India, he became actively involved in the organisational work of these societies, serving on the Councils of the RAS and Royal Society.[58] Like Playfair, Strange served as a juror at the Paris Exhibition in 1867. The two men shared a deep concern for the future of scientific progress in Britain. Strange eagerly participated in the January 1868 conference on improving the status of science in British educational institutions organised by the Society of Arts in response to Playfair's request for an inquiry on the matter. While others who became involved in this educational reform movement focused on the problem of constructing an institutional framework within which to situate a practical and technical science education curriculum, Strange believed that, to effect scientific progress, there first had to be an infusion of financial support for education from the Government.[59] This was a radical proposal in a time when many still had greater faith in unencumbered individual enterprise.

Strange presented an appeal for State support for science in an address before the Mathematics and Physics Section of the BAAS at its Norwich meeting in 1868.[60] The BAAS was moved to appoint an investigatory committee, a committee on which Huggins served, as did Strange, Stokes, Thomas Henry Huxley (1825–95) and other scientific luminaries. The committee reported that there was widespread belief among members of the scientific community that they lacked sufficient resources to carry on their research at a competitive level. Unsure about what recommendations to make to locate the source of the problem, let alone rectify it, the committee suggested that a Royal Commission be appointed to launch a thorough investigation.[61]

8.4 The Devonshire Commission

Thus it was that in 1870 the Government appointed a special commission composed of eight eminent men with the proper political credentials and an avowed interest in scientific research, science education and its practical applications. The Commission was chaired by William Cavendish, 7th Duke of Devonshire (1808–91), a Second Wrangler in the Mathematical Tripos at Cambridge who possessed all the characteristics listed above.[62] The Devonshire Commission, named in honour of its chair, was tasked to solicit a range of expert opinion on the matter of Britain's current scientific health, to analyse the spectrum of comments and recommendations with which it was presented, to identify and assess what was already in place, to suggest improvements that would emend any failings, and to propose means by which these changes could be implemented. The panel heard testimony aimed at stimulating progress in a wide range of scientific fields. That the hearings served both as a vent and a showcase for the simmering diversity within the astronomical community is shown by the lengthy testimony presented by individuals concerned about the development of astronomical physics.

The hearings began in June 1870. The published transcriptions of the testimony – particularly that presented in the spring of 1872 – expose the tension among practising astronomers, professionals and amateurs alike. At the monthly RAS meeting on 12 April 1872, Strange prepared his fellow astronomers for his forthcoming testimony before the Commission by presenting a statement, 'On the insufficiency of existing national observatories', which contained both his assessment of the current state of astronomical research and his recommendations for improving it.[63]

The way Strange saw it, the research agenda pursued by mathematical and physical astronomers until the second half of the nineteenth century had narrow and well-defined goals related to navigational needs and interests. Timekeeping was a central concern, and routine, repetitive measurement of positions of celestial bodies was the principal method. By the 1870s reliable marine chronometers had largely eliminated sailors' need to starshoot with sextants, and the mission of the Royal Observatory had gone beyond the mere cataloguing of celestial positions. In fact, Strange contended, astronomy itself had changed so much in recent years that 'Greenwich cannot, as at present constituted, be expected to contribute systematically to the advancement of the new branch of astronomy ... now generally recognised as the Physics of Astronomy.'[64] Or, as the *Astronomical Register* more bluntly paraphrased his words, it would be 'impossible for Greenwich to follow up the numerous branches into which the science has ramified'.[65] Not that the Astronomer Royal was incapable, but his many duties to the existing research programme at Greenwich prevented him from attempting anything else. In fact, Strange took pains to remind his audience that Airy had recently recommended establishing a special observatory given solely to observing Jupiter's satellites because there was no longer any room in the daily routine at the Royal Observatory for that project.[66]

Private individuals had initiated a variety of projects of their own design incorporating methods and instruments commensurate with the new astronomical agenda. But Strange considered this sort of decentralised approach to be inefficient. He believed that astronomical physics required both long-term and large-scale research projects, and he warned that the continuity so essential to the collection of useful data would be lost if individuals working out of their own private observatories were relied upon too heavily. As Strange explained: 'It is certain that we cannot count upon such extended continuity from private energy, which however supreme it may be, must die with its possessor.'[67] Besides, he argued, the sheer quantity of observations needed in order to make possible some reasonable analysis was simply 'too heavy a work for individuals'. Such an effort 'required co-operation and assistance, or it would never be completed'.[68]

Strange pointed to solar physics as a case in point. He was particularly anxious that routine and thorough study of the Sun be undertaken because:

what we now call the uncertainties of climate, are connected with the constant fluctuations which we know to be perpetually occurring in the Sun itself. The bearing of climatic changes on a vast array of problems connected with navigation, agriculture, and health, need but be mentioned to show the importance of seeking in the Sun, where they doubtless reside, for the causes that govern these changes. It is indeed my conviction, that of all the fields now open for scientific cultivation, there is not one which . . . promises results of such high utilitarian value as the exhaustive, systematic study of the Sun.[69]

He believed that the only way to make possible the long-term and large-scale research required by the new astronomy was through active Government support of astronomical 'laboratories'.

The kind of laboratory Strange had in mind was exemplified by that operated by De la Rue, Balfour Stewart (1828–87) and Benjamin Loewy (1831–92) at Kew Gardens where – weather permitting – the solar surface was photographed daily.[70] Unfortunately, as Strange noted in his report to the RAS and emphasised again in his testimony before the Devonshire Commission, the Kew operation was coming to a close owing to lack of funds.[71] After Kew's demise, he claimed, there would be no regular photographs being taken of the Sun's surface anywhere in the British Empire. In his opinion, 'Any prolonged loss of continuity in the series will be a most deplorable circumstance. No time . . . should be lost in averting such an evil.'[72]

Strange hoped to alert his colleagues in the RAS to the tremendous opportunity afforded them by the ongoing investigation being conducted by the Devonshire Commission. He recognised this as a rare chance to influence science policy directly. He believed it was a matter of extreme importance that the RAS provide the Commission with information on the current state of astronomical research as well as 'any recommendations which . . . may seem to it advisable'.[73] Such testimony would probably determine the shape of astronomical research opportunities in Britain 'for a long time to come'.[74]

After Strange's presentation, an animated exchange ensued among the Fellows present. Airy was quick to point out with a humorous anecdote the reluctance of the public to support scientific endeavour with their tax money.[75] De la Rue wondered if meteorological stations throughout the Empire could not be converted into solar observatories by supplying them with the necessary equipment and resources to prepare daily photographs of the Sun. Another Fellow announced that he had been engaged in recording the solar surface for the last nine years and that, should he live long enough, he planned to complete one full solar cycle of eleven years.[76]

Perhaps to demonstrate the mutually beneficial and cost-effective cooperation that was possible when forced to rely on one's wits and a shoestring budget (but more likely to underscore the importance of the routine work done at Greenwich), Airy described how his efforts to transmit time signals throughout England had been assisted by the South-Eastern Railway Company. Additionally, he wished to point out that Government-supported agencies were assigned only to collect data, not interpret them. He saw no value in 'groping about for causes' as, for example, those inherent in the relationship being suggested between sunspot cycles and terrestrial weather patterns.[77] In Airy's view, 'It was the place of a Government not to establish philosophical institutions, but working bodies.'[78] His outburst triggered a final, impassioned response from Strange:

The Government is not remarkable for scientific knowledge, and the nation is absolutely destitute of science, because it has not had the teaching requisite for carrying out objects unintelligible to the ignorant ... The Government is ignorant, and the people more so if possible. Whose business is it to teach them? We must teach [the Government and the people] to understand that not only is [scientific investigation] good in an intellectual sense, but that in time it will bring forth utilitarian results likewise. This is why I dwell in my paper upon the necessity for the study of our great luminary [the Sun], as nothing can be more calculated to influence the material prosperity of the people.[79]

Twelve days later, Strange testified before the Devonshire Commission and submitted to the public record his views on the importance of Government support for national laboratories. He repeated much from his statement made before the RAS, arguing for the importance of systematic solar observations to nurture the growth and development of meteorology.[80]

Strange's testimony is crucial to understanding the development of William Huggins's research agenda at this time. He presented the example of Huggins and the Grubb telescope as evidence that the current system of awarding individuals money or equipment through the Royal Society was not only inequitable, but subject to abuse by individuals who used it to further their own personal research goals in lieu of those that serve the national interest.

He enumerated three reasons why it would not be desirable for Greenwich to undertake such observations: insufficient room for the instruments necessary to undertake such research, unsuitability of the mathematical 'order of mind' to the pursuit of physical astronomy and incompatibility of mathematical astronomy with the study of the physics of astronomy. Expecting Greenwich to perform astrophysical observations was only 'undesirable', not 'impossible'. He knew better than to declare George Biddell Airy unfit for any astronomical assignment.

That Strange should name three difficulties in support of his claim against Greenwich is worth noting. Surely several would be necessary to persuade the Government to remove some astronomical responsibility from Greenwich and place it in new hands, and listing three reasons before a Commission jealous of its time implied there were likely to be more. But only one of the difficulties listed was based on purely physical limitations. The other two were strictly matters of opinion regarding method, motivation and aptitude in astronomical research. Including them in his list of concerns signalled Strange's desire to mark a boundary, however hazy in these early stages of separation, between astronomers who map the stars and those who feel they have the key to understanding the cosmic links between celestial and terrestrial phenomena.

When pressed by the chairman, William Cavendish, on the work in the new astronomy being done by individuals not directly connected with Greenwich Strange responded critically:

It is rather difficult to get from private individuals exact information upon this point, but, so far as I know, those studies are not systematically conducted, and not fully conducted by any university, or by any private individual. One section of such observation was conducted some time ago at the Kew Observatory ... in an irregular, confused sort of manner, as most things connected with science are conducted in England, partly by means of grants from the Government, partly, I think, by means of grants from the British Association, and partly by means of private liberality. But that series of researches has come to an end ... Therefore, at present, that particular method of studying the sun is not now pursued, I believe, by any institution or any individual systematically in the British dominions.[81]

He had particularly harsh words for Huggins, who, in his opinion, used the privilege of the loan of the Royal Society's telescope to pursue his own personal research agenda, rather than apply it to projects like regular solar study, which Strange believed would benefit the nation more generally. For Strange, Huggins's case exemplified the inadequacy and the inefficiency of counting on individuals to carry out the essential work of the new astronomy. And there was the all-important matter of public accountability in scientific endeavour to consider:

With respect to ... studying the sun by use of the spectroscope, there are several private individuals who are pursuing that, and one by the aid of the Royal Society, Dr. Huggins, that is to say, the telescope is suitable for the spectroscopic examination of the sun, but I understand that Dr. Huggins employs it not so much for that purpose as for researches on stellar, nebular, and cometic spectra. The Royal Society some three or four years ago, when I was myself on the council, voted a sum of money for the construction of a large telescope to be placed in the hands of Dr. Huggins, who undertook on his part to erect an observatory for its shelter, and to apply himself sedulously to using it. I think it is as well that I should mention here that though I voted with the rest of the council for this grant of money, being anxious that every aid that was possible should be given to science, still, I objected then, as I object now, to an arrangement which I did not think a satisfactory one. It did not promise such continuity, nor was there present that element of responsibility which I think is essential in such matters. It is possible, on Dr. Huggins's death, for instance, that there might be some difficulty with his executors regarding this double arrangement; and, moreover, I did not think that a great undertaking like that should devolve upon a private individual; I thought it a matter great enough for the nation to take up, and I still hold those views.[82]

Two days later, Airy delivered his own testimony before the Commission.[83] Where Strange had found disorganisation and waste, Airy found untapped potential. He admitted that adding new observations to the burden currently borne at Greenwich would give the Astronomer Royal 'a good deal of work on his hands'. Nonetheless he firmly believed that any project capable of routinisation could be handled by the Royal Observatory in a satisfactory way.[84]

Airy regarded the problem of slowed scientific progress in Britain as being rooted in the inadequate mathematical education provided by British institutions of higher learning.[85] A young man possessing the proper mathematical foundation in conjunction with sufficient curiosity about the world and self-motivation could pursue scientific research on his own in a manner most convenient to himself. In Airy's view, restricting financial assistance for scientific research was not a hindrance, but a way to weed out the unworthy and ensure that only the best ideas would survive. He believed that any individual who believed strongly enough in his own proposed research agenda would pursue it at all cost. If, after some exploration of the matter, the project demonstrated unusual promise, an application for financial assistance could be submitted to the Government Grant Committee of the Royal Society, a committee with much experience in allocating funds to individuals to support their private scientific endeavours. There could then be no complaints from the public about tax money ill-spent.

On the face of it, Airy's testimony bears the unmistakable earmark of a liberal political bent with a firm belief in free market science. But Airy was himself a beneficiary of Government support for science and had been for over thirty-five years. The kind of research programme he directed at Greenwich would not have been possible without continued long-term financial support. As Robert Smith has aptly pointed out, Airy's experience as a recipient of Government funding had made it clear to him that the budgetary pie was finite in size, particularly that portion available for scientific endeavour.[86] While Strange may have imagined that he could persuade the Devonshire Commission to recommend larger apportionments for astronomical research, and that the Government would be moved to take immediate action, Airy knew that offering any slice of the Government's already meager pie to develop a solar observatory elsewhere, would simply reduce Greenwich's portion. The Greenwich programme could not weather any cuts.

8.5 Dissension in the ranks

By the RAS meeting on 10 May 1872, the first after Strange and Airy had testified before the Commission, the Society's Council found itself at odds over the issue of the establishment of, and Government support for, new national observatories independent of the Astronomer Royal's direction and given to the study of the physical and chemical nature of celestial bodies. One group – a minority of the Council who backed Strange's proposals – argued for the creation of a Government-sponsored solar observatory separate from Greenwich. The other group, which included William Huggins, allied themselves with Airy. They saw greater value in continuing to place principal responsibility on the shoulders of Greenwich to launch a programme of regular astrophysical observations, while relying on private observers to follow their own personal research interests with support coming to them through small grants awarded on the basis of individual merit.

Huggins was sympathetic to Airy's position for a number of reasons. For one thing, he knew the Astronomer Royal was preparing to begin regular spectroscopic work at Greenwich. In fact, he wrote to Airy as early as 28 February 1872 to express his pleasure on learning that solar spectroscopy would be undertaken at Greenwich. He offered his help in this enterprise, promising Airy that he would 'think for a few days as to the getting of a suitable spectroscope'.[87] Thus, from a purely practical standpoint, it made no sense for the Government to allocate funds or expend any additional effort on what could very well turn out to be duplicative of what Greenwich was already tooling up to do.

On a moralistic note, Huggins always viewed his own reputation as a careful observer and the custodial award of the Royal Society's telescope as a prize earned by virtue of his own hard work. There is no question that he would have been seriously and unforgivingly offended by Strange's personal attack in his testimony before the Devonshire Commission. He was unprepared to be singled out for such hard-hitting criticism. He was not, after all, taking unmerited advantage of the Royal Society's generosity, and could not conceive of good science being done by anyone who was. Though his research agenda conformed with his own personal interests, those interests were synonymous with the public promotion of natural knowledge.

Huggins believed, and would always believe, that he was his own man. He was not entirely naive or self-deceptive in holding this belief. The eclecticism and opportunism that had marked his personal research agenda to this point in 1872 characterised much of his later work as well. But after accepting responsibility for the Society's Grubb telescope, his choices of both subject and method were unconsciously, yet deeply, guided by his strong desire to remain worthy of the Society's trust and the growing realisation that, as the telescope's custodian, he was publicly accountable for his use of it. Recall, for example, the intense personal anxiety that he had expressed to Robinson over his worthiness to be selected as the telescope's custodian even while the matter was still under consideration by the Royal Society's Council in 1868 and 1869.

In choosing to side with Airy, Huggins may also have reckoned that Airy's view would ultimately prevail and that, in terms of Government support of science, matters would stand much as before. It is always wise to ally oneself with the winning side, and in this sense it can be said that Huggins made a prudent choice. Although the Devonshire Commission largely endorsed the establishment of a national astrophysical observatory of the type Strange had advocated, it was several years before any action was taken in this direction.[88] In the meantime, Huggins gained additional stature as chief advisor on spectroscopic matters to Airy and other members of the Greenwich staff.

At the RAS meeting on 10 May, Strange moved that the president of the Society submit a statement to the Devonshire Commission that would formally represent the Society's belief in the importance of astronomical physics and the need for more research in that area. The motion gave rise to a debate that continued into the next regular meeting. With the matter still unresolved, the Council called a special meeting at the end of June 1872. Needless to say, the question of the statement's wording generated considerable heat.[89]

Huggins, for his part, proposed a three-part resolution to be submitted to the Devonshire Commission by Society President Arthur Cayley.[90] He urged greater support for the 'cultivation of the Physics of Astronomy', but requested that support be given in the form of increased assistance to existing national observatories such as that at the Cape of Good Hope, and/or the establishment of a new observatory somewhere in the Empire where the climate would be suitable for making routine observations. The third clause recommended that no separate solar observatory sponsored by the Government be established. Its strong wording and the fact that it was ultimately approved by the majority of the Council made it clear that, in their minds at least, Greenwich was the preferred site for prosecuting solar research. This point was underscored by the well-timed announcement that action had already been taken at Greenwich to bring about such research.

Strange countered with an amendment to Huggins's statement recommending the establishment in England of an 'Observatory, with a Laboratory and Workshop of moderate extent attached to it'.[91] He also wanted a network of similar observatories to be set up throughout the Empire to collect daily Sun photographs and to 'investigate the effect of the Earth's Atmosphere on Physico-Astronomical Researches in different geographical regions, and at different altitudes'.[92] This amendment was not approved, however, and Huggins's original statement was submitted to the Devonshire Commission, virtually unchanged, in the form of a letter in lieu of personal testimony from the RAS President.[93]

8.6 The Lockyer factor

A. J. Meadows has attributed the negative response given to Strange's counter-proposal to the fact that Strange allied himself with Norman Lockyer (see Figure 8.2).[94] Shortly after joining the RAS in 1862, Lockyer gained considerable attention and prestige from his colleagues for his lunar observations as well as his drawings of Mars during its favourable opposition in that year. But it did not take long for him to drop these telescopic research projects and take up spectroscopy. He was inspired in great part by Huggins's success in applying spectrum analysis to celestial objects. The recognition Huggins had received in consequence of his spectroscopic study of the stars and nebulae appealed to Lockyer's

Figure 8.2 Joseph Norman Lockyer (1836–1920). (*PRS* **104**, opposite p. i)

nature. He wished to establish his own area of spectroscopic expertise. Thus, he turned to study of the Sun. By 1872, Lockyer had made quite a name for himself.

One might assume that having such a promising young solar observer on board would have assisted Strange in his efforts to convince his colleagues of the need for a national observatory devoted to making daily observations of the Sun. But, in his quest for recognition and advancement, Lockyer had managed to alienate himself from a number of his colleagues in the RAS with what they perceived as brash, impulsive and self-promoting behaviour.[95]

Many people are irritating in their own unique way, but their redeeming qualities can make their less admirable characteristics endurable, or even charming quirks of personality. Huggins, for example, sufficiently annoyed his companions in his role as leader of the solar eclipse expedition to Oran in 1870 that even his friend William Crookes was prompted to write in his diary:

On returning to the ship we had a committee meeting. Little Huggins's bumptiousness is most amusing. He appears to be so puffed up with his own importance as to be blind to the very offensive manner in which he dictates to the gentlemen who are co-operating with him, whilst the fulsome manner in which he toadies to Tyndall must be as offensive to him (Tyndall) as it is disgusting to all who witness it. I half fancy there will be a mutiny against his officiousness.[96]

In spite of this and the fact that the expedition was clouded out, there was no mutiny, only private derision as evidenced in a later diary entry by Crookes describing a particularly rough sea one evening as their ship made its way home:

Eating was almost impossible, for nearly all one's attention was required to keep the meal, &c., on the plate, and ourselves on the benches. Huggins being small and not very careful, disappeared once, plate and all, under the table.[97]

But by 1870 Lockyer had antagonised enough of the Fellows in the RAS to be denied the three-quarters majority required to be awarded the Society's Gold Medal in January 1871.[98] His colleagues also rankled at his treating the data he collected at the 1871 solar eclipse as though it were a marketable commodity.[99] Suspicion grew that Lockyer was Strange's choice to head his proposed solar observatory. Unfortunately for Strange, Lockyer's personality was one upon which it was difficult for those who knew him to remain neutral.

In the months following the presentation of President Cayley's (read Huggins's) letter to the Devonshire Commission, feelings in the RAS on both sides of the question of the need for national observatories intensified. The animus towards Lockyer is evident in the emotional correspondence published in the *Astronomical Register* in the wake of what was termed by the journal's editor as 'a very stormy meeting' of the Society in February 1873. The furore was triggered by a circular that Strange sent around in advance of the meeting suggesting that the current RAS Council be replaced by men he believed would discharge their duties more competently. In all this, Strange's opponents saw the figure, influence and ambition of Lockyer. Letters published in a subsequent number of the *Register* railed against him, particularly for his lack of participation in Society activities, his lack of contribution to the Society's journals and his apparent lack of regard for the Society's aims of gathering and disseminating within its ranks any useful knowledge gained by an individual observer. In their view, such selfishness should not be rewarded with the directorship of a state-supported observatory.[100] Thus, the Lockyer factor appears to have done much to crystallise individuals' opinions on the question of establishing an independent solar observatory.

***

William Huggins's designation as custodian of the Royal Society's Grubb telescope embroiled him in the vigorous and often political exchange that ensued. By allying himself with the Astronomer Royal in this heated controversy, he strengthened his ties within the social and institutional boundaries of traditional astronomical research and facilitated the introduction of spectrum analysis into the agenda of positional astronomy. Nevertheless, the tumultuous events spawned by Strange's initiatives, and Huggins's need to be more publicly accountable as custodian of the Royal Society's telescope, influenced his own personal research programme.

Notes

1. Strange, 'Testimony before the Devonshire Commission, 24 April 1872', Royal Commission on Scientific Instruction and the Advancement of Science, *Minutes of Evidence, Appendices, and Analyses of Evidence*, vol. 2 (1874), p. 76.
2. 'Visitors to the Observatory, 1868', Notebook 2, WCL/SC. For more on Thomas and Howard Grubb, and the Great Melbourne telescope, see Glass, *Victorian Telescope Makers* (1997), ch. 2.
3. Extract from 'Minutes of Council of the Royal Society' (19 March 1868), and 'Report of Committee, Council of the Royal Society', *Correspondence Concerning the Great Melbourne Telescope* (1871), Part 3, pp. 9–12.
4. William Lassell to Sabine, 28 February 1868, in *Correspondence Concerning the Great Melbourne Telescope* (1871), Part 3, pp. 12–13.
5. W. Huggins to Robinson, 20 April 1868, Add MS 7656.TR69. GGS, CUL.
6. Jackson, *Spectrum of Belief* (2000), pp. 176–9.
7. J. G. Hofmann (fl. 1850–1875) was a German-born optician working in Paris, who had developed the direct-vision spectroscope in 1862. J. G. Hofmann, 'Sur un nouveau modèle de prisme pour spectroscope à vision direct' (1874).
8. The assembled telescope weighed in at 18,170 lb, of which 3,500 lb was due to the primary mirror. There were two such mirrors to be shipped. Robinson and Thomas Grubb, 'Description of the Great Melbourne Telescope' (1869). The telescope was shipped from Dublin to Liverpool in April 1868. It left Liverpool in July and arrived in Melbourne in November, but did not become operational until the following June. See Hyde, 'The calamity of the Great Melbourne Telescope' (1987).
9. King, *History of the Telescope* (1955), pp. 251–2. W. Huggins to Robinson, 20 April 1868, Add MS 7656. TR69, GGS, CUL.
10. The telescope was made in 1861 with a 7.25-in objective with a 48-in focal length. Glass describes this as an 'unusual refractor with two cemented doublets'. Glass, *Victorian Telescope Makers* (1997), p. 255.
11. W. Huggins to Robinson, 6 May 1868, Add MS 7656.TR70, GGS, CUL.
12. *Ibid.*
13. 'Instruments for Sale', *MNRAS* **28** (1868), p. 224.
14. Six months later, Huggins again referred to his special arrangement with Cooke: 'I had been before on treaty for a smaller instrument (abt. 11 inches) which they [Cooke & Sons] agreed to make under the usual price for *me* . . .' W. Huggins to Robinson, 6 November 1868, Add MS 7656.TR72, GGS, CUL.
15. W. Huggins to Robinson, 26 October 1868, Add MS 7656.TR71, GGS, CUL.
16. W. Huggins, 30 September 1868, 7h 30, Notebook 2, WCL/SC. 'My intensely beloved mother died. From the above date to Nov 5 I was quite unable to observe.'
17. W. Huggins to Robinson, 26 October 1868, Add MS 7656.TR71, GGS, CUL.
18. W. Huggins to Robinson, 12 November 1868, Add MS 7656.TR73, GGS, CUL.
19. 'Minutes of Council', 21 January 1869, RSL.
20. W. Huggins to Robinson, 6 November 1868, Add MS 7656.TR72, GGS, CUL.
21. *Ibid.*

22. For the notice of Oliveira's election, see *PRS* **3** (1835), p. 338. For notice of his 1854 gift to the Society, see Royal Society, *The Record of the Royal Society of London for the Promotion of Natural Knowledge* (1940), p. 142.

23. Initially, there was hope the Oliveira Bequest would bring the Society as much as £1,800, but claims on the money by other scientific societies reduced this to £1,350. See 'Minutes of the meeting of the Council of the Royal Society', 21 January 1869. The letters between Huggins and Robinson show that the decision to expend this money on the proposed telescope took several months.

24. W. Huggins to Robinson, 18 January 1869, Add MS 7656.TR74, GGS, CUL.

25. W. Huggins to Robinson, 23 January 1869, Add MS 7656.TR75, GGS, CUL.

26. W. Huggins to Robinson, 31 January 1869, Add MS 7656.TR76, GGS, CUL.

27. H. Grubb to Stokes, 6 February 1869, Add MS 7656.G473, GGS, CUL.

28. W. Huggins to Robinson, [no date, but probably 19 February 1869], Add MS 7656.TR79, GGS, CUL.

29. W. Huggins to Robinson, 27 February 1869, Add MSS 7656.TR77, GGS, CUL.

30. Robinson to W. Huggins, 27 February 1869, Add MSS 7656.TR78, GGS, CUL.

31. W. Huggins to Robinson, 6 March 1869, Add MSS 7656.TR80, GGS, CUL.

32. W. Huggins to Robinson, 20 March 1869, Add MSS 7656.TR81, GGS, CUL.

33. Glass, *Victorian Telescope Makers* (1997), pp. 65–7.

34. 'Numerous observations on the sun with the spectroscope, and also with coloured solutions made up to August 1869 on which day the telescope, some micrometers & observing chair were removed by Mr. Corbett, to whom they had been sold . . .' W. Huggins, undated entry, sometime between February and November 1870, Notebook 2, WCL/SC. According to Corbett's obituary, he became interested in astronomy about the time he purchased Huggins's instruments. See 'Charles Joseph Corbett', *MNRAS* **44** (1884), p. 131.

35. W. Huggins to Airy, 6 November 1869, RGO 6.271/7/2/304–8, CUL. See also W. Huggins to J. F. W. Herschel, 22 September 1869, HS.10.55, JFWH, RSL.

36. W. Huggins to Robinson, 10 November 1869, Add MS 7656.TR82, GGS, CUL.

37. Sabine, 'President's address' (1869), pp. 105–6.

38. W. Huggins to Robinson, 13 December 1869, Add MS 7656.TR83, GGS, CUL.

39. Glass, *Victorian Telescope Makers* (1997), p. 67.

40. Schellen, *Spectrum Analysis and its Application to Terrestrial Substances, and the Physical Constitution of the Heavenly Bodies* (1872). Huggins refers to his having undertaken this task in a letter to Airy, 28 September 1870, RGO 6.382/6/7/401, CUL. His notebook entries indicate that he visited Lassell at Maidenhead in March of both 1871 and 1872. See 25 and 26 March 1871; 31 March 1872, Notebook 2, WCL/SC.

41. The eclipse expedition was clouded out, a major disappointment for Huggins who had hoped to use the occasion to examine spectroscopically the Sun's outer atmosphere. I shall discuss this expedition in more detail in Chapter 9.

42. W. Huggins to Airy, 5 August 1870, RGO6.131, file 2, f. 52, CUL.

43. W. Huggins, 12 November 1870, Notebook 2, WCL/SC. In the notebook entries which have survived from this period, Huggins does not mention whether he was using the refractor or the reflector, although he preferred refractors for visual observation.

44. W. Huggins, 14 November 1870, Notebook 2, WCL/SC.

45. The eclipse took place on 22 December 1870. Huggins and his observing party left England on 5 December and returned on 5 January 1871.

46. W. Huggins, undated entry following that for 15 November 1870, Notebook 2, WCL/SC.

47. Sabine, 'President's address' (1869), p. 105.

48. In 1860, the Fellows numbered 380. By 1870, this had increased to 509. See Dreyer and Turner, *History of the Royal Astronomical Society* (1987), p. 245.

49. One of the central issues animating Meadows's fourth chapter in *Science and Controversy* is the contentious process by which recipients of the RAS's prestigious Gold Medal were selected during the 1870s. On three occasions between 1870 and 1880, no Medal was awarded because agreement could not be reached on who deserved it. On one other occasion, after much indecision regarding possible British recipients, Italian astronomer Schiaparelli was selected. Throughout this period there was much discussion on how to amend the process of Medal selection. See Meadows, *Science and Controversy* (1972), pp. 95–6 and pp. 108–9; Dreyer and Turner, *History of the Royal Astronomical Society* (1987), pp. 172–3, p. 176 and p. 206.

50. Cardwell, *The Organisation of Science in England* (1972), pp. 111–26; M. E. W. Williams, 'Astronomy in London: 1860–1900' (1987), pp. 13–6; MacLeod, 'Resources of science in Victorian England' (1972), pp. 111–66; Alter, *The Reluctant Patron* (1987), pp. 13–74.

51. Cardwell, *The Organisation of Science* (1972), p. 111.

52. Cardwell cites the report of British engineer John Russell Scott: 'It was not that we were equalled, but that we were beaten, not on some points, but by some nation or another at nearly all those points on which we had prided ourselves.' Cardwell, *The Organisation of Science* (1972), pp. 111–12.

53. *Ibid.*

54. Alter, *The Reluctant Patron* (1987), pp. 100–1; Meadows, *Science and Controversy* (1972), pp. 76–7.

55. Cardwell, *The Organisation of Science* (1972), pp. 75–7.

56. The key elements of this episode are outlined in Dreyer and Turner, *History of the Royal Astronomical Society* (1987), pp. 173–8.

57. Dunkin, 'Lieutenant-Colonel Alexander Strange' (1877).

58. Crowther, *Statesmen of Science* (1966), pp. 237–69; Trotter, 'Lieut.-Col. Alexander Strange'(1898).

59. Many of those who were active in this movement were chemists like Playfair: Henry Roscoe from Owens College in Manchester, for example, and Edward Frankland from the Royal College of Chemistry in London.

60. Strange, 'On the necessity for State intervention to secure the progress of physical science' (1868).

61. Cardwell, *The Organisation of Science* (1972), pp. 119–21; M. E. W. Williams, 'Astronomy in London: 1860–1900' (1987), pp. 14–15; Meadows, *Science and Controversy* (1972), pp. 82–3; MacLeod, 'Resources of science in Victorian England' (1972), pp. 123–5; Alter, *The Reluctant Patron* (1987), p. 80.

62. Crowther, *Statesmen of Science* (1966), pp. 213–33; Meadows, *Science and Controversy* (1972), pp. 82–3.

63. Strange, 'On the insufficiency of existing national observatories', *MNRAS* (1872). For a paraphrase of Strange's remarks at the meeting of the RAS, with excerpts of the discussion which followed, see Strange, 'On the insufficiency of existing national observatories', *AR* (1872).

64. Strange, 'Insufficiency of existing national observatories', *MNRAS* (1872), pp. 238–9.

65. Strange, 'Insufficiency of existing national observatories', *AR* (1872), p. 113.

66. *Ibid.* See also Airy, 'Proposed devotion of an observatory to observation of the phenomena of *Jupiter's* satellites' (1872).

67. Strange, 'Insufficiency of existing national observatories', *MNRAS* (1872), p. 239. In the *Astronomical Register*, Strange was paraphrased in this way: 'The labours of private individuals are liable to be interrupted by their deaths or the loss of the possession of the observatory.' See 'Insufficiency of existing national observatories', *AR* (1872), p. 114.

68. Strange, 'On the insufficiency of existing national observatories', *MNRAS* (1872), p. 240. Strange's own words were that this type of work would be so laborious that 'individuals can never complete [it] within any reasonable period'.

69. *Ibid.* The suspected relationship between sunspot numbers and terrestrial weather patterns was of growing interest in Britain owing to the desire to anticipate droughts in India and hence prevent famine. See Porter, *The Rise of Statistical Thinking* (1986), pp. 274–9.

70. More than efficient organisation was needed to insure a successful data gathering effort. It also helped to have a ready supply of patient and attentive assistants to hand, as indicated in the following note: 'The Kew photographs are now taken by Miss Beckly, the daughter of the mechanical assistant of Kew; and it seems to be a work peculiarly fitting to a lady. During the day she watches for opportunities of photographing the Sun with that patience for which the sex is distinguished, and she never lets an opportunity escape her. It is extraordinary that even on very cloudy days, between gaps of cloud, when it would be imagined that it was almost impossible to get a photograph, yet there is always a record at Kew.' *MNRAS* **26** (1866), p. 77.

71. The Duke of Devonshire's questions provided Strange with the opportunity to put the story behind the demise of the Kew's solar observations in the public record. Strange recounted how the British Association had, until recently supported, the Kew operation. When this money was cut off, James Gassiot had offered a large sum of money to the Royal Society to maintain many projects at Kew, but solar observations had not been included. See Strange, 'Testimony before the Devonshire Commission, 24 April 1872', Royal Commission on Scientific Instruction and the Advancement of Science, *Minutes of Evidence, Appendices, and Analyses of Evidence*, vol. 2 (1874), p. 76.

72. Strange, 'On the insufficiency of existing national observatories', *MNRAS* (1872), p. 240.

73. *Ibid.*, p. 241.

74. Strange, 'On the insufficiency of existing national observatories', *AR* (1872), p. 114.

75. Airy, 'Response to Col. Strange' (1872).

76. It is interesting to note that the very next day after this particular RAS meeting, Huggins spent time observing the Sun. The three notebook entries that Huggins made during the next week indicate that he observed the Sun twice again. On these occasions, however, he was not observing the Sun's surface to gain a better understanding of the structure of the Sun itself. Instead, he subjected the solar spectrum to exacting scrutiny in order better to gauge the shift he had perceived in the spectral lines of Sirius. Where individuals like De la Rue, Strange, and

Lockyer saw the Sun as an object with intrinsic scientific interest, Huggins viewed it as a valuable calibration device for assisting in the study of other celestial bodies in which he had a greater personal interest.

77. The question of the relationship among such apparently disparate phenomena as sunspot numbers, fluctuations in the Earth's magnetic field, and terrestrial weather patterns was an old sore point for Airy. Airy had directed the keeping of daily records of geomagnetic variations at Greenwich in the 1830s and 1840s, but later analysis of these records failed to convince him that any discernible pattern existed. Both the accuracy of the Greenwich measures and the correctness of Airy's interpretation of them was challenged by Edward Sabine and other magnetic observers who reported evidence for a ten-year cycle in geomagnetic variations. Airy did not ordinarily respond favourably to such criticism, but in this case, as Meadows has shown, Airy's reaction was magnified by his intense personal dislike of his chief critic, Edward Sabine. Clearly then, Strange pressed some of Airy's more sensitive alarm buttons with his forceful arguments in favour of initiating solar observations based on what Airy considered unproven connections between solar and terrestrial phenomena. See Meadows, *Greenwich Observatory, Volume 2* (1975), pp. 96–100.

78. Airy, 'Response to Col. Strange' (1872), p. 118.

79. Strange, 'On the sufficiency of existing national observatories', *AR* **10** (1872), p. 119.

80. Strange, 'Testimony before the Devonshire Commission, 24 April 1872', Royal Commission on Scientific Instruction and the Advancement of Science, *Minutes of Evidence, Appendices, and Analyses of Evidence*, vol. 2 (1874), p. 75.

81. *Ibid.*, p. 76.

82. *Ibid.*

83. Airy, 'Testimony before the Devonshire Commission, 26 April 1872', Royal Commission on Scientific Instruction and the Advancement of Science, *Minutes of Evidence, Appendices, and Analyses of Evidence*, vol. 2 (1874), pp. 93–100.

84. *Ibid.*, p. 94.

85. *Ibid.*, p. 95.

86. R. W. Smith, 'A national observatory transformed' (1991), p. 6.

87. W. Huggins to Airy, 28 February 1872, RGO 6.271/7/2/327, CUL. It was only two weeks later, on 15 March 1872, that Airy requested information from George Stokes on what he called 'das Doppelsche Princip, or some such term.' This episode is discussed in Chapter 7.

88. Meadows, *Science and Controversy* (1972), pp. 94–5 and p. 105.

89. *Ibid.*, pp. 96–8.

90. *Ibid.*; see pp. 97–8 for a discussion of what took place at this special RAS meeting and a transcript of Huggins's resolution.

91. 'Minutes of the RAS Council Meeting, 28 June 1872', cited in Meadows, *Science and Controversy* (1972), p. 97.

92. *Ibid.*

93. 'A. Cayley (RAS President) to Lockyer (Secretary to the Royal Commission on Scientific Instruction), 2 July 1872' in Royal Commission on Scientific Instruction and the Advancement of Science, *Minutes of Evidence, Appendices, and Analyses of Evidence*, vol. 2 (1874) Appendix VII, p. 30. Indicative, perhaps, of the dissension out of which this resolution was born, Cayley stated in his letter that he was submitting to the public record a resolution approved by the Royal Astronomical Society's Council in accord with his duty as the Society's President, but that he had no 'wish to offer any evidence to the Commission' in person.

94. Meadows, *Science and Controversy* (1972), p. 95.

95. Plotkin, *Henry Draper* (1972), pp. 87–100.

96. From William Crookes's diary entry for Friday, 16 December 1870, cited in D'Albe, *The Life of Sir William Crookes* (1923) p. 154.

97. William Crookes's diary entry for Monday, 2 January 1871; *ibid.*, p. 172.

98. Meadows, *Science and Controversy* (1972), pp. 95–6.

99. 'Socius Nauseatus' to the Editor, 'The Royal Astronomical Society and a recent fiasco', *AR* **11** (1873), pp. 100–2.

100. See 'Letters to all Fellows of the Royal Astronomical Society' (from J. Browning, T. W. Burr, E. B. Denison, W. Noble and R. A. Proctor; and A. Strange); as well as 'Letters to the Editor' (from Proctor; and 'Socius Nauseatus'), *AR* **11** (1873), pp. 93–102. For a summary of the tumultuous sequence of events, see Meadows, *Science and Controversy* (1972), pp. 99–102.

9

Solar observations

... after our failure but one idea seemed to possess all, and that was to get
away from Oran and on our homeward voyage as quickly as possible.
                                                                  – William Crookes[1]

By the 1870s, when William Huggins assumed responsibility for the Great Grubb
Equatorial, the discipline of astronomical physics was developing on many fronts. Which
would open up the most fruitful line of investigation? No one knew. Unabashedly eclectic in
his research interests and methods, Huggins shrewdly ventured down many paths that
promised discovery and recognition. Sometimes – but not always – he encountered new
opportunities to press the spectroscope into service.

Concern for priority with its attendant thrill of the chase occasionally provoked him to
examine the Sun, just as solar observers like Lockyer were drawn to study stars and nebulae.
Huggins was not always happy with the results of his solar investigations, however, and they
are conspicuously missing from his retrospective account, 'The new astronomy'.[2] Because
his biographers and historians of science have relied heavily on this essay for details of his
career, his contributions to the rapidly growing body of knowledge about the Sun and its
atmosphere have been all but forgotten.[3]

The unpublished record brings light and life to these episodes once again. Huggins
learned a great deal about discovery's delicate dance from his setbacks and failures, and
so can we. Two of his solar projects are particularly instructive: his attempts to observe the
Sun's so-called red flames without an eclipse in the late 1860s, and his participation in an
expedition to view the total eclipse of 22 December 1870 in Oran, Algeria, which we alluded
to in the previous chapter. His correspondence, notebooks and published accounts of these
efforts give valuable insight into how Huggins modified his research methods and instrumentation when faced with serious obstacles and outright failures in achieving his goals.

To better understand Huggins's interest in these two projects, we begin this chapter with a
brief survey of the events and discoveries during the total solar eclipse of 1868, the so-called
'Great Indian Eclipse'.

9.1 The 'Great Indian Eclipse'

In February 1867, Major James Francis Tennant (1829–1915) of the Royal Engineers
alerted the RAS to a total solar eclipse coming in August 1868. Although still a full year

and a half away, he believed it warranted the Society's immediate attention.[4] For one thing, the eclipse promised totality of an extraordinary duration.[5] But, more importantly for British astronomers, its projected path crossed a number of favourable sites in the imperial domain, from the Aden Protectorate on the Arabian Peninsula to India, then to the Tenasserim Province in Lower Burma and beyond – offering them politically, as well as physically, accessible locations with reasonable prospects of good weather, comfortable accommodations, and – in some cases – telegraph service. In short, it promised to be the first eclipse since July 1860 to be worth the time, effort and expense required to mount a British expedition.

Tennant made his case shortly before a partial eclipse swept across the British Isles on the morning of 6 March, and it is easy to imagine that many of those blessed with clear skies on that day were left wishing to see more. Eager to serve in the role of 'old India hand' to his home-based colleagues, he soon followed up with a lengthy report offering information and strategies for planning expeditions to India including a map of the eclipse path, a table of expected times and geographical limits of totality in the region, weather prognostications, location recommendations and detailed schedules for observing the eclipse from beginning to end at three selected sites.[6]

In Chapter 6 we saw that, during the 1860 eclipse, Warren De la Rue had demonstrated the power of photography to freeze the fleeting moments of these rare celestial events and bring them home. There, undistracted by the awful frenzy that invariably turned each precious moment between second and third contact into a blinding blur of sensory overload, cooler heads had time to examine and re-examine the photographs at will, measure and remeasure the features they recorded, consider and reconsider what each one represented. Tennant suggested that all observing stations be equipped to photograph the Sun throughout the eclipse, particularly during totality. Furthermore, to enable analysts to put these photos into a broader context, Tennant urged that the solar disc be photographed for a week before and after the eclipse.

Since 1860, the tools, methods and expertise of celestial photographers had improved. And the astronomer's toolkit had expanded to include the spectroscope. Heeding Tennant's clarion call, RS Secretary George Stokes began gathering his colleagues' opinions on the role the spectroscope might play in analysing the red flames' light during the eclipse. John Herschel, for one, feared the prominences might prove too feeble as light sources to produce a discernible spectrum. On the other hand, if their light turned out to be sufficiently bright, it should be possible to isolate them using 'a long tube *without an object-glass*, closed by a screen with a narrow slit'. After passing through the slit, the flames' light could then be dispersed and examined. In fact, he presciently opined, such a method might be tried even without an eclipse, to compare the spectrum of 'the light at the extreme border of the limb of the sun itself with that of the central portion'.[7] Huggins urged placing a skilled observer in charge of an equatorially mounted telescope of 3¾-in aperture capable of being moved by hand. But, in his opinion, even a competent observer, given the limitations of expense imposed on a Government-sponsored eclipse expedition, would find it difficult to obtain more than the 'general character of the spectrum of a red prominence, that is, whether it be continuous, or consist of bright lines'.[8]

9.1.1 Aden

As the August eclipse approached, teams of observers from around the world established camps along the long eclipse path.[9] Photographers from Berlin, led by photographer and chemist Hermann Wilhelm Vogel (1834–98), arrived in Aden well in advance of the eclipse date. There they rendezvoused with an Austrian group headed by astronomer Edmund Weiss (1837–1917).[10] Despite its oppressive August heat, the nearly rain-free British Protectorate of Aden was an ideal place to be. If all went as expected on the day of the eclipse, the Sun would rise a gleaming crescent. Over the next half hour, the crescent would reduce to a sliver, then a mere glint, and finally disappear for the next three minutes providing astronomers and photographers alike with a glimpse of the red flames and corona.

Do these enigmatic solar phenomena change over time? Like Tennant, Vogel hoped an answer would be found through careful comparison of photographs taken from different points along the eclipse path. His goal was to take as many photographs as possible during the brief interval of totality at Aden to establish the baseline. To that end, he and his colleagues divided the labour of preparing, inserting, exposing and developing the plates to form a smooth and efficient assembly line, the operation of which they 'regularly practised, as artillerymen do their cannon'.[11] While they mastered their routine, they also accustomed themselves and their methods to conditions in which their 'perspiration flowed ... in streams: it ran from the tips of the fingers, dropped from the face' often spoiling 'a well-cleansed or prepared plate'.[12]

The Austrians, meanwhile, planned to concentrate on spectroscopic observations of prominences and the corona. Weiss, for example, intended to spend part of his time before, during and after totality studying the structural appearance of the corona for signs of any correlation between irregularities in the erose lunar rim and the distribution of coronal streamers.

9.1.2 Jamkhandi

An hour after leaving Aden, the shadow of the Moon would enshroud a group of land-based observers in India's Bombay Presidency led by Lieutenant John Herschel, RE. Based in the Bangalore Cantonment, a British military garrison adjoining the Mysore capital, young Herschel had been working throughout southern India since 1864 on the Great Trigonometric Survey. He happened to be in England in summer 1867, just when the Royal Society was mobilising its resources to mount an eclipse expedition to the sub-Continent. On 19 June, he paid a visit to Huggins's Tulse Hill observatory.[13] Unfortunately, Huggins did not record the details of this or any subsequent meeting he might have had with Herschel at the time. Thanks to RS President Edward Sabine, however, we know he gave Herschel valuable instruction in the art of observing the spectra of celestial bodies.[14] Huggins's expert guidance prepared Herschel for making his own spectroscopic observations of solar phenomena during totality, and for training others before the event so that they could carry out similar studies elsewhere in the region.

John F. W. Herschel, the young lieutenant's illustrious father, had counselled against using a telescope to observe prominence spectra, but the RS ignored this advice and provided his son with a 5-in aperture 'equatorially mounted telescope, furnished with a spectroscope and clock-movement'. They also made four handheld spectroscopes available to him to distribute among his colleagues.[15]

Once back in India, the industrious lieutenant set to work on the all-important task of identifying a site on or near the centre line with good prospects for clear skies in August. It was a challenging task. Each year from roughly June to October, the southwest monsoon brings clouds and rain to much of the targeted region. Nevertheless, Herschel found a promising spot: Jamkhandi, a princely state nestled within the boundaries of the Bombay Presidency. Situated in the rain shadow of the Western Ghats, the mountain range that parallels India's coast along the Arabian Sea, Jamkhandi routinely endures parched Augusts that only mad dogs and English solar astronomers could love. So, with hope of fair weather and kind offers of assistance from Jamkhandi's ruling chief, Ram Chandra Rao Gopal (1833–97), Herschel chose it as the place from which he and his team would view the upcoming eclipse.

Clear nights in Bangalore from March to the start of the monsoon in mid-May 1868 gave him the opportunity to practise what he had learned from Huggins while examining the spectra of some fifty nebulae in the Southern sky with the RS instruments. Unfortunately, the telescope's finder proved to be quite useless for locating faint objects. He would have replaced it, of course, but 'India is not England', he noted wryly, 'and Bangalore is not London'.[16] Having to aim the telescope by trial and error sorely tried his patience and luck.

In June, everything expedition participants would need to set up a working field observatory was carefully packed for transport to Jamkhandi. Because of its military importance, the Bangalore Cantonment had been connected by railway with Madras (Chennai) since 1864. But there was no such convenient means to carry freight or passengers from Bangalore to Jamkhandi. The nearly 400-mile overland journey the convoy faced was an arduous undertaking even in the best of weather. The monsoon's road-eating rain turned the trek into a month-long slog.

The equipment reached Jamkhandi on 7 August. Herschel arrived a week later to find the sky 'thick with passing cloud'.[17] If his 'exceedingly disagreeable' journey had put him in a dismal mood, the gloomy sky in Jamkhandi probably did little to improve it. At least, as promised, it was not raining. He received local assurances that the overcast skies were unusual and could not continue for long.[18] But he could not afford to dwell on the weather. He had only four days to make sure all the instruments were in working order, that he and all members of the expedition knew exactly what to do during the slightly more than five minutes of totality they hoped to witness, and that everyone was well-practised in how and when to do it.

Unfortunately, the persistent clouds threatened to spoil any chance of viewing the eclipse. To spread the risk, Herschel had already planned to have several small groups observing the eclipse simultaneously from different locations. Lieutenant William Maxwell Campbell, who had been working with Herschel on a survey project in Bangalore, set up an observing station some distance away from the main camp. There, he and his team would employ a polariscope to analyse the light of the corona and prominences during totality. In addition, Herschel had trained Campbell to use one of the RS's handheld spectroscopes so that he could contribute to what was shaping up to be a global study of prominence spectra.[19]

Captain Charles Thomas Haig (1834–1907) and Captain Henry Charles Baskerville Tanner (1835–98) travelled from their station in Pune to view the eclipse at Bijapur. Like Campbell, Haig was entrusted with an RS-supplied handheld spectroscope and charged with recording the characteristics of any prominence spectra he observed. Tanner was to measure the heights of the prominences and make quick sketches of all he saw throughout the eclipse, particularly sunspots on the solar disc, and the corona and red flames during totality.

9.1.3 Guntur

The RAS sponsored a second British expedition under Major Tennant's leadership. He had calculated that totality would last nearly six minutes at Masulipatam on India's Bay of Bengal.[20] Perhaps the prospect of clouds at that coastal location encouraged him to look further inland, for he ultimately settled on Guntur as the site from which to view the eclipse.

His team spent months mounting a full-scale assault on the Sun during totality using state-of-the-art instruments supplied by the RAS. They planned a battery of observations with spectroscopes and polariscopes to analyse the solar corona and prominences. While in London, Tennant visited Huggins's observatory. Warren De la Rue trained one of Tennant's men to take a set of photographs of the eclipsed Sun similar to those he had captured back in 1860.

In March 1868, the instruments and photographic equipment that had been shipped from England began arriving in India. The equatorial spectroscope appeared none the worse for wear, but the badly damaged photographic apparatus required extensive repair. The silvered glass mirror that would be used to capture the photographic images was delivered in May. Fortunately, the damage it suffered in transit was not too bad and by June all the equipment was in working order.[21]

Guntur was also the site selected by Pierre Jules Janssen (1824–1907), France's pre-eminent solar observer. Situated between the mist-topped Eastern Ghats and the foggy coast, Guntur offered Janssen one additional advantage: the welcoming hospitality provided by a settlement of French merchants. He and his team trained hard in advance to handle the mental and physical pressures of the work to be done during totality.[22]

9.1.4 Observations

In Aden, Vogel and his team successfully exposed three plates producing six images of distinctive prominences around the lunar disc.

The clouds at Jamkhandi cleared sufficiently during totality to allow Herschel a glimpse of the spectrum of a prominence.[23] The sight of the bright lines sparked an idea for a method of monitoring prominences on a regular basis, a method relying on intervening absorptive media very similar to that devised independently by Huggins nearly two years earlier. He wrote to his father:

It has occurred to me that, since the whole light of the 'flames' is of three refrangibilities only (or nearly so), dark glasses (could they be formed) which allowed these *only* to pass, would so enormously diminish the light *from the Solar disk*, as to enable the flames to be seen without the interposition of an opaque body, *i.e.* without an eclipse . . . Is the idea a practical one? If it is, these flames may come to be studied *at leisure!*[24]

Hindered by cloud cover during the second half of totality at Guntur, Tennant and his team obtained six faint, but, in their opinion, satisfactory, photographs of the solar atmosphere during totality which showed considerable detail in the structure of the visible prominences. Tennant's spectroscopic observations of the prominences were thwarted by a recalcitrant clock drive, although he noted seeing several bright lines in their spectra which he believed were coincident with some of the Fraunhofer lines in the solar spectrum.

Before clouds obscured his view of the eclipsed Sun, Janssen subjected two large prominences to spectroscopic examination. He, too, concluded they were gaseous in nature. He also surmised, when he saw how vividly bright the lines were, that they might remain visible under spectroscopic scrutiny even in full sunlight, just as Huggins and Herschel had suggested. Janssen tested his hypothesis the next day. Remembering the approximate location of the larger prominences on the solar limb, he quickly detected their bright lines.[25]

9.2 Viewing the red flames without an eclipse

In February 1868, months before the Indian eclipse, Huggins disclosed in his annual observatory report that he had tried to view solar prominences without an eclipse.[26] He did not mention his attempt back in November 1866 (discussed in Chapter 6) to make the red flames visible by filtering the Sun's rays through a stack of coloured glass plates. Instead, he described a spectroscopic method he had devised based on his experience with nebular spectra. If prominences are gaseous, he reasoned, they ought to display an array of bright emission lines when viewed spectroscopically. Because these colourful lines suffer little additional dispersion after passing through a prism (regardless of its refractive power), they should remain conspicuously brilliant compared with the fainter, broadly dispersed light of the background sky or solar limb.

He claimed to have been engaged in making such observations for the past two years.[27] Although this was a bit of an exaggeration, it sufficed to establish him as a serious contender for priority in this limited arena, and, without access to his private notebooks, his contemporaries could neither confirm nor refute his claim. His principal rival was none other than Norman Lockyer, whose growing interest in solar chemistry and physics had led him naturally into spectroscopic study.[28] In January 1867, with £40 awarded to him by the RS's Government Grant Committee, Lockyer commissioned Thomas Cooke to make a spectroscope with dispersing power sufficient to render solar prominence spectral lines visible without an eclipse.[29] But the famed instrument maker was otherwise engaged and the project languished. In January 1868, Lockyer handed the job over to John Browning (1835–1925).[30] Unfortunately, Browning's slow work pace and Lockyer's ill-health conspired to prevent the latter from trying his new instrument until October, two months after the Indian eclipse.[31]

Meanwhile, Lockyer became acquainted with some of the eclipse observers' cabled reports. Although he seems to have been ignorant of Janssen's discovery, he knew that others had confirmed the suspicion that the prominences produced bright-line emission spectra. The qualitative nature of these observations did not provide Lockyer with sufficient information to know which spectral lines to look for, but at least he was given hope that some bright lines would be readily visible.[32] On 19 October 1868, just days after obtaining his new spectroscope, he observed carefully the bright-line spectrum of a prominence and he immediately sent out letters describing his preliminary findings.[33] Because of communication delays, Janssen's announcement arrived at the Académie des Sciences in Paris at virtually the same time as Lockyer's. The two observers were awarded a special medal by the French Government in honour of their independent discoveries. No official recognition, however, was bestowed on either man by the British scientific societies.[34]

While Lockyer awaited the completion of his new spectroscope, Huggins did have an instrumental edge. But he did not see the bright spectral lines produced by a solar prominence until 19 December 1868, fully two months after Lockyer's success. He had a method, and he had the instrumentation to render them visible. In fact, when he viewed the solar prominences for the first time, he tells us that he did so with his trusty spectroscope that boasted two dense 60° prisms by Hofmann and Ross:

Applied 2 prism spectroscope to edge of sun. Saw at once the three bright lines. The red very strong. The F line ... broader at the base. The line near D apparently double, not certain of this, a little more refrangible than D. Tried red glass, but could not see prominence.[35]

Colleagues were puzzled as to why he had been unable to see the spectral lines until then. In November 1868, De la Rue wrote to Stokes:

It is curious that Huggins should have failed in discovering the 'red flame' – because he is a very skilful observer: – had he been successful he would have anticipated Janssen and, by so doing, rendered great assistance to the Eclipse observers.[36]

Agnes Clerke later concluded that Huggins 'devised various apparatus for bringing them into actual view; but not until he knew where to look did he succeed in seeing them'.[37] Hers was a harsh verdict for an observer who had achieved considerable respect for his care and perspicacity. Recall that during this critical period Huggins was distraught over the death of his mother and extremely anxious about arrangements to obtain a new telescope. A lack of concentration resulting from these two factors may have been partially responsible for Huggins's lack of success, but Clerke's assessment is, by and large, on the mark.

Huggins knew from his own experience that 'fishing around the limb of the sun', as Lockyer so aptly put it, for elusive emission lines was a hit or miss activity.[38] An individual who had seen the lines during an eclipse would be better prepared, but 'not knowing what kind of lines would appear in the prominences',[39] or where to expect them, he remonstrated, 'it would have been by accident only if I had succeeded in obtaining a view of the flames'.[40] His insinuation that Lockyer had succeeded principally because he had some knowledge of the recent eclipse observations was not lost on his rival who angrily denied he had been helped in any way by the findings of the 1868 eclipse observers. Lockyer pointed to his 1866 *Proceedings* paper as evidence that he had been working toward just such a discovery on his own for two years. Allies John Browning and Balfour Stewart, Director of the Kew Observatory, came to his defence.

Huggins was unconvinced and unrepentant.[41] He felt that, once eclipse observers had definitively resolved the question of the character of prominence spectra and described where to look, actually seeing the spectral lines without an eclipse was merely a technical exercise, not a major discovery. Clearly, Janssen had beaten everyone to the visual discovery. But any skilled observer with the right information and sufficient instrumental power could have repeated it. What really counted was figuring out that it could be done in the first place. And, on that score, Huggins believed he had beaten them all. Lockyer may have had plans to examine the prominences spectroscopically in October 1866 – but what were they?

Lockyer continued to seethe long after the fact.[42] And Huggins, through his role in the dispute, became more aware of the need to establish and preserve his priority whenever he engaged in some research project he believed to be original. Astronomical physics was evolving rapidly, and priority, once claimed, required constant reassertion. Investigators were provided with neither clear boundaries nor well-defined research guides. What might seem peripheral at one time could prove to be central at another. Thus, in February 1869, Huggins submitted a paper to the RS describing his earlier thermometric work to preempt others' claims.[43] And, when De la Rue published what he thought was a new method of observing the whole of a prominence without an eclipse by using only absorptive filtering, Huggins quickly advertised that he had already thought about that 'three or four years since'.[44]

Having a successful method to render solar prominences visible to observers on any cloudless day was extremely important for solar observers. Now they could monitor prominences on a regular basis just like sunspots, without waiting for an eclipse, without begging for funds to sponsor an expedition, and without praying for clear skies during totality. Prominence activity could also be examined on the visible surface of the Sun instead of just at its limb. When reports from all the 1868 eclipse observing stations had been gathered, there were hints that prominences were active solar features that could and did change in structure and appearance over very short periods of time. Near the close of the RAS meeting in January 1869, which was devoted almost entirely to discussion of the 1868 eclipse, De la Rue observed: 'It comes to this, that the spots are the least frequent occurrences. The changes of the faculae [bright spots] and prominences are much more numerous.'[45]

Astronomical physics itself was changing even more quickly. In September 1869, the *Astronomical Register* opened with the statement, 'The age is essentially a *fast* one ... [W]ith regard to our present subject [i.e. spectroscopic observations of the Sun] ... the race for fresh discoveries is so evident and the competition among observers so keen.'[46] It was becoming almost necessary for investigators, like *The Looking-glass*'s Red Queen, to run merely to stay in place.[47] The lack of rigidity in Huggins's research programme served him well here. His own early attempts to view solar prominences without an eclipse, despite his lack of success, gave him a central role in the ensuing priority dispute. Instead of sitting on the sidelines as a nebular and stellar specialist, Huggins chose to jump into the fray with the solar observers.

9.3 The eclipse expedition to Oran

The next total solar eclipse took place in 1869. It swept across North America from the Territory of Alaska to the Carolina coast. No British expedition was sent to observe it.

Where attention had been focused on the solar prominences during the 1868 eclipse, observers participating in the 1869 expeditions were interested in studying the solar corona – the soft white glow around the limb of the Moon during the total phase of a solar eclipse. On some occasions, observers had described the corona as reaching out like asymmetric fingers into the darkened sky. Other times it escaped any notice at all.

A solar eclipse was predicted for December 1870 that would be visible from sites nearer to European observers. This time, the Moon's shadow would make landfall briefly at Cabo de São Vicente, Portugal, and again at Cadiz, Spain, then move on to Gibraltar before crossing the Mediterranean to Algeria where it would pass conveniently near the cities of Oran and Tunis. After leaving Africa, the centre line would cut across southern Sicily falling midway between the cities of Catania and Syracuse.

Huggins led a small expedition to Oran to view the 1870 eclipse with the intention of examining the solar corona spectroscopically. It was the only eclipse expedition in which he participated as an observer. The experience exposed him to the challenges of obtaining financial support from a tight-fisted liberal Government, providing effective leadership under stress, eliciting group cooperation from a collection of observers accustomed to working independently, and coping with the frustration of a clouded-out eclipse.

In 1870, there was still fertile disagreement as to the nature of the solar corona. No one was sure whether the corona was physically connected to the Sun, a meteorological effect comparable to a mock Sun or a tangential arc, or simply an illusion of contrast. Interest in the solar corona was greatly enhanced after the 1868 eclipse in part because of the bounty of information obtained on the nature of solar prominences. Another important factor was the promise held out to solar observers by the new investigative tools of photography, spectroscopy and polariscopy for achieving an understanding of the nature of other features of the near solar environment. In addition, there was a suggestive report offered by Edmund Weiss, who had participated in the Austrian eclipse expedition team sent to Aden in 1868. Where others in his party had observed the prominences, Weiss had concentrated on the corona. His results, which he reported in person at the April 1869 meeting of the RAS, startled and intrigued the Fellows present.

Solar enthusiasts had expected that, if the corona were of solar origin, it would either be shown to consist of photospheric light reflected off tiny suspended particles swarming around the Sun's visible surface (in which case it would show a continuous spectrum interrupted by the same absorption lines seen in the normal solar spectrum), or a glowing gas of very low vapour density (in which case it would show bright emission lines like the solar prominences). Weiss told the RAS that he had observed a truly continuous spectrum in his examination of the coronal light, a result which he admitted he could only report but not explain. The observation suggested to him another possible explanation for the corona, namely that it comprised a cloud of incandescent solid or liquid particles, a suggestion clearly at odds with current views on the Sun's temperature gradient.[48]

The corona was quickly proving to be a tangle of challenging puzzles for observers interested in the physics of the Sun. The observations made by the American eclipse teams during the 1869 eclipse complicated the matter even more. William Harkness (1837–1903) of the US Naval Observatory, for example, described the coronal spectrum as continuous – that is, showing no absorption lines – and crossed by one bright green line. Charles Augustus

Young of Dartmouth College also noted the presence of a green emission line in the coronal spectrum and determined its location at '1474' on Kirchhoff's scale (530.3 nm on modern scales), very close to one associated with terrestrial iron. If the prominences were predominately hydrogen, the lightest of the known elements, how could a layer of vaporous iron remain for very long above it? How could temperatures be sufficiently high to cause the incandescence of solid particles at such great distances from the solar surface? In general, the coronal spectrum observed in 1869 stretched the interpretive sophistication of spectroscopic experts beyond the limits for which their experience had prepared them.

Reports from this North American eclipse stimulated great interest in Britain – among professional and amateur astronomers alike – in the upcoming Mediterranean eclipse. Airy, for example, was impressed by the potential for spectroscopic observation to assist celestial mechanicians in determining the precise time of events such as the first contact of the leading edge of the Moon's shadow with the solar limb, events over which visual observers had disputed in the past. In early November 1869, Airy began to press Huggins to view the 1870 eclipse in Gibraltar.[49] Amateur interest may have been aroused by a popular article on the American observations of the 1869 eclipse contributed by William Crookes to the January 1870 number of the *Quarterly Journal of Science*. This article was accompanied by a remarkable full-colour illustration of totality.[50] The next month, a letter from 'A constant reader' appeared in the *Astronomical Register* asking if any expeditions to view the 1870 eclipse were being planned by the RAS. If so, the writer continued, 'I and certain friends of mine would proceed there.' He hinted at a potential windfall for the Peninsular and Oriental Company should they offer reduced fares on transport to Gibraltar since many others like himself would then be encouraged to make private arrangements to see the eclipse. 'A total solar eclipse is such a rare phenomenon', he pointed out, 'and as none can be seen in Great Britain in the lifetime of anyone now alive, I am sure many lovers of astronomy would wish to proceed where this treat can be obtained.'[51]

9.4 Planning the expedition

In March 1870, the Council of the RAS began to consider plans for an expedition.[52] The Royal Society had already formed a Solar Eclipse Committee in January. In April, the two societies joined forces to constitute a Joint Eclipse Committee (JEC) with each society contributing £250 to get things started. Much of the discussion at the RAS meeting in April 1870 centred on eclipse preparation and planning.[53] RAS President, William Lassell, announced that it seemed probable the Admiralty would furnish a ship to transport a British expeditionary team to a favourable eclipse site, perhaps Oran, Syracuse, or Jerez north of Cadiz. Considerable discussion followed on the relative merits of the proposed observation sites. Gibraltar was given a great deal of support by Airy and others with experience observing there during the winter months, although Airy warned that travel to the Mediterranean region through the Bay of Biscay was likely to be a treacherous experience and so might limit the numbers of skilled observers inclined to participate.

He was delegated to request assistance from the Admiralty in the form of two ships for transport, one to Spain and one to Sicily.[54] He did not relish the task. But, in recent years, the Admiralty had made ships available to eclipse expeditions, as well as other projects. And his enquiries made earlier in the spring had received an optimistic response, so there was some hope for success.[55]

Huggins and others offered their expert advice during the meeting. In order to observe both the prominences and the corona spectroscopically, Huggins argued that a total of five spectroscopes would be required at each observing station: 'two for the prominences, two for the corona, and one for the outer part of the latter light'. Three individuals would be required to operate each of the first four of these instruments to best advantage: one to observe through the instrument, one to keep the desired object on the slit of the spectroscope, and one to record. The instrument trained on the outer region of the corona would not require guidance, so only an observer and recorder would be necessary. Thus, each station, in Huggins's opinion, would need a minimum of fourteen spectroscopic observers![56]

Aware that there were many, like 'a constant reader' and his friends, who might be tempted to sign on to the Society's expedition to obtain a glimpse of the eclipse for their own pleasure and amusement, the discussion ended with the caveat that it be 'understood that no mere idle spectators were wanted in the expedition and that those who volunteered would take the parts assigned to them without grumbling'.[57] De la Rue declared that participants 'must remember they will have no time to look about'.[58] Indeed, just before the eclipse parties were scheduled to depart, Huggins wrote a last-minute letter to Airy to express his lingering concern for the intrusion of what he referred to as a whole 'host of men looking through a telescope for the first time'. In his opinion, the interests of science would best be served if each eclipse party were kept as a small, well-trained group. Airy tartly responded in agreement, 'I absolutely protest against taking hunters of the picturesque.'[59]

In April, a circular was distributed to all Fellows of the RAS. The response was overwhelming. The Assistant Secretary of the RAS, John Williams, sent Greenwich a list of over fifty aspirants interested in joining the eclipse expedition.[60] The JEC was left with the task of paring down the names. By June, Huggins, as Secretary of the RAS, sent Airy a memo listing principal observers in four categories. The photographic work would be organised by Warren De la Rue, John Browning and photographer Alfred Brothers (1826–1912) of Manchester. Spectroscopy would be in the hands of Huggins, Lockyer and chemist John Hall Gladstone (1827–1902). Polarisation observations would be conducted by Airy, Stokes and Pritchard. A general survey of the eclipse would be directed by Lassell, Strange and lens maker John Henry Dallmeyer (1830–83).[61]

But, as the summer progressed with no word from the Government concerning financial and material assistance, hopes for the expedition began to dim. In fact, they all but evaporated in mid-July when the French declared war on Prussia. The British were left to watch nervously on the sidelines after their eleventh-hour diplomatic efforts failed to broker a peaceful resolution to the impending hostilities. If Britain were drawn into the conflict, thousands of troops would have to be shipped to the Continent on a moment's notice.

In early August, when Airy received news that it was likely that the Admiralty would refuse the JEC's request, it looked as though all the preliminary planning had been for naught.[62] 'It seems to me doubtful if the exped can be carried out under the present political circumstances', Huggins wrote dolefully to Airy.[63] Indeed, the Astronomer Royal was already thinking it best, as he wrote to De la Rue, to 'abandon the whole scheme'.[64] De la Rue replied with an alternative suggestion:

I am inclined to think with you that a regular expedition is no longer possible; while, at the same time, if two or more sets of observers are still willing to go out & make specific observations I think that it would be legitimate to aid them out of the sums placed at the disposal of the Committee. The observations most needed are spectroscopic observations of the entities usually confounded with the corona – also of the corona proper at distances of 1′, 2′, 3′ from the moon's limb, also polariscope observations of the corona.[65]

A worried Huggins wrote to Stokes: 'I fear we shall get no assistance from the Admiralty for the Eclipse Exp. which perhaps will have to be given up. I have called a meeting of the Council R. Ast. S. on Friday, & will write to you after that day.' Following his signature, he added: 'Could you let me know if, in your opinion, no assistance from Government being obtained, part of the funds of R.S. voted for Exp. could be applied to assist in sending out any observers.'[66] He even tried to persuade Lockyer to attend this special meeting to decide 'what steps are to be taken under these circumstances'.[67]

But Lockyer was planning his own full-scale attack on the problem from the bully pulpit of his new weekly science journal, *Nature*.[68] 'A RUMOUR is current', he announced in the 18 August number, 'that the Government have refused both ships and assistance' for the planned expeditions. 'We can hardly believe that the Government will thus venture to brave the opinion of all men of science and culture ... [W]e are a nation of Philistines!'[69] On 8 September, the rumour confirmed, Lockyer loudly denounced the Government's 'astounding' decision. 'The American Government, more enlightened than our own, are making extensive preparations' to observe the eclipse firsthand. '[U]pon the results of their labours and those of the Continental Governments', he warned, 'Englishmen must therefore fall back, in a research which is eminently English.'[70] Two weeks later, he reminded his readers of the great advances in astronomical knowledge that had been made by participants in past British expeditions, all thanks to Government support.[71] The upcoming eclipse promised a rich opportunity to continue this important work. Passing it up, he cautioned, would 'bring shame upon the scientific repute of England'.

On 6 October, Lockyer ratcheted the debate up a notch or two in a scathing rebuke of the Government's refusal to aid 'astronomers anxious to observe so rare a phenomenon as a total eclipse'. In his view, the root cause of such short-sighted action lay in Britain's lack of a proper scientific mindset. By contrast, he declared, the 'Prussians, whatever their other qualities, are emphatically a scientific people'. Indeed, their recent military triumphs attest to the fact that 'the spirit of science possesses the entire nation'.[72] 'Where is our science?' he demanded.

To make his case, he pointed to the recent (many would say, scandalous) loss of the HMS *Captain* and crew off the coast of Cape Finisterre, Spain.[73] The masted turret ship was an experimental design 'built by an amateur in spite of the demonstration of our professional adviser that she must be unsafe'. With its deck too close to the waterline and its centre of gravity too high, the *Captain* blew over 'in the first gale'. Lockyer concluded, 'We fear it is a terrible truth that the absence of scientific method is as conspicuous with us as its presence is with the Germans.'

The next week, he advertised that the American Government had appropriated the munificent sum of £6,000 to support their eclipse expeditions. 'It will be in the recollection of our readers', he wryly added, 'that our own Government have refused to give either a single ship or a single shilling in aid of our own observations.'[74]

The situation continued to appear hopeless until 21 October when, at long last, the Government let it be known that the 'question might be reconsidered'.[75] Unfortunately, as De la Rue explained to the Chief Hydrographer, many who had originally expressed interest in participating back in the spring had made other plans. If only the Government had given more timely word that the question was still open, the expedition might have been saved. At this late date, in his view, it was better to simply 'let the matter drop'.[76]

Nevertheless, the matter did not drop. Lockyer, for his part, continued to embarrass the Government with exuberant reports of plans for the well-funded American expeditionary efforts.[77] His pen-and-paper threats of impending competition for scientific hegemony were becoming harder to ignore. Indeed, they appeared to be realised when, in late October, London fell victim to an 'invasion' by what Simon Newcomb later described as 'an army of American astronomers'.[78] Newcomb, then an astronomer and professor of mathematics at the US Naval Observatory, was one of those invaders. He and his colleagues from the Naval Observatory, as well as expeditionary teams under the direction of Benjamin Peirce (1809–80), Superintendent of the US Coastal Survey, were passing through London on their way to their posts in the Mediterranean.[79]

Peirce was the first to reach London and had already paid a visit to Huggins's Tulse Hill observatory by the time Newcomb arrived on 1 November.[80] Newcomb recalled being surprised to learn of the 'parsimony' of the British Government which had 'declined to make any provision for the observations of the eclipse'. His English colleagues hoped, since the eclipse 'was visible on their own side of the Atlantic', that 'their government might take a lesson from ours'. He offered an account of a meeting with Prime Minister William Ewart Gladstone (1809–98) in which Peirce 'expressed his regret that her Majesty's government had declined to take any measures to promote observations of the coming eclipse of the sun by British astronomers'. According to Newcomb, Gladstone replied he 'was not aware that the government had declined to take such measures' and assured Peirce 'that any application from our astronomers for aid in making these observations would receive respectful consideration'.[81]

Nevertheless, concerned that assistance might not be forthcoming, Peirce wrote to Lockyer inviting him and 'other eminent physicists of England' to join his expedition either to Spain or to Sicily.[82] Meanwhile, the JEC hastily prepared their own direct appeal in reply to the Government's recent overture. Now that it seemed likely neither France nor Germany would be able to send their own teams of observers, they 'hoped that Mr. Gladstone may take a more liberal view of the subject'.[83]

Unfortunately, the execution of this appeal became mired in a muddle of miscommunications and misdirected letters. Lockyer wrote a searing editorial in *Nature* blasting the JEC for its inept handling of the whole affair. He singled out Huggins – though not by name – whose inability to keep members of the Committee informed on matters shortly to come before them, was, in Lockyer's opinion, emblematic of the larger problem of mismanagement in the JEC. Lockyer concluded his tirade by pointing out, 'There is still ample time to organise an expedition which shall do much good work, though perhaps it is too late to send out and erect the largest class of instruments.'[84]

Finally, on 11 November, word was received that the Government would provide a ship after all to Spain, Gibraltar and Oran as well as £2,000 for the Sicilian expedition's land transportation. The news was announced at the RAS meeting that very evening. Newcomb later reflected, 'I suspect that our coming [to England], or at least the coming of Peirce, really did help them a great deal.'[85]

An anxious three weeks followed as individuals scurried to prepare themselves for the journey. On 6 December, Huggins, his eclipse party, and others planning to view the eclipse from sites en route, set sail from Portsmouth on the HMS *Urgent*, an aptly named ship, provided by the Government.[86] Used in the past to transport troops, the ship had been fitted for the expedition's voyage in just under a week.

Known by those who had sailed her before for a propensity to roll violently, the *Urgent* did not disappoint. Thanks to several members of the party who kept diaries of their voyage, we know that, just as Airy had feared, a fierce storm at sea caused serious damage to the ship and tossed its not-too-seaworthy occupants about their cabins. Based on his journal, William Crookes composed a vivid account of the 'Gale in the Bay of Biscay' for *The Times* of London.[87] 'The day after we started the wind freshened, the sea rose, and about noon, when the Bay of Biscay was entered, the motion of the ship was such as to send below all who were not good sailors.' When the ship reached Cape Finisterre on 9 December, 'there was a general sigh of relief at seeing the southern extremity of the much dreaded Bay of Biscay'. Unfortunately, it was there they encountered a real gale. The rolling of the ship became 'so excessive that even practised sailors could with difficulty keep their feet'. The passengers found themselves worrying their ship might 'join the ill-fated [HMS] *Captain* at the bottom of the Atlantic'. In the middle of the night, Crookes 'dressed and went on deck to observe the extent and appearance of the storm'. Overcome by 'the spirit of investigation', he found a 'suitable post of observation' from which he attempted to gauge the roll of the ship, the height of the waves and their spacing from crest to crest.

Another diarist, Crookes's colleague physicist John Tyndall (1820–93), had joined the expedition as 'a general stargazer'. He too 'wished to see the storm' and climbed onto deck to observe and document its fury.[88] By contrast, fellow passenger Mary Caroline Hassler Newcomb (1840–1921), who was accompanying her husband to Gibraltar, spent almost the entire trip in her quarters.[89] On 7 December, one day after the *Urgent* put to sea, she wrote, '[I tried to go below] grasping the rail, but it broke and flung me to the bottom. I was up in a minute but my head ached severely in consequence.' She 'kept in bed' the next day. On 9 December, 'I was up a little, but worse.' Finally, on 10 December, she 'Had so much pain and was so tired of the confinement ... Really three miserable days ...'

Following stops at Cadiz (12 December) and Gibraltar (13 December) to drop off the other two observing parties, Huggins and his group continued on to Oran, which port they reached on 16 December.[90] The next day, Huggins and a few of the group went ashore to locate an appropriate site for the eclipse observatory. They found an ideal spot near the railway station which provided them with ready access to telegraph services. Stormy weather raised concern over the prospects for the eclipse, but in a few days the weather cleared giving the observers an opportunity to work with their instruments.

9.5 A registering spectroscope

Crookes complained that his spectroscope had been 'roughly put together' and his telescope, a 4½-in Grubb refractor, had no clockwork. To ready his instrument for the spectroscopic observations he was to make during the eclipse, he removed the telescope's eyepiece
and inserted the spectroscope. This spectroscope was structured somewhat like Lockyer's
sunspot spectroscope, an instrument which had become more or less standard for solar
spectroscopists by that time. But it had one important new feature: it had been specially
designed by Huggins to allow its user to register the observed positions of spectral lines
automatically.

Reports from observers of recent eclipses impressed solar spectroscopists with the need to
automate the process of recording the locations of spectral lines during totality. Much time
was routinely lost removing one's eye from the eyepiece of the spectroscope to note the
reading on the head of the micrometer screw as each spectral line passed over the cross-
hairs. The eye had to accommodate each time from distance to near vision in order to make
an accurate reading. In addition, viewing a faint spectrum in the spectroscope required a
satisfactory level of dark adaptation. Unfortunately, reading the micrometer exposed the
observer's eyes to light. During a normal spectroscopic observation, an object could be kept
in view for an extended period of time with the principal hazard being eye fatigue from
repeatedly going back and forth between the eyepiece and the micrometer. There is no such
leisure during eclipses. Not only that, but with the corona as the object of prime interest, the
spectrum to be observed was expected to be extremely faint. Uninterrupted dark adaptation
was an absolute necessity to heighten the observer's visual acuity and thus ensure that every
detail possible was recorded.

Just days before the eclipse, Lieutenant A. B. Brown, RA, a member of Lord Lindsay's
party [James Ludovic Lindsay, 26th Earl of Crawford and 9th Earl of Balcarres
(1847–1913)], a privately sponsored expedition team that planned to observe the eclipse
from Maria Luisa Observatory in Cadiz, became concerned about the need for some way to
make an automatic record of spectroscopically observed phenomena during totality. Brown
immediately 'set to work to construct a recorder, and fitted it to the spectroscope, which, as
will be hereafter seen, worked very well. Indeed, I know of no other mode of measuring the
position of lines quickly that I would adopt during an eclipse.'[91] Unfortunately, Brown
provided little information in his eclipse report as to his recorder's design, or method of use.
The only clue given suggests that he generated a comparison strip on which were marked the
principal Fraunhofer lines. This strip was positioned in such a way as to be visible in the
spectroscope. As Brown described it, following first contact and the occultation of a number
of sunspots, 'I fitted and adjusted my spectroscope, and carefully recorded the position of the
solar lines . . . both above and below the spaces left for corona and prominence observations
by means of my automatic recording arrangement.'[92] During totality, Brown observed only
a faint continuous spectrum with no absorption lines or emission lines despite looking
especially hard for signs of Young's green '1474' line.

Huggins had also anticipated the special challenge of spectroscopic observation under the
pressure of a total eclipse, but he had done his thinking about the problem before he left
England and so was able to provide the spectroscopists on his eclipse team with a fairly

sophisticated automatic recording device attached to their instruments. On consultation with Howard Grubb, Huggins contrived an ingenious but simple device capable of permitting the observer to note the position of each spectral line observed without ever removing his eye from the eyepiece.[93] Crookes, who spent considerable time before the eclipse practising with this new device, described it in this way: 'At the eye end of the spectroscope is an arrangement for rapidly bringing a pointer on to any line in the spectrum and pricking its position on a card.'[94]

Huggins's practical observing experience told in the design of this recording instrument. There were two needles that could be operated separately, allowing the observer to indicate if the particular line being noted was an emission line (by pressing only one needle) or an absorption line (by pressing both needles). Each card was made so that five separate spectra could be recorded in succession before replacing it with a new one. This device was later improved upon by others,[95] so Huggins's initial contribution has been somewhat obscured over time, particularly as he and his party were unable to put the instrument to good use during the 1870 eclipse.

9.6 22 December 1870

The day of the eclipse, each observer set up his equipment and readied himself for the rush of totality. Crookes again practised using the recording device on his spectroscope. He consulted with Huggins and Revd Howlett (who would be guiding Crookes's telescope during the eclipse) as to just how they would proceed with the coronal observations during the little more than two minutes they would have of totality. The aim was to record three critical views of the coronal spectrum: placing the spectroscope radially on the trailing limb of the Moon (just before totality), placing the spectroscope tangential to the Moon's limb in the lowest region of the corona (during totality), and finally aiming the spectroscope at some region of the corona some distance away from the Moon in order to get a view of the streamers (once again during totality).

Crookes planned to cover his eyes just before totality to increase his light sensitivity. During these important coronal observations, he was to be on the alert for Fraunhofer lines, extra black lines, and the bright green '1474' line noted by Young and Harkness. Like modern Olympian lugers, these men committed their motions to autonomic kinesthetic memory. Timing, coordination and precision were everything.[96] At around 11 a.m., Crookes spotted the small bite out of the solar disc which signalled the start of the eclipse. Unfortunately, it also signalled the start of an invasion of clouds which obliterated all hope of making any but the most general observations!

Their painstaking practice all for naught, Huggins and the other observers reluctantly left their instruments and concentrated instead on gross changes in light and colour in their surroundings. Shortly after totality, Crookes and Captain William Noble (1828–1904) sent telegrams to the *Daily News* and *The Times* of London with the report of their failure. 'As soon as the telegrams had been sent', Crookes wrote in his diary, 'we returned to the tents, and after standing for one or two photographs commenced to take the instruments down and pack them up, for after our failure but one idea seemed to possess all, and that was to get away from Oran and on our homeward voyage as quickly as possible.'[97]

Lockyer had no better luck with the weather at his station in Catania, Sicily. But a number of others had more success including Charles Young, who reported viewing, for the first time, the bright lines of the Sun's reversing layer.[98] Photographs showing the corona were obtained from Cadiz by Lord Lindsay; from Jerez de la Frontera by O. H. Willard, a photographer from Philadelphia and member of the eclipse party mounted by Harvard Observatory Director, Joseph Winlock; and from Syracuse by Alfred Brothers.

Unfortunately, the observations made during the eclipse of 1870 by those individuals blessed with clear skies during totality contributed but little to solving the corona puzzle. For one thing, the increasing numbers of people reporting their observations of the same celestial event, and the disparity of their representations, brought into greater relief the risks of relying on visual observations alone. The instrumentation, the interpretive skills of the fledgling corps of astronomical spectroscopists, and the photographic capabilities available at the time were simply not up to the task of recording the faint corona. Nevertheless, speculation as to its nature and structure flourished, and encouraged the development of improved mechanical observing and recording techniques.[99]

In the final analysis, the Report of the Council to the RAS in February 1871 contended that:

Although the gain to our knowledge of Solar physics is much less full and decided than doubtless it would have been if the very efficiently equipped and competent observers had been favoured with a cloudless sky, the new information which comes to us from the eclipse is very valuable, and well repays the large amount of thought, time, and money, which were so freely bestowed upon the preparations.[100]

However, when it came time to provide support for a British team to observe the eclipse in India in December 1871, the RAS committee appointed to weigh the question showed considerably less enthusiasm. Lockyer, the only one to continue to press for support, finally received what he wanted through the intervention of the British Association.[101]

***

Organising eclipse expeditions was a risky business that required a great deal of money and ambition. Ambition was never in short supply, but few individuals had the money to mount their own expeditions. Huggins certainly did not. In 1870, he planned an expedition and helped organise a band of fellow eclipse enthusiasts to urge the Government to underwrite it. But even the Government could not control the weather during the eclipse. When clouds prevented him and his eclipse party from making their planned observations, he was left to wonder at the efficacy of expending so much capital on such hit-or-miss ventures.

Indeed, soon afterward, as we saw in Chapter 8, he took the side of those who questioned Government support for the running of national astrophysical laboratories. He never participated in another eclipse expedition. In fact, his interest in solar observation returned to its old occasional pace, leaving the frustration he had experienced in the wake of his clouded-out eclipse to ferment quietly and creatively for the rest of the 1870s. When his interest in solar eclipses was rekindled in 1882 (see Chapter 11), he came back to it hopeful and ready to become actively engaged, not in organising a new eclipse expedition, but in developing a new and innovative way to gather the same information on a daily basis without having to invest everything in a risky venture.

Notes

1. William Crookes, cited in d'Albe, *Life of William Crookes* (1923), p. 161.
2. W. Huggins, 'The new astronomy' (1897), pp. 907–29.
3. A notable exception is A. J. Meadows in *Early Solar Physics* (1970). Although Meadows chose not to include any of Huggins's papers in the appended collection of original papers on solar research, Chapter 2 on 'The new astronomy' contains numerous references to his solar work ranging from early examination of sunspots through efforts to photograph the solar corona without an eclipse. Meadows also discusses Huggins's role as one of the first to observe solar prominences using a spectroscope in *Science and Controversy* (1972), pp. 54–7. Anstis briefly discusses Huggins's work on the solar corona (see Anstis, 'Scientific work of William Huggins' (1961), pp. 194–6). Hufbauer lists Huggins as one of a handful of early pioneers in solar physics, but provides no further information on his contributions (see *Exploring the Sun* (1991), p. 60). Pang omits the plans and efforts of Huggins's solar eclipse team which was dispatched to Oran in 1870, perhaps because the expedition was confounded by clouds. See *Spheres of Interest* (1991) and *Empire and the Sun* (2002).
4. Tennant, 'On the solar eclipse of 1868, August 17' (1867).
5. Tennant, 'On the eclipse of August 1868' (1867).
6. *Ibid.*
7. J. F. W. Herschel to Stokes, 5 May 1867, in Larmor, *Memoir and Scientific Correspondence*, vol. 1 (1971), p. 211.
8. W. Huggins to Stokes, 15 May 1867, Add MS 7656.H1112, GGS, CUL.
9. Schellen, *Spectrum Analysis* (1871), pp. 308–31.
10. H. Vogel, *The Chemistry of Light and Photography* (1883), pp. 171–2. A lengthy excerpt from Vogel's initial report appeared soon after the eclipse in 'Spectrum observations connected with the recent eclipse of the Sun', *CN* **18** (1869), p. 228.
11. *Ibid.*, p. 174.
12. *Ibid.*, pp. 172–3.
13. W. Huggins, 'Visitors to the Observatory', Notebook 2, WCL/SC.
14. Sabine, 'President's address' (1868), p. 138.
15. *Ibid.*, pp. 138–9.
16. J. Herschel, 'Account of the solar eclipse of 1868, as seen at Jamkhandi in the Bombay Presidency' (1868), p. 111.
17. *Ibid.*, p. 112.
18. In fact, Herschel later discovered that it was not uncommon for the region to experience a two-week long 'cloudy season' related to the south-west monsoon. It was the timing that was unusual.
19. W. [M.] Campbell, 'Lieut. Campbell's report' (1868).
20. Tennant, 'On the eclipse of August 1868' (1867), p. 177.
21. Tennant, 'Report of the observations of the total solar eclipse of August 1868' (1869), p. 36.
22. Janssen later wrote: 'The decisive moment drew near, and we waited for it with great anxiety. This however, did not affect our intellectual powers: they were rather over-excited, and this feeling was amply justified by the grandeur of the phenomena nature had prepared for us, and by the knowledge that the fruits of our great preparations and a long voyage depended entirely upon the observation of some moments' duration.' See Janssen, 'Summary of some of the results obtained during the total solar eclipse of August 1868, Part I' (1869), pp. 108–9.
23. Herschel was in a very excited frame of mind during this eclipse. The anticipation of the great responsibility he had assumed, the thrill of the first moments of totality and the intermittent cloud cover which threatened the success of the entire operation resulted in measures being made of only three of the bright lines he observed in the solar prominence. See his remarks in 'The great eclipse of the Sun, I', *The Engineer* **6** (1868), pp. 345–6.
24. Lieutenant John Herschel to J. F. W. Herschel, 2 September 1868, cited in J. F. W. Herschel, 'On a possible method of viewing the red flames without an eclipse' (1868).
25. Janssen to the Permanent Secretary of the Académie des Sciences, 19 September 1868, translated and reprinted in Meadows, *Early Solar Physics* (1970), pp. 117–18; Janssen, 'Summary of some of the results obtained during the total solar eclipse of August 1868, Part II' (1869), p. 131.
26. W. Huggins, 'Mr. Huggins' observatory' (1868).
27. Recall Huggins's first attempt to observe the solar prominences out of eclipse on 10 November 1866. At that time he remarked that he was trying to accomplish this feat using a new method that he had been thinking about for several weeks. Given his phrasing, there can be no question that it describes the beginning of his efforts on this particular project.

28. Recall Lockyer's enthusiastic suggestion, made in October 1866, that the spectroscope might 'afford us evidence of the existence of the "red flames" which total eclipses have revealed to us in the sun's atmosphere' and which 'escape all other methods of observation at other times'. J. N. Lockyer, 'Spectroscopic observations of the Sun' (1866), p. 258.

29. 'Account of the appropriation of the sum of £1000 annually voted by parliament to the Royal Society (the Government Grant), to be employed in aiding the advancement of science', *PRS* **19** (1870), pp. 135–45; p. 142.

30. Meadows provides a good discussion of this sequence of events although he mistakenly ascribes Lockyer's change of instrument makers to Thomas Cooke's death. Cooke died on 19 October 1868, months after the project had been handed over to Browning, and ironically the very day Lockyer first succeeded in seeing the prominences out of eclipse with his new instrument fashioned by Browning. See Meadows, *Science and Controversy* (1972), pp. 52–3; C. Pritchard, 'Thomas Cooke' (1869).

31. 'Spectroscopic observations of the "red prominences" without an eclipse of the Sun', in 'Report of the Council', *MNRAS* **29** (1869), pp. 162–3.

32. Meadows, *Science and Controversy* (1972), pp. 54–5. As Meadows points out, this information may have been critical since Lockyer had recently begun to express doubt in the gaseous nature of the prominences. See Stewart and J. N. Lockyer, 'The Sun as a type of the material universe, I and II' (1868), p. 254.

33. Lockyer made the important discovery that the bright yellow line attributed by some observers to the close pair of solar D lines (associated with sodium) was not in fact coincident with them. After considerable study, he later contended that this line, the so-called D_3 line, was produced by an as yet undiscovered element. Convinced of his assertion, he named the unknown element, choosing to call it 'helium' because the source of the D_3 spectral line seemed abundant in the Sun. Its presence had thus far escaped detection in terrestrial samples, however. As we shall see in Chapter 14, helium was identified years later by the chemist William Ramsay as a natural emanation from the mineral clèveite. Lockyer also announced the discovery of a new thin layer of solar atmosphere which he called the 'chromosphere' for its bright pink colour. Both of these claims aroused considerable discussion.

34. As mentioned earlier, Lockyer – though placed in nomination for the Royal Astronomical Society's prestigious Gold Medal in 1871 for his work in solar physics – failed to receive the three-quarters majority required.

35. W. Huggins, 19 December 1868, Notebook 2, WCL/SC.

36. De la Rue to Stokes, 23 November 1868, Add MS 7656.D200, GGS, CUL.

37. Clerke, *History of Astronomy* (1885), p. 215.

38. For Lockyer's remarks, see the published notes from the animated discussion on solar prominence observation which took place at the 13 November 1868 RAS meeting following William Huggins's presentation: 'On a possible method of viewing the red flames without an eclipse' in *AR* **6** (1868), pp. 263–5; p. 264.

39. *Ibid.*, p. 265.

40. W. Huggins, 'On a possible method of viewing the red flames without an eclipse' (1868).

41. See discussion following Lockyer, 'Note on a paper by Mr. Huggins', *AR* **7** (1869), p. 42.

42. Consider this excerpt from a letter from Lockyer to Stokes in August 1869: 'As I know how far & wide what you say will be read I take the liberty of sending you a *tirage à part* [he meant "*tiré à part*", an offprint] of one of my papers, showing that Hydrogen was not so obviously indicated by the Eclipse observers as Mr. Huggins stated it to be when he attempted to take a good part of the credit due to Janssen & myself working without an eclipse away from us.' J. N. Lockyer to Stokes, 14 August 1869, Add MS 7656.L558, GGS, CUL.

43. Huggins's announcement of his intention to publish a paper on this topic prompted Edward Stone to remark that he had begun a similar programme of research. Stone was unaware of Huggins's earlier efforts and had, in fact, recently purchased some expensive apparatus to carry it out. See W. Huggins, 'Note on the heat of the stars' (1869); Stone, 'Approximate determinations of the heating-powers of Arcturus and α Lyrae' (1869).

44. Here again, Huggins exaggerates somewhat in dating his work on solar prominences by inferring it took place somewhere between 1865 and 1866 rather than at the end of 1866. See De la Rue, 'On some attempts to render the luminous prominences of the Sun visible without the use of the spectroscope' (1869); W. Huggins, 'Note on Mr. De La Rue's Paper' (1869).

45. 'Meeting of the Royal Astronomical Society, January 8, 1869', *AR* **7** (1869), pp. 35–45; p. 44. The suggestion that prominences change with great rapidity was soon withdrawn, however, when it was discovered that the changes observed were due to faulty alignment of sequential eclipse photographs.

46. 'Spectroscopic observations of the Sun. I', *AR* **7** (1869), pp. 193–6.

47. 'Well, in our country,' said Alice, still panting a little, 'you'd generally get to somewhere else – if you ran very fast for a long time as we've been doing.' 'A slow sort of country!' said the [Red] Queen. 'Now, here, you see, it takes all the running you can do, to keep in the same place. If you want to get somewhere else, you must run at least twice as fast as that!' Carroll [Dodgson], *Through the Looking-glass* (1946), p. 32.

48. Weiss, 'Presentation of results of 1868 solar eclipse expedition' (1869), pp. 100–3.
49. Airy to W. Huggins, 5 November 1869, RGO 6.271/7/2/303, CUL.
50. Crookes 'The total eclipse of August last' (1870), p. 29.
51. 'A constant reader', February 1870, Letter to the Editor, *AR* **8** (1870), pp. 58–9.
52. Dreyer and Turner, *History of the Royal Astronomical Society* (1987), p. 169.
53. 'Discussion of the plans for the eclipse expedition in December 1870', *AR* **8** (1870), pp. 99–105.
54. *Ibid.*
55. In 1860, the Admiralty furnished the *Himalaya* to transport De la Rue, Airy and others to the eclipse in Rivabellosa, Spain. See R. W. Smith, 'The heavens recorded' (1981), p. 7; Pang, *Spheres of Interest* (1991), pp. 28–30. In 1868, the *Lightning*, a paddle steamer, was made available to William Benjamin Carpenter (1813–85) and Charles Wyville Thomson (1830–82) by the Admiralty for a deep-sea dredging project in the North Atlantic. In 1869, the *Porcupine* was provided to them for the collection of biological specimens off the coast of Ireland and Shetland. See M. B. Hall, *All Scientists Now* (1984), p. 175.
56. 'Discussion of the plans for the eclipse expedition in December 1870', *AR* **8** (1870), p. 103.
57. For a discussion of the social aspects of eclipse expeditions and their planning, see Pang, *Spheres of Interest* (1991), Chapter 3, pp. 85–146.
58. 'Discussion of the plans for the eclipse expedition in December 1870', *AR* **8** (1870), p. 104.
59. W. Huggins to Airy and Airy to W. Huggins, 19 November 1870, RGO 6.132/5/130–1, CUL. Airy had used nearly the same words ten years earlier to ward off a would-be eclipse observer, one Joseph Beck, who wished to accompany De la Rue on his expedition to Spain in 1860. See Pang, *Spheres of Interest* (1991), p. 29.
60. J. Williams to E. J. Stone, 19 May 1870, RGO 6.131/1/14, CUL.
61. W. Huggins to Airy, 23 June 1870, RGO 6.131/1/38, CUL.
62. Admiral G. H. Richards to Airy, 4 August 1870; cited in Pang, *Spheres of Interest* (1991), p. 38.
63. W. Huggins to Airy, 5 August 1870, RGO 6.131/2/52–3, CUL.
64. Airy to De la Rue, 5 August 1870, RGO 6.131/2/51, CUL.
65. De la Rue to Airy, 6 August 1870, RGO 6.131/2/54–5, CUL.
66. W. Huggins to Stokes, 16 August 1870, Add MS 7656.H1124, GGS, CUL.
67. W. Huggins to Lockyer, 16 August 1870, JNL, UEL.
68. The first number of *Nature* was published on 4 November 1869.
69. 'Notes', *Nature* **2** (1870), p. 320.
70. 'Notes', *Nature* **2** (1870), p. 379.
71. 'The Government and the eclipse expedition', *Nature* **2** (1870), pp. 409–10.
72. 'Scientific administration', *Nature* **2** (1870), p. 449.
73. The HMS *Captain*, designed by Captain Cowper Phipps Coles (1819–70), foundered and sank shortly after midnight on 7 September 1870. Only seventeen of the nearly 500 aboard survived. News stories about the disaster filled the pages of *The Times* of London for weeks, particularly as discussion heated up over its cause. For a sampling of these articles, see the following from *The Times* of London: 'Loss of the Captain', 10 September 1870, p. 5d; 'Her Majesty's Ship Captain', 12 September 1870, p. 12a; 'Her Majesty's Ship Captain', 13 September 1870, p. 3c; 'Captain Coles and Her Majesty's Ship Captain', 14 September 1870, p. 4e; 'The loss of Her Majesty's Ship Captain', 14 September 1870, p. 9f; and Sherard Osborn, 'Loss of Her Majesty's Ship Captain', 14 September 1870, p. 10b.
74. 'Notes', *Nature* **2** (1870), p. 475.
75. Airy to the Editor of the *Daily News*, 14 November 1870, published in *AR*, **8** (1870), pp. 257–8; p. 257.
76. De la Rue to Airy, 25 October 1870, RGO 6.131/2/65, CUL.
77. 'The American Government eclipse expedition', *Nature* **2** (1870), p. 517 and 'Notes', *Nature* **2** (1870), p. 520.
78. It is important to keep in mind that Newcomb was recalling these events nearly thirty years after the fact. S. Newcomb, 'Reminiscences of an astronomer. I' (1898), p. 245. The essays in this series were collected and reprinted five years later in *The Reminiscences of an Astronomer* (1903). See p. 274.
79. Newcomb was persuaded by Airy to establish his observation station at Gibraltar. Three others from the Naval Observatory – Asaph Hall, William Harkness and John Robie Eastman (1836–1913) – were assigned to observe the eclipse from Syracuse. Peirce was also headed to Sicily. He observed the eclipse near Catania. The other team from the US Coastal Survey included director of the Harvard Observatory, Joseph Winlock (1826–75) and Charles Young from Dartmouth who observed the eclipse from Jerez, Spain.
80. 'Profr Peirce of Washington' and 'Mr. Lassell' are listed as visitors to the observatory on 31 October 1870. Newcomb visited on 7 November. Charles Young and Samuel Pierpont Langley (1834–1906), then director of Allegheny Observatory, are recorded as having called on Huggins on 23 November.
81. Newcomb, 'Reminiscences' (1898), pp. 245–6; *Reminiscences* (1903), pp. 274–7.

82. 'Notes', *Nature* **2** (1870), p. 520.

83. *Ibid.*

84. 'Notes', *Nature* **3** (1870), p. 13. See also Lockyer's tirade written after the Government agreed to provide assistance to the expedition: 'Notes', *Nature* **3** (1870), pp. 52–3.

85. Newcomb, 'Reminiscences' (1898), p. 245; *Reminiscences* (1903), p. 274.

86. The HMS *Urgent* carried three eclipse parties. The first, headed by the Revd Stephen Joseph Perry (1833–89) planned to view the eclipse at Cadiz, Spain. A second party, headed by Captain Robert Mann Parsons (1829–97) was going to Gibraltar. The third party, led by Huggins (including Admiral Erasmus Ommaney, John Tyndall, Revd Frederick Howlett, James Carpenter, Captain William Noble, John H. Gladstone, William Crookes and others) was travelling on to Oran in Algeria. The Sicilian group headed by Lockyer planned to travel over land to Naples where they would then proceed by ship to Sicily. See 'English Government Eclipse Expedition', *Nature* **3** (1870), pp. 87–9.

87. For excerpts from Crookes's diary, see d'Albe, *Life of William Crookes* (1923), pp. 135–73. Crookes's article in *The Times* of London, 'A gale in the Bay of Biscay', was attributed to 'A Scientific Correspondent' and printed on 18 January 1871, p. 7a. It was later condensed and illustrated in *The Graphic, An Illustrated Weekly Newspaper*, 18 March 1871, pp. 255–7.

88. J. Tyndall, 'Voyage to Algeria to observe the eclipse. 1870' (1897), pp. 142–74,

89. Diary of Mrs Newcomb, Simon Newcomb papers, Box 2, Library of Congress. [Box 1 contains a partial typewritten transcript of this particular diary which documents her first trip to Europe.]

90. Crookes recounted a humorous anecdote about Huggins to make the point that Huggins was vulnerable to visual suggestion. As the *Urgent* approached what the crew believed to be the city of Cadiz, Huggins took out a small telescope and observed a pilot ship approaching them. Crookes wrote: '"There's Lord Lindsay," cried Huggins . . . The man came alongside, when, instead of Lord Lindsay, he turned out to be a seedy-looking pilot who could not speak English.' See d'Albe, *Life of William Crookes* (1923), p. 151.

91. Brown, 'Report of observations of the solar eclipse' (1871), p. 52.

92. *Ibid.*, p. 54.

93. W. Huggins, 'On a registering spectroscope' (1871).

94. d'Albe, *Life of William Crookes* (1923), p. 156.

95. John W. Draper, an American physician and amateur astronomer, came up with an idea for registering spectral lines which he believed was, as he described it, 'infinitely beyond' the card pricking system devised by Huggins. The elder Draper strongly suggested to his son, Henry, that he develop this invention quickly since he considered '*this the most important improvement in the spectroscope that has yet been made*'. John William Draper to Henry Draper, 15 August 1875, HD, NYPL. I wish to thank Howard Plotkin for bringing this letter to my attention.

96. This sort of preparation tempered natural excitability during totality and united human control with instrumental capabilities for optimum performance. In an article on the 1869 solar eclipse in Burlington, Iowa, Crookes himself had recounted the testimony of Dr Mayer, a photographer from Philadelphia: 'I had nothing but an *instrumental* consciousness, for I was nothing but part of the telescope, and all my being was in the work I had to perform.' Unfortunately, Mayer's efficiency left him with 50 seconds of totality to spare and no more photographic plates at the ready! See Crookes, 'The total solar eclipse of August last' (1870), p. 43.

97. d'Albe, *Life of William Crookes* (1923), p. 161.

98. The so-called solar 'reversing layer' had been predicted, but not observed. Theoretically, the layer of the Sun's upper atmosphere which produces the dark Fraunhofer lines in sunlight coming from the solar disc should produce an emission spectrum which precisely mirrors the Sun's absorption spectrum when observed spectroscopically at the limb of the Sun. The brilliance of the solar disc and the thinness of this layer contribute to the difficulty in seeing it. During the 1870 eclipse, for example, Young reported having seen the reversing layer's bright-line spectrum for only two seconds.

99. For a summary and discussion of the range of speculations being put forth shortly after the 1870 eclipse, see Proctor, 'Theoretical considerations respecting the corona, Parts I and II' (1871).

100. 'The total solar eclipse, Dec. 22, 1870', Report of the Council, *MNRAS* **31** (1871), pp. 111–17; p. 113.

101. For a discussion of the events leading up to the eclipse of 1871, see Meadows, *Science and Controversy* (1972), pp. 68–9.

10

An able assistant

... I had the great happiness of having secured an able and enthusiastic
assistant, by my marriage in 1875.

– William Huggins[1]

With each passing year, Huggins became increasingly involved in observations
requiring assistance. Until the untimely death in 1870 of his neighbour, chemist
William Allen Miller, Huggins relied on him to confirm important telescopic observations and assist in spectroscopic comparisons. On occasion, he invited others to
work with him at Tulse Hill.[2] But he could not long continue as a solitary observer if
he wished to maintain his position on the cutting edge of research in astronomical
physics.

By the mid-1870s, he faced a growing field of able competitors in London and abroad,
who vied with him for the same prize discoveries: to decipher the spectral code of the
nebulae, to reduce the varieties of stellar spectra to a seemly and sensible order, to bring the
full potential of the spectroscope's analytic power to bear on the solar surface and its
immediate environs, and/or to be the first to observe some new as yet unimagined celestial
phenomenon. He had already experienced a loss of priority to Lockyer's and Janssen's
independent claims to have found a spectroscopic method for viewing solar prominences
without an eclipse. He would have to work hard to ensure that he did not lose such an
opportunity again.

Another difficulty gradually arose as astronomical photography became an accepted,
even expected, part of the serious amateur's toolkit. All of Huggins's observations, including his work on stellar motion in the line of sight, were made visually. To remain a leader, he
would need to incorporate photography in his research and adapt his methods and
instruments to the idiosyncrasies and special needs of photographic celestial spectroscopy.

It is in the context of these critical career challenges that we meet Margaret Lindsay
Murray, a woman nearly a quarter century younger than William Huggins, and in whom he
found both a lifelong and devoted companion as well as an interested and capable collaborator. As we shall see later in this chapter, Margaret was more than an able assistant,
amanuensis and illustrator, whose work conformed to her husband's research interests. Her
very presence and expertise not only strengthened but also shaped the research agenda of the
Tulse Hill observatory. Both their observatory notebooks and their extensive
correspondence with scientists around the world vividly describe on-going daily activity

in the Hugginses' laboratory and observatory, and make possible, for the first time, a more definitive assessment of Margaret's role in the work at Tulse Hill.

10.1 The solitary observer

Recall that after the sale of his 8-in Alvan Clark refractor in 1869, Huggins was without a large telescope for well over a year.[3] While waiting for his new instruments to arrive, he devoted time to other projects, but these activities did not prevent him from worrying a great deal about the personal responsibility he had assumed in accepting custody of the Royal Society's instruments. He probably worried about increased competition, too, as the circle of observers trying their hand at celestial spectroscopy widened to include novices seeking guidance, but lacking the necessary personal connections to obtain it informally. In March 1870, for example, 'W. P.' wrote to the *Register* suggesting an article be written on 'Spectroscope construction' to 'enable an amateur like myself, by the purchase of suitable prisms, and the exercise of a little ingenuity, to construct a simple spectroscope suitable for a telescope of, say, from two to three inches aperture'.[4] About the same time, instrument maker William Ladd (1815–85) began selling coloured representations of the solar spectrum showing as many as 500 absorption lines visible with the aid of a magnifying glass for just 7s. 6d.[5]

Huggins knew from his own experience that it was possible for novices like W. P. to make noteworthy contributions to this fledgling scientific enterprise. Maintaining his advantage would become less problematic once the Grubb telescope was in place. With his 8-in telescope, Huggins had found that the feeble light of a nebula faded to invisibility when dispersed by the power of a highly refractive prism. He anticipated that the 15-in, with nearly four times the light-gathering power of his old telescope, would free him to use prisms of greater dispersive power for observing such faint objects.

Howard Grubb designed a unique spectroscope for him (see Figure 10.1). It consisted of four compound prisms (2, 3, 4 and 5) and two semi-compound prisms (1 and 6) contained in an airtight box and arranged so that, after traversing the entire system once, an incoming light beam would be reflected back again (at prism 7), thereby doubling the instrument's effective dispersion to the unheard of equivalent of ten compound prisms, or 90°![6] But Grubb knew that dispersive power was not everything, particularly when extremely faint light sources were being analysed. So he made each individual component removable to give Huggins control over the spectroscope's total dispersion. He also incorporated an innovation recently introduced by John Browning that made it possible to adjust the system automatically to the minimum dispersion angle of any selected wavelength using a system of levers attached to fixed centres at the back of the spectroscope. Amateurs with 2-, 3- or even 8-in aperture telescopes would be no match for Huggins and his new apparatus.[7]

Nevertheless, Huggins's enthusiasm for receiving these fine instruments was tempered by the recognition that taking possession of them meant increased responsibility and challenge to stay in the forefront of research in the new astronomy. He needed an assistant! There was always Miller, of course, whenever something new presented itself, instrumentation required an extra hand, or an observation needed confirmation.[8] Unfortunately, on

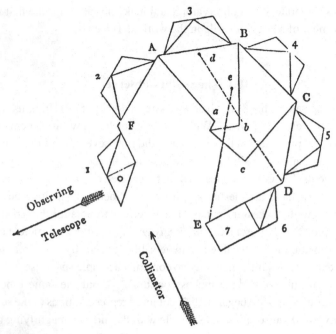

Figure 10.1 Grubb's automatic spectroscope. (*MNRAS* **31**, p. 37)

30 September 1870, Miller died unexpectedly at the age of fifty-two after collapsing on his way to the BAAS meeting in Liverpool.[9] Thus Huggins lost the trusted hands and eyes of his principal collaborator just months before Grubb installed the new instruments.[10] Because the public record provides no clues to the degree of his reliance on Miller's assistance after 1864, no explanation has been sought heretofore, or even thought necessary to account for the fact that when Huggins resumed his astronomical research in February 1871 he did so virtually alone with only occasional visits from others.

Huggins may simply have preferred his ad-hoc system of intermittent invitational assistance over engaging someone to help him on a regular basis. But it is more likely that it was the best he could do given his personal circumstances. Recall for a moment his concern, confided in a letter to Robinson, that hiring an assistant would 'cripple' him and limit his observational opportunities. Such concern reflected both his financial constraints and his wish to maintain his independence.[11] Around 1870, Robinson was paying his own assistant at the Armagh Observatory an annual salary of £100.[12] This was comparable to the salary paid to the lowest echelon assistants at Greenwich.[13]

There is no direct information on Huggins's source of income following the sale of the family silk business, although there are occasional references in his correspondence to a reliance on rents. He did not consider himself rich. We may estimate that he took in approximately £500 per annum given the home he lived in and the fact that he employed two live-in servants.[14] A paid assistant would be a sizeable expense, and one he felt he could not afford. In time, however, he would have to balance his reluctance to engage an assistant against the growing pressures inherent in assuming custody of the Royal Society's telescopes.

One pressure Huggins soon felt was a sense of accountability to the RS. Where his notebook entries had become somewhat more detailed following his election to Fellowship in the Royal Society in 1865, in 1871 they became more frequent.[15] In fact, the impression given by the notebook entries after his acquisition of the Grubb telescope is one of nearly continuous and complete record keeping. He began to include excuses for gaps in the observatory record. We find, for example, that in early March 1871, a wind storm severely damaged the roof of his observatory and the necessary repairs made it impossible to observe until 20 March.[16] Then, on 25 March he was obligated to send the 'eye-end' of his new telescope back to Grubb in Dublin for adjustment, thus preventing observations until 12 April.[17] In addition to these mishaps and delays, weather frequently interfered with observation, as did lengthy trips abroad.[18]

Still in his prime, Huggins nevertheless found the new work pace exhausting. His notebook entries after 1871 contain frequent references to fatigue. On 22 April, for example, he spent the early, lighter hours of the evening measuring some eyepieces and then turned to observing Uranus and Jupiter spectroscopically. But he had to stop, he tells us, because 'The sky became a little overcast & with three hours work I found my eye becoming fatigued.'[19] On 1 May, after spectroscopic observations of Venus, Jupiter and Mars, he turned to examine the spectral lines of Uranus. He wrote:

> very difficult to see with sufficient distinction the lines in the red … [One particular] line was very difficult to manage. After great trouble & repeated attempts, I am not certain – It appeared very near indeed to air-line – If anything *probably* less refrangible. By this time, my eyes weary, & I felt it prudent to give over.[20]

Comments like these appear with some regularity throughout his 1871–4 notebook entries, most often after four- or five-hour stints in the observatory, or when the observations proved repetitive and strenuous.[21]

Two unfortunate incidents during the next few years may have heightened his sense of need for a regular and knowledgeable assistant in the observatory. In early 1873, while in the midst of examining the spark spectrum of a gas-filled tube loaned him by the young German-born physicist Arthur Schuster, Huggins left the room briefly to consult a book in his library. Upon his return to the laboratory, he 'found that the black cloth covering the spectroscope was on fire from a small lamp on the table'.[22] Less than a year later, he sprained his ankle in a fall on the ice outside Burlington House. This injury prevented him from observing for a period of time.[23]

For Huggins, the trust which custody of the Royal Society's telescope represented carried with it a sometimes overwhelming obligation to use it. When problems arose and he found he was unable to meet this obligation, he questioned his worthiness to keep the telescope. In December 1874, exhaustion, exacerbated by technical problems with the new telescope's clock drive, prompted him to contemplate surrendering it. 'I feel I ought not to retain the telescope unless I can work with it', he wrote to RS Treasurer, William Spottiswoode (1825–83). 'In consequence of overwork my eyes really needed rest & during the last year, with the exception of the comet, I have not been working. My eyes are certainly stronger, & I may shortly find myself able to commence work again.'[24] The Society's prompt response offering to pay the cost of any repairs necessary to correct the problem in the clock drive demonstrated that officials still viewed him as the one individual in whom they wished to entrust the care of their investment.[25]

10.2 An able assistant

Accounts of the work of William Huggins always mention that he was assisted in his research by his wife.[26] Mills and Brooke, for example, tell us that 'William Huggins did not allow his marriage to interfere with his work', rather that he 'derived great benefit from his wife's able assistance'.[27] They cite Margaret's own account of her contributions to the work of the Tulse Hill observatory which, in addition to observation, included arranging instruments, working batteries, mixing chemicals, dusting and washing up the laboratory, doing 'small things' and being 'generally handy'.[28] Her words lead the reader to understand that, while these tasks provided essential support to the work at hand, they were nonetheless subordinate to the research agenda designed and directed by her husband. They represent a symptom of the invisibility of vital support personnel in astronomical observation: neighbours, friends, assistants, instrument makers, relatives, wives – people who are routinely written out of the published accounts of work done in the observatory.[29]

Figure 10.2 William Huggins at the star-spectroscope, c. 1904. The photograph first appeared in Traill and Mann, *Social England*, p. 691. Huggins proclaimed it to be 'a good one of a difficult subject'. (Frontispiece, W. Huggins and M. L. Huggins, *Scientific Papers*)

Figure 10.3 Margaret Lindsay Huggins (1848–1915). (W. Huggins and M. L. Huggins, *Scientific Papers*, xv)

The invisible people in scientific endeavour have in some cases been obscured by those who sought to gain the greater glory for themselves. Other times they are not so much invisible as simply forgotten. But there are some invisible participants who have cloaked themselves in socially appropriate ways in order to be allowed to play the game – who not only embrace their supportive, helpmeet role, but fashion and nurture that image in the public record. Wives of astronomers form a special subset of that group.[30] Their relationships to their husbands' research programmes run the gamut from 'telescope widow' to one half of a 'two-person career'.[31] The range and complexity of the possible relationships of husband and wife research teams is such that great care must be taken in interpreting the historical record.[32] Indeed, to see these invisible people, we must look beyond the public record. Only there will we find the soul of the complementary collaborative couple.[33]

The historical figure of William Huggins, crafted in part by his wife, has loomed large in retrospectives on the origins of what came to be known as astrophysics. This image is reflected in the classic photograph that serves as the frontispiece for *The Scientific Papers of Sir William Huggins* (see Figure 10.2).[34] Here we see the solitary stellar explorer seated alongside his star-spectroscope in his private observatory free of the distracting bureaucratic entanglements as well as the debilitating methodological and theoretical restrictions which, in Huggins's view, encumbered institution-bound observers at Greenwich, South Kensington and Kew.

10.3 Margaret Lindsay Murray

Margaret Lindsay Murray was born in Dublin on 14 August 1848 (see Figure 10.3).[35] She was the second child and elder daughter of one John Murray, a solicitor, and his wife, the former Helen Lindsay.[36] Though her parents made their home in Ireland, they were of Scottish ancestry and her father was educated at the Edinburgh Academy.[37]

Obituaries and memorial essays written by her friends help us construct a plausible, but somewhat apocryphal, story of Margaret's childhood. From them we learn she was only eight years old when her mother died in January 1857 at the age of thirty-one. Margaret then moved with her family to a Georgian townhouse overlooking the sea at 23 Longford Terrace in Kingstown (now Dun Laoghaire).[38] She spent a good deal of time with her paternal grandfather, Robert Murray, a wealthy bank officer, and attended private school in Brighton for some unspecified period of time. Few girls in Ireland attended school in the 1850s. The majority were educated at home: wealthier families hired tutors while mothers provided the educational needs of their daughters in middle-class and poorer homes. Margaret may not have received any formal education until after her mother's death.[39]

10.4 Interest in astronomy

In her adulthood, Margaret's enthusiasm for astronomical research was a marvel to those who knew her. After her death, her friends' reminiscent memorial essays drew on childhood anecdotes – many no doubt related to them personally by Margaret herself. They speculated on the extraordinary circumstances that might have predisposed her to engage in this unique vocation. Some attributed her early astronomical training to her grandfather, others to her reading an article on spectroscopy in a young people's magazine. A number even suggested that the author of this magazine article was none other than William Huggins himself. A typical example is the obituary for Margaret written by Wellesley Astronomy Professor, Sarah Frances Whiting:

Before [Margaret Huggins] reached her teens she worked with a little telescope making drawings of the constellations and sunspots. Later, inspired by anonymous articles in the magazine, *Good Words*, she became interested in the spectrum, and made a little spectroscope for herself by which she detected the F[r]aunhofer lines. It was the romance of her life that she afterwards became the wife of the astronomer who wrote the papers, and with him made many discoveries with the magic instrument.[40]

Some version of this story is repeated with authority in a variety of widely read sources. However, a search through the volumes of *Good Words* published between 1860 – the year the magazine was first published – and 1875 – the year of Margaret's marriage – uncovered no articles by Huggins and only one series of anonymous articles on astronomy. Given the style and content of the articles and his own eclectic research interests at the time, it is clear Huggins was not their author.[41]

Articles on a variety of astronomical subjects written by such prominent astronomers of the day as John Herschel, Richard Proctor and Charles Pritchard appeared in the magazine during Margaret's youth. In fact, Herschel offered the first description of the basic principles of spectrum analysis to be found in *Good Words* as part of a short series of articles to acquaint its young readers with the properties and behaviour of light. 'The analysis into its prismatic elements of the colour of any natural object', he wrote, 'is readily performed by examining through the refracting angle of a prism of perfectly colourless glass a rectilinear band or strip of the colour to be analysed.' And he included instructions for making a simple instrument to try such observations:

An exceedingly convenient arrangement for this purpose is to fasten across one end of a hollow square tube of metal or pasteboard blackened within, of about an inch square and twelve or fourteen inches long, a metal plate having in it a very narrow slit parallel to one side, quite straight, and very cleanly and sharply cut. At the other end within the tube is to be fixed a small prism of highly dispersive colourless flint glass, having its refracting angle parallel to the slit, and so placed that when the tube is directed to the sky, or rather to a white cloud, the slit shall be seen dilated into a clear and distinct prismatic spectrum . . . The use of this little instrument, at once simple, portable, and inexpensive, will be found to afford an inexhaustible source of amusement and interest . . .[42]

Herschel gave no hint of the spectroscope's astronomical applications, choosing instead to emphasise its aesthetic powers:

To the water-colour painter, the study of the prismatic composition of his (so fancied) simple washes of colour and the effects of their mixture and superposition; – to the oil painter, that of the various brilliantly coloured powders which mixed with oil form the material of his artistic creations, are all replete with interest and instruction.[43]

Margaret would have been sixteen years old when these words appeared in print. If they encouraged her to make her own spectroscope, she may have done so to further her artistic interests.

There is one article in *Good Words* by Charles Pritchard, then RAS President, which is worth noting. Titled 'A true story of the atmosphere of a world on fire', it appeared in the April 1867 number.[44] The 'world on fire' Pritchard described was the recently discovered nova in Corona Borealis (see Chapter 6). After recounting the flurry of interest that this object generated throughout the international community of astronomers, Pritchard introduced W. A. Miller and 'Mr. Huggins' of the Tulse Hill observatory. According to Pritchard, these gentlemen were experts in spectroscopy, and thus able to analyse the nature of this unusual star.[45]

Pritchard's use of such kinetic phrases as 'sudden bound', 'strong impulse in a new direction', 'aspires to a loftier aim' and 'no longer . . . restricted' conveyed the advantages he felt spectroscopy brought to astronomical research. He provided simple directions for constructing and using a spectroscope complete with suggestions to guide the reader's expectations and ensure successful observations.[46] If Margaret's interest in astronomical spectroscopy was indeed piqued by reading one particular article in *Good Words*, Pritchard's is a likely candidate. But several articles in later volumes of the magazine discuss stellar spectroscopy and mention specifically the contributions of William Huggins.[47] Thus, it seems more likely her inspiration came from a combined influence of many articles read over an extended period of time, perhaps even one like the following on Caroline Herschel written by Pritchard in 1869:

During the forty years of her brother's astronomical labours, whether at Bath, or Datchet, or Slough, she acted as his one and only assistant. She always sat up with him at night, writing down the observations from his verbal dictation as they occurred, and reading off and noting the clock at each observation requiring a register of the time. In such calculations she was perfectly indefatigable, and when she had not work enough to do for her brother, she struck out a course of calculation for herself . . . Nothing seemed to her a sacrifice, or a hardship, or a privation, if made in furtherance of his objects.[48]

10.5 The 'two star-gazers'

It is not yet clear when or how William and Margaret met. One romantic account tells how 'the two star-gazers stopped their investigations long enough to "exchange eyes"'.[49] A plainer, but more detailed, version recalls that the pair were first introduced at the London home of the Montefiore family where Margaret was a frequent visitor. This initial encounter was followed, with the help of Howard Grubb's mediation, by further meetings in Dublin during Huggins's visits to inspect Grubb's progress on his new telescope.[50]

No documents or personal letters have thus far been found to substantiate these stories. It is unfortunate, because the question of how they met and chose to marry is of historical interest in the context of collaborative couples.[51] Social, intellectual and cultural barriers virtually prohibited women with scientific aptitude or interest from pursuing independent scientific research. Victorian social critic John Ruskin (1819–1900) wrote:

I believe, then, with this exception, that a girl's education should be nearly, in its course and material of study, the same as a boy's; but quite differently directed. A woman in any rank of life, ought to know whatever her husband is likely to know, but to know it in a different way. His command of it should be foundational and progressive; hers, general and accomplished for daily and helpful use . . . [S]peaking broadly, a man ought to know any language or science he learns thoroughly – while a woman ought to know the same language, or science, only so far as may enable her to sympathise in her husband's pleasure, and in those of his best friends.[52]

No doubt, many gave up their scientific interests entirely or channelled their creative energies into more socially acceptable activities. Those inclined to persevere – if they wished to make contributions to scientific enterprise instead of simply reading about the exploits of others or attending public lectures on scientific topics – required a tolerant male partner with whom to collaborate and through whom to connect to the larger community of scientists. Was Margaret such a woman? Or did her scientific interests emerge in consequence of an offer of marriage from a gentleman in search of an observing partner whom he could train to assist him in his own research programme?

Margaret and William Huggins were married on 8 September 1875 at the Monkstown Parish Church near Margaret's family home.[53] She was twenty-seven years old and he was fifty-one. She soon became involved in the work of the observatory, and her presence changed both the kind of work done there and its organisation. William's terse notebook jottings were replaced by Margaret's lengthy and detailed entries. More importantly, photography suddenly appeared as a new method of recording what had previously been purely visual spectroscopic observations. Evidence gleaned from the notebooks points to Margaret as a strong impetus behind the establishment of a successful programme of photography-based research at Tulse Hill.

10.6 Celestial photography

In November 1873, Lockyer presented an illustrated lecture at the Society of Arts on the application of photography to spectrum analysis. He drew attention to the pioneering work being done by two Americans: Lewis M. Rutherfurd and Henry Draper.[54] Lockyer assured his audience that the spectral photographs these men had taken would prove to be of great assistance to those who would understand the chemistry and physics of the Sun: 'I do think we have in photography not only a tremendous ally of the spectroscope, but a part of the spectroscope itself.'[55] In addition, Lockyer shared news he had just received in a letter from Draper that, after several months of experimentation, he [Draper] had obtained a photograph of the spectrum of α Lyrae (Vega) which showed four distinct lines.[56] 'Now that', Lockyer proclaimed to his audience, 'is of the very highest importance, because the sun is nothing but a star, and the stars are nothing but distant suns.'[57]

A few days after his lecture, Lockyer wrote to Draper to congratulate and encourage him in his efforts: 'However difficult this is the most important thing to be done in the present state of science.'[58] But Draper was a busy man. He had astronomical projects other than photographing stellar spectra to occupy his limited time: photographing a normal solar spectrum, for example, and planning for the transit of Venus.[59]

In February 1874, just before the accident in which he sprained his ankle, Huggins wrote to Draper for advice and to let him know he was not the first to try his hand at spectrum photography:

Is the process [of photography you used] delicate? Would it be possible to apply it to the stars? I am thinking of trying to apply photography to stellar spectra. The few I took some years ago [1863] showed that there is a prospect of success.[60]

Draper's response carried a congenial, but challenging, catch-me-if-you-can tone:

I am very glad to learn that you think of continuing your former experiments in applying photography to stellar spectra. I have made some new trials in that direction with my silvered glass reflector of 28 inches aperture and find that I can get the great bands of Vega readily and even the spectrum of alpha Aquilae [Altair]. It is a very difficult subject and requires so favorable a series of circumstances that a number of observers might work at it a long time before fine results were achieved.[61]

There were numerous difficulties to be overcome in prosecuting this line of investigation particularly when long exposures were required for faint objects. In 1875, one intrepid astronomical photographer wrote in the *British Journal of Photography*:

As all our work requires to be done in the shortest possible time, my first experiment was to get a collodion which should combine the greatest amount of sensitiveness with stability ... as it would sometimes be a week or more between the working nights ... In order to make reliable tests for sensitiveness it is necessary to have some means by which to compare the results with considerable nicety. To do this I constructed a photometer ... I have made nearly one hundred trials for sensitiveness, and have kept a record of how each was made and of its qualities.[62]

The Hugginses began preparing themselves and their observatory to meet these challenges. In late July and early August 1876, they proudly showed two American visitors around their observatory. The first, Captain Richard Samuel Floyd (1843–90), visited Huggins on 27 July.[63] Floyd was not an astronomer, but had recently been named President of the Board of Trustees for the new mountaintop observatory being built in California with funds donated by the wealthy American land baron James Lick (1796–1876). Floyd was on a tour of Europe to meet with and gather useful information from some of the world's most renowned astronomers and instrument makers.

A few days later, the Hugginses hosted astronomer Edward Singleton Holden, an assistant to Simon Newcomb at the US Naval Observatory. Holden had been sent to London to study scientific instruments.[64] He had tried unsuccessfully to get an update on his friend Draper's photographic work before he saw Huggins. After his visit to Tulse Hill, Holden wrote to Draper 'your note ... did not give me what I wanted: that is not details about your work. I want to tell Huggins how much you have done – for strictly between ourselves I think he is afraid of you *now*.'

Like an intelligence officer reporting from the field, Holden laid out for Draper the situation he witnessed at Tulse Hill:

[Huggins] has taken his refractor down & put up his reflr with everything ready for photographing – & last & best for him he has his clock newly overhauled by Grubb & a 2dary control put in from his normal clock. The idea of this control is new & *extremely* ingenious & I think there is no difficulty in applying this idea to any one.

Huggins is very pleasant & everything about him is *thorough* – his Obsy & working places are part of his house & every bit of apparatus in them works like a charm – smoothly & easily. He has a wife now, & she is devoted to him & science & altogether they seem to have work in them.[65]

Huggins's first paper devoted solely to spectral photography done at Tulse Hill was published in the *Proceedings* in December 1876. William was named the sole author. Margaret's clear involvement in the photographic work detailed in the paper raises the natural, but unanswerable, question of the degree of her participation in writing and editing it as well. Their first co-authored article did not appear until 1889.[66] It mattered little whether or not William wrote the papers he submitted between 1875 and 1889 entirely by himself. Every effort would have been made to present them as if he had. Margaret would have participated in the process knowing that she was effectively writing herself out of the published accounts.

Because he was not known for photographic work when this paper appeared, Huggins established his authority by invoking the revered name of his late collaborator, Miller, who was skilled in the art. For purposes of priority, he reminded readers that '[i]n the year 1863 Dr. Miller and myself obtained the photograph of the spectrum of Sirius'.[67] During the course of their 1863 investigations, they had indeed twice captured an image of Sirius's spectrum on a wet collodion plate. The results were disappointing: '[T]he spectrum, though tolerably defined at the edges, presented no indications of lines'. Their attempt to photograph the spectrum of α Aurigae (Capella) resulted in a similar line-less image. Professing at the time to be undaunted by these initial failures, they remarked: 'Our other investigations have hitherto prevented us from continuing these [photographic] experiments further; but we have not abandoned our intention of pursuing them.'[68]

Nevertheless, it seems they did not continue their photographic experiments on stellar spectra. At least, Huggins left no record of such work anywhere in the sections of the observatory notebooks that date from the early 1860s.[69] It is possible that Miller took the notes, or that the two collaborators viewed the photographs themselves as the only necessary record of their activity. Perhaps they judged their results too poor to warrant comment.

In his 1876 paper, Huggins compressed the intervening thirteen years into a moment's hesitation, boldly announcing 'I have recently resumed these experiments . . . '. The delay in resuming photographic work, he explained, 'has arisen from the necessity . . . of a more uniform motion of the driving-clock'.[70] In fact, back in December 1874, he informed Spottiswoode that the telescope's clock drive had been giving him some trouble since the spring.[71] And Grubb installed a new one in January 1876, just before the Hugginses began their programme of photographic experimentation.[72] However, the observational obstacles he deemed worthy of mention in his notebook during that period have more to do with eye fatigue and his sprained ankle than mechanical glitches.

Some twenty years later, in 1897, the problem with his clock drive a long-forgotten annoyance, Huggins reminisced that he and Miller 'did not persevere in our attempts to photograph the stellar spectra' because the available photographic methods were unsuitable for such work. Wet collodion, he claimed, was inconvenient and dry plates were not sensitive enough.[73]

The wet collodion process, though popular, was not the only method available to photographers in 1863. Miller discussed in considerable detail the principles and practice of such varied photographic processes as Talbotype, Daguerreotype and chrysotype in the 1860 edition of his textbook *Elements of Chemistry*.[74] Throughout the 1860s, photographers were mixing substances like honey, glycerine and beer with the collodion to prolong the exposure time available to users of wet plates.[75] They also experimented with ways of increasing the light sensitivity of various types of dry plates. But because he and Huggins were attempting something quite new and untested in applying photography to their infant programme of observing stellar spectra, it seems reasonable to assume that Miller's previous experience with the wet collodion process in his own spectroscopic research would have made that his process of choice.

Balky clock drives and clumsy photographic processes served Huggins more as rhetorical devices in his latter-day accounts – symbols those few who had made similar efforts would readily recognise as representative of the multitude of technical difficulties pioneers like themselves faced in the decade separating the first visual observations and the first successful photographs of stellar spectra. That Huggins bothered to give any explanation for his lack of use of photography during this period demonstrates his awareness that some excuse was necessary. Others in London who were actively engaged in celestial photography – Warren De la Rue and Captain William de Wiveleslie Abney (1843–1920), for example – saw astronomical needs and photographic capabilities as interdependent, one driving the improvement of the other. The gelatine dry plate, a stable photographic plate which, although not as light sensitive as wet collodion, could bear lengthy storage and long or multiple exposures, was introduced in 1871 and quickly improved upon.[76] Although Huggins continued his observations of stellar motion in the line of sight in the early 1870s, and compared nebular spectra against those of various terrestrial metals, he pursued this taxing research using only visual observations. Why was he not motivated to adapt and apply photography to his research needs? Perhaps he preferred to wait until photographic methods had matured sufficiently to meet the demands of his astronomical research agenda. Perhaps he needed someone with skill and interest to guide his efforts.

10.7 Photography at Tulse Hill

Enter Margaret Huggins. The observatory notebooks make it clear that her photographic expertise made possible an important shift in the research agenda at Tulse Hill. Many accounts credit her with having learned the basic principles of photography at some time in childhood or adolescence. One close friend went so far as to say that Margaret's skills in photography were self-taught and that she mastered them before she made her spectroscope.[77]

How fashionable was it for a young woman in the 1860s to be handling smelly photographic chemicals, or managing cumbersome tripods and other photographic equipment? One survey of photographic portrait studios in Britain in the mid-nineteenth century revealed that 22 of the 750 individuals engaged in portrait photography between 1841 and 1855 were women.[78] The number of women who achieved some degree of renown for their work during the early history of photography was small, but not inconsequential, and included such accomplished photographers as Julia Margaret Cameron (1815–79) and Lady Clementina Elphinstone Hawarden, the Countess of Rosse (1822–65).[79] Queen Victoria was an early patron of the Photographic Society of London. This royal enthusiasm for photography may have encouraged women with both the leisure and the financial means to experiment with the emerging art form. There were also increasing employment opportunities for working-class women in photography. By 1873, one-third of all photographic assistants were women. One observer at the time expressed hope that this ratio might soon increase to one-half: 'It is an occupation exactly suited to the sex.'[80]

Margaret's interest in photography identifies her as one of an adventurous group of young women of her day. Novelist and clergyman Edward Bradley (1827–89), writing under the penname of 'Cuthbert Bede', advertised photography as the ideal entrée for a young lady into the science of chemistry. Remarkable effects could be produced with little or no understanding of basic chemical principles. And photography was a socially acceptable topic for polite, if somewhat 'mystical', ballroom conversation. Bradley cautioned his readers that experimentation with photography had its inherent risks. A humorous illustration depicts the consequences of one such mishap (see Figure 10.4).[81]

A PHOTOGRAPHIC POSITIVE.

LADY MOTHER (loquitur) "*I SHALL FEEL OBLIGED TO YOU, MR. SQUILLE, IF YOU WOULD REMOVE THESE STAINS FROM MY DAUGHTER'S FACE. I CANNOT PERSUADE HER TO BE SUFFICIENTLY CAREFUL WITH HER PHOTOGRAPHIC CHEMICALS AND SHE HAS HAD A MISFOR-TUNE WITH HER NITRATE OF SILVER. UNLESS YOU CAN DO SOMETHING FOR HER. SHE WILL NOT BE FIT TO BE SEEN AT LADY MAYFAIR'S TO-NIGHT.*"

Figure 10.4 'A Photographic Positive'. Humorous depiction of the perils faced by young women who engaged in photographic experimentation. (Bede [Bradley], *Photographic Pleasures*, opposite p. 50)

Photographic work at that time was a complex and often frustrating activity even for those with experience and ability. In fact, until the 1880s when ready-made photographic plates became widely available, few were prepared to invest the time, money and energy required to make photography an avocation. Although the facts behind Margaret's training in photography remain unclear, there is sufficient evidence available in the laboratory notebooks to demonstrate that her practical and technical photographic expertise was considerable by early 1876, when she assumed the task of making the notebook entries.

In her first notebook entry on 31 March 1876, she wrote (see Figure 10.5):

Photographed Sirius. Wet Plate, 9 minutes exposure. Photograph on the edge of the plate in consequence of want of adjustment. 3 lines across refrangible end of spectrum.[82]

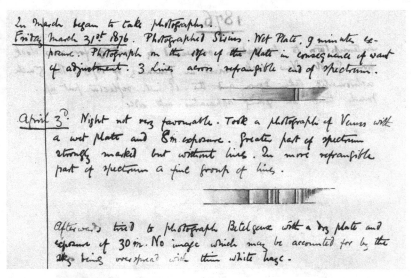

Figure 10.5 First notebook entries by Margaret Huggins. 31 March and 3 April 1876, Notebook 2. (WCL/SC)

On 3 April she recorded:

Took a photograph of Venus with a wet plate and 8 m. exposure ... Afterwards tried to photograph Betelgeuse with a Dry plate and exposure of 30 m. No image[,] which may be accounted for by the sky being overspread with thin white haze.[83]

Nearly every entry thereafter contains some mention of photographic work. The process employed at the start was wet collodion. On 7 May, she documented their comparison of the wet and dry process:

The dry plate gave best results ... These results were so good that thought I might endeavour to photograph the spectrum of Venus using the same narrow slit I had from the Solar Spectrum.[84]

Here we see her first use of the first person, a practice that could be attributed to the fact that she was merely transcribing her husband's personal notes about the work that was done. There is no way to check this independently, and it cannot be ruled out as a possibility.[85] However, by December 1876, Margaret was using the first person plural.[86] In July 1879, Margaret began to use the initial 'W' to single out her husband's contribution to the work at hand, while she referred clearly to her own efforts in the first person singular.[87]

Margaret and William became acclimated to working together. At the same time, they expended considerable time and energy learning photography's limitations, moulding its capabilities to match their astronomical research agenda, and testing new emulsions for sensitivity, stability and reliability.

Early on, Margaret's entries reveal her knack for innovation and interest in experimental design. On 9 May 1876, for example, she wrote that she 'took one or two photographs of Solar spectrum with a view to determining how wide I might open the slit and still obtain lines'.[88] By June, she was demonstrating her expertise in improving and adapting both instruments and methods to the new photographic tasks at hand (see Figure 10.6):

I had a new and much smaller camera made to use in connection with the above described apparatus . . . I was occupied upon all favourable days in testing and adjusting this photographic apparatus upon the solar spectrum: at the same time testing different photographic methods with a view to finding, relatively to different parts of the spectrum the most sensitive, and relatively to the whole spectrum the quickest method for star spectra.

I found that although otherwise desirable wet collodion processes are open to serious objection on account of oblique reflection – a second spectrum in greater or less degree being invariably present. This arose from a second reflection from the back of the plate the light having passed through . . .

After this I used in turn Emulsion, Gelatine, and Captain Abney's Beer plates and obtained some excellent photographs of the solar spectrum both by direct sunlight reflected by a Heliostat and by diffused daylight.[89]

Indeed, the entire summer of 1876 seems to have been devoted to experimentation with different types of light-sensitive plates. Her last mention of wet collodion was on 17 August. After that, they used only dry or gelatine plates.

She made substantial improvements to the observatory's equipment. On 19 September 1876, she wrote:

Finding it impossible to feel certain whether the apparatus was perfectly axial in the telescope, I had a small brass tube made & placed as marked in the diagram. This tube being furnished with cross wires placed very accurately at right angles: it could be ascertained by observing when the angles of the cross wires coincided whether the apparatus was perfectly axial . . .

This rendered it possible to have more accurate adjustments and saved danger of throwing the apparatus out of adjustment in other respects. Instead of a scale of black lines on the silver plate,

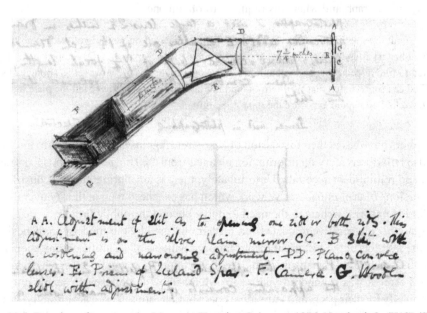

Figure 10.6 Drawing of a camera by Margaret Huggins. 7 August 1876, Notebook 2. (WCL/SC)

I placed two wedge shaped indicators one at centre of each half of slit, it being always intended to bring the stars into the lower portion of the slit.[90]

Her interest in problem selection comes out in a note she added to a letter of William's to David Gill in 1879:

If only a few nights sufficiently clear come, I want to try photographing a nebula. The difficulty would be to keep it on the slit: – but difficult as it would be I am most eager that we should try & get some result. It would be valuable.[91]

By 1887, she was sufficiently confident in her own interpretive skills to note, 'I cannot feel sure there is anything on the nebula plate but William fancies there is. Well if there be anything it's practically useless it's so faint.'[92]

In 1888, her interest turned from obtaining photographs of nebulae to photographs of their spectra. After receiving a photograph of the Andromeda Nebula from Isaac Roberts, the Liverpool building contractor and celestial photographer *par excellence*, Margaret exclaimed:

It would be of special interest we think to supplement this remarkable photograph with some photographs of the spectrum of this neb. Mr. Roberts' work gives the body: if we can get good spectra we should have the soul.[93]

Initial unsuccessful trials in her quest for the nebula's 'soul' led her to suggest keeping the photographic plate in over several nights to collect sufficient light: 'I persuaded W[illiam] to gently close the shutter and leave the plate in the camera to go on with the next fine night.'[94]

As the years went by, she continued to take the initiative whenever photography was employed at Tulse Hill. By 1893, if not before, she had her own ideas about what counted as a quality photograph and what was required to obtain one:

I was . . . unable to be in the Observatory but W[illiam] insisted on working alone. Again tried [the globular cluster] Messier 15, giving exposure from 6.10 to 9 p.m. . . . Developed next day and delighted to find a spectrum good enough to tell us something. It is not however as strong as I should have liked & I regret much that W[illiam] would not take my counsel & have left the plate in so that it might have had continued exposure the next fine night.[95]

***

Biographers have based their discussion of Margaret Huggins's contributions to the work of the Tulse Hill observatory on information gleaned from the Hugginses' published scientific papers and reminiscent accounts. Unfortunately, there is an alluring internal consistency in these versions of their collaborative work which has enhanced their authority over the years and blinded past researchers to the need to delve beyond the public façade.

Margaret's collaborative role has been further obscured by our modern predisposition to see in the inkblots of the historical record familiar patterns of scientific organisation around which to structure our analysis. Today, collaboration in scientific research is far more common than it was a hundred years ago. Members of large research teams are ranked hierarchically by their peers and individual contributions are assessed accordingly. Principal investigators are rewarded for their originality and insight into the theoretical and practical problems encountered in on-going research; support personnel's success is measured on the basis of how well they tend instruments, follow instructions and work cooperatively as part of a team.[96]

Reliance upon such standards for historical analysis, however, unnaturally constrains discussion of complementary collaborative efforts in the nineteenth century. Such a scheme makes little sense when describing the work of Robert Bunsen and Gustav Kirchhoff, for example, or that of George Liveing (1827–1924) and James Dewar (1842–1923) – pairs of men with comparable levels of professional training and technical skill. However, when investigative partners are also husband and wife, there is a risk that a hierarchical evaluative structure may appear applicable to their joint scientific work.

This risk is enhanced when the body of published papers contains what appear to be clear signatures of something like today's hierarchical teamwork structure: husband as observer, wife as instrument tender; husband as principal interpreter of data, wife as recorder; husband as analyser of measurement error, wife as corroborator; and so on. Thus, it may seem fitting to evaluate the husband's work in terms of its originality and theoretical insight, and the wife's contributions in terms of the peripheral support it supplies the research effort.

The Tulse Hill observatory notebooks show that Margaret and William did not work together as principal and secondary researchers; rather, they worked as complementary collaborative investigative partners. The constant give-and-take on which such a work relationship is based blurs the usual markers that distinguish an idea or plan's originator from its implementor. Their notebooks reveal Margaret's initiative in such diverse activities as problem selection, instrument design, methodological approach and data interpretation thus providing a clearer sense of the nature, extent and value of her scientific contributions to the work done at Tulse Hill.

Thanks to Margaret, William Huggins soon found himself moving quickly into the forefront of spectroscopic astrophotography. In their first decade of working together as complementary collaborators, the Hugginses became engaged in two long-term research projects, both of which sorely tested their skills in state-of-the-art photography. The first involved photographing the solar corona without an eclipse, an elusive and often frustrating challenge that will be the focus of the next chapter. We will turn to their second major project of the period in Chapter 12, namely analysing the spectra of different nebulae to resolve the nature of what came to be known as the 'chief nebular line', a green emission line that William had noted, but not identified, some years earlier. Margaret contributed actively to both of these projects. In doing so, she, like William, became embroiled in the ensuing controversy over methods, instruments and interpretation of received data.

Notes

1. W. Huggins, 'The new astronomy' (1897), p. 926.
2. The men most often mentioned are Sidney Bolton Kincaid and Arthur Finch, both of whom lived nearby.
3. Huggins's 8-in Alvan Clark was removed in August 1869. The new 15-in Grubb telescope became operational on 20 February 1871.

4. W. P., 'Spectroscope construction' (1870).

5. G. J. W., 'Solar spectrum' (1870).

6. H. Grubb, 'Automatic spectroscope for Dr. Huggins' Sun observations' (1870).

7. J. Browning, 'On a spectroscope in which the prisms are automatically adjusted . . . ' (1870).

8. Miller's collaboration with Huggins was highly valued, though neither regular nor frequent: he was present for only about 10% of the observations Huggins recorded in Notebook 2 before 1870.

9. Trotter, 'Miller' (1870), p. xxvi.

10. In later life Huggins so successfully projected an image of himself as an independent and solitary observer during the years before his marriage that his biographers routinely wrote Miller out of the story after 1864. Dyson, for example, cited the 'pressure of other duties' as responsible for Miller's discontinuing his productive collaboration with Huggins, leaving Huggins to pursue 'his researches single-handed' until the time of his marriage. See, Dyson, 'Sir William Huggins' (1910), p. iv.

11. W. Huggins to Robinson, 18 January 1869, Add MS 7656.TR74, GGS, CUL.

12. Bennett, *Church, State and Astronomy in Ireland* (1990), pp. 146–7.

13. Meadows, *Greenwich Observatory, Volume 2* (1975), p. 8.

14. For a discussion of the range of middle-class incomes and the accoutrements traditionally associated with those income levels, see May, *An Economic and Social History of Britain 1760–1970* (1987), pp. 201–4. Huggins's 1871 census report lists only a cook and a female house servant living in the home. 1871 Census Return for 90 Upper Tulse Hill Road, Lambeth, RG 10/684/10. It is worth noting that George Airy, the Astronomer Royal, received an annual salary of £800.

15. William Huggins recorded observations on 51 occasions in 1866, 56 in 1867, 51 in 1868, 6 in 1869, and 3 in 1870. So few were recorded in 1869 and 1870 because he was without a telescope. In 1871, He noted a record 63 observations, but in 1872, this dropped to 54, with only 27 in 1873, and 15 in 1874. He recorded no observations in 1875.

16. W. Huggins, 20 March 1871, Notebook 2, WCL/SC.

17. W. Huggins, Note before entry on 12 April 1871, Notebook 2 WCL/SC.

18. Huggins's first entry after the new Grubb telescope was installed to his satisfaction ended with, 'Clouds came over' (20 February 1871). In the next two months we find remarks like, 'perfectly clear up to 6 o'ck then became hopelessly clouded' (25 March 1871); 'some clear patches of sky, but generally a haze more or less thick over the whole sky' (26 March 1871); 'Having been at work for four hours, & haze over sky shut up observatory' (13 April 1871); 'Clouds & haze between observations' (17 April 1871); 'From April 17 weather bad tonight' (Note before entry on 21 April 1871); 'Evening fine with drifting clouds. Soirée R. S. but remained at home to make use of the night' (22 April 1871); 'From April 22 evenings all bad until tonight when practically fine with clouds & masses of haze' (27 April 1871). In a note written between 16 May and 25 August 1871, he tells us, 'alterations to reading telescopes, bad weather & journey to Scotland prevented observation up to Aug 20'. [All entries in Notebook 2, WCL/SC.]

19. W. Huggins, 22 April 1871, Notebook 2, WCL/SC.

20. W. Huggins, 1 May 1871, Notebook 2, WCL/SC.

21. Complaints about fatigue during this period appear pretty consistently in about 1 out of every 8 entries. It must be kept in mind that Huggins was nearing 50 years of age during this period.

22. W. Huggins to Arthur Schuster, 3 April 1873, Sc.95, AS, RSL. Schuster, then just 21 years old, had studied spectroscopy under Roscoe at Manchester and Kirchhoff at Heidelberg. See Kargon, 'Arthur Schuster' (1970).

23. W. Huggins to Christie, 20 February 1874, RGO 6.174/3/8/85, CUL.

24. W. Huggins to Spottiswoode, 30 December 1874, MC 10.186, RSL.

25. Stokes to Spottiswoode, 1 January 1874, MC 10.188, RSL. As this letter is clearly in reference to Huggins's letter to Spottiswoode of 30 December 1874, Stokes must have written the letter on 1 January 1875 and by mistake written in the previous year's date.

26. See, for example, W. Huggins, 'The new astronomy' (1897), p. 926; Mills and Brooke, *Sketch of the Life* (1936), pp. 37–42; Maunder, *Sir William Huggins and Spectroscopic Astronomy* (1913), p. 64; Ogilvie, 'Marital collaboration' (1987), pp. 111–15.

27. Mills and Brooke, *Sketch of the Life* (1936), p. 38.

28. *Ibid.*, pp. 38–41.

29. See Pang, *Spheres of Interest* (1991), p. 147; Shapin, 'The invisible technician' (1989).

30. See Rossiter, '"Women's work" in science, 1880–1910' (1980); Pang, 'Gender, culture, and astrophysical fieldwork: Elizabeth Campbell and the Lick Observatory-Crocker eclipse expeditions' (1996); McKenna-Lawlor, *Whatever Shines Should be Observed* (1998); Ogilvie, 'Obligatory amateurs' (2000); M. T. Brück, *Women in Early British and Irish Astronomy* (2009).

31. Sheryl J. North, 'The telescope widow syndrome' (1990). The concept of a two-person career describes those occupations in which the spouse of an employee is expected, or even required, to assume certain central responsibilities tied directly to the success of that employee's career. These include such occupations as the ministry, career military or diplomatic corps, high level business administration or civil service. Traditionally, it has been the husband whose salaried career is supported by his wife's performance of necessary but unpaid duties. See, for example, Papanek, 'Men, women, and work' (1973); Seater, 'Two person career' (1982) and Beeson, Jr, *Influences on the Identification of Wives with the Air Force Organization* (1986).

32. On Elizabeth Campbell, wife of Lick Observatory Director William Wallace Campbell, and Mabel Loomis Todd (1856–1932), wife of American astronomer David Peck Todd, see Pang, *Spheres of Interest* (1991), pp. 147–203. On Angeline Stickney Hall (1830–92), wife of American astronomer Asaph Hall (1827–1907), see A. Hall, *An Astronomer's Wife* (1908), pp. 72–3. Isobel Sarah Gill (1845–1919), whose husband David Gill (1843–1914) served for many years as Her Majesty's Astronomer at the Cape of Good Hope, wrote a book describing her experiences while travelling with her husband's expedition to Ascension Island in the summer of 1877 to measure the parallax of Mars: Mrs Gill, *Six Months in Ascension* (1880).

33. B. J. Becker, 'Margaret and William Huggins at work in the Tulse Hill Observatory' (1996); M. T. Brück, *Women in Early British and Irish Astronomy* (2009), ch. 11.

34. Frontispiece, W. Huggins and M. L. Huggins, *The Scientific Papers of Sir William Huggins* (1909).

35. Concerning biographical accounts of Margaret Huggins: M. F. Rayner-Canham and G. W. Rayner-Canham include a few paragraphs on her in their paper, 'Pioneer women in nuclear science' (1990). Unfortunately, there are a number of factual errors in this essay. Perhaps the most detailed examination of Margaret Huggins's early life has been carried out by Maire Brück and Ian Elliott. Much of the information in this section was made available to me through the kindness of Drs Brück and Elliott before their published accounts appeared. See M. T. Brück, 'Companions in Astronomy' (1991); M. T. Brück and Elliott, 'The family background of Lady Huggins (Margaret Lindsay Huggins)' (1992); M. T. Brück, 'An astronomical love affair' (1997); Elliott, 'The Huggins' sesquicentenary', *IAJ* **26** (1999), pp. 65–8; and M. T. Brück, 'The new astronomy' in *Women in Early British and Irish Astronomy* (2009), pp. 161–83. See also McKenna-Lawlor, *Whatever Shines Should Be Observed* (1998), pp. 75–123.

36. 9 September 1875, *Irish Times*.

37. M. T. Brück to the author, 17 October 1991.

38. Elliott to the author, 3 October 1991.

39. Burstyn, *Victorian Education and the Ideal of Womanhood* (1984), pp. 30–47.

40. S. F. Whiting, 'Lady Huggins' (1915), p. 854. See also S. F. Whiting, 'Margaret Lindsay Huggins' (1915); Hodgkins, 'Lady Huggins: Astronomer' (1915); Newall, 'Dame Margaret Lindsay Huggins' (1916); Ogilvie, 'Marital collaboration' (1989), p. 110.

41. Anonymous, 'God's glory in the heavens', *Good Words* **1** (1860), p. 23, p. 116, p. 161, p. 225, p. 289, p. 465, p. 513, p. 577, p. 625, p. 729. This series included such articles as, 'The Moon, is it inhabited', 'The approaching total eclipse of the Sun', 'Comets – Their history', and 'The structure of the planets'. *Good Words* had a readership of 70,000 and cost only 6 pence per number.

42. J. F. W. Herschel, 'On light' (1865), p. 363.

43. *Ibid.*

44. C. Pritchard, 'A true story of the atmosphere of a world on fire' (1867).

45. *Ibid.*, p. 250.

46. *Ibid.*, p. 251.

47. C. Pritchard, 'Perceiving without seeing' (1869); 'Historical sketch of solar eclipses' (1871); Proctor, 'A giant Sun' (1872); Carpenter, 'Spectrum analysis' (1873).

48. C. Pritchard, 'Stars and lights; or, the structure of the sidereal heavens – V. The arrival of Herschel's faithful assistant' (1869), pp. 613–14.

49. Hodgkins, 'Lady Huggins' (1915), p. 1417.

50. Mills and Brooke, *Sketch of the Life* (1936), pp. 33–4.

51. A collection of historical case studies concentrating specifically on the opportunities and impediments faced by women who made their principal scientific contributions in the context of collaborative scientific couples can be found in Pycior *et al.*, *Creative Couples in Science* (1996).

52. Ruskin, 'Of Queens' Gardens' (1905), pp. 117–18.

53. Marriage Record, 8 September 1875, Monkstown Parish Church, Dublin County.

54. For a general overview of astronomical photography, see Lankford, 'The impact of photography on astronomy' (1984); for a discussion of Lewis Rutherfurd's photographic contributions, see Warner, 'Lewis M. Rutherfurd' (1971); for Henry Draper's contributions, see Plotkin, *Henry Draper* (1972), pp. 33–59.

55. J. N. Lockyer, 'On Spectrum Photography' (1874), p. 255.

56. Draper, 9 August 1872, Research Notebook XI, MAH, SI.

57. J. N. Lockyer, 'On spectrum photography' (1874), p. 255.

58. J. N. Lockyer to H. Draper, 26 November 1873, HD, NYPL.

59. Plotkin to the author, 4 March 1992.

60. W. Huggins to Draper, 10 February 1874, HD, NYPL.

61. Draper to W. Huggins, 5 March 1874, HD, NYPL.

62. D. C. Chapman, 'Astronomical photography' (1875), p. 631.

63. W. Huggins to Holden, 26 July 1876, LOA.

64. At the time, Holden perceived himself to be the leading candidate for directorship of the new observatory. He took advantage of Floyd's presence in London 'to ingratiate himself with the new trust president and to check on Floyd to see whether Newcomb's misgivings about him were valid'. Osterbrock *et al.*, *Eye on the Sky* (1988), pp. 24–34.

65. Holden to H. Draper, 2 August 1876, Boxes 1 and 2, HD, NYPL.

66. W. Huggins and M. L. Huggins, 'On the spectrum, visible and photographic' (1889).

67. W. Huggins, 'Note on the photographic spectra of stars' (1876), p. 445.

68. W. Huggins and Miller, 'On the spectra of some of the fixed stars', *PRS* **13** (1864), p. 244; 'On the spectra of some of the fixed stars', *PTRSL* **154** (1864), p. 428.

69. At the top of one page in his first notebook, there is a heading dated 18 March 1859 which reads, 'Photog^c: memoranda – ...' But the single entry under the heading has been partially excised due to the removal of a drawing of Jupiter on the reverse side of the page. See verso of page containing notebook entries for 11, 13, 14 and 23 February 1859, Notebook 1, WCL/SC. Given the date of this heading, it was probably intended to mark a page devoted to records of his lunar photographs.

70. W. Huggins, 'Note on photographic spectra of stars' (1876), p. 446.

71. W. Huggins to Spottiswoode, 30 December 1874, RSL.

72. M. L. Huggins, undated note entered before 31 March 1876, Notebook 2, WCL/SC. See also Stokes to Spottiswoode, 1 January 1875, MC. 10.188, RSL, for Stokes's approval of the expenditure of £40 for improving the clock drive. The informal and extremely positive character of Stokes's brief note to Spottiswoode provides some sense of the confidence which Stokes placed in Huggins's assessment of what was needed to maintain the Society's telescope at peak performance.

73. W. Huggins, 'The new astronomy' (1897), p. 914; W. Huggins, 'Address of the president' (1891), pp. 31–2. See also Clerke, 'Sir William Huggins' (1910).

74. In fact, Miller devoted an entire chapter of his textbook to the chemistry of photographic processes. See 'Influence of light on affinity – Photography', in *Elements of Chemistry* (1860), pp. 825–46.

75. Ackland, 'The collodio-albumen process' (1856); Turnbull, 'A few words on the beer and albumen process' (1874); Clarke, 'The beer and albumen process' (1875).

76. Newhall, *The History of Photography from 1839 to the Present Day* (1964), pp. 47–57, pp. 83–95; R. Hunt, *A Manual of Photography* (1853), pp. 100–3 and pp. 276–87; Ackland, 'The difficulties of the dry processes' (1860); Dawson, *A Manual of Photography Founded on Hardwich's Photographic Chemistry* (1873), pp. 98–130.

77. See, for example, 'Lady Margaret Huggins', *Who Was Who: 1897–1916*, (London, 1935); Donkin, 'Margaret Lindsay Huggins', *The Englishwoman* (1915), p. 152.

78. Heathcote and Heathcote, 'The feminine influence' (1988), p. 260.

79. *Ibid.*, p. 269. T. Powell, *Victorian Photographs of Famous Men & Fair Women* (1973).

80. Hughes, 'Photography as an industrial occupation for women' (1873).

81. Bede [Bradley], *Photographic Pleasures* (1855), pp. 49–52.

82. M. L. Huggins, 31 March 1876, Notebook 2, WCL/SC.

83. M. L. Huggins, 3 April 1876, Notebook 2, WCL/SC.

84. M. L. Huggins, 7 May 1876, Notebook 2, WCL/SC.

85. In the correspondence uncovered thus far, it would appear that Margaret rarely took dictation from William, but when she did, she made this point clear to the reader. On 6 December 1895, for example, Margaret wrote a letter to David Peck Todd which was dictated to her by William. At the end of the letter, she explained, 'My husband at present is *unable to write* owing to a chill having affected his hand neuralgically ... so I am giving myself the pleasure of being his secretary.' W. Huggins to D. P. Todd, 6 December 1895, DPT, YUL.

86. On 7 December 1876 (Notebook 2, WCL/SC), Margaret noted 'Tried a lens newly ground by Hilger to 10 inches focus. Found definition better than that of the lens *I* have been using.' [Emphasis added.] On 14 December 1876, she recorded: 'A faint spectra of β Pegasi; lines faintly discernible. The faintness of Hydrogen spectra *we* think due to the Battery having gone down.' [Emphasis added.] In a typical notebook entry for this period, Margaret wrote on 22 December 1876: 'Began work at 6.15 . . . . Moon half full. Took a photograph of its spectrum' leaving it ambiguous as to who precisely was doing the work.

87. M. L. Huggins, 28 July 1879, Notebook 2, WCL/SC.

88. M. L. Huggins, 9 May 1876, Notebook 2, WCL/SC.

89. M. L. Huggins, 'June' 1876, Notebook 2, WCL/SC.

90. M. L. Huggins, 19 September 1876, Notebook 2, WCL/SC.

91. Note added by M. L. Huggins. W. Huggins to D. Gill, 7 October 1879, DG, SAAOA.

92. M. L. Huggins, 21 March 1887, Notebook 2, WCL/SC.

93. M. L. Huggins, 4 December 1888, Notebook 2, WCL/SC. A number of astronomers, including William and Margaret Huggins, believed Roberts's photograph of the Great Nebula in Andromeda (M31), as well as others taken by him of other nebulous objects, provided conclusive proof of the nebular hypothesis of Laplace. After seeing a photograph Roberts had taken of the Dumbbell Nebula (M27), Huggins described it to George Stokes as showing 'for the first time to the eye of man its true nature. A solar system in the course of evolution from a nebulous mass! It might be a diagram to illustrate the Nebular Hypothesis! I never expected to see such a thing. There are some 6 or 7 rings of nebulous matter already thrown off, & in some of them we see the beginning of planetary condensation & one exterior planet fully condensed. The central mass is still larger, to compare it with the solar system, say as large as the orbit of Mercury. The rings are all in one plane & the position is such that we see it obliquely.' W. Huggins to Stokes, 27 November 1888, Add MS 7656.H1230, GGS, CUL. For a discussion of others' reactions to Roberts's nebular photographs, see R. W. Smith, *The Expanding Universe* (1982), pp. 4–5. For a discussion of the influence of Roberts's photograph of M31 on his contemporaries' understanding of the structure of that nebula, see Vaucouleurs, 'Discovering M31's spiral shape' (1987).

94. M. L. Huggins, 5 December 1888, Notebook 2, WCL/SC. This plate was left in place and exposed on a number of occasions from the beginning of December until the end of February 1889, for a total of about 2 1/2 hours exposure. No nebular spectrum was recorded on this plate, however.

95. M. L. Huggins, 12 November 1893, Notebook 5, WCL/SC.

96. See, for example, Price, *Little Science, Big Science* (1963), pp. 86–91; Hagstrom, *The Scientific Community* (1965), pp. 105–58; Weinberg, *Reflections on Big Science* (1967), pp. 47–53; *idem*, 'Scientific teams and scientific laboratories' (1970); Hargens *et al.*, 'Research areas and stratification processes in science' (1980).

11

Photographing the solar corona

... though it is very easy to obtain a corona-like image, one may readily be deceived in such matters.

– William H. Pickering[1]

William Huggins's work on solar prominences, and his expedition's failure to observe the eclipse of 1870, encouraged him to attempt a bold plan for photographing the solar corona without an eclipse. His initial impression of success in this project led him to pursue it for many years with great interest and drive. The inconclusiveness of his results tested the strength of his persuasive power and encouraged him to try to build an international network of confirmatory witnesses. The evidence Huggins believed he needed to argue successfully for the validity of his method was not forthcoming. Nevertheless, his correspondence contains the details of the verbal and visual rhetorical process by which he was able, as a relative outsider to solar observation, to shape the development of methods of observation in the emerging discipline of solar research, the types of questions being asked about the solar atmosphere, the kind and form of observation that counted as real and conclusive evidence, and finally the direction in which solar observation was taken by the growing network of solar observers up to the turn of the century.

11.1 The Egyptian eclipse

Despite growing threats of local unrest in Egypt, Arthur Schuster travelled to that country in May 1882 to observe a total solar eclipse. He was a member of an expedition led by Norman Lockyer, whose principal object of interest was, once again, the solar corona. Totality was expected to last only a little more than one minute. Nevertheless, the fact that the Moon's shadow crossed over arid Egypt virtually guaranteed 100% visibility.[2] A strict division of labour and diligent practice would ensure success.[3]

By 1882, Lockyer had four eclipse expeditions under his belt, and the advantage of having observed totality over the course of nearly one complete eleven-year sunspot cycle.[4] With the seventy-four seconds of totality promised in May, he wanted another crack at proving his theory that there was, in fact, no so-called 'reversing layer' in the Sun's atmosphere, contrary to American astronomer Charles Young's claims. Lockyer was also anxious to verify a theory he had developed following the 1878 eclipse that related coronal shape and structure with sunspot numbers. Following a few years in which almost no sunspots were seen (1878 having the least), 1882 had produced a relative bumper crop, and at eclipse time there were over twenty spots visible.[5] Lockyer predicted the corona in 1882 would bear a strong resemblance to the complex and highly textured corona he had seen in 1871, the last time he had observed an eclipsed Sun with many sunspots on its surface.

On 17 May, the day of the eclipse, England and France sent a fleet of ships to Alexandria to restore order in the wake of a popular uprising against the Khedive, the Egyptian ruler whose pro-European sympathies were under attack. The next day there were numerous reports in *The Times* of London under the headline 'Latest intelligence' with news of 'The eclipse expedition' and 'The crisis in Egypt' given from many different national perspectives.[6] Henry Roscoe wrote to Schuster: 'We only trust that you have escaped safely from Egypt where it seems that things are likely to turn out pretty awkward.'[7]

Not only did the expedition 'escape' and return to England without mishap, they came back satisfied that the venture had been well worth the trouble. For although Lockyer was not altogether successful in his attempt to rid the Sun of its reversing layer, he was thrilled by the appearance of the solar corona. In addition, Schuster obtained many fine photographs of the corona, including a first-time photograph of the coronal spectrum.[8]

Today, astronomers attribute the corona's visible light to three principal sources. One component, a faint continuous spectrum, is produced by photospheric light scattered from free electrons in the corona. Because it mimics the visible solar spectrum in relative intensity of individual wavelengths, it displays a maximum in the yellow-green region of the spectrum around 490 nm. Another component is an emission spectrum generated by high-temperature ions in the corona. It contains a few discrete bright lines, many of which are only visible during periods of unusual solar activity. The most prominent coronal emission line is Young's green '1474' line. The third component of the visible light seen in the corona, an extremely faint replica of the Sun's absorption spectrum, is produced by photospheric light reflecting off debris in the ecliptic plane. Known as the zodiacal light, this component of the visible corona is unrelated to physical processes in the solar atmosphere.

But at the time of the Egyptian eclipse, the corona remained a mystery. Young, by then an eminent solar observer and professor of astronomy at Princeton, hinted in his new book, *The Sun*, that the advent of the spectroscope had put astronomers on the verge of unlocking the corona's secrets. Nevertheless, he confessed, 'The corona as yet has received no explanation which commands universal assent. It is certainly truly solar to some extent, and very possibly may be also to some extent meteoric.'[9]

11.2 Photographing the corona

Meanwhile, that spring William and Margaret Huggins had been busy trying to photograph the spectrum of the Orion Nebula. On 7 March, they succeeded.[10] But their elation soon turned to dismay upon learning that their work may have been anticipated by Henry Draper, who announced in May his own success in taking two photographs of the Nebula's spectrum.[11] The Hugginses, anxious as always to hold on to their priority in this difficult area of celestial photography, intensified their efforts in nebular spectroscopy.

Indeed, priority was their major concern at this time. In April, William had written to Holden, then Director of the Washburn Observatory in Madison, Wisconsin. He wished to correct what he viewed as a serious error published in Simon Newcomb's book *Popular Astronomy*, which stated that he (Huggins) and Father Angelo Secchi had simultaneously discovered the gaseous nature of the nebulae. In fact, Huggins argued, Secchi had no inkling of the existence of such lines until he heard about them from Otto Wilhelm von Struve. Huggins wrote, 'I had an account *from Struve himself* . . . Struve visited Secchi & told him of my discovery. Secchi at first would not believe it, but . . . in Struve's presence [Secchi] pointed [his instrument] to a nebula & saw one or more lines.' Margaret added at the bottom, 'And in the interests of morality I denounce Sechi's [*sic*] above proceeding as shameful thievery.'[12]

In consequence of these priority concerns, the Hugginses' interest in solar observation was at a very low ebb at the time of the Egyptian eclipse. That is, until 19 May, when they read Schuster's eclipse report in *The Times* of London announcing that he had found the violet region of the spectrum, near lines H and K, to be the brightest part of the coronal spectrum.[13] Margaret and William later claimed that they quickly recognised this information offered a means for photographing the corona without an eclipse and accordingly set to work.

The notebooks contain no entries contemporaneous with any of their efforts to photograph the solar corona in 1882. In fact, there are no regular entries in their observatory notebooks between June 1882 and April 1886! However, on 15 December 1882 – two days after William submitted his first paper on their corona project to the RS – Margaret devoted over seven handwritten pages of the notebook to a personal history of their work over the past several months. It is a notable exception that bridges a curious chronological gap in the record.[14]

'It at once occurred to us', she wrote, to try 'the principle propounded by William in 1866–8 in his endeavours to see the prominences without a prism – viz that using absorbing media for all except a limited part of the spectrum.'[15] Guided by Schuster's claim that the coronal light 'was clearly brightest through some little *range* about H and K', they cemented three or four pieces of violet glass together with some castor oil to prevent light loss from reflection.

Margaret does not mention any attempt to glimpse the corona by eye using this filter. Even with its aid, the corona's light, while technically within the limits of human visual perception, would have been nearly impossible to differentiate from the background noise of the luminous atmosphere. Besides, as William noted in his paper, detecting the subtle structural details and variations within the corona's luminosity – so crucial for establishing a practical record of daily coronal observations – was beyond his own eyes' capability, and probably that of most other observers.[16] The eye of the camera, especially when applied with skill, offered a more promising alternative. The Hugginses positioned the stack of cemented coloured glass in front of the image plane of their photographic apparatus.[17]

At the Royal Society's soirée on 21 June, plates from the Egyptian eclipse expedition were shown. It was at this time that Margaret and William were actually able to see the evidence from which Schuster had deduced that the violet region of the coronal spectrum was the brightest. Margaret wrote: 'We then set to work and on every fine day when we could be sure of the Sun for a few moments free from clouds we took photographs of it with varying exposures.'[18]

Throughout their trials, Margaret and William continually tinkered with their instruments and methods to improve the images they captured. To say the technical challenges were daunting is an understatement. They identified sources of stray light in the system that could confuse or even falsely represent the reality they wished to record and they did what they could to eliminate them. Early on, for example, flaws in the violet glass encouraged them to change the absorbing material to a smooth glass container filled with a strong solution of potassium permanganate. However, sunlight caused small light-scattering particles to form in the solution. They then tried iodine in carbon disulphide, but this solution suffered from the same problem. Ultimately they dispensed with a filtering medium altogether and relied instead on the narrow range of light sensitivity found in photographic plates coated with a silver chloride emulsion.[19]

Concern about chromatic aberration in their camera's lenses prompted them to employ a reflecting telescope for this project. Rather than tie up their 18-in Grubb reflector, the Hugginses chose to invest in an old telescope by James Short which had a 6-in aperture and a 3-ft focal length.[20] They shifted the position of the shutter along the optical path and varied exposure times to capture the sharpest images. They worked at perfecting their darkroom technique to find the optimal rate for developing the photographic plates that would enable them to stop the chemicals' action at the moment the corona's image came into view. This last effort proved particularly frustrating. The Hugginses took the last of their first set of corona photographs on 28 September.

In December, Margaret wrote, 'We were often struck when developing with what seemed to be a flashing out of the Corona at a certain point in development. But we found it very difficult to stop development so as to secure this effect alone.'[21] She boasted:

In certainly 20. of our plates, a form peculiarly coronal is to be seen. In the overexposed plates it is most distinct in some respects than in the under exposed ones. But in the under exposed plates the inner corona can be made out . . . [C]areful measures of the average height of the outer & inner corona in the Egyptian plates and in ours' agree in the relative proportions they bear to the diameters of the suns in the two sets of plates.

According to Margaret, they planned to continue these experiments using a black disc the same size as the Sun's image in front of the absorbing material in order to further reduce glare.

In November William shared these photos with Stokes, hoping, it seems, to get some supportive response from the experienced spectroscopist and esteemed Secretary of the Royal Society. But Stokes apparently missed key features in them that the Hugginses believed were obvious. William quickly dispatched a lengthy letter offering Stokes guidance in 'seeing' the photographs:

I write now in the hope that you will once more look at the photographs as you do not seem to have seen what we see. If we had merely seen the little difference of illumination immediately about the sun's image, I should at once have considered it of too slender importance, & I would not have troubled you with the photographs.

My wife & I, independently, see in all the photographs (in which we see the appearance at all) a *definite form* of difference of illumination which is *essentially the same* in all the phots. taken at different times, with different plates, & different absorbing media . . .

I have just looked at one of the *ten good plates* I have here, & I see it perfectly. My wife has done her best to make a drawing of this form (necessarily a little exagerated [*sic*] in distinctness) relatively to the shape of the plate. It should correspond on the other plates except in the cases in which the plates were placed at right angles to the usual position . . .

Will you kindly examine one of the photographs carefully in different lights until you are able to see the definite form such as is shown in the enclosed sketch. When you have seen it in one photograph, I think you will not have much difficulty to discovering exactly the same *complex & peculiarly coronal* form of outline in all the others.[22]

In case that did not provide sufficient help and encouragement, he added in a postscript:

My wife has been careful not to put more in than she sees. I confess *I* think *I* can see rather more *structure* & can trace the appearance rather farther from the sun's limb. My wife has added notes of her own.

The appearance is more distinct *on one side*, this is not out of harmony with what has been seen at eclipses.

The drawing corresponds with the photographic plate when laid over it, with the gelatine side uppermost. As the little camera is not permanently attached to the telescope, it is possible it was not always attached quite accurately in same position.

At some point in the next few weeks, Huggins must have received the positive response he wished from Stokes,[23] for he soon informed Schuster of his apparent success.[24] He also invited William de Wiveleslie Abney, the photographer whose plates were used during the Egyptian eclipse, to his home to examine the photographs. Abney was impressed with the similarity in appearance of what Huggins claimed to be the corona in his photographs and the images Abney himself had captured during totality in Egypt. Not only were the general features the same, Abney wrote to Huggins, 'but also . . . details such as rifts & streamers have the same position and form'.[25] Abney went so far as to claim, 'If in your case the coronal appearances be due to instrumental defects, I take it that the eclipse photographs are equally untrustworthy.'[26] Margaret later recalled that Abney told William, 'As surely as I stand here, you did photograph the Corona.'[27]

On 13 December, Huggins submitted a paper to the RS on his efforts to photograph the solar corona out of eclipse. He emphasised that his success in this endeavour, while unquestionable, was nevertheless hard won due to the 'unpropitious' English climate. A few months later, in a report to the RAS, he suggested trying his method in a location with clearer skies, perhaps at a higher elevation to avoid the obscuration of the thicker layers of Earth's atmosphere. If the method worked, Huggins claimed, 'the corona may be success-fully photographed from day to day with a definiteness which would allow for the study of the changes which are doubtless always going on in it'.[28]

The Hugginses replaced the Short reflector with a 7-ft Newtonian lent them by the late William Lassell's daughter. Lassell's telescope with its fine 7-in speculum mirror produced larger and less distorted solar images. The Hugginses modified this instrument to create a prototype coronagraph (see Figure 11.1). To limit the telescope's prodigious light-gathering ability, they fitted the open end of its tube with a cover having a hole on one side to admit light from the Sun and its immediate surroundings into the tube. Because the Hugginses had removed the small secondary mirror that is a signature feature of all Newtonian reflectors,

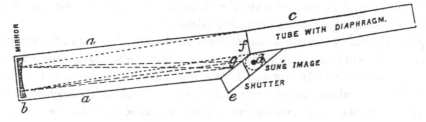

Figure 11.1 William Huggins's prototype coronagraph. (W. Huggins, *RBAAS, Southport*, 1883, p. 347)

there was nothing to divert the light after it had struck the primary mirror, thus allowing it to pass through to a second small hole in the cover and onto a photographic plate. William and Margaret used the instrument to photograph the Sun at every possible opportunity. Unfortunately, they were hampered by wet weather the following spring.[29]

11.3 The Caroline Island eclipse

The Hugginses were particularly interested in obtaining some good photographs of the eclipseless corona on or about 6 May 1883, the date on which a solar eclipse of exceptionally long duration had been predicted to cross the Pacific Ocean making landfall on Caroline Island, a small coral island in the Marquesas.[30] Despite the difficulty in getting there and the uncertainty of the weather, a number of observers from around the world travelled to view this particular eclipse. Among them were two representatives from the Royal Society, H. A. Lawrance and Charles Ray Woods (1859–1920), both assistants to Lockyer at South Kensington, who were charged with photographing the corona during the eclipse. Huggins anticipated that a comparison of their photographs with those he planned to take on the same day back in London would conclusively demonstrate the validity of his method.

While conditions were favourable for eclipse observers on Caroline Island, bad weather in London that day prevented the Hugginses from taking any photographs at Tulse Hill. Fortunately, they had captured what they believed were good images of the corona on three other dates (3 April, 31 May and 6 June) close enough to 6 May, in their view, to justify using them for comparison. Huggins gave these photographs to the talented illustrator and RAS Assistant Secretary, William Wesley. Wesley had had considerable experience over the previous decade converting photographic plates of solar eclipses into reproducible drawings for the RAS's ambitious volume 41 of its *Memoirs* detailing the history of solar observation.[31] At Huggins's request, Wesley also made drawings of the coronal photographs taken at Tulse Hill.

Upon their return from Caroline Island, Lawrance and Woods first saw Wesley's drawings from the Tulse Hill plates, and then the plates themselves. They were impressed with the similarities between their results and the images Huggins had captured without an eclipse. Particularly notable in all the images, in their opinion, was evidence of a rift in the corona near the Sun's north pole. Lawrance wrote to Stokes, 'As a result of the comparison I should say that Dr. Huggins' coronas were certainly genuine as far as 8 minutes from the limb.'[32] It is possible that the order in which the visual evidence was presented to Woods and Lawrance shaped their opinion of the reality of the effect depicted.

Despite Lawrance's positive assessment, Huggins was not fully satisfied that the trial had proven the efficacy of his method. To try it again meant waiting for the next convenient solar eclipse. A good candidate, one that would cross New Zealand, was coming up in September 1885. But Huggins did not want to wait that long, nor did he want to rely again on the whims of the weather to make a direct comparison possible. Instead, he devised a new plan, one that involved obtaining a vast number of coronal images in broad daylight over a long period of time. A high degree of internal consistency in the appearance of these images would, in his view, establish his method's validity with certainty. To collect so many photographs efficiently, however, would require working in a more favourable and less fickle environment.

In December 1883, at other Fellows' urging, he sent a note to the RS's Government Grant Committee requesting a sum of about £200 to finance a small expedition to a place that was both easily accessible and free of lower atmospheric obscuration.[33] If a party of experienced photographers – the men who went to the Caroline Island eclipse, for example – could spend two or three months on a mountain in Switzerland, say, they could gather a sufficient number of photographs to permit a judgement to be made.

Hoping to gain an ally, Huggins outlined his plan in a short note to Stokes. As a member of the Grant Committee himself, Huggins recognised the delicacy of the situation and the possible questions of conflict of interest that could be raised given his own personal stake in the results of such an expedition. He assured Stokes 'I do not want to have *any* personal grant. If it is done, it must be by means of a committee who would be (*not I*) responsible for the use of the money.'[34]

Huggins was excited about how things were going thus far: Abney and Stokes had given him written statements attesting to their belief in the authenticity of his original coronal photographs, Lawrance and Woods had compared his more recent coronal photographs quite favourably with those they had taken during the eclipse at Caroline Island, and now he was being encouraged to apply for a grant from the Royal Society to obtain even more persuasive evidence of the validity of his new method.[35]

While waiting for the Committee to consider his request, Huggins continued to search for ways to extract more evidence from the three photographs he had taken around the time of the 6 May 1883 eclipse. Seeing Wesley's drawings of them had done much to heighten his own conviction that the images showed the corona. He began to calculate probable changes in appearance of the solar corona based on known rates of solar rotation and the assumption that the corona moved with it. He wrote to Stokes 'In *confidence* for the present' in January 1884:

In three of the drawings from my plates there is a very marked V shaped rift which is the most conspicuous feature of the Eclipse photograph . . . Now if we take the synodical rotation of Sun at 27 days (there is a little uncertainty about Sun's rotation say 25.2 days), the successive positions of the rift are not far from where we should expect them to be. The rift itself seems to widen out either in reality or from perspective. It is least distinct on April 3, & widest & strongest on June 6. Of course it may be that the rift has not rotated, but shifted toward the axis from the first plate on April 3.

When I get back the plates I may be able to find the rift on the other side of the axis on some other plates. Mr. Wesley has concentrated his attention on the plates he selected to draw from. He is about to take the drawings to Prof. Bonney to have them engraved. It seems to me that it would be desirable for me, when I get back the plates, to see how much can be got out from them & to put the results in a little paper for the Proceedings. If I do this it would be necessary to have copies of the plates. No doubt the B. Ass. would allow the R.S. to have copies taken. I think they will be engraved in steel. The drawings contain much more detail than those you saw. The plates have been strengthened & Mr. Wesley has had Captain Abney's assistance. I was careful not to influence Mr. Wesley myself.[36]

He included sketches of an apparent rift in the corona that he claimed was visible in the three photographs (see Figure 11.2). He expanded upon these claims in a lengthy letter to Holden in June and conjectured that the persistence of the rift over several rotations of the Sun was indicative of something relatively permanent in the structuring of physical forces in the near solar environment.[37] Margaret, clearly seeing their coronal work opening new avenues of solar research added, 'Astronomical prospects grow wider and wider . . .'

In May, as the Committee's vote approached, Huggins wanted Wesley's drawings (see Figure 11.3) to be displayed at the upcoming Royal Society conversazione, the so-called 'black soirée' held each May to which all serious men of science were invited. He wrote to Wesley on the all-important matter of the drawings' presentation. He thought it best to place them under glass, so they could not be touched, and in a location where they could be illuminated properly. It would be advisable, in his opinion, to accompany the drawings with a 'good sized label calling attention to the rift' to guide viewers' inspection of them.[38] He also believed it wise to include some notice of the drawings in the soirée's printed programme to draw them to people's attention. Huggins concluded by asking Wesley to bring the drawings to the Royal Society a little early in order to 'choose the best position *for light*, so much depends upon this. They should *have a lamp to themselves.*'[39] Shortly after the event, Huggins presented the collection of Wesley's drawings to the RAS. He did not wish to have them in his possession out of concern for later accusations that he had altered them.[40]

11.4 The Riffel expedition

The Royal Society generously allocated £250 to send an experienced photographer to Switzerland to replicate the Hugginses' method of photographing the solar corona without an eclipse at extremely high elevation. They deemed the 8,500-ft Riffelberg near Zermatt to be the ideal location. The photographer selected was Ray Woods, whose experience in photographing the solar corona at both the recent Caroline Island eclipse and the Egyptian eclipse of 1882 made him the logical choice.[41]

Before leaving for Switzerland, Woods consulted with both Huggins and Abney in order to master Huggins's special photographic technique. Meanwhile, over a ton of equipment was sent ahead by boat and rail to the town of Visp in Switzerland's Rhone valley. The fourteen-crate shipment contained heavy but delicate apparatus, supplies of photographic plates and chemicals, as well as building materials to construct a sturdy, weather-resistant observing tent and dark-room. Two crates were required just to pack the components of the coronagraph that Grubb built specially for the expedition modelled after Huggins's prototype.

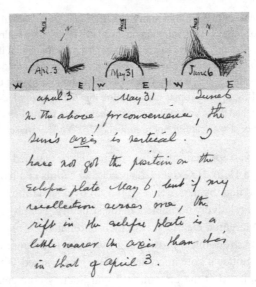

Figure 11.2 Sketches by William Huggins of the solar corona without an eclipse as it appeared on photographs taken at Tulse Hill on 3 April, 31 May and 6 June 1883. (W. Huggins to Stokes, 2 January 1884, Add MS 7656.H1168, GGS, CUL)

Woods documented his expedition in a series of five lively and informative letters that were published between 1 August and 5 December 1884 in *The Photographic News*.[42] When Woods and all the crates finally arrived at Visp, he arranged for their transport from the valley up to the Riffel. Baedeker's *Handbook for Travellers* to the region describes the route from Visp to Zermatt as 'easy and attractive'.[43] Woods found the path steep and narrow noting that it required long carts barely two feet wide to transport the equipment on that leg of the journey. Pack mules and sturdy porters were needed after Zermatt. In fact, Woods had to disassemble the heavier pieces of apparatus and personally supervise several trips up and down the mountain to ensure their safe arrival at the Riffel Hotel, where, at an elevation of 8,430 feet, Woods planned to set up his observatory. [44] The hotel was popular with English tourists anxious to view the surrounding mountain peaks. Baedeker's hailed it as 'an admirable starting-point for glacier excursions'.[45] Its proprietor Alexander Seiler, who had owned and operated many of the inns and hotels in the region for decades, knew how to make his guests comfortable.

Woods arrived on the mountain in July and set immediately to work erecting his observatory and dark-room on a stony outcrop near the hotel. He engaged in a strenuous battle with the alpine winds to secure the structure's roof, but when the job was done he was thrilled to think that his handsome canvas and cardboard observatory was the highest in the world.[46] From then until 21 September, whenever the weather permitted (except on Sundays) he photographed the Sun using Huggins's method. Woods complained that meteorological conditions on the Riffel varied widely from moment to moment. He had little to occupy his time in that isolated place while waiting out a spate of wind, mist or rain. But, as he soon discovered, the summer of 1884 proved to be an inauspicious time to be testing such a delicate and exacting process. The transparent sky normally found on clear

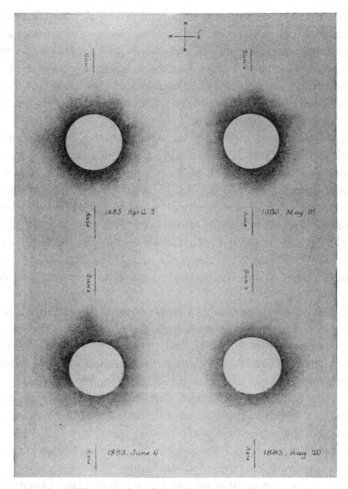

Figure 11.3. Drawings by William Wesley of the solar corona without an eclipse from photographs taken by William Huggins on 3 April, 31 May, 6 June and 20 August 1883. (Plate X, *RBAAS, Southport*, 1883, opposite p. 348)

days at the Riffel's elevation was lost by a chain of cataclysmic volcanic eruptions in Indonesia the previous year.

Beginning in May 1883, a cloud of vapour issued out of Mt Krakatoa, the 2,700-ft high centrepiece of an uninhabited volcanic island roughly the size of Manhattan and located just off the coast of Java. It was estimated that the cloud may have reached a height of nearly seven miles. In mid-June and again in July, the mountain showed signs of activity. On 26 August in the early afternoon, a continuous series of volcanic eruptions, earthquakes and tsunamis rocked the region for nearly twenty-two hours. The next morning, the island of Krakatoa exploded with a blast that was heard nearly half-way around the world. All told, close to 40,000 people are believed to have perished in the disaster.

Volcanic eruptions in the late twentieth century have demonstrated the profound impact that large quantities of fine debris injected into the upper atmosphere can have on astronomical observations.[47] The material from Krakatoa quickly encircled the globe bring-ing brilliant sunsets for artists and poets, but wreaking havoc on the plans of astronomers in need of transparent skies, even as long as a year later. As Huggins later complained, 'It is most certainly matter in the wrong place so far as astronomical observations are con-cerned.'[48] Stimulated by the accounts of seamen located close enough to the island at the time to have witnessed the eruption's destructive force firsthand, the RS formed a committee to study the Krakatoa event and its effects.[49]

Woods frequently reported seeing a great aureola around the Sun. One Sunday, for pleasure, he climbed the Cima di Jazzi. From its 12,000-ft summit he marvelled at the view of the mountains that surrounded him. 'But the most interesting sight to me . . . was the remarkable haze round the sun', he reported.[50] 'A pink glow extended for some twenty degrees around the sun, and at the extremity of this glow was a vivid and well-defined red ring', a clear reference to an atmospheric effect now known as Bishop's Ring, first noted on 5 September 1883, barely ten days after Krakatoa's eruption, by the Revd Sereno Edwards Bishop (1827–1909) of Hawai'i.[51] In Huggins's circle, Abney attributed the aureola to ice crystals, although others, whom Huggins called the 'Krakatowa-ites', blamed it on the volcano.[52]

'If ever I admired my deadliest enemy', Woods continued, 'I did then; and it is my deadly enemy . . . On every clear day we have had here, this peculiar haze has been more or less apparent; and the more strongly it has appeared, the more difficult has it been for me to get on my plates anything that one could say with positive certainty is truly coronal.'[53] Despite this impediment, during his two months on the Riffel – approximately two solar rotations – Woods obtained about 150 photographs, half of which appeared to show the general form of the solar corona.[54] A few even displayed what he considered to be indications of detailed coronal structure, but these were hard-won prizes.[55]

In the fourth instalment of his *Photographic News* series, Woods confessed that the challenges of this photographic project left him unsure 'that the negatives I had obtained were to be relied on, even as a partial success; that much was false I knew'.[56] He worried about expressing too much enthusiasm for his early results out of fear they would later be found incorrect. Huggins, he noted, had done much, of course, to improve his original method before the Riffel expedition began. To reduce one troubling source of unwanted scattered light, for example, Huggins eliminated altogether the need to filter the incoming light using coloured glass or potassium permanganate. Instead he relied on selective sensitivity in the photographic plates themselves to register the presence of light only from the desired narrow region of the solar spectrum. Woods also backed all of the photo-graphic plates with asphaltum dissolved in benzole, a coal-gas tar distillate, in order to prevent halation, an undesirable halo effect caused by stray light reflected off the glass's anterior surface. Although some critics questioned the efficacy of the asphaltum backing, Woods believed the large numbers of consistent images of the corona he had obtained demonstrated its value.[57]

But all these means of improving the quality and reliability of the corona photographs paled in comparison to the effect achieved by the insertion of a small disc, just a little larger than the Sun's image, between the shutter and the photographic plate. Woods wrote, 'The results then obtained were more delicate and more trustworthy than formerly, and in spite of the evil influence of the haze that has prevailed, may yet throw some light on the nature of the corona when a searching examination has been made of them.'[58] After the disc was added, Woods noted happily, the general appearance of the corona photographs remained the same, thus offering strong evidence for the reliability of Huggins's method. Furthermore, because the disc had made his own job so much easier, there was great hope for its future incorporation into observatory routine.

Huggins was pleased with the results achieved despite the 'quite exceptional state of the sky.'[59] In his excitement, he wrote to Holden in America: 'It seems of the first importance that the method be adopted at some elevated observatory.'[60] Where that might be, and under whose sponsorship, was unclear at that point.[61] Perhaps Huggins's mind was on the new Lick Observatory being built on the 4,209-ft high Mt Hamilton in California when he wrote these words. He had long been aware of – and to some extent even involved in – the plans for this observatory financed by James Lick, the wealthy and eccentric American investor, who a decade earlier had set up a trust to build it to house 'a powerful telescope, superior to and more powerful than any telescope ever yet made'.[62]

As we shall see in Chapter 13, Huggins's correspondence with Holden and other American astronomers increased in the early 1880s. Now, he wanted them all to know of his breakthrough in photographing the solar corona without an eclipse. In December 1882, he wrote to Edward Charles Pickering (1846–1919) of the Harvard Observatory.[63] In January 1883, Margaret wrote to Holden.[64] When Benjamin Apthorp Gould (1824–96), an American astronomer then directing the Cordoba Observatory in Venezuela, visited London to receive his Gold Medal from the RAS, Huggins made sure to show him the photographs he had taken.[65]

Selecting a favourable site was crucial to the success of any future efforts to photograph the solar corona without an eclipse, but the method's sensitivity demanded the expertise of a specially trained photographer. In Huggins's view, Ray Woods was that individual. Even before Woods had returned from the Riffel, Huggins began working behind the scenes to play matchmaker between the intrepid young photographer and the right observatory director so that his corona work could continue uninterrupted. To that end, he wrote to Gill at the Cape Observatory, encouraging him to apply for money from the RS's Government Grant Fund to cover Woods's salary, passage to South Africa and other costs of prosecuting the corona project.[66] Although the observatory was situated near sea level, it averaged as many as 350 days of clear sky each year.[67] With prospects like that, the Cape could prove to be the ideal place to conduct the kind of coronal study Huggins desired.

Privately, however, Huggins worried that if Woods went to the Cape, his attention would be diverted to other projects deemed of greater national interest. He deftly played both sides of the Atlantic, confiding in Holden that Gill was hoping to lure Woods to work with him there. He suggested that Holden, in planning for his future staff at the Lick, might make Woods a counter-offer, of say, £250 to £300, in order to tempt him to spend some time there as soon as the new observatory was complete.[68] It would be valuable to have a trained photographer like Woods where the skies were clear so that a programme of solar corona photography could be carried out on a regular basis.[69]

But in November, the Royal Society awarded Gill £300 – £50 more than he had requested – from the Government fund explicitly for Ray Woods to take a 'series of daily photographs of the Solar Corona' at the Cape Observatory.[70] The grant is testimony to Huggins's success in persuading the Committee that the Cape was the place – and Woods the only man – for the job. Its amount reflects Woods's dogged negotiation of a favourable contract with Gill.[71]

Soon after receiving word of the Committee's decision, Huggins placed an order with Howard Grubb to build a new coronagraph for the Cape.[72] Both he and Gill hoped it would be in place by the time Woods got there so the next phase of the corona project could begin in earnest as soon as possible. Woods 'arrived by the mail' on 20 February 1885. The very next day he began familiarising himself with his new surroundings and routines.[73] The corona-graph, however, was still in Grubb's shop. Woods would have to earn his generous salary assisting Gill in other ways for the time being.

11.5 The Bakerian lecture

February 1885 marked a high-point in William Huggins's scientific career. Not only was he feeling extremely optimistic about the future of his method for photographing the corona, but just one week after he celebrated his sixty-first birthday, he became one of only a handful of astronomers to be awarded a second Gold Medal by the RAS.[74] The lengthy address in his honour delivered by RAS President Edwin Dunkin (1821–98) detailed a stunning array of achievements in a wide variety of research areas. Dunkin praised Huggins for having given 'practical life' to the study of stellar radial motion, for his study of nebular spectra, for his pioneering work in observing solar prominences and for his 'recent labours' on the photo-graphic spectra of stars, nebulae and comets. Dunkin hoped that Huggins's photographs of stellar spectra would eventually serve as a foundation for a complete atlas of stellar spectra. He argued there was still much work to be done.

Dr. Huggins's successful application of photography to the subject of these inquiries is now so far acknowledged to be an important astronomical achievement that it is hoped that others will follow his example. But any attempts to follow in his footsteps will certainly end in failure unless the observer is in possession of instruments of the highest class as well as the necessary zeal for the work.[75]

In Dunkin's view, Huggins's great virtue in all this was perseverance. As a pioneer, he had encountered difficulties and occasional failure. Despite this, Huggins invariably turned setbacks into success because he did not give up. William Huggins exemplified the motto: 'The anxious inquirer after knowledge conquers all difficulties at last.' Perhaps this was meant as tacit encouragement for the success of his, as yet unsubstantiated, method of photographing the solar corona, the one area of Huggins's research which received no mention at all in Dunkin's address.

The very next week, Huggins addressed the Royal Institution on the subject of the solar corona. With the Cape work not yet started, he had nothing new to report. So he summarised observations of the corona from recent eclipses and described briefly some of the more popular theories proposed to explain those observations. He pointed out the obvious benefits of having the opportunity to subject the corona to daily scrutiny and provided details of his own method for accomplishing this, which, he was quick to point out, had received the approbation and encouragement of a number of experienced photographers and eclipse observers.

Here, for the first time, Huggins speculated publicly on the nature of the corona. He drew the attention of his audience to the analogous appearance of the structural features of the solar corona to the 'luminous streamers and rifts and curved rays' seen in the tails of comets. Photographs taken during the Egyptian eclipse clearly showed the image of a comet with its tail pointing radially away from the Sun as though pushed by a powerful repulsive force. This evidence when added to reports from numerous visual observations of comet tails made over the years had led to increasing speculation that comets were subjected to 'electrical disturbances' as they approached the Sun, perhaps similar to those responsible for aurorae and lightning in Earth's atmosphere. High electrical potential at the level of the photosphere could account for both the movement and the glow of oppositely charged material far above the Sun's surface, particularly if that material was extremely tenuous. The morphological variety presented by coronal features could be easily accommodated within the bounds of such a theory.

Huggins reminded his audience that, despite such speculation, many questions about the corona remained unanswered. It was still unknown, for example, by what physical mechanism the corona emits its light, or what happens to the ejected coronal matter. Huggins closed his talk with a little demonstration. An evacuated glass bulb lent him by William Crookes contained a small metal ball which was connected to the cathode of a battery. When sufficient potential difference was applied, the metal ball became surrounded by a 'corona of blueish-grey light which was sufficiently bright to be seen from all parts of the theatre'. Huggins believed this simulation of physical conditions in the near solar environment provided a convincing demonstration of electricity's probable role in producing the form and structure of the solar corona. But he refused to speculate further, claiming he was sure that David Gill's regimen of daily coronal photographs would reveal much of additional interest to solar investigators.[76]

One month later, he sent Stokes a copy of his Royal Institution lecture hoping he would 'be disposed to approve of the speculation about the corona'.[77] Stokes must have acquiesced, for, just four days later, Huggins sent off an animated and enthusiastic note:

I am very much pleased to find that you think so well of the corona theory. On this subject I value your opinion more highly than that of any other living man, & your approval is a source of much gratification to me.[78]

Stokes questioned Huggins about the demonstration he had performed with Crookes's apparatus, wondering whether the glow produced by the current in the evacuated tube could rightly be considered a simulation of the electrical effects producing the solar corona. In truth, no one at the time claimed to understand what caused the eerie glow in Crookes's apparatus. And Huggins did not really care about such details. He was satisfied that the glow represented a 'very *attenuated matter luminous in connection with an electric discharge*, which was the point I wished to illustrate'.[79]

In May, the Council of the Royal Society selected him to deliver the annual Bakerian lecture, established by the Society in 1774 in conformity with the terms of the will of Fellow Henry Baker (1698–1774), an antiquarian and naturalist.[80] It was a great honour to be selected. And Huggins was quick to express his deep appreciation to Stokes. 'I heard of the action of the Council in conversation from two or three members of the Council', he wrote. 'I was too much taken aback to say anything then.'[81]

Huggins had already sent Gill good news from Grubb that the coronagraph was finally on its way to the Cape.[82] Now he wrote to let Gill know about the Bakerian lecture.

[I]t is of the *extremest importance* that I should be in a position to say that you have been succeeded [*sic*] at the Cape. Indeed, that your success is the *one thing* now needed to make the proof complete. When this reaches you, you will have had the instrument for some three weeks or so. Under the circumstances will you *telegraph to me* . . . putting your telegram in such words that I can put them in my paper, or lecture . . .

What I want is to be able to say, that *you have succeeded*, if you can say more, so much the better, but the chief point is that the method has succeeded in your hands. If from your plates you can say in confirmation or otherwise any of the points I have raised in my lecture, pray do so in your telegram.

I feel almost overburdened with responsibility in giving this lecture, & I am very anxious to be able to include the commencement of work at the Cape in the lecture . . .

Pray let me have a telegram as soon as you are successful.[83]

Meanwhile, he confessed privately to Stokes that, other than what he had already said in his recent Royal Institution lecture, he felt he had nothing new to say about the Sun – unless, of course, some last-minute confirmatory report should arrive by telegraph from Gill. Huggins explained that he was certain his method of photographing the corona without an eclipse would work, '& yet considering the extreme delicacy of the process, & the amount of doubt existing in some minds, I should have greatly preferred receiving some further confirmation before writing, or speaking again about it'.[84]

His tone betrayed the depth of his anxiety. He was as concerned about his ability to measure up to the high quality expected by the Royal Society in this effort as he had been almost fifteen years earlier when he contemplated the responsibility of accepting custody of the Society's great telescope. He worried whether the skies at the Cape had sufficient clarity to capture the corona's image on film. He worried whether Gill would notify him of positive results in a timely manner.[85] He even worried whether the Council wanted him to show lantern slides to illustrate his lecture. But most of all, he worried whether his lecture would be '*worthy of the Society*'.

Much of this concern can be ascribed to overactive conscientiousness. But Huggins was perturbed by a brief, but public, confrontation over the validity of his claim to have photographed the solar corona without an eclipse. The clash had begun in April 1885, when a letter critical of his method was published in the American journal *Science*.[86] The letter was written by American astronomer and photography expert William Henry Pickering (1858–1938). Not yet thirty years old, Pickering was teaching at the Massachusetts Institute of Technology (MIT) and assisting in the photographic research projects of his well-known older brother, Edward, Director of the Harvard College Observatory. Young Pickering's letter detailed his own attempts to duplicate Huggins's coronal photographs. 'Very corona-like effects were certainly produced', Pickering admitted, but 'unfortunately no two of the pictures were alike'. He concluded that 'though it is very easy to obtain a corona-like image, one may readily be deceived in such matters'. In the event, William Pickering sent a copy of his *Science* letter to Huggins, who was, as might be expected, incensed by the insinuation that he was lacking in certain rudimentary methodological and interpretive skills.[87]

Complicating matters somewhat was the fact that Huggins had recently criticised Edward Pickering for what, in Huggins's view, were 'wildly wrong' measures for the spectral lines in α Lyrae. He wrote to the elder Pickering privately to urge him to re-examine his data because 'it would be better that you should yourself discover the error and publish the correction, than that the matter should be pointed out by others'. Huggins added that he had also sent 'the substance of this note to Prof. Young as his name is on one paper'. He did not, however, tell Pickering that he had taken the liberty to share his views with Edward Holden in a letter that concluded, 'Such papers [as Pickering's] throw back the progress of science, and are much to be regretted, as so many astronomers & book-makers have not the necessary knowledge to distinguish between what is sound & what is not.' Huggins later suspected that his criticisms of Edward Pickering's work contributed to his younger brother's desire to return the favour.[88]

Before Huggins was able to rally his forces in England, Charles Young attempted, albeit with limited success, to cast doubts on the validity of Pickering's own methodology.[89] Meanwhile, Huggins drafted a reply which he sent to Stokes for editorial guidance. The original letter has not been uncovered so we cannot know what changes Stokes may have suggested, if any, to soften its tone, but the letter as it appeared in *Science* was both caustic and personal. In it, Huggins made clear that he was willing to admit failure of his method due to some unforeseen physical limitation or theoretical constraint, but he could not tolerate criticism of his method based on accusations of unskilled ingenuousness.[90]

Pickering's initial letter appeared in print again, this time in the *Photographic News*, which added further insult to the injury Huggins already felt. Fearing a rash of inquiries would be directed to the RAS in search of some explanatory response from him, he sent William Wesley a copy of his reply to Pickering so that the RAS Assistant Secretary could 'answer any point raised' should the need arise.[91]

Huggins did not want to be distracted by a dispute with Pickering, or anyone else, at that time. He had to prepare his worthy address. Nevertheless, he could not ignore the fact that delicate new methods, such as the one he was proposing, would always draw fire. Pickering's criticism alerted him to the potential for similar attacks from his own country-men in consequence of his forthcoming lecture.

Thus, with the lecture date just three weeks away, he wrote again to Stokes. He wished to include something in his lecture about the photographs taken by Ray Woods in Switzerland, because 'I know that astronomers in other countries are much interested in these plates'.[92] However, he worried that discussing the plates would place him in the position of having to state his own opinion of them, and he wished very much to avoid that.

In retrospect, it is clear Huggins harboured personal doubts about Woods's Riffel photo-graphs, doubts that he could not then openly disclose to Stokes. He assured Stokes, in confidence, that he believed the Riffel plates showed the corona, but, he added, 'there are [a] good many markings about which I do not know what to say. Some of these may be, and probably are truly normal, many others I fear are instrumental, or due to stains in the film.' He had considered using Woods's own drawings made from the photographic plates, but he rejected that out of hand, because, in his view, Woods's drawings were unreliable as evidence. Their value, he told Stokes, was 'only as a sort of index to what is on the plates. [Woods] put nearly everything in his drawings, and perhaps too strongly.'

He suggested that Stokes sit down for an hour with Airy's assistant William Christie, Abney and De la Rue, all reputable photographers, and examine the Riffel photographs. Together they could formulate a collective written opinion of the plates: 'If this could be done, so that I could quote the verdict in general terms, it would add much to the interest of the lecture.'[93] Stokes must have demurred, for Huggins made only brief allusions to the plates in his lecture, and introduced supportive evidence not from the plates themselves, but from Woods's interpretive remarks concerning them.[94]

Meanwhile, still stuck in transit limbo, the coronagraph sat idly by in its packing crate.[95] On 1 June, Gill sent his sincere regrets for being unable to 'send the reply you wished relative to the Coronagraph for insertion into the Bakerian Lecture'. He complained, 'Grubb has somehow mismanaged the transport and always seems to do so.' Nevertheless, Gill had 'every hope of complete success when we get the apparatus'.[96]

Huggins's Bakerian lecture was, in many ways, a more elaborate and sophisticated version of the one he gave to the Royal Institution in February. He carefully laid out the theoretical foundation for the preponderance of violet light in the corona which made possible his method of photographing it. He described his photographic method in detail, giving particular attention to the many precautions he had taken to prevent the appearance of false effects – a direct result of his earlier bout with Pickering. He painstakingly built up the case for the acceptance of his interpretation of the coronal images that had been taken by his new method. He emphasised the persuasive value of the sheer numbers of successful plates obtained, the corroborative interpretations of these plates by expert witnesses, and the positive outcomes of the confirmatory tests that had been tried thus far. In his view, adding this new investigative method to the toolkit of solar observers would go far toward answering many of the perplexing questions that remained about the solar corona.

In addition to discussing the details of his photographic method, Huggins elaborated further on the electrical theory of the solar corona which he believed was strongly supported by the photographs taken using his method. He repeated the evidence he had cited in his Royal Institution lecture adding that the fact that coronal streamers could be seen in a number of the photographs pointing in the direction of Mercury and Venus indicated some kind of electrical attraction between the coronal material and the planets. But much of the detailed support for his argument in favour of the electrical theory of the solar corona was not in place at the time the lecture was given.

11.6 The Cape Observatory

At long last, on 23 June, Gill announced in a letter to Margaret 'The Coronagraph has just arrived and looks a very big beast.'[97] (See Figure 11.4.) He hoped 'to have it fairly started in operation in course of a few days'. By 8 July it was 'fully mounted & ready for work'[98] but cloudy skies intervened. More than two weeks later, Gill informed William, 'we have only had one fine day in the Coronagraph sense when Woods secured 8 fine pictures'. Eager for success, he told his friend, 'I think there is little doubt that we have got the real thing, but', he

Figure 11.4 An improved coronagraph built by Howard Grubb (1885) for the Cape Observatory. Contemporary photograph of the instrument employed by David Gill and C. Ray Woods in their attempts to photograph the corona without an eclipse according to Huggins's method. (Ian Glass, SAAO)

added, 'I reserve a definite opinion till we get another similar set. I know that you wish me only to telegraph a mature conclusion and I will not lose an hour when such opinion has been arrived at. I need not say anything to you about what a delicate matter it is to decide upon. I find it more so even than I expected.'[99]

Throughout the summer of 1885, while awaiting definitive news from Gill at the Cape, Huggins worked on refining the text of his Bakerian lecture for publication. He carried on an extensive and what seems to have been for him a theoretically challenging correspondence with George Stokes on the central question of the electrical theory of the corona.[100] It is unfortunate that we only have one side of this correspondence. We are left to speculate on the content of Stokes's expert guidance on the basis of Huggins's queries and comments in his letters to Stokes and from the notes added to the lecture's text after the date it was delivered.

Refining his argument for the electrical theory was not the only difficulty facing Huggins in readying his lecture text for publication. He received more questions from William Pickering concerning his claims to have photographed the corona without an eclipse.[101] Pickering challenged Huggins to photograph the Moon at midday, a feat Pickering claimed to be impossible.[102] Huggins was unsure how to respond. On the one hand, he did not wish to create the appearance that he was shying away from criticism. 'If there were reason to suppose that a mistake had been made, and that the corona had not, & could not be photographed, in that case I should wish to be myself the first to say so. My only anxiety is for the *truth*.'[103] On the other hand, he was concerned about the 'tone' of Pickering's letter, the sense of 'personal animus' that made the matter a delicate one. He sought Stokes's advice on how to handle the situation diplomatically without damaging his own case.

He also asked Stokes what he should do about the lack of supporting evidence from the Cape. Huggins had had high hopes that the Cape photographic regimen would clinch his case. Thus far, it had not. It was not that Gill had been unable to photograph the Sun. He had. In spite of obtaining images that seemed likely to represent the solar corona, Gill could not say he was '*quite sure* about success' given the atmospheric conditions. It was this lack of certainty that plagued Huggins.

Indeed, Gill had sent photographic plates to Huggins for review. Margaret examined them carefully and offered her own advice on how to achieve improved results in the future. 'All depends *now* on Woods' photography', William firmly stressed.[104] The '*first* thing to do', he instructed Gill (and Woods), is to make every effort to avoid instrumental effects. 'The *second* thing is the discrimination of air-effect from the true corona.' Use only 'the *most perfect plates*' and control both exposure and development. Try shorter exposures and short development accompanied by slower intensification. 'If Woods does not take kindly to this', he declared, 'let him show a better way. It is of *but little use multiplying such plates as you have got*. It seems to me you are within a hair-breadth's of definite success, & *this you must get*.'

It was difficult to direct the delicate process from quarterway round the globe. When words failed to convey the visual and kinesthetic secrets of his method, Huggins repeated, underscored and double-underscored the text of his lengthy letters to Gill. '[I]f you only knew exactly what to look for you might be able *even in your present plates* to be sure that it is there ... If you know exactly what to look for you might find it.'[105]

In early October, Gill replied:

we had a glorious clear day on Saturday when Woods carried out your suggestions. He has made some excellent chloride plates which develop very clearly and after short exposure and development and after intensification they do show something very delicate that looks like the corona shape – something certainly much nearer to the real thing than any we have previously secured ... More than this I could not say until we get more pictures of like quality ...[106]

In November, when another hopeful, but this time less certain, letter from Gill arrived, Huggins once again despaired of gaining the kind of absolute proof he desired, especially since even the comparison of Gill's photographs with those taken at the long-awaited New Zealand eclipse of September 1885 was inconclusive.[107]

The excitement and enthusiasm Huggins had felt in early 1885 turned to real exasperation by the beginning of 1886. In January, Arthur Cowper Ranyard (1845–94), a barrister and avid amateur solar observer, read a paper before the RAS, in which he argued that the limitations on photographic plates were such that there was simply no 'chance of photographing the corona under ordinary daylight conditions'.[108] Ranyard had tested Wratten and Wainwright 'extra sensitive' plates as well as commercial dry plates from various makers by exposing them in steps: opening the shutter at selected time intervals and exposing limited portions of each plate. In these experiments, the plates distinguished, at most, only nineteen gradations in intensity whereas contemporaneous studies of human sensory response showed that humans can detect differences in illumination of one in sixty with certainty.[109]

Huggins became increasingly discouraged by what he called the 'rough ways of most professional & amateur photographers', which were hindering any hope of success. The reputations of astronomers, he asserted, were at stake and dependent on the skill, or lack thereof, of these artisans. He cited Pickering and Ranyard as examples of individuals who have an 'obvious incapacity even to understand properly the conditions of the method', while Abney, clearly an exceptional photographer, had been able to detect photographically differences in illumination of only 1 in 120 and was aiming to reach 1 in 200![110]

Just then, encouragement came from an unexpected source: a letter arrived from French astronomer Étienne Léopold Trouvelot (1827–95) of the observatory at Meudon announcing that he had observed the corona when the Sun was behind a cloud.[111] The news provided Huggins with another method to ask Gill to try.[112] But Gill was busy trying to meet Pickering's challenge to photograph the new Moon in front of the corona. He had his own administrative responsibilities and research projects to attend to as well. Feeling overwhelmed, Gill extended the first of several invitations to Huggins to come to the Cape and supervise the delicate photographic work himself. 'You wd have the finest months of the whole year – September & October – when the climate is the finest in the south and you and your wife would have a home and a right kindly welcome.'[113]

In May, having received no response from Huggins on the matter, Gill sent a second, more earnest, plea: 'let me beg you & Mrs Huggins to come to us. It is quite possible that the utmost has not been done here and no one but yourself can say . . . [I]t is very desirable you should try [that] which can be done in our climate.'[114] But Huggins declined the invitation. 'Apart from all personal difficulties, the Committee has not too much money, and I am sure they would not feel justified (I a member of it would not) to give £100 for one to go to do, what Woods, if he is worth his salt, should be able to do.'[115] Huggins was frugal, especially with other people's money. But a letter to Stokes at just that time hints that it is likely he was feeling more sanguine about the prospects for obtaining conclusive results without the need for making such an epic journey: 'Wesley tells me that he can make several *satisfactory* drawings from Gill's plate, & that *he* has no doubt of the reality of the coronal appearances, at least on some of the plates.'[116]

Gill persevered. He wrote to Margaret:

My opinion is still the same that you should come. Woods cannot. I cannot do what Dr. Huggins cd do. Think of his special experience & special knowledge on this and co-related matters, his long preoccupation with similar ideas. You can hardly realize the difference between such special powers and special conditions for such work, as there exists between a man like him, armed and ready at all points, and another like myself who have but recently taken it up, and whose life and thoughts are full of masses of work of varied kinds and part of which only is specially my own. You can rest assured that all I can do I will do, but there is no false humility in my saying that in such a matter as corona-photography, there is between his best and my least a great gulf fixed which I can by no means pass over. I have only one desire in the matter, and that is to arrive at convincing proof that this matter is sound and practical as a method, and to learn what can be ascertained as truth about the Suns Corona.[117]

And he wrote separately to William:

I have already said that I have little doubt the Corona is there – but what alone is useful now is not an expression of opinion but an unquestionable claim of evidence that must carry conviction to all persons competent to form a judgement on such matters … It is not a matter of sensational character … but simply a question of scientific fact to be earnestly, soberly, and quietly worked out … I am sure that the best thing you could do would be to come out and see us – come and try what you yourself can do on the occasion of the Eclipse, in photographing the Sun [*sic*; read, Moon] on the background of the Corona. I cannot be certain that the utmost has been done in our favourable circumstances till your own master hand has tried the work.[118]

But he could not convince the Hugginses to make the trip.

Both Huggins and Gill had already invested time and energy on plans for the 29 August eclipse to which Gill referred in his letter. It was slated to be a long one – over six and one-half minutes at its center point. But its watery path did not present observers with many good land-based viewing options. Before embarking on its trans-Atlantic journey, the Moon's shadow was to cross the tiny Caribbean island of Grenada where totality would last four minutes. It would not make landfall again until it reached the port of Benguela on the west coast of Africa (in modern-day Angola) where the duration of totality was expected to be about forty seconds longer. Gill had thoroughly investigated the weather prospects there. After weighing the limited chances for clear skies against the high cost in time and money, he concluded he could not afford to mount an expedition.[119]

The RS and RAS, meanwhile, pinned their hopes and plans on Grenada. If the weather there cooperated, which was questionable, eclipse observers could reap much valuable information in the time available. Huggins had arranged for another coronagraph to be made for Captain Leonard Darwin (1850–1943), son of naturalist Charles Darwin, to capture images of the corona before, during and after totality. In particular Huggins wished to confirm the presence of coronal streamers in the direction of Mercury and Venus as observations gathered in two previous eclipses had suggested.[120] Unfortunately, the weather was temperamental on the day of the eclipse and, although observers did see the Sun during totality, the corona did not present a particularly pronounced appearance.[121] Huggins was disappointed that efforts to photograph the corona during the partial phases of the eclipse using his method were unsuccessful.[122]

Having opted to stay at the Cape during the eclipse, Gill launched a full-scale assault on the corona during the partial phases visible there:

I cleaned the mirror in the morning with a slice of lemon and washed off with water and afterwards pure alcohol – centred it carefully – and Woods blackened all diaphragms. I looked after everything myself, saw from time to time that the focussing was right, took the chronometer times and checked off the numbering of the plates after each batch of 6 or 8. We began 10 minutes before first contact, and ended 10 minutes after last contact, and working deliberately took 59 plates in all.[123]

Even before Woods finished developing all the plates, Gill reported 'We are unable to … detect the slightest trace of the moon on any of them' despite the fact that the 'pictures seem fairly good'. Further examination might identify 'some faint trace of the moon', he ventured, 'but I am now at least certain that it is quite useless to attempt to study the Corona by photography except in atmospheric conditions very much more favourable than can anywhere be obtained near sea level ... I still think that the thing *should* be possible', he added, 'but I believe *now* that it is only possible at very great altitudes and in a dustless atmosphere'. He closed with an uplifting yet sobering message: 'We must not be discouraged – what is truth is truth, and truth is the noblest and best result.'

Although the description of the corona's appearance provided by visual eclipse observers conformed to his expectations in the sense that luminous regions in the near-solar environment were described as being generally in line with the inner planets, Huggins was disheartened by his method's weak performance. He agreed with Gill that, given the current weakened state of the corona, the method was especially ineffective at sea level.[124] He confided to Stokes, 'I am doubtful as to how I should look at the early favourable results 1880–1883.'[125] Indeed, he began to suspect he had been wrong to eliminate the layer of coloured absorbing material from the coronagraph. Despite problems with the homogeneity of the coloured glass and maintaining the clarity of the potassium permanganate solution, perhaps, he now worried, he had been too hasty in abandoning some sort of filtering agent altogether.[126]

Howard Grubb suggested a new instrumental arrangement to Huggins that promised to make it possible to observe individual images of the Sun that were '*approximately* homogeneous' in wavelength. By 'placing suitable diaphrams [*sic*] in front of prism, & observing with a similar prism, reversed in position, next the eye[,] I got well-defined images of the sun of different colours'.[127] And Stokes continued to provide him with both encouragement and suggestions for further improving his method.[128]

During the autumn of 1887 and spring of 1888, Huggins renewed his efforts, interspersing daytime coronal work with evening nebular observation. In September 1888, he once again accumulated enough photographs which were 'certainly coronal in appearance' to send them on to Stokes for inspection. He feared, however, 'there is not enough evidence to come to any decision upon the point at present'. He also wrote to Gill to request his cooperation in taking yet another round of photographs based on Stokes's suggested improvements to serve as comparisons.[129]

The continuing lack of conclusive evidence in support of his coronal method made Huggins long to be able to show Stokes 'one of the very early trials – *the one*, which made me feel pretty certain about the matter, and *the one*, which when Abney saw he exclaimed: – "It is the corona."' But, he explained, that particular photograph had deteriorated with time and 'I have never since got any photo, showing distinct forms in the same manner.'[130] In strict confidence, Huggins raised the same concerns about the quality of the Riffel photographs that he had expressed when preparing his Bakerian lecture, only this time he carefully began distancing himself from them. In a portion of a letter to Stokes marked '*Private*', Huggins admitted:

It ... seemed desirable [in 1884] to compare drawings of the plates, and a number of drawings were made by Mr. Woods under Capt. Abney's direct superintendence. When we compared these among themselves we (my wife especially, who had spent much time over them, & whose eyes are artistically trained,) began to doubt very much if the features which made one photo differ from another were really due to the corona. We then had here the original negatives, a further examination of these confirmed us in the view, that these features were almost certainly due to instrumental causes of some kind.

We took them back to South Kensington & there went over some of them with Capt. Abney, who still expressed himself in favour of a true coronal origin of these forms, or marked features.

The result of the prolonged examination in our own minds was a strong conviction that all the more marked features seen upon the plates were instrumental, & that any true coronal effect was confined to a very short distance from the limb, & even then was too much mixed up with atmospheric & instrumental effects for any conclusions as to the corona and its possible changes to be got from the plates.[131]

To make matters worse for the confirmation process, Gill's latest photographs arrived from the Cape. Huggins proclaimed his profound exasperation to Stokes:

I think the truth of the matter is summed up in what I said to my wife on my return home on Thursday that 'All the photos are spoilt through *bad* photography.' This, of course, in *confidence*, but it would be well to point out to Gill, if you write, (Gill himself has no knowledge of photography) that the photographs for the corona need much greater delicacy of treatment, that the development must be stopped with greater judgment, & that the chemical processes must be so conducted that a solarized image of the sun is almost quite white. It is badly damaging to the method to send photos which show so little intelligence in their production . . .

I looked at the prints hastily, but I think there are no whites anywhere. All that Woods has to do is follow the processes on the enclosed paper, & use *great cleanliness*, avoidance *of light* coming upon the paper. What is wanted above all is, what I may call, *intelligence* in the darkroom, namely to see when to stop the development, & also to use the solutions weak so that they be slow in their action, & thus enable the operator to stop at the exact moment.[132]

Huggins's strong words indicate more about his own state of mind than Gill's photographic abilities. Gill was one of the foremost celestial photographers of his day, having already begun collaborative work with the Dutch astronomer Jacobus Cornelius Kapteyn (1851–1922) on the *Cape Photographic Durchmusterung*, a prodigious photographic catalogue of nearly a half-million stars in the southern sky.[133]

Huggins showed even greater impatience and disappointment the following summer, when Woods, who had taken most of the photographs Huggins used to corroborate his method, visited London on a vacation from his duties at the Cape. Woods unexpectedly called on Huggins to deliver some more photographs. Huggins was taken aback by what he saw. He complained loudly to Stokes that the photographs showed 'nothing, or next to nothing, which can be supposed to be coronal'. He quizzed Woods on his developing technique and found, much to his dismay, that Woods had, in his view, been processing the coronal photographs incorrectly. 'I have urged him to experiment in London until he can develop properly.'[134] Huggins ceased to have any confidence in further efforts at the Cape and turned elsewhere for evidentiary support.

In late 1889, William Huggins again complained to Stokes, 'There has been only one day in which there seemed to me any chance of the Corona, but I have been so *overwhelmed with anxious work* (up to yesterday) over my paper on Orion . . . Now I will watch everyday.'[135] He sought some positive word from Americans Edward Holden and Charles Young.[136] He eagerly solicited the help of the new Director of the Athens Observatory, Demetrius Eginitis (1865–1934), who advertised the skies above Athens as 'very pure & black', 'preeminently the place' for attempting Huggins's corona method.[137] But three years later, in July 1892, we find Huggins still awaiting news of success from the Athens Observatory.[138]

In 1894 and 1895, the method received considerable, though critical, attention from young American astronomer George Ellery Hale. It was a frequent subject of the correspondence between the two men into 1896.[139] In December 1895, Huggins wrote to David Peck Todd (1855–1939) of Yale University expressing his delight that Todd was going to give the method another trial.[140] Yet, when, in 1897, Huggins wrote his retrospective essay, 'The new astronomy', he did not mention these many efforts to photograph the corona without an eclipse. The perception that he had developed a successful method to photograph the solar corona in broad daylight had inspired tremendous enthusiasm in 1883. But well over a decade later, when the enthusiasm had subsided, Huggins himself helped sweep any memory that remained of his pioneering role into the shadows of astronomy's history.

***

Huggins's efforts to photograph the solar corona without an eclipse show that failure, rather than stifling his research efforts, motivated him to improve his research methods and instrumentation. It forced him to hone his rhetorical and technical skills in order to persuade his colleagues of the soundness of his observations. His struggle to convince others of the validity of his coronal photographs renders visible the ordinarily more tacit discussion of how a scientific community achieves consensus on what counts as conclusive evidence. In an important way, his confidence that the outer atmosphere of the Sun could be studied without an eclipse helped to shape the methods of solar observation that were developed, the types of questions being asked about the solar atmosphere, and the direction in which solar observation was taken up to the turn of the century.

Notes

1. W. H. Pickering, 'An attempt to photograph the corona' (1885), p. 266.
2. J. N. Lockyer, *Chemistry of the Sun* (1887), p. 359.
3. Clerke, *A Popular History of Astronomy* (1885), p. 226.
4. Lockyer had thus far travelled to observe eclipses in 1870, 1871, 1875 and 1878.
5. Clerke, *A Popular History of Astronomy* (1885), p. 226.
6. *The Times* of London, 18 May 1882, p. 5.
7. Roscoe to Schuster, 2 June 1882, Sc. 158, AS, RSL.
8. Clerke, *A Popular History of Astronomy* (1885), pp. 226–7.
9. Young, *The Sun* (1881), p. 19.
10. W. Huggins to Stokes, 7 March 1882, Add MS 7656.H1148, GGS, CUL; W. Huggins to Christie, 7 March 1882, Correspondence of the Society, RASL. The notebook entry recorded by Margaret Huggins and dated 7 March 1882 was clearly written some time after the fact and contains much pointed criticism of Henry Draper and his efforts to obtain the same result. See M. L. Huggins, 7 March 1882, Notebook 2, WCL/SC.
11. Draper, 'On photographs of the spectrum of the nebula in Orion' (1882).
12. W. Huggins to Holden, 10 April 1882, MLSA, LO.
13. 'The Eclipse Expedition', *The Times* of London, 19 May 1882, p. 5d.
14. When the entries begin again, they are written in William's hand. Margaret did not resume her role as note taker until November 1886. There is no explanation given for this in the notebook.
15. M. L. Huggins, 15 December 1882, Notebook 2, WCL/SC.
16. W. Huggins, 'On a method of photographing the solar corona without an eclipse' (1882), p. 410.

17. Flaws in the violet glass encouraged Margaret and William to change the absorbing material to a smooth glass container filled with a solution of potassium permanganate.
18. M. L. Huggins, 15 December 1882, Notebook 2, WCL/SC.
19. W. Huggins, 'On some results of photographing the solar corona without an eclipse' (1883), pp. 347–8.
20. In early March 1882, Huggins had written to Edward Ball Knobel (1841–1930), then Secretary of the RAS, to say that he was in need of a reflecting telescope of about 4-in aperture for some new experiments. He wanted to know if the Society might have such an instrument available for loan and requested the address of the individual who had offered such an instrument for sale at the last RAS meeting.
21. M. L. Huggins, 15 December 1882, Notebook 2, WCL/SC.
22. W. Huggins to Stokes, 12 November 1882, Add MS 7656.H1152, GGS, CUL.
23. The letter from Stokes does not survive, and hence cannot be firmly dated. However, Huggins cited a brief excerpt from it as evidence in support of his interpretation of the photographs in his *Proceedings* paper: 'The appearance is certainly very corona-like, and I am disposed to think it probable that it is really due to the corona.' See W. Huggins, 'On some results of photographing the solar corona without an eclipse' (1882), pp. 412–13.
24. W. Huggins to Schuster, 2 December 1882, Sc. 96, AS, RSL.
25. Abney to W. Huggins, 15 December 1882, inserted in 15 December 1882 entry, Notebook 2, WCL/SC. This letter is reproduced with a few alterations in W. Huggins, 'On some results of photographing the solar corona without an eclipse' (1882), p. 414.
26. Huggins substituted the word 'causes' for 'defects' when citing this letter in his *Proceedings* paper.
27. This anecdote is included in a series of undated notes added to Margaret's lengthy 15 December 1882 entry in Notebook 2 and which, given the dates mentioned in them, were probably written in 1888. Its inclusion at this late date is indicative of the continued weight the Hugginses wished to place upon Abney's initial response to seeing their photographs in light of his later scepticism that the photographs genuinely showed the corona.
28. Christie, 'Dr. Huggins' method of photographing the solar corona without an eclipse', (1883), p. 232.
29. W. Huggins to Stokes, 13 May 1883, Add MS 7656.H1159, GGS, CUL; W. Huggins to Holden, 14 September 1883, MLSA, LO.
30. This eclipse, with totality lasting just under six minutes, is a member of a notorious family of long-duration eclipses (Saros 136). More recent eclipses in this Saros cycle – 11 July 1991 (6 m 53 s) and 22 July 2009 (6 m 39 s) – were seen by millions of people. See Clerke, *A Popular History of Astronomy* (1885), pp. 230–3; Ottewell, *The Under-standing of Eclipses* (1991), p. 87.
31. For a discussion of how this remarkable volume came to be written, see Pang, *Spheres of Interest* (1991), pp. 334–41.
32. 'Dr. Huggins' method of photographing the solar corona without an eclipse', *MNRAS* **44** (1884), p. 203. See also H. A. Lawrance to Stokes, 14 September 1883, included in a letter from W. Huggins to Holden, 9 October 1883, MLSA, LO.
33. Huggins described this note in a letter to Stokes: W. Huggins to Stokes, 12 December 1883, Add MS 7656. H1167, GGS, CUL.
34. *Ibid.*
35. For some indication of Huggins's excitement, perhaps even a hint at the 'bumptiousness' Crookes had noted in his personality a decade earlier, see Huggins's letters to William Wesley and T. G. Bonney regarding the engravings to be made of his photographs: W. Huggins to Bonney, 20 December 1883, MM 7.4, RSL; W. Huggins to Wesley, 22 December 1883, Correspondence, RASL.
36. W. Huggins to Stokes, 2 January 1884, Add MS 7656.H1168, GGS, CUL.
37. W. Huggins to Holden, 2 June 1884, MLSA, LO.
38. W. Huggins to Wesley, 5 May 1884, Correspondence of the Society, RASL.
39. *Ibid.*
40. W. Huggins to Knobel, 9 May 1884, Correspondence of the Society, RASL.
41. 'Dr. Huggins's method of photographing the solar corona without an eclipse', *MNRAS* **45** (1885), pp. 258–9.
42. Woods, 'Photo-astronomy at the Riffel' (1884). For a terser, more 'scientific' report, see Woods, 'Photographing the solar corona' (1884).
43. Baedeker, *Switzerland, and the Adjacent Portions of Italy, Savoy, and the Tyrol* (1883), p. 302.
44. Woods, 'Photo-astronomy at the Riffel, No. I', pp. 490–2.
45. Baedeker, *Switzerland, and the Adjacent Portions of Italy, Savoy, and the Tyrol* (1883), p. 306.
46. Woods, 'Photo-astronomy at the Riffel, No. II' (1884), p. 533.
47. The eruption of Mexico's El Chichon in March and April 1982, for example, put so much material into the upper atmosphere (particularly in the northern hemisphere) that the total lunar eclipse in July of that year was

notable both for its unusual darkness and the asymmetry of its appearance. Mt Pinatubo's eruption in June 1991 caused colourful sunsets the world around and lengthened the average evening twilight for many months as sunlight scattered from particles located as much as sixteen miles above the Earth's surface. See Meinel and Meinel, *Sunsets, Twilights, and Evening Skies* (1983), pp. 39–61.

48. W. Huggins, 'On the solar corona' (1885), p. 209.

49. Symons, *The Eruption of Krakatoa and Subsequent Phenomena* (1888).

50. Woods, 'Photo-astronomy at the Riffel, No. III' (1884), p. 582.

51. Such unusual atmospheric effects are most often observed when fine volcanic ash has been injected into the upper atmosphere. See Meinel and Meinel, *Sunsets, Twilights, and Evening Skies* (1983), pp. 79–81, and Plate 8–1.

52. W. Huggins to Holden, 29 October 1884, MLSA, LO.

53. Woods, 'Photo-astronomy at the Riffel, No. III' (1884), pp. 582–3.

54. Woods, 'Photo-astronomy at the Riffel, No. IV [*sic*; read V]' (1884), p. 772.

55. 'Dr. Huggins's method of photographing the solar corona without an eclipse', *MNRAS* 45 (1885), pp. 258–9.

56. Woods, 'Photo-astronomy at the Riffel, No. IV' (1884), p. 611.

57. Asphaltum in benzole was a popular material for backing photographic plates because of its ease of use. But photographers debated for decades whether it solved the problem of halation or simply made it worse. Naysayers pointed out that applying asphaltum to the back of a glass plate effectively turned the plate into a black mirror that reflected rather than absorbed the short wavelength rays that struck it. See, for example, Penlake [Salmon], 'The question of backed plates' (1911).

58. Woods, 'Photo-astronomy at the Riffel, No. IV' (1884), p. 612.

59. W. Huggins to Stokes, 4 October 1884, Add MS 7656.H1171, GGS, CUL.

60. W. Huggins to Holden, 29 October 1884, MLSA, LO.

61. By the time Woods returned from Switzerland, Abney had spent all the money appropriated for this project by the Royal Society. If Wesley was to be asked to convert the photographs to drawings, more money would have to be requested. Huggins wrote to Stokes to ask for an additional £20 for this purpose from the donation fund. See W. Huggins to Stokes, 4 October 1884, GGS, CUL.

62. W. W. Campbell, *A Brief Account of the Lick Observatory* (1927), pp. 3–17.

63. W. Huggins to E. C. Pickering, 22 December 1882, ECR, HUA.

64. Though still Director of the Washburn Observatory in Madison, Wisconsin, Holden was the designated future Director of the Lick. Holden met the Hugginses during a trip to London in the summer of 1876 (see Holden to Draper, 2 August 1876, HD, NYPL). Afterwards, the Hugginses maintained a friendly correspondence with him.

65. Huggins reported to Stokes that after viewing these photographs, Gould was 'convinced that the appearance was really the corona'. See W. Huggins to Stokes, 13 May 1883, Add MS 7656.H1159, GGS, CUL.

66. W. Huggins to D. Gill, 26 August 1884, RGO15.136/115, CUL; 4 September 1884, RGO15.136/116, CUL; D. Gill to W. Huggins, 17 September 1884, RGO15.136/119, CUL.

67. Krisciunas, *Astronomical Centers of the World* (1988), pp. 195–6.

68. Huggins had originally written £200 to £250, but scratched these out and inserted the higher amounts in the space above.

69. The Lick Observatory was, by this time, some ten years in the planning and building. The 12-in Clark refractor was installed within its small dome in time for the transits of Mercury and Venus in 1881 and 1882 respectively, but those were singular events. No routine observations were carried out until all the construction and instrument-making was completed in 1888. Its official scientific operation began on 1 June 1888. Even if Woods had wanted to forgo his opportunity to work with Gill at the Cape, there would have been little to do at the Lick for almost four more years! See W. W. Campbell, *A Brief Account of the Lick Observatory* (1927), pp. 6–11; Osterbrock *et al.*, *Eye on the Sky* (1988), pp. 38–43 and pp. 53–63.

70. 'Dr. Huggins's method of photographing the solar corona', *MNRAS* 45 (1885), pp. 258–9.

71. Woods did not endear himself to either Gill or the Hugginses with his seemingly cavalier demands for such amenities as first class passage and generous time off to visit family and friends in England. Gill was insulted by Woods's worry that he would be '*buried*' by accepting an assignment at such a distant outpost (W. Huggins to D. Gill, 25 November 1884, RGO15.136/127, CUL). Outraged, Gill replied, 'If his mind is full of holidays, if he considers himself a martyr or "buried" here – the less we have to do with him the better. I can stand no nonsense of that kind' (D. Gill to W. Huggins, 23 December 1884, RGO15.136/135, CUL). One month later, Margaret Huggins wrote to restore Gill's equanimity with a letter headed '*Private about Mr. Woods* . . .': 'Unless we felt satisfied that Mr. Woods is really well qualified to be a Photographic Assistant to so able and energetic a Director as yourself, rest assured we should have dissuaded you from entering into negotiations with

him, – not have done our best to help on the negotiations. Our opinion of Mr. Woods is, that like many another who has some skill in a certain direction but who is very ignorant in most others, he completely mistakes his own position and fancies himself on a level with men who are as gods compared to such as he. As a photographer he is clever: inspire him & he will carry out your aspirations: give him work and he will work hard *with all his heart*.' (M. L. Huggins to D. Gill, 22 January 1885, RGO15.136/147, CUL).

72. W. Huggins to D. Gill, 17 December 1884, RGO15.136/133, CUL.

73. D. Gill to W. Huggins, 23 February 1885, RGO15.136/155, CUL.

74. Recall that Huggins had shared the prestigious award in 1867 with his collaborator, W. A. Miller. Other illustrious recipients of this honour up to 1885 include J. F. W. Herschel (1826, 1836), F. Baily (1827, 1843), G. B. Airy (1833, 1846) and U. J. J. Leverrier (1868, 1876).

75. Dunkin, 'President's address on presenting the Gold Medal of the Society to William Huggins' (1885), p. 292.

76. He addressed the Royal Institution on 20 February 1885, a few days before Woods's arrival at the Cape. W. Huggins, 'On the solar corona' (1885).

77. W. Huggins to Stokes, 25 March 1885, Add MS 7656.H1172, GGS, CUL.

78. W. Huggins to Stokes, 29 March 1885, Add MS 7656.H1173, GGS, CUL.

79. Huggins sent Stokes a proposed replacement for his earlier description of the effect observed in Crookes's apparatus. He hoped this new paragraph would satisfy Stokes's criticisms. Unfortunately, he explained, James Knowles, the editor of the popular magazine *Nineteenth Century*, had requested that Huggins convert the text of his lecture on the corona into an article suitable for his readership. Huggins had already submitted the manuscript for publication, and was thus unable to make any further adjustments to it. See W. Huggins, 'The Sun's corona' (1885), p. 689.

80. For a discussion of the Bakerian lecture, see Lyons, *The Royal Society 1660–1940* (1968), p. 192; and M. B. Hall, *All Scientists Now* (1984), p. 224, n. 13.

81. W. Huggins to Stokes, 4 May 1885, Add MS 7656.H1175, GGS, CUL. The three members who revealed the Council's decision to Huggins were Christie, De la Rue, and Abney.

82. W. Huggins to D. Gill, 23 April 1885, RGO15.136/178, CUL. 'Grubb sent off the apparatus some little time ago, & you have it doubtless by this time.' Unfortunately, the coronagraph would not arrive until the end of June.

83. W. Huggins to D. Gill, 30 April 1885, RGO15.136/192, CUL.

84. W. Huggins to Stokes, 4 May 1885, Add MS 7656.H1175, GGS, CUL.

85. *Ibid.* Huggins had 'begged' Gill to telegraph him immediately, 'which he can do free of expense, in case of scientific urgency'.

86. W. H. Pickering, 'An attempt to photograph the corona' (1885), p. 266.

87. W. Huggins to Stokes, 20 April 1885, Add MS 7656.H1174, GGS, CUL.

88. See W. Huggins to E. Pickering, 12 March and April/May 1884, ECP, HUA; W. Huggins to Holden, 2 June 1884, MLSA, LO; W. Huggins to Stokes, 4 September 1885, Add MS 7656.H1188, GGS, CUL.

89. Young, 'An attempt to photograph the corona' (1885). Young's letter was dated 8 April.

90. W. Huggins, 'An attempt to photograph the solar corona' (1885).

91. W. Huggins to Wesley, 26 April 1885, Correspondence of the Society, RASL.

92. W. Huggins to Stokes, 19 May 1885, Add MS 7656.H1176, GGS, CUL.

93. *Ibid.*

94. W. Huggins, 'On the corona of the Sun – The Bakerian lecture' (1885), pp. 115–16. It is possible that Stokes merely let the issue drop. In preparing the written text of his lecture for publication, Huggins indirectly raised the question to Stokes again: 'I presume it is not desirable to say more about the Riffel plates, than the short quotation I have given of Mr. Woods's own words.' W. Huggins to Stokes, 5 July 1885, Add MS 7656.H1181, GGS, CUL.

95. Huggins explained to Gill that the delay in delivering the instrument was due to the steamer having been diverted for government use. W. Huggins to D. Gill, 2 June 1885, RGO15.136/213, CUL.

96. D. Gill to W. Huggins, 1 June 1885, RGO15.136/211, CUL.

97. D. Gill to M. L. Huggins, 23 June 1885, RGO15.136/217, CUL.

98. D. Gill to W. Huggins, 8 July 1885, RGO15.136/222, CUL.

99. D. Gill to W. Huggins, 27 July 1885, RGO15.136/225, CUL.

100. See W. Huggins to Stokes, 15 June, Add MS 7656.H1180; 5, 12, 17, and 22 July, Add MS 7656.H1181–1184; 19 and 24 August 1885, Add MS 7656.H1185–1186; 4 September 1885, Add MS 7656.H1188; 18 September, Add MS 7656.H1191; and 2 October 1885, Add MS 7656.H1192; all in GGS, CUL.

101. W. H. Pickering, 'An attempt to photograph the solar corona without an eclipse' (1885), p. 132.

102. Huggins claimed he was able to obtain four images of the Moon between 11:30 a.m. and noon. Thus, he believed that 'Pickering is fully disposed of.' To bolster his own view that Pickering had no practical or

theoretical foundation for his claims, Huggins consulted his photographic expert, William Abney, who had become quite an authority on the ability of various photographic processes to detect slight gradations in the intensity of light emitted by the object being photographed. See W. Huggins to Stokes, 4 September 1885, Add MS 7656.H1188, GGS, CUL.

103. *Ibid.*

104. W. Huggins to D. Gill, 8 September 1885, RGO15.136/229, CUL.

105. W. Huggins to D. Gill, 8 September 1885, RGO15.136/231, CUL.

106. D. Gill to W. Huggins, 6 October 1885, RGO15.136/233, CUL.

107. See W. Huggins to Wesley, n.d. and 14 November 1885, Correspondence of the Society, RASL; W. Huggins to Young, 20 November 1885, CAY, DCL.

108. Ranyard, 'On the connection between photographic action, the brightness of the luminous object and the time of exposure' (1886).

109. *Ibid.*, p. 305.

110. See, for example, W. Huggins to Stokes, 22 and 26 January 1886, Add MS 7656.H1194–1195, GGS, CUL.

111. Trouvelot's letter has not been uncovered. It is referred to in W. Huggins to Stokes, 7 February 1886, Add MS 7656.H1197, GGS, CUL.

112. Gill was also trying to photograph the new Moon in front of the corona. See D. Gill to W. Huggins, 2 March 1886, Add MS 7656.H1199a, GGS, CUL.

113. D. Gill to W. Huggins, 9 March 1886, RGO15.136/267, CUL.

114. D. Gill to W. Huggins, 12 May 1886, RGO15.136/277, CUL.

115. W. Huggins to D. Gill, 18 May 1886, RGO15.136/282, CUL. Gill continued to urge the Hugginses to visit and oversee the corona photography work there.

116. W. Huggins to Stokes, 16 May 1886, Add MS 7656.H1201, GGS, CUL.

117. D. Gill to M. L. Huggins, 12 June 1886, RGO15.136/288, CUL.

118. D. Gill to W. Huggins, 12 June 1886, RGO15.136/296, CUL.

119. D. Gill to W. Huggins, 12 May 1886, RGO15.136/277, CUL. The 'possibility of unsatisfactory conditions' and 'the certainty of upsetting useful work here' led Gill to conclude 'there is a very great risk of utter failure'.

120. Just after Huggins delivered his Bakerian lecture, and before it was published, Tennant calculated the positions for Mercury, Venus and Mars at the moment of totality during the 1871 and 1882 eclipses. Huggins was able to relate these positions, particularly those of Mercury and Venus, to the arrangement of notable streamers in the solar corona. See, W. Huggins, 'On the corona of the Sun – The Bakerian lecture' (1885), p. 132.

121. Clerke, *Popular History of Astronomy* (1885), pp. 235–6.

122. W. Huggins, 'Photography of the solar corona' (1886). Huggins wrote this note to *Nature* and a similar letter to *The Times* of London (13 September 1886) because he wished 'to be the first to make known this untoward result'.

123. D. Gill to W. Huggins, 1 September 1886, RGO15.136/336, CUL.

124. He was encouraged in this view by Andrew Ainslie Common (1841–1903), an ardent celestial photographer. See W. Huggins, 'Photography of the solar corona' (1886), p. 470.

125. W. Huggins to Stokes, 23 September 1886, Add MS 7656.H1202, GGS, CUL.

126. W. Huggins to D. Gill, 15 September 1866, RGO15.136/341, CUL.

127. W. Huggins to Stokes, 22 July 1887, Add MS 7656.H1215, GGS, CUL.

128. Stokes was particularly interested in Huggins's speculation that the solar corona was an electrical phenomenon. He had proposed similar ideas relating terrestrial aurorae, geomagnetic fluctuations, so-called 'earth-current' (spontaneously induced current in insulated wire), and solar phenomena in a lecture at South Kensington in April 1881. The lecture, 'Solar physics', was published in two parts in *Nature* **24** (1881), pp. 593–8 and pp. 613–18.

129. W. Huggins to Stokes, 18 September 1888, Add MS 7656.H1221, GGS, CUL.

130. W. Huggins to Stokes, 19 October 1888, Add MS 7656.H1224, GGS, CUL.

131. W. Huggins to Stokes, 7 November 1888, Add MS 7656.H1229, GGS, CUL.

132. W. Huggins to Stokes, 27 December 1888, Add MS 7656.H1231, GGS, CUL.

133. B. J. Becker, 'David Gill', in Lankford, *History of Astronomy* (1997), p. 235.

134. W. Huggins to Stokes, 13 July 1889, Add MS 7656.H1248, GGS, CUL.

135. W. Huggins to Stokes, 27 April 1889, Add MS 7656.H1243, GGS, CUL.

136. See, for example, W. Huggins to Holden, 12 April 1889; Holden to W. Huggins, 4 May 1889; W. Huggins to Holden, 20 May 1889 and 13 July 1889; M. L. Huggins to Holden, 16 July 1889; Holden to W. Huggins, 28 August 1889, MLSA, LO. Also, W. Huggins to Young, 12 April 1889, and W. Huggins to Young, 14 July 1889, CAY, DCL.

137. W. Huggins to Stokes, 13 July and 28 September 1889, Add MS 7656.H1252, GGS, CUL.
138. W. Huggins to Stokes, 1 July 1892, Add MS 7656.H1268, GGS, CUL.
139. Hale had serious, but generally friendly, disagreements with Huggins over the validity of the theoretical basis of Huggins's photographic method. See Hale, 'On some attempts to photograph the solar corona without an eclipse' (1894), pp. 664–6; 'Note on the exposure required in photographing the solar corona without an eclipse' (1895) and 'Note on the Huggins method of photographing the solar corona without an eclipse' (1895).
140. W. Huggins to Todd, 6 December 1895, DPT, YUL.

12

A scientific lady

I have added the name of Mrs. Huggins . . . .

– William Huggins[1]

As focused as the Hugginses were during the 1880s and 1890s on demonstrating the feasibility of their method of photographing the solar corona without an eclipse, they also remained alert to and actively involved in other projects. The astrophysics playing field was becoming crowded with new players – some joining in to assist, others to compete. The rules and even the game itself were constantly changing.

William and Margaret found it both exciting and discomfiting to be involved. To participate successfully required no small measure of vigilance and alacrity: vigilance to defend one's claims to priority of discovery and reputation as an observer; alacrity to keep abreast of the latest investigatory opportunities, technical improvements and methodological innovations. William Huggins had always possessed these characteristics, but as both his career and astrophysics matured, he and Margaret had to remain alert and ready to act to preserve his role as patriarch in the field and maintain what both of them believed was William's rightful place in the history of the new science.

Like their corona work, their investigation of the so-called 'chief nebula line' was a project in which Margaret was actively involved. It embroiled the Hugginses in controversy over methods, instruments and interpretation of received data. Their findings formed the basis of the first paper on which Margaret Huggins appeared as co-author.[2] This paper, which appeared in 1889, was a benchmark in the Hugginses' collaborative relationship.[3]

As we shall see in this chapter, the Hugginses' correspondence and notebook entries during the time they investigated the spectrum of the Orion Nebula bring to light Margaret's essential role in this challenging endeavour. They also provide clues as to why this work and not some other was chosen as the subject of their first joint publication.

12.1 ' . . . zeal and perseverance . . . '

There are, as we have seen, unexplained gaps in the Tulse Hill observatory records including the lengthy hiatus that extended from December 1882 to April 1886.[4] The Hugginses were hardly idle during that period. William's publications along with his active (and Margaret's occasional) correspondence document their tireless efforts to photograph the solar corona and William's struggle to write his Bakerian lecture. The letters also shed some light on the Hugginses' desultory observations of an apparent new star in Andromeda, later dubbed S Andromedae.[5] The puzzling brighter-than-usual condensation of light was first noted by German-born observer Carl Ernst Hartwig (1851–1923) on 20 August 1885. It appeared to be located within (or in the direct line of sight with) the nucleus of the Andromeda Nebula. Some (not Huggins) speculated that the two were physically connected – perhaps a nova in a cluster so distant its individual stars could not be resolved.[6] For his part, Huggins exhibited surprisingly little interest in the object, observing it only twice.

Two brief notes he jotted off to Stokes on 4 September tell us what really consumed his attention at the time. In the first, he reported with some frustration that he and Margaret 'tried hard on the spectrum of this star-like nucleus, but there is scarcely light enough to do much. It gave a continuous spectrum with an abnormal brightness in the orange. I could not be sure if this was due to a bright line, or lines, or to a greater brilliancy of that part of the spectrum.' At almost any other time in his career, Huggins would have dropped everything to take on the new star's teasing challenge. But not even an intriguing new star could distract him from his corona work in 1885. Later the same day, he sent a second note to Stokes to complain that despite 'a good [deal] of anxiety' over Pickering's criticisms of his corona photography 'I do not see I can do anything but wait for Dr. Gills further results.'[7]

The reappearance of William's characteristically terse entries on 3 April 1886 marks the first time in ten years that he, not Margaret, recorded the observations made at Tulse Hill.[8] They document a month of intensive work on Sirius: taking photographs of its spectrum in order to measure and remeasure the star's velocity in the line of sight. The diversion is striking in light of his continued focus on the solar corona. He had not devoted any time to stellar radial velocity since 1872. Besides, Airy had made line-of-sight measures part of the normal routine at Greenwich in 1874 under the supervision of William Christie and Walter Maunder. Although the Greenwich observers did not deem their instruments and method-ology to be sufficiently reliable at first, by 1876 they believed they had worked out the kinks in the system. Huggins admitted there were still 'discordances' between Greenwich's measures and his own for some of the stars' velocities, which would have to be resolved, but he was encouraged by the degree of agreement for brighter stars like Sirius, Castor and Vega. In June 1876, for example, Christie reported a recessional velocity of 25 miles per second for Sirius, compared with Huggins's value of 18–29 miles per second.[9]

But in 1884 Airy alerted the RAS to the curious fact that recent measures of Sirius's radial motion appeared to indicate that the star was slowing down and – possibly – beginning to reverse its direction.[10] 'But that', he conceded, 'is a point that requires a great deal of confirmation.' The next year, Maunder confirmed that, since 1882, Sirius's spectral lines had shifted slightly toward the blue, a signal that the star had begun to *approach* the Earth. Now, in 1885, he announced, the star's 'mean motion of approach' is 22 miles per second! It was, he exclaimed, 'a very extraordinary thing, and no satisfactory explanation of it can be offered at present'.[11]

At the time, Huggins was too overwhelmed with corona work and the Bakerian lecture to follow up. But, a year later, in April 1886, he made time to investigate the radial motion of Sirius for himself.[12] He introduced a new tool to the effort: photography. Instead of relying on visual comparisons of the celestial and terrestrial spectra, Huggins photographed them. It was an ingenious step in the rapidly evolving method of detecting and measuring the small shifts in spectral lines due to radial velocity.[13]

On 3 April, he photographed both the spectrum of Sirius and that of the sky for the purpose of comparison. After inspecting the images by eye, he thought the star's lines appeared shifted slightly toward the red. He then viewed them under a microscope to measure the shifts precisely. A calculation revealed the star to be receding from the Earth at a velocity of 26 miles per second, a value within the range he had obtained previously but completely at odds with Maunder's most recent finding that Sirius is *approaching* the Earth at a rate of 28 miles per second.

Huggins tinkered with the apparatus and battled the weather until 23 April. Then, over the next two weeks, he took numerous photographs. A careful examination of the best of these images, one captured on 24 April, produced the same results as the 3 April plate. On 8 May, he concluded 'This would make Sirius to have a recession as great, as at the time of my early measures, & is inconsistent with M. Maunder's measures at Greenwich, which show a motion of approach.'[14]

He neither published these results, nor mentioned them in any public forum. Furthermore, there are no hints that he discussed them privately. Huggins never shied away from correcting colleagues when he believed they had erred. Even when he was not sure, but strongly suspected, that someone had come to a wrong conclusion, he seldom hesitated to voice his concerns with trusted and more knowledgeable friends. His silence in this case suggests that, despite his scepticism concerning the reliability of Greenwich's radial velocity measures, he lacked sufficient confidence in his own to raise the issue with anyone.[15]

After Margaret's return as record keeper on 16 November, the two focused their attention on stellar and nebular spectroscopy. In her first entry, she offered a hint of how much she had grown in her collaborative role during the lengthy absence, stating bluntly, 'My husband thought the nebula [M57] was brighter in the achromatic. I thought not – but probably I was wrong.'[16] A few months later, she described a disagreement between them over the interpretation of a nebular photograph.[17]

William and Margaret took advantage of the season to observe and photograph the Great Nebula in Andromeda. 'With low powers', she wrote, 'one readily believes there are stars in the central portion.' The illusion is dispelled, she noted, with higher magnification which reveals 'the supposed stars becoming more and more plainly simply *bits of greater brightness* – not stars'. Indeed, as Margaret spent more time examining the nebula, 'it seemed to me more than ever unlikely that the Nova of last year [S Andromedae] was physically connected with it. William ... said to me that he was not inclined to believe in the physical connection' although he now admitted the nebula's central region had changed:

He used to see, a little removed from the nucleus, a spot or little patch of considerable brightness relatively to the rest of the nebula. Now he was conscious of the central portion having *three* or four – three he felt sure of – little bright patches. The Nova of last year was quite different to these: it was distinctly a point – a star.[18]

12.2 The Henry Draper Memorial

Recall that in March 1882, before they became engrossed in their solar corona project, William and Margaret had been hard at work photographing the spectrum of the Orion Nebula. They returned to it with some interest in March 1887. But in May of that year, Huggins was suddenly and brutally awakened to the reality of competition on an unprecedented and, for him, unimaginable scale. This challenge materialised as an indirect consequence of Henry Draper's death some five years earlier.[19] The cause of Huggins's anxiety was the announcement that Harvard College Observatory had just received a large endowment from Draper's widow, Anna, to begin a long-range programme there of photographing, measuring and classifying stellar spectra.[20] The news dredged up unpleasant memories of priority disputes with Draper over photographs of the spectra of Vega (1874–7), the Orion Nebula (1879) and of Comet *b* (1881).

Suffice it to say that, while Huggins was genuinely grieved by Draper's untimely death in November 1882, he was nonetheless relieved to be free of the tension arising from the fact that, in his view, the ambitious young American's research agenda was coming to parallel his own far too closely.

Huggins believed that, during a visit to the Tulse Hill observatory in the summer of 1879, Draper had stolen his research plans and techniques for photographing stellar spectra. Soon after Draper's death, William pled his case in an impassioned letter to Charles Young which included strict instructions to '*burn this note*' at once lest its contents 'add to the grief of Mrs. Draper and his friends'.[21]

Young appears to have destroyed the original letter as instructed. Nevertheless its contents have survived thanks to a copy Margaret sent (with William's permission) to Lick Director, Edward Holden. 'You cannot imagine the pain this Draper matter has caused us', she lamented, confessing 'that for a time I was bitterly angry that my gentle noble-breasted husband should have been so used'. Although 'I do not wish to say hard things', she continued, 'Truth is Truth, and it should be helped to prevail.' In this case, she was certain 'The facts speak for themselves.' After all, 'Who ever heard to any purpose of Dr. Drapers star results before his visit to Tulse Hill?' Margaret acknowledged that 'D. was (and is) your friend. But', she emphasised, 'we are friends too.'

The letter to Young that Margaret passed on to Holden is marked '*Private*'. In it William reminded Young that 'the *idea* of taking photographs of the more refrangible parts of the spectra of stars originated with Dr. Miller and myself . . . in *1863*'. After recounting a brief exchange with Draper concerning photographs of Vega's spectrum in 1876 and 1877, Huggins confided:

In the summer of *1879* Draper and his wife paid us a visit here. I had then nearly completed my work for my paper in the Philosophical Transactions given into R.S. Decem 11th 1879. As my wont is I was willing to show Draper everything. He was greatly surprised at my spectra, and I told him the main points I had made out, and that my paper was in course of preparation. I then offered to show him my special apparatus and arrangements; he immediately said 'I should like to see everything but I have given up star spectra and I do not intend to do any more, you need not hesitate to show me your apparatus.' This statement he repeated *certainly twice*, and I think three times. It was unasked for on my part as I believed him to be a man of honour. Towards the close of his visit he expressed a wish to have a spar prism and quartz lenses like mine 'but I am not going to take star spectra.' I immediately offered to see Hilger and use my influence to get them done in time for him to take with him. He took them with him and *immediately* set to work, got much better spectra than he had before, and tried to anticipate my paper by his paper in Nature, & in American Journal of Science Nov 27th 1879.[22]

Huggins was further aggrieved that Draper subsequently approved wording of a published description of his work that implied Huggins had simply duplicated his accomplishment. Huggins implored Young 'to read carefully my paper in Phil. Trans. 1880 p. 669 and say whether you consider my position in reference to Dr. Draper to be honestly and truthfully put, as "a person repeating his experiment".' He was indignant. 'This is the way I am spoken of in all American accounts I have seen.' He had viewed himself as Draper's friend and mentor. 'I have always been willing to show everything', he declared. 'I have always considered that scientific workers should be above all others, men of honour and of perfect truth.' Draper had betrayed his trust.

Huggins concluded his anguished epistle by asking Young a special favour: 'Still it may be possible for you after a time to bring about quietly without any direct reference to Dr. Draper a more truthful appreciation of my work on the photographic spectra of stars than exists in America if I may trust to popular prints.'[23] Margaret boldly marked this sentence for Holden's attention, noting 'provided you feel justified in [fulfilling William's request] ... we shall be content and the matter need never again be mentioned. We wish to forget in a large and kindly spirit.'

But they could not forget. The announcement of the Draper endowment revived and enhanced Huggins's old worries about his ability to keep apace, let alone make any future contributions of value to the rapidly changing field of astronomical physics. 'A *grave question* is upon my mind', he wrote his trusty confidante, George Stokes:

I have just received a paper from Harvard Observatory & there, through the large endowment of Mrs. Draper the photography of star spectra is to be carried on upon a magnificent scale. Three large instruments are to be kept at work all through the night by relays of photographers & photographs to be enlarged by special methods & measured by other men.

The question is, is it worth my while to continue working in this direction now that it is being done under circumstances with which no zeal & perseverance on my part will enable me to be in an equal position.

It is scarcely worth while to do what will be done well, no doubt, elsewhere –

I do not at this moment see clearly any entirely *new* direction of work ...

As the telescope belongs to the R.S. I am the more anxious to do the best work I can with it.[24]

Having vented his concerns, Huggins soon turned his 'zeal & perseverance' to a new project, one which 'at present has not been attempted at Harvard', namely, photographing the ultraviolet region of stellar spectra.[25] As he was currently 'suffering unfortunately from reduced rents', Huggins claimed he was unable to afford the necessary improvements to the Royal Society's telescope to carry out this programme of research and asked Stokes privately to ask the Council of the Society for a grant.[26]

12.3 The 'meteoritic hypothesis'

In October 1888, William and Margaret returned to their study of the Orion Nebula with a passion, this time to identify once and for all the cause of what was known as the 'chief nebula line'. William had noted the green emission line (500.7 nm) some years earlier in the spectra of several nebulae.[27] It is located tantalisingly close to, but not precisely coincident with, spectral lines associated with several terrestrial elements, most notably, magnesium. The true identity of the line's source remained a mystery until 1927 when, in the wake of recent advances in atomic theory, American physicist and astrophysicist Ira Sprague Bowen (1898–1973) identified it as the mark of a so-called 'forbidden' transition in doubly ionised oxygen (OIII). It was a physical process wholly unimagined – indeed, unimaginable – by pioneer spectroscopists in the 1880s.[28]

On 12 October, Margaret described their plans:

We are very anxious to try and determine whether the Mg line which Lockyer asserts is coincident with the 1st Nebula line, and the Mg line which he also asserts to be coincident with our new Nebula line, – really are so coincident. To try and throw light on this very important point – for much may turn on it – we wish to examine by eye various nebulae for the 1st line, with the 15″ and compare directly with the nebula line the spectrum of Mg.[29]

Norman Lockyer had made quite a name for himself in the scientific community by 1888. He was editor of *Nature*, the renowned weekly science journal he had founded in 1869. He had recently been named Professor of Astronomy at the Normal School of Science (later the Royal College of Science) at South Kensington and given complete authority over the Solar Physics Observatory, an astrophysical research facility funded by the Government.[30] He was also becoming an arch-rival of William Huggins.

The clash of the two men's personal styles coupled with the growing similarity of their research interests had long since placed them on a collision course. They had vied to be the first to view solar prominences without an eclipse, they had bickered over planning the 1870 solar eclipse expeditions and they had taken opposing stands on the national observatory debate. The unresolved question of the identity of the chief nebula line provided one more opportunity for them to meet head-on.

But there were no hints of trouble in the autumn of 1887 when Lockyer began laying down the principles of an exciting and truly universal view of the nature and structure of celestial bodies, a view that had been taking shape in his mind for a decade. He sketched its outline in a lengthy report to the Solar Physics Committee based on years of extensive and intensive laboratory analysis of the spectra of meteorites and metallic elements.[31] Careful comparisons of the positions and characteristic appearances of their spectral lines with those generated by a wide variety of celestial bodies convinced Lockyer that elements commonly found in meteorites such as oxygen, manganese and magnesium are also the main ingredients of stars, nebulae, comets and novae (see Figure 12.1a). He concluded that, despite their apparent physical and structural differences, 'all bodies in the heavens shining by their own light, except stars like the Sun and Sirius, are produced by meteorites in various aggregations and at different temperatures'.[32]

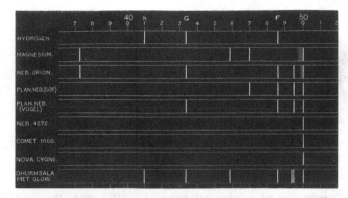

Figure 12.1a Spectra of nebulae compared with spectra of hydrogen, cool magnesium and meteorite glow. (J. N. Lockyer, *PRS* **43**, p. 134)

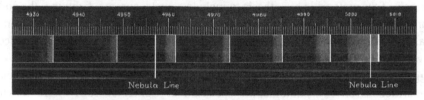

Figure 12.1b Comparison of visible magnesium spectrum with that of nebula. (W. Huggins and M. L. Huggins, *PRS* **46**, p. 50)

Lockyer interpreted nebular bright-line spectra differently from Huggins, who saw them as conclusive evidence of the gaseous nature of their sources (see Figure 12.1b). In his report, Lockyer acknowledged, but did not comment upon, Huggins's contrasting view.[33] For his part, Huggins seems to have had little to say about Lockyer's report. In January, he calmly told Stokes that he had 'no a priori objection to Mg in nebula'.[34] After all, he and Miller had 'found this element, together with iron, hydrogen & sodium, to be widely present' in their spectra. Nevertheless, as always, he preferred to remain cautious: 'there are some important points which have to be determined before any conclusion can be come to on this question'.

In April 1888, Lockyer delivered the Royal Society's prestigious Bakerian lecture.[35] He used the opportunity to flesh out what he called his 'meteoritic hypothesis', beginning with the important role of meteorites in determining the structure and evolution of nebulae and stars.[36] Once again, he pointed to Huggins's assertion that bright-line nebular spectra signal the source's gaseous nature. But this time he listed Huggins's view as the last in a series of outdated attempts to account for the nature and structure of nebulae. Lockyer explained that Huggins's early belief in gaseous nebulae was both understandable and excusable. The observations needed to arrive at the correct conclusion 'were not available to Dr. Huggins when his important discovery of the bright-line spectrum of nebulae was given to the world'.[37]

Ordinarily, Huggins would not have let a comment like that go unchallenged. But, if his correspondence is any indication, he might have been too consumed by angst over Gill's failed efforts to photograph the corona to notice, let alone respond.

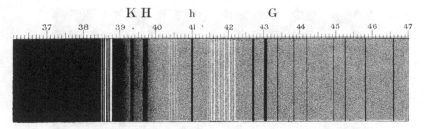

Figure 12.2 Drawing of the spectrum of Comet *b* 1881 by Margaret Huggins. (W. Huggins, *PRS* **33**, opposite p. 2)

He was finally pushed to take action in January 1889 after seeing the text of an addendum to the Bakerian lecture Lockyer had submitted in November for publication in the *Proceedings*.[38] In this lengthy and detailed follow-up report, Lockyer extended the explanatory reach of the meteoritic hypothesis to include comets, aurorae, lightning, the zodiacal light and binary stars. It seems that, while reviewing the existing literature on cometary spectra, Lockyer was led to question the accuracy of a diagram Huggins had included in a paper written years earlier on Comet *b* 1881 (see Figure 12.2).[39]

Huggins was incensed by Lockyer's remarks. He complained to Stokes that Lockyer's recent 'criticisms on my photographic spectrum of Comet of 1881' were 'obviously founded upon an erroneous interpretation of my diagram and paper'.[40] The diagram in question, William explained to Stokes, 'was drawn by Mrs. Huggins & there is a slight error in the relative strength of H & K, but', he emphasised, 'the diagram was not intended to be a *picture* of the solar spectrum, but simply to show the relative positions of the solar lines & the new lines'.

Margaret, it should be pointed out, defended her illustration – if only to herself – by adding the following comment to her original notebook entry made in June 1881:

In making the diagram I was in consultation with my husband. The measures upon which the diagram was laid down were made for the most part by both of us one checking the other and were many times repeated ... But no matter how much care is taken in a matter of this sort, accuracy can only be approximate. The characters and intensities of the lines and groups are I think fairly truthfully represented – but the number of the lines given in the faint group between H and h is only a guess ... But while acknowledging these guesses, I wish to state that they are not guesses at haphazard, but are guesses founded upon careful examination of our comet plates compared with others concerning the interpretation of which there can be no doubt.[41]

Huggins informed Stokes that he had explained all this to Lockyer weeks ago, hoping to clear up the misunderstanding before the report appeared in print. He even showed Lockyer the original plate on which the diagram was based, at which time Lockyer had 'expressed himself strongly, as perfectly satisfied'. But now, 'I see by the proof, that he [Lockyer] has left all the statements uncorrected, & makes matters worse by the note he has added, which, seems to me, to support to some extent the statements in the text'. Huggins wrote once again to Lockyer urging him to correct the error. The reply he received made it clear 'that he does not intend to alter his statements'. Unsure how to proceed, Huggins asked Stokes for advice. '[T]ell me if you think I may, *without a sacrifice of truth by the consent of silence*, allow the matter to pass without sending a short note of explanation to the R.S. I am, for several reasons, very unwilling to send a note, unless it is a *positive duty.*'[42]

Stokes appears to have dealt with Huggins's complaint, if not entirely to Huggins's satisfaction. A week later, he thanked Stokes for his 'kind attention to the matter'.[43] He was grateful 'to be relieved from the necessity of sending any note' to the RS and entrusted the resolution to Stokes's best judgement. Without Lockyer's original manuscript, we cannot tell what changes, if any, were made to the final published version of the paper. Certainly Lockyer's criticisms of the diagram in Huggins's paper on Comet *b* 1881 remained, but there is no note like the one Huggins described among the addenda to the paper.[44]

12.4 The 'chief nebula line'

The confrontation with Lockyer over cometary spectra occurred as William and Margaret were ramping up their own investigations of the spectra of nebulae. Lockyer's noisy claims that nebulae are composed of swarms of meteorites intensified the Hugginses' desire to obtain conclusive evidence to support their own view that bright-line nebulae are gaseous. Recall that when he first observed the chief nebula line in 1864, William had suggested that its proximity to a pair of lines associated with nitrogen indicated the possibility that nebulae contained some exotic form of that element.[45] Subsequent observation dissuaded him of this view, however, and by 1889, he and Margaret were of the opinion that nebulae might be composed of some new, and as yet undiscovered, material. 'Nebulium', a name apparently coined by Agnes Clerke, was the most popular given to this mystery element. William Huggins called it *nebulum*, while Margaret suggested *nephelium* or *nephium* in order to keep in line with the Greek nomenclature used in naming two other recently discovered new gases, helium and argon.[46] Hence, they were inclined to argue that the nebular line, while very close, was distinct from that of any associated with magnesium.

Like all other empirical spectroscopists of their day, they were limited to the familiar and, in most instances, adequately probative method of matching unknown spectra to those of known laboratory samples. They set out to compare several nebular spectra with that of burning magnesium.

The persuasive power of Huggins's reports in the past stemmed largely from the fact that his apparatus permitted direct comparisons of telescopic and laboratory spectra. To conduct this particular investigation to the Hugginses' satisfaction required perfect alignment of the spectrum observed with that of the comparison apparatus. Margaret and William achieved this alignment using Fraunhofer's *b* group, a triplet of green lines in the solar spectrum (516.7 nm, 517.3 nm and 518.4 nm) associated with neutral magnesium, as a visual guide. They both considered this calibration to be of critical importance and they took much care in its execution. As Margaret explained:

It has taken much trouble to get everything satisfactorily arranged ... First, we directed the telescope & spectroscope to the sky ... Then Mg was flashed in as required. We did not leave the apparatus until we both felt satisfied that coincidence between the dark *b* [group] of daylight & the bright *b* [group] of the burning Mg was perfect ... We considered the apparatus as now ready for use.[47]

Throughout the autumn and winter, the Hugginses directed their attention to the problem of the nebular line comparison. In February 1889, William began keeping a separate record of his own observations in an old notebook.[48] The occasional overlap in Margaret's and William's notebook entries during this period provides sparse but valuable insight into their individual research interests, methods and concerns.

On 6 March, after a number of visual observations and one attempt to secure a photograph of the nebular spectrum in direct comparison with that of burning magnesium, William sketched what he had observed. With the crosswire centred on the magnesium line, the chief nebular line appeared to him to be just a bit to the left, or more refrangible side of the magnesium line.[49]

In Margaret's 6 March entry, she mentioned that they were thinking of sending a paper to the Royal Society about their work. To that end, they rechecked the calibration of their apparatus on 9 March.

W[illiam] then put the spectroscope on the Moon bringing in the *b* group so that I might observe whether the Mg lines coincided exactly with those of the *b* group.
     I thought *not* decidedly . . . [50]

William's own 9 March entry makes no mention of the recalibration, but rather concentrates again on confirming his previous observation of the relative positions of the nebular and magnesium lines.[51]

Two days later, Margaret tells us they checked the calibration once again:

W[illiam] thought the bright line did fall coincident. Then I observed. I found a difficulty in getting good observations. A number, did rather give me the impression that it was after all coincident: but one thoroughly good observation showed me as distinctly as I saw it on Saturday night that the bright line was not truly coincident but on one side. I left off feeling certain on the point.[52]

Despite the care they had taken earlier in calibrating their instruments, Margaret noticed that the telescopic spectrum was no longer coincident with that of the comparison apparatus. Although William was initially unable to confirm her observation, she felt 'certain' of it. He became convinced of the misalignment on the night of 11 March, but apparently remained unconcerned, remarking casually in his notebook entry that 'this state of adjustment is satisfactory for comparison of nebulae, and can be allowed for'.[53]

William may not have been aware of it at the time, but Margaret's discovery of the lack of alignment in the comparison apparatus averted what would have been for him an unspeakable calamity.[54] His constant references in his published papers to the care and accuracy of his observations were underscored by his use of four significant digits in reporting his results. If the reliability of the Hugginses' 1888/1889 data were ever brought into question, it would have meant not only the loss of the debate with Lockyer on the question of the chief nebular line, but it would have damaged Huggins's credibility in the wider community of astronomers as well.

By the time the Hugginses recognised the misalignment, the season for observing the Orion Nebula was coming to an end. There was no time to recalibrate the instruments and take chances on the weather providing enough clear nights to make a second set of observations. Besides, their only successful photograph had been taken just a week before the misalignment was discovered. Difficulties in getting good photographs of the nebular spectrum had already persuaded them to abandon further attempts for the moment and concentrate their energies on visual comparisons.[55] Still, the importance of having some photographic evidence to support their visual observations was uppermost in the Hugginses' minds.[56] They simply would have to make do with the photograph they had. Trusting in Margaret's assessment that their spectroscope was 'not *shifty*' – that the observed disparity was constant over all their observations – they based their paper on data gathered when the instruments were ever-so-slightly out of alignment.

The separation between the nebular line and that attributed to magnesium was clearly small. Despite the slight displacement of their equipment which reduced the separation of these lines even more, both Margaret and William were confident they each had observed it. This, Margaret believed, added even greater strength to their argument. She noted:

Now this observation is important for it showed that our arrangements really *displaced the Mg line slightly to the left towards the neb. line*, thus making it more difficult to observe the doubleness by reducing the true separation between the neb. line & the Mg oxide one ...

This observation that our Mg lines are displaced slightly towards the *left*, surely gives great force to our observations showing duplicity ...[57]

In their paper on the spectrum of the Great Nebula in Orion, the Hugginses artfully converted this potentially disastrous turn of events into a forceful argument in favour of their view on the nebula line. They perfunctorily declared the serendipitous misalignment to be a purposeful one – a conscious instrumental adjustment made as part of an experimental design chosen to give their opponents every conceivable advantage:

Indeed, to prevent any possible error in the observation of apparent want of coincidence of the nebular line ... the arrangement was purposely made that the lines of magnesium were seen to fall ... a very little on the more refrangible side of the middle of those lines ... if under such circumstances, the nebular line was seen on the more refrangible side of that of magnesium the observation would be much more trustworthy ...[58]

The day after submitting the paper to the Royal Society, William wrote to Charles Young with an almost smug confidence:

I have just sent into R.S. a long paper on the Neb. in Orion in which I give a good deal of *quite new* matter, which I am sure will interest you greatly ... I took the great labour of comparing Mg-flame spectrum band in the telescope directly with the nebula line. Of course it was not coincident ... [The paper] will probably be 'boycotted' by 'Nature'.[59]

Unfortunately, his confidence was shaken soon after he posted this letter. In several separate comparisons of another nebula's spectrum against that of magnesium, Huggins found the 'end of Mg band, came almost exactly, or very dangerously near the place where I had put the neb. line'.[60] Very worried, he hastily penned a postcard to Young with the distressful message: 'I have just made some observations which will *may* lead me to greatly modify, the statements I made about Mg. Please *burn* the letter I sent you. I will explain in a few days.'[61]

Margaret kept no records during this period. William summarised his efforts to resolve the problem in a long entry dated simply 'Evening, April'. It appears to have been written after he had satisfied himself fully of the outcome. 'During the first half of the observations, I felt *no doubt* in my mind, but I was greatly disturbed by the later observations', he confessed privately. Still, he noted optimistically, 'In no case did the Mg. appear to the left of the neb. line. If the obsers. had been measures, and a mean taken the result would have been quite satisfactory.' He concluded with a word to the wise: 'it is a safe rule to doubt observations when there is not the accordance that may be expected'.[62]

A few days later, a calmer, much relieved Huggins sent yet another note to Young: 'I was too hasty in sending you the post card. It was a moments hestitation only, & the results are more satisfactory than before. All I said in the *letter* you may accept as my views.'[63]

Nevertheless, the Hugginses could not shake their insecurity as they awaited the reading of their paper before the Royal Society on 2 May. On 23 and 24 April, they checked and rechecked their observations. 'Madge compared very carefully', William wrote as he finally declared himself 'satisfied' that the *b* group of the comparison magnesium spectrum was not coincident with the chief nebula line.[64]

On 27 April, William confided to Stokes, the 'Orion paper has caused me immense labour & anxiety', nevertheless with so much that was new to report, 'it was scarcely possible, nor indeed desirable, for me to remain absolutely silent, as to the views of Mr. Lockyer on the presence of "cool magnesium" in the nebula'.[65] Although he had 'refrained up to this time from expressing my opinion' on Lockyer's interpretation of nebular spectra, he wanted Stokes to understand 'I have not accepted that view, because I considered the careful comparisons which I made from 1870 to 1874 of the nebular lines with terrestrial lines, were sufficient to show that "the remnant of the Mg fluting" could not be considered, with, though very near, the nebular line.' He described the 'labourious & anxious task' of directly comparing the spectrum of burning magnesium with that of a nebula. But this difficult work, he added with pride, 'I and Mrs. Huggins, who is now a very trained observer of such things have done to the utmost of our ability, and with the greatest possible care.'[66]

12.5 'I have added the name of Mrs. Huggins ...'

Margaret and William had been working together as a team for over thirteen years when this, their first co-authored paper appeared in the Royal Society's *Proceedings*. Until then, William had neither noted her presence in the observatory nor credited her numerous contributions in his published accounts of previous investigations. Given his awareness of this particular paper's potential for stimulating controversy, he might have introduced Margaret on a gentler slope.[67] Why did the Hugginses choose to submit their first co-authored paper on the challenging question of the chief nebula line? In this instance an array of extraordinary factors came together to push William over the conventional brink.

In the paper's introduction, Huggins proclaimed: 'I have added the name of Mrs. Huggins to the title of the paper, because she has not only assisted generally in the work, but has repeated independently the delicate observations made by eye.'[68] These 'delicate observations' required making repeated direct visual comparisons of the nebular spectrum, a light so faint it tests the threshold of human visual sensitivity, against that of the blinding light of burning magnesium. Needless to say, their observations were exhausting.

In her notebook entries, Margaret complained that the dazzling magnesium light tired her eyes 'even with dark glasses'.[69] She worried about her observations' reliability under these conditions. Always working as collaborative partners during each observing session, Margaret and William served alternately as observer and apparatus tender to allow their fatigued eyes to rest without interrupting the course of the evening's investigation. While William might have been able to overlook Margaret's earlier contributions, these efforts would have been harder to ignore.

In a different vein, it is possible William chose to add Margaret's name to this particular paper in order to diffuse or even deflect blame should a problem arise when others tried to replicate their observations. Lockyer's recent criticisms of the diagram in his paper on Comet *b* 1881, for example, would probably have been on his mind at the time he and Margaret were preparing their paper.

But that was not all. Lockyer had devoted over twenty pages of the 'Appendix' to a discussion of the aurora.[70] He believed spectrum analysis demonstrated auroral light to be the glow of meteoritic dust falling into the Earth's upper atmosphere. Huggins had witnessed what he described as a 'Brilliant Aurora' on the night of 4 February 1874.[71] He recorded both visual impressions and spectroscopic observations of it, but did not publish his findings at the time.[72] Now, fifteen years after the fact and in the midst of a marathon research effort on the spectrum of M42, Huggins felt compelled to respond to Lockyer's claims. He reviewed his old notes and cobbled together a small paper on aurorae for the *Proceedings*.[73]

The page and a half of notes on which Huggins based his paper served him as a worksheet in the true sense of the word (see Figure 12.3). Through the clutter, one can discern the usual spectroscopist's hieroglyphics: a background of text in Huggins's neat and orderly hand; a simple line sketch of a spectrum; numbers, most representing wavelengths, sprinkled throughout. Deciphering these notes as originally entered would be challenge enough. But they are muddled by an overlay of barely legible marginal notes, scratched out text and numbers, with alternatives squeezed in above, below and to the side. All belie at least one – perhaps more – wholesale review, reconsideration and reconstruction of the original text.

The story is further complicated by the published record. Two weeks after submitting his *Proceedings* paper, but before it was read, Huggins amended it with a lengthy comment declaring 'Mr. Lockyer's recent statement that: – "The characteristic line of the aurora is the remnant of the brightest manganese fluting at [λ]558," is clearly inadmissible, considering the evidence we have of the position of this line.' Lockyer responded a month later with a forceful defence of his auroral line claims.[74] He drew critical attention to Huggins's counter-claim that the principal auroral line was seen 'to fall about midway between two strong lines in the spectrum of tin, λ5564 and λ5587'. Lockyer pressed him to elaborate on this assertion since – as Huggins himself admitted – 'further details of this comparison are not given in my note-book'. Emotions ran high on both sides as the battle spilled into their correspondence and – thanks to vocal Lockyer critic, Captain William Noble, writing under the pseudonym of 'F.R.A.S.' – flooded the pages of the *English Mechanic*, where, for weeks, comments from readers and Noble's blunt replies laid bare the intensity of popular pro- and anti-Lockyer sentiment.[75] On 17 May, Noble awarded himself the final word on the subject: '"Lockyer dixit" may satisfy some folk; but I happen to be an astronomer, and, rightly or wrongly, astronomers do not take his utterances as inspired.'[76]

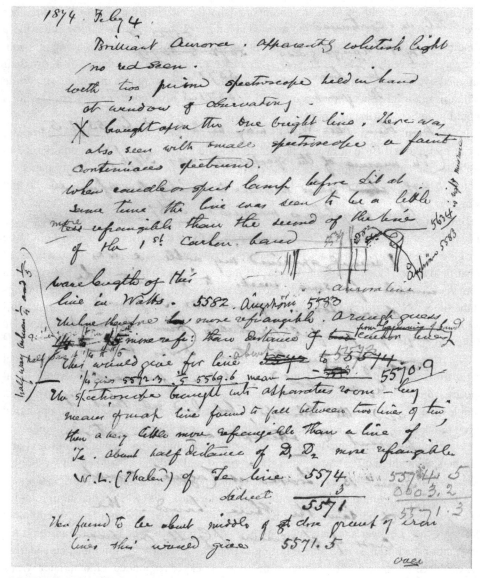

Figure 12.3 Observations of a 'brilliant aurora' by William Huggins. 4 February 1874, Notebook 2. (WCL/SC)

Lockyer ultimately out-manoeuvred Huggins in this skirmish and forced him publicly to acknowledge that the published description of his observations was unsupported by his notebook account:

I find, on reference to my note-book, that the words written at the time are 'map showed reading between lines of tin.' It was in expanding this sentence for my paper that the words 'about midway' came in. I noticed when the paper was in proof that the expression was not strictly accurate, but I allowed it to pass, as the sentence was obviously merely a rough indication of position for the selection of the metal for comparison.[77]

Huggins hoped to close the episode on a conciliatory note. 'I am much gratified that you have spared me from any controversy', he wrote to Lockyer, 'but it was not from any personal feeling but from my strong wish always to do & say what *I believe* to be the Truth that I felt I could not allow the paragraph to pass without correction.' He expressed regret 'that a difference of opinion on several points exists between us but I am anxious that we should remain friends however much we may differ in opinion'.[78]

Perhaps William determined it would be useful to have Margaret in a more visible position this time to help shoulder the burden of proof should Lockyer eventually turn up some damning evidence to counter their nebular line work.

12.6 A scientific lady

Today, it is tempting, but probably mistaken, to suggest that Margaret's new status as co-author was the consequence of her husband's pioneering support of women engaged in scientific work. Consider William's reaction in November 1906 to the news that Hertha Ayrton (1854–1923), a pioneer in the investigation of such disparate phenomena as electric arcs and sand ripples, had just been voted the recipient of the prestigious Hughes Medal at a meeting of the Royal Society's Council, a meeting from which he had been absent.[79] It seems that Margaret, whether out of innocent concern for her husband's well-being, or personal interest in the outcome of the Council's vote, assessed his health on the day of the meeting and judged him too ill to attend. Given the turn of events, Huggins regretted having missed his opportunity to vote in opposition to the award. He complained to Joseph Larmor, the Society's Secretary:

The papers will teem with publications from all the advanced women! I suppose the P[resident, Lord Rayleigh] will invite her [Ayrton] to the dinner, and ask her to make a speech. As the only lady – I should say woman – present, the P. will have to take her in, and seat her on his right hand! And all this comes from what appeared as the *pure accident* of my taking a chill on Wednesday ... Was it Providence on her behalf or was it 'the D – taking care of his own' – which? Can we now refuse the Fellowship to a Medallist?[80]

William Huggins preferred ladies who pursued scientific interests to women who sought scientific recognition.

The transition from sole author to co-author was a difficult one for Huggins. After all, it had been twenty-five years since he was last in that role. The image of Margaret as a subordinate rather than a collaborative partner is more clearly projected in the published paper itself, which is written principally in the first person singular. The first person plural appears less than twenty times in twenty pages, and nearly half of the time it connotes a more general usage. Margaret's name appears twice to credit her independent corroborations of William's observations. However, it should be pointed out that the paper does not mention the nature or quality of these confirmatory observations even though details of corroborating evidence supplied by several other astronomers are given. Even more important, no attention is ever drawn to her methodological or interpretive contributions.

Changing one's accustomed mode of expression takes time, and subsequent papers co-authored by the Hugginses do show an increased use of the first person plural. Most, if not all, readers of this first co-authored paper, unfamiliar with the degree of Margaret's involvement in the research effort, would have had no reason to assume her role to have been anything more than titular.

***

William and Margaret Huggins lived and worked together for thirty-five years as complementary collaborative investigative partners. This new interpretation of their working relationship differs from that presented in the published record and reminiscent accounts. Given the rich store of extant primary source material providing insight into their lives and work, why has the full extent of their collaboration only recently come to light? Why has the image of William as the principal investigator and Margaret as his able, but subordinate, assistant persisted for so many years? It may be argued that the correspondence is too widely scattered, or that the notebooks are not readily accessible to the historian wishing to examine these documents in tandem with the more visible and available published record. These are indeed serious obstacles. But there is a more formidable barrier which must be overcome, and that is the power of the Hugginses' historical image itself.

The traditional and romanticised image of the Hugginses' collaborative efforts is largely their own creation. It has endured because it has been verified and amplified by the published accounts, and because it has fitted the needs and expectations of those who have retold the tale. William and Margaret worked hard to present themselves as classic representations of Ruskin's ideal Victorian couple:

[The man] is eminently the doer, the creator, the discoverer, the defender. His intellect is for speculation and invention ... But the woman's ... intellect is not for invention or creation, but for sweet ordering, arrangement, and decisions.[81]

The strength of this legendary image is captured in the often-reproduced photograph of William seated alone beside his star spectroscope (see Figure 10.2). The absence of Margaret is telling.

In October 1910, after the death of her husband, Margaret Huggins wrote to Joseph Larmor, his former friend and confidante:

No doubt you know about my Pension. £100 a year has been granted me, 'for my services to Science by collaborating with' my Dearest. This I *could* accept without *any* reflection on the memory of my Dearest – & with *honour* to myself as well as to *him*. I do regard the Pension as an honour *to him* though it is honourable also to me, & I humbly hope, – *really earned* for the 35 years of *very hard work*. None of you know *how* hard *we* worked here just our two unaided selves.[82]

Here we catch her in a rare moment of candor which offers us a touching glimpse of her own longing for acknowledgement as neither woman nor lady, but as a scientist in her own right.

Notes

1. W. Huggins and M. L. Huggins, 'On the spectrum, visible and photographic' (1889), p. 40.
2. *Ibid.*
3. See, for example, Ogilvie, 'Marital collaboration' (1989), pp. 111–14.
4. As in the case of the curious absence of notebook entries between 1862 and 1865, we are left to wonder if records were kept in another notebook that has not yet been uncovered.

5. Variable star cataloguer Seth Carlo Chandler (1846–1913) listed the mysterious object as number 224 in his 'Catalogue of variable stars', *AJ* **8** (1888), pp. 81–94; p. 84. At that time, he simply called it '_ Andromedae'. Five years later, he listed it as S Andromedae in his 'Second catalogue of variable stars', *AJ* **13** (1893), pp. 89–110; p. 92. American Association of Variable Star Observers [AAVSO] archivist Michael Saladyga writes, 'I found an interesting note in Chandler's working papers for his catalogues, for this star: "Disc. Hartwig 1885" and below that "Only candidate: hence can call S Androm. or leave blank until Schonf. [German astronomer, Eduard Schönfeld (1828–91)] decides." This seems to suggest that he listed the star as "S And" in his 1893 catalogue on his own initiative; it also suggests that Schonfeld had not officially assigned the name "S And" at least as late as 1888 (?).' Michael Saladyga to the author, 2 September 2009.

6. S Andromedae was the first extragalactic supernova ever observed and one of the most brilliant stars ever viewed from Earth. Its maximum brightness probably occurred on 17 August 1885 and faded from sight in February 1886. See R. Burnham, *Celestial Handbook*, vol. 1 (1978), pp. 1443–6; Jones, 'S Andromedae, 1885: An analysis of contemporary reports and a reconstruction' (1976).

7. All of Huggins's published and unpublished accounts refer solely to two observations of S Andromedae, one on 3 September 1885 and the other on 9 September. W. Huggins to Stokes, 4 September 1885, Add MS 7656.H1187, and 11 September 1885, Add MS 7656.H1190, GGS, CUL; W. Huggins, 'The new star in Andromeda' (1885); 'On the spectrum of the Stella Nova visible on the Great Nebula in Andromeda' (1885).

8. The Hugginses give no explanation for the temporary shift in record-keeping duties.

9. W. Huggins, 'Letter in reply to Father Secchi's letter on the displacement of stellar lines' (1876), p. 73.

10. 'Meeting of the Royal Astronomical Society, January 11, 1884', *AR* **22** (1884), pp. 31–2.

11. 'Meeting of the Royal Astronomical Society. Friday, March 13, 1885', *OBS* **8** (1885), pp. 109–10; Maunder, 'The motions of stars in the line of sight' (1885).

12. W. Huggins, 3, 5, 9, 12, 13, 14, 24, 26, 27, 30 April and 6, 8, 11 May 1886, Notebook 2, WCL/SC.

13. Hermann Carl Vogel (1841–1907), working at Potsdam, is credited as the first to introduce photography in radial velocity measures in April 1887.

14. W. Huggins, 8 May 1886, Notebook 2, WCL/SC.

15. For a summary of the challenges faced by nineteenth-century spectroscopists in their attempts to measure stellar radial velocities, see Holberg, *Sirius: Brightest Diamond in the Sky* (2007), pp. 90–2.

16. M. L. Huggins 16 November 1886, Notebook 2, WCL/SC.

17. M. L. Huggins, 21 March 1887, Notebook 2, WCL/SC.

18. M. L. Huggins, 29 November 1886, Notebook 2, WCL/SC.

19. Draper died in November 1882 at the age of 45 from pleurisy contracted while on a lengthy hunting expedition in the Rocky Mountains. See Plotkin, *Henry Draper* (1972); Whitney, 'Henry Draper' (1970).

20. Plotkin, 'Harvard College Observatory' (1984).

21. See 'Copy of portion of a letter from Dr. Huggins to Professor Young. Private', in M. L. Huggins to Holden, 31 January 1883, MLSA, LO.

22. Draper, 'On photographing the spectra of the stars and planets' (1879).

23. M. L. Huggins to Holden, 31 January 1883, MLSA, LO.

24. W. Huggins to Stokes, 10 May 1887, Add MS 7656.H1211, GGS, CUL.

25. W. Huggins to Stokes, 7 June 1887, Add MS 7656.H1213, GGS, CUL.

26. *Ibid.* Huggins's reference to 'rents' provides one of the only clues as to his source of income after he sold the family silk shop in the mid-1850s. He repeated this complaint a year later as a rationale for requesting more financial assistance from the Royal Society (see W. Huggins to Stokes, 27 October 1888, Add MS 7656.H1225, GGS, CUL). There is no information given as to where his leased properties were located, or what they may have been. In the 1861 census, he and his mother were listed as 'fundholders' under the column for 'Rank, Profession, or Occupation'. Beginning with the 1871 census, however, Huggins merely enumerated his honorary degrees. See Census Reports for Lambeth: 1861, R. G. 9/364/23; 1871, R. G. 10/684/10; 1881, R. G. 11/615/10.

27. W. Huggins, 'On the spectra of some of the nebulae' (1864).

28. Bowen, 'The origin of the nebulium spectrum' (1927).

29. M. L. Huggins, 12 October 1888, Notebook 2, WCL/SC.

30. The Solar Physics Committee (SPC) was established in 1878 after pressure was brought to bear by the Duke of Devonshire on the Committee of Council on Education to provide financial support for the development of astronomical physics. The SPC was tasked to explore new methods of observation, keep tabs on solar research in other countries and to direct the collection and reduction of new solar observations in England. See Meadows, *Science and Controversy* (1972), p. 109.

31. J. N. Lockyer, 'Researches on the spectra of meteorites' (1887); *idem*, 'Researches on meteorites' (1887).

32. J. N. Lockyer, 'Researches on the spectra of meteorites' (1887), p. 118.

33. *Ibid.*, p. 150.

34. W. Huggins to Stokes, Add MS 7656.H1216, GSS, CUL.

35. J. N. Lockyer, 'Suggestions on the classification of the various species of heavenly bodies – The Bakerian lecture' (1888).

36. The phrase 'meteoritic hypothesis' can be found on p. 5 of Lockyer's Bakerian lecture. Lockyer soon brought the meteoritic hypothesis to wider public attention in his book *The Meteoritic Hypothesis* (1890).

37. J. N. Lockyer, 'Suggestions on the classification of the various species of heavenly bodies – The Bakerian lecture' (1888), p. 5.

38. J. N. Lockyer, 'Appendix to Bakerian lecture' (1889).

39. *Ibid.*, pp. 200–1. See W. Huggins, 'Preliminary note on the photographic spectrum of Comet *b*, 1881' (1881).

40. W. Huggins to Stokes, 22 January 1890, Add MS 7656.H1235, GGS, CUL.

41. M. L. Huggins, n.d., but probably early 1889; added under 30 June 1881 entry, Notebook 2, WCL/SC.

42. W. Huggins to Stokes, 22 January 1890, Add MS 7656.H1235, GGS, CUL.

43. W. Huggins to Stokes, 28 January 1890, Add MS 7656.H1236, GGS, CUL.

44. Lockyer, 'Appendix to Bakerian lecture' (1889), p. 184; pp. 188–9; pp. 215–17.

45. W. Huggins, 'On the spectra of some of the nebulae' (1864), p. 444.

46. See M. L. Huggins, '. . . Teach me how to name the . . . light' (1898); Hirsh, 'The riddle of the gaseous nebulae' (1979), p. 203.

47. M. L. Huggins, 12 October 1888, Notebook 2, WCL/SC.

48. W. Huggins, 18 February 1889, Notebook 3, WCL/SC.

49. W. Huggins, 6 March 1889, Notebook 3, WCL/SC.

50. M. L. Huggins, 9 March 1889, Notebook 2, WCL/SC.

51. W. Huggins, 9 March 1889, Notebook 3, WCL/SC.

52. M. L. Huggins, 11 March 1889, Notebook 1, WCL/SC.

53. W. Huggins, 11 March 1889, Notebook 3, WCL/SC. Unfortunately, Margaret stopped making her separate entries in the notebook at this point and did not resume them until September, 1889. Of course, she may have kept her notes elsewhere. William, on the other hand, continued his entries with some regularity into May of the following year. These entries provide additional opportunities to explore the Hugginses' collaborative relationship as the controversy over their nebular line work intensified.

54. This would become clearer to them in the year following the appearance of their paper as Lockyer raised the issue of what he viewed as William Huggins' excessive claims to measurement accuracy. See J. N. Lockyer, 'On the chief line in the spectrum of the nebulae' (1890).

55. See M. L. Huggins, 6 March 1889, Notebook 2, WCL/SC.

56. At the very outset of this research effort, Margaret wrote, 'If we could have such plates then we might have *photographs* of the complete neb. spectrum and photograph the Mg lines afterwards on the same plates as the neb. spectrum is on . . . At present our results are not too satisfactory: but I hope and believe we shall succeed in doing what we wish. It would be very important not to have to depend on *eye* observations in anything so difficult and important as that identity of Mg lines question.' M. L. Huggins, 24 October 1888, Notebook 2, WCL/SC.

57. M. L. Huggins, 9 March 1889, Notebook 2, WCL/SC.

58. W. Huggins and M. L. Huggins, 'On the spectrum, visible and photographic' (1889), pp. 48–9.

59. W. Huggins to Young, 12 April 1889, CAY, DCL. At the top of the letter is written in William's hand: 'Please consider statement about paper confidential until it is read.'

60. W. Huggins, Evening, April 1889, Notebook 3, WCL/SC. This note seems to describe events that occurred between 12 and 18 April. The next dated entry was recorded on 23 April.

61. W. Huggins to Young, 15 April 1889, CAY, DCL.

62. W. Huggins, Evening April, 1889, Notebook 3, WCL/SC.

63. W. Huggins to Young, n.d. [probably around 18 April 1889 given the date of receipt stamped on the card], CAY, DCL.

64. W. Huggins, 23 April 1889, Notebook 3, WCL/SC. The rivalry – for that is what it was – between the Hugginses and Norman Lockyer over the identity of the chief nebular line subsided during the remainder of 1889, only to flare up again in early 1890. The details of this unpleasant episode will be discussed in Chapter 13. See Meadows, *Science and Controversy* (1972), pp. 183–6, and Osterbrock, *James E. Keeler* (1984), pp. 92–103.

65. W. Huggins to Stokes, 27 April 1889, Add MS 7656.H1243, GGS, CUL.

66. *Ibid.*

67. There is no way to know with confidence if William's adding Margaret's name to this paper was an action taken on his own initiative, or if it resulted from her insistence on recognition for her contributions. Thus, I have used the word 'introduce' here in a very literal sense, to refer to the fact that Margaret's inclusion in the title of this paper was made possible by William's Fellowship in the Royal Society.

68. W. Huggins and M. L. Huggins, 'On the spectrum, visible and photographic' (1889), p. 40.

69. See M. L. Huggins, 24 October 1888 and 9 March 1889, Notebook 2, WCL/SC.

70. J. N. Lockyer, 'Appendix to Bakerian Lecture' (1889), pp. 217–41.

71. The auroral display was also seen by others. See 'Auroral display', *Nature* **9** (1874), p. 303.

72. W. Huggins, 4 February 1874, Notebook 2, WCL/SC.

73. W. Huggins, 'On the wave-length of the principal line in the spectrum of the aurora' (1889).

74. J. N. Lockyer, 'On the wave-length of the chief fluting seen in the spectrum of manganese' (1889).

75. W. Huggins to J. N. Lockyer, 6 May 1889; J. N. Lockyer to W. Huggins, 7 May 1889; and W. Huggins to J. N. Lockyer, 7 May 1889, JNL, UEL. 'The spectrum of the aurora . . . ', *EM* **49** (5 April 1889), p. 108; 'The meteoritic theory and "F.R.A.S."', *EM* **49** (3 May 1889), pp. 195–6; 'Scientific News', *EM* **49** (10 May 1889), p. 217; and 'The meteoritic theory and "F.R.A.S."', *EM* **49** (17 May 1889), pp. 236–7.

76. 'The meteoritic theory and "F.R.A.S."', *EM* **49** (17 May 1889), p. 237.

77. J. N. Lockyer, 'On the wave-length of the chief fluting seen in the spectrum of manganese' (1889), p. 38.

78. W. Huggins to J. N. Lockyer, 7 May 1889, JNL, UEL.

79. Hertha Ayrton was married to physicist William Edward Ayrton (1847–1908). Her interest in fluid vortices led her to conceive of a mechanical device for dissipating clouds of poison gas during World War I, which came to be known as the Ayrton fan. Mason, 'Hertha Ayrton (1854–1923) and the admission of women to the Royal Society of London' (1991), p. 206.

80. W. Huggins to Larmor, 2 November 1906, Lm 948, JL, RSL. For additional discussion of this incident, see Mason, 'Hertha Ayrton (1854–1923) and the admission of women to the Royal Society of London' (1991), pp. 214–16.

81. Ruskin, 'Of Queens' Gardens' (1905), p. 107.

82. M. L. Huggins to Larmor, 17 October 1910, Lm.790, JL, RSL.

13

Foes and allies

No doubt Dr. Huggins can give you some pointers. You know he is the
founder of the science of astronomical spectroscopy.

– James Keeler[1]

By 1890, William Huggins had acquired considerable renown and prestige. He also faced
serious threats from all sides: refutations and criticisms from Lockyer, reports that old
problems were being conquered with improved instrumentation at other observatories,
and lack of recognition from men too young to remember his pioneering role in the
development of observational techniques they took for granted.

How did he manage to keep himself on the forefront of discovery and hold his detractors
at bay? One way he did this, as we shall see in this chapter, was by cultivating and nurturing
personal alliances with prominent American astronomers. He had discerned early on that the
locus of cutting-edge astronomical research was shifting from the Old World to the New.
Visionary plans were being drawn up and executed on the other side of the Atlantic thanks to
eccentric tycoons like James Lick and Charles Yerkes (1837–1905) with egos and fortunes
large enough to cover the cost of erecting monumental observatories equipped to face the
challenges of the new astronomy. It is emblematic of the eclectic and dynamic nature of
Huggins's investigative interests and methods that he embraced the work of these new
American observatories and made use of their resources to further his own research goals.

13.1 Controversy

In June 1889, the Hugginses published a routine update on earlier observations of the spectra
of Uranus and Saturn.[2] But, for the most part, after submitting their paper in April on M42's
spectrum, they returned to their on-going solar corona project until autumn and winter skies
brought their favourite target nebula back into view.

Lockyer, meanwhile, continued to focus on explaining and promoting his meteoritic
hypothesis. It was a task he had been pursuing with considerable enthusiasm since delivering his Bakerian lecture in April 1888. That August, he launched a series of 'Notes on
meteorites' in *Nature* to reach the wider community of interested laymen.[3] In February
1889, he showed the scientific elite in the RS how the meteoritic hypothesis and its
interpretive framework could guide spectrum analysts in dividing stars he classified as
Group III into two distinct subcategories: those that are growing hotter (forming), and those

that are cooling (dying).[4] In May, he used his meteoritic hypothesis to criticise Huggins's recent paper on aurorae.[5] In November, he demonstrated its power to explain complex patterns of variability in the brightness and colour of certain stars.[6] He published an update to the Bakerian lecture and its Appendix in December in which he boldly concluded that 'It is now universally agreed that comets are swarms of meteorites' and confidently asserted that the data he had presented thus far 'show that nebulae, bright-line stars, stars with mixed flutings, and the aurora have spectra closely resembling those of comets, and are therefore probably also meteoritic phenomena'.[7] He followed this with a brief report offering further support for his claim that nebulae and 'many of the so-called stars' are really 'sparse groups of meteorites'.[8]

Lockyer submitted one other paper to the RS that December. It was a lengthy and contentious response to the Hugginses' charge in their paper on M42 that 'I [Lockyer] am wrong in my identification of the origin of some lines in the spectrum of the nebulae.'[9] He began with a polemic on the futility of declaring five-figure accuracy when instrumental and human observational limitations place practical restrictions on the precision attainable by even the most experienced observers working with the best of instruments. The Hugginses positioned the 'chief nebula line' at $\lambda5004.6$ to $\lambda5004.8$. Lockyer, by contrast, put it simply at $\lambda500$. The seemingly different values were all one and the same to him. Any apparent 'discrepancy' was much smaller than the margin of error due to tracking inconsistencies, imperfect focus or involuntary eye movements. To drive his well-taken point home, he drew attention to the variations among Huggins's five-figure wavelength measures and those of other observers that Huggins accepted as reliable. These differences would vanish if only three figures were used.

His real disagreement with the Hugginses stemmed from their insistence that the chief nebula line is 'perfectly sharp and well-defined', a feature central to their view that nebulae are gaseous.[10] Lockyer, on the other hand, had been struck in his many observations by a lack of clarity in the line, particularly along its more refrangible edge. Indeed, it was the gradual way in which the line's brightness appeared to taper off on that side that had led him to conclude it was the 'remnant of a magnesium fluting' common to the spectra of cool burning meteorites. He was quick to point out that others – even Huggins – had described the line as 'nebulous at the edges' in the past. Could it be that Huggins has become blind to evidence that contradicts his own theoretical preconceptions?

Finally, although Lockyer deemed his adversary's claim that nebulae are hot and gaseous to be 'untenable', he was pleased to note that Huggins had abandoned his old view that nebulae possessed 'a structure, and a purpose in relation to the universe, altogether distinct and of another order from' all other celestial bodies. Instead, Lockyer cheerfully observed, Huggins was now acknowledging a key element in the meteoritic hypothesis, namely that nebulae 'represent an early stage in the evolutionary changes of the heavenly bodies'.[11] He considered it a small but significant victory.

The RS permitted Lockyer to read his paper at the Society's 16 January meeting, but the Committee of Papers ordered its publication 'suspended for further consideration'.[12] In the meantime, the Council appointed chemist George Downing Liveing and physicist Arthur William Rücker (1848–1915) to conduct an independent and disinterested inquiry into the claims from both sides and to referee the papers the litigants had submitted to substantiate them.[13]

Liveing visited Tulse Hill on 9 February to use the Hugginses' own apparatus and see for himself how the positions of M42's spectral lines compared with those of burning magnesium. He wrote his impressions directly in their observatory notebook declaring that he 'repeated the observations several times ... came to same conclusion, namely that the edge of the fluting was less refrangible than the nebular line. Observed the nebular line with various widths of slit ... The same appearance as to sharpness on the more refrangible side was observable ...'[14]

He returned on 29 April to check 'Dr. Huggins' arrangement for throwing in a comparison spectrum attached to his telescope, with a view to see if there could be any error arising from the direction in which the light is thrown on to the slit of the spectroscope'.[15] Again he described his observations in their working notebook, noting 'The pointer was first placed on b_1 [in the magnesium triplet], and then I carefully observed this line while Dr. H. moved the burning Mg from side to side. This was done several times but I could not detect any shift of the line upon the pointer in consequence of the movement.' After repeating the test with the pointer on 'the brightest part of the band of Magnesia', Liveing attested 'I could not detect any shift; and I came to the conclusion that there is no sensible shift due to moving the burning magnesium.' Finally, he compared the magnesium band with a narrow line in the spectrum of lead which Huggins had used as a fiducial mark to gauge the chief nebula line's position. Satisfied with what he found, Liveing reported 'So far as my memory will serve the distance from the edge of the MgO fluting at which the nebular line appeared when I observed it on Feb 9 was not far short of the distance now observed between the lead line & the edge of the MgO fluting.'

13.2 American allies

Not content to leave the controversy's resolution solely in RS hands, Huggins looked across the Atlantic for support. As early as 1874, when the Lick Observatory was still just a glint in its namesake's eye, Huggins had noted that construction of this 'giant telescope for St. Francisco (Slick or Slick's money)' would mean greater competition for instrument makers' limited time and energy.[16]

In 1877, in the early stages of his rivalry with Draper, Huggins wistfully told Holden 'though you Americans have "l'*avenir* du monde" you cannot rob us of the past'.[17] Huggins had just returned from the French city of Le Havre where he had gone on 'a pilgrimage to places of interest in connection with our conquerors (& yours too) the Normans'. In a literal sense, then, the 'past' he referred to was Europe's event-filled millennia. America, by contrast, was an infant looking forward to a history that remained to be written. But figuratively, given his concerns about Draper's recent photographic advances, he was reminding Holden that celestial spectroscopy had a venerable 'past', too. He was prepared to fight to protect that past and his contributions to it from usurpation by young upstarts.

By 1890, the Lick Observatory was up and running with Holden as Director. Huggins, ever the pragmatist, was ready and willing to avail himself of its unparalleled power to probe the heavens. In late January, less than one week after Lockyer read his critical paper, 'On the chief line in the spectrum of nebulae', William sent a plea to Holden: 'If you should have made any observations on the spectrum of the Orion nebula, I should be very grateful to you if you would let me know any results as soon as possible.'[18] Holden obliged by putting his top observer, James Edward Keeler, on the task.

Huggins was delighted by Holden's 'ready kindness to examine the Nebula of Orion'.[19] To prepare the eyes and minds of his proxy observers at the Lick for making these '*very delicate*' observations, he emphasised that Liveing 'was "quite sure", to use his own words, that the 1st line is where we put it, a very *little* more refrangible than the terminal line of the MgO band'. He reminded Holden of the 'extreme precautions' required to avoid 'shift – flexure of spectroscope, &c.' Having Liveing's confirmation did not '*make us less anxious to receive the report of your observations*, which we cannot but believe will be confirmatory'.

He advised Holden to be alert to the mottled appearance of the nebula which 'may give to an incautious eye the appearance of a serrated line, but the *apparent* increased breadth of the blotches extend equally on *both sides of line*. It is *apparent* only, and we can see no indication whatever of a *flare*, or a fluting'. Besides, he insisted, 'From the chemical side it is all *but impossible* that the line can be the residue of the fluting of magnesic oxide, – when we know hydrogen is freely present.'

On 3 April, Keeler informed the Hugginses that four observations he had made of M42 with the Lick's 36-in telescope supported their interpretation of its spectrum. 'One thing that struck me particularly', he added, 'and that there could be no doubt of, was the perfect sharpness and fineness of the nebular lines under the very considerable dispersion used. There is not the least doubt in my mind that they are all of gaseous origin – not "remnants of flutings".'[20]

William shared Liveing's and Keeler's confirmations with Charles Young at Princeton. He hoped the good news would allay any concerns that might have beset Young upon learning that 'Lockyer has sent a long paper (abominable in tone & in misrepresentation) asserting that the neb. line is coincident with Mg & is a fluting.'[21]

In a separate letter, Margaret further strengthened their alliance with Young by complimenting his supportive work on M42's spectrum. It had been a great help:

It is not easy work – the direct comparison of it with the MgO flame: but we have spared no pains to try & find the truth about the matter. Night after night we have worked at this matter; and we have discussed and *criticised* our results *with the most anxious care*. I have an unspeakable dread of wrong observations. In the long run, of course Truth must prevail: but even one wrong observation may be in the way of Truth for a time. So much do I feel all this that in spite of loving Science, she robs me in a way, of peace of mind and I think her great text is 'Truth hath the plague in his house.'[22]

Two days later, William wrote to Young once again. 'Lockyer's paper [on the chief nebula line] is more unfair than probably is clear to you without going into minute details', he warned. 'The attention of the Council R.S. has been called to it [here he inserted "*not by me*"] and it is uncertain at present whether [Lockyer's paper] will be printed or not.'[23] He included a near-verbatim account of Liveing's testimonials to underscore the support they have received. But he confessed to being worried by a reference in Keeler's confirmatory letter to 'some source of constant error which could not be discovered for want of opportunity to apply the proper tests', a problem Keeler recognised after comparing his results with those the Hugginses had published. Huggins interpreted Keeler's remark to imply the source of error lay in the Tulse Hill apparatus or method. 'We cannot detect a suspicion even of error in our observations', he anxiously confided in Young while soliciting his thoughts (and assurances) on the matter.[24]

William concluded his long letter with words of consolation for Young on *Nature*'s tepid review of his new book, *The Elements of Astronomy*.[25] 'You must not mind the criticisms in "Nature"; everybody here knows that Fowler [spectroscopist, Alfred Fowler (1868–1940)] is Lockyer's assistant, and writes to order, to praise or blame just in proportion as L's views are accepted or not.' Clearly he and Young were confederates beleaguered by a common enemy.

13.3 Irreconcilable differences

Liveing and Rücker turned in their report to the RS on 27 May. They had read the papers being considered for publication – three from Lockyer; two from the Hugginses.[26] And they had carefully reviewed the additional information the two parties had offered in support of their differing views. They recommended all be published in the *Proceedings*, although they requested Lockyer first revise the one he had read in January which had prompted the Society's intervention. As for deciding whether the nebulae were gaseous or meteoritic in nature, they believed 'it would be unwise for the Royal Society to accept either alternative as fully proved'.[27]

Liveing and Rücker had found it difficult to untangle the knotty issues at the heart of the controversy. 'Both methods used appear to us well-suited to the end in view. Both observers tested their arrangements by comparing the position of the lines given by burning magnesium with *b* in the spectrum of the moon. Unfortunately their results differ.'

The negotiations taxed their diplomatic skills. 'The matter could best be settled by friendly cooperation', they declared. Happily, in some cases 'our criticism appears to have been due to a misunderstanding caused by some ambiguous expression, & when this was modified the ground of our objection was removed'. But they could not always coax the two rivals to work toward reconciliation. Lockyer, for example, 'gave Dr. Huggins three invitations to inspect & criticise his apparatus & methods'. Regrettably, 'Dr. Huggins did not avail himself' of that opportunity. Furthermore,

The notes sent in by Dr. Huggins were of such a character that it would have been quite impossible to have shewn them to Mr. Lockyer. We have therefore been compelled not only to assume the responsibility of deciding which of Dr. Huggins' criticisms were pertinent, but of ourselves stating them in our own terms to Mr. Lockyer. This appears to us to be fair to neither party. Dr. Huggins might repudiate our interpretation of his notes. Mr. Lockyer might hereafter complain that some of Dr. Huggins' criticisms were not placed fairly before him. We have had alternately to state & to decide upon the case of one of the disputants.

They concluded that resolving controversies 'in private under the auspices of the Council of the Royal Society' is not only impractical, but imprudent. In the future 'it would be better if in such cases both parties were made clearly to understand that all documents would be open to the inspection of both sides'.

In the summer of 1890, Keeler sent further confirmation of Huggins's claim that the chief nebula line was not located in or near the magnesium *b* group.[28] But Lockyer was already focused on a different comparison: matching lines in the spectra of nebulae and comets with those of carbon. Huggins had to change course to follow Lockyer's new tack. This sort of shift in the subject of debate is a feature common to scientific controversies, particularly those, like this one, that involve phenomena with a seemingly endless array of independent variables and limited theoretical structure to guide investigators in selecting the salient ones among them.

Huggins responded by pressing Holden to put the Lick's powerful instruments to work again on his behalf. It was bad enough that Lockyer 'has been trying to find Carbon lines in the spectra of the bright line nebulae as well as Mg (which is *not* there)', Huggins complained, 'but he has also asserted that the carbon bands (or one of them) are to be seen in the *continuous* spectra of nebulae'.[29] If Lockyer were successful in making this connection, he could then include nebulae like M31 in his grand meteoritic scheme of the cosmos. 'He now asserts that his assistants (who I fear *must* see & find what he orders) have seen the carbon band in *Neb. 4058* [NGC 5866].' 'The chief thing at this moment', he urged Holden, 'is *a thorough examination of spectrum of 4058.*'

Lockyer's colleagues were losing their patience with his continued use of the RS as a forum to promote his meteoritic hypothesis, especially now that he had published a popular book devoted to the subject.[30] Reports from referees of a paper he submitted in November 1890 on spectroscopic observations of novae hint at efforts to rein him in. William Christie wrote:

This paper is merely a very lengthy discussion of the observations of the three Novae of 1866 [T Coronae], 1876 [Nova Cygnus] and 1885 [S Andromedae], which as far as they go have an important bearing on the author's meteoritic hypothesis. But the observations are necessarily subject to much uncertainty and I doubt whether trustworthy inferences can be drawn from them in regard to identity of lines in the spectra or changes of colour.[31]

Another reviewer was physicist George Howard Darwin (1845–1912), son of Charles Darwin, who had been moved by Lockyer's earlier spectroscopic evidence regarding the abundance of meteoritic material in luminescent celestial bodies to test the physical and mathematical likelihood that meteorite swarms could play a role in the origin of stars and planetary systems like our own.[32] Darwin was hardly opposed to Lockyer's general hypothesis, but he urged Lockyer to remain more circumspect in his claims. '[W]e are said to *know* that comets, nebulae and bright line stars are swarms of meteorites', Darwin wrote. 'Now although I am quite prepared to admit the probability of this view, *know* seems too strong a word.'[33] The paper, revised as recommended, was soon published in the *Transactions*.[34]

In December, Lockyer submitted another pointed critique aimed at the Hugginses.[35] In it, he questioned Keeler's supposed confirmation of their claims and pointed to Greenwich's recent measures of the radial motion of nebulae, which appeared to cast doubt on all previous statements about the true positions of nebular spectral lines.[36]

The attack prompted the Hugginses, in turn, to draft a forceful refutation.[37] The hand-written text shows that it was written by both William and Margaret. In the sections written by Margaret, she used the pronoun 'we' to attribute their work and conclusions. In each case, however, 'we' was later stricken and replaced by 'I'. William would take on Lockyer by himself. Both his and Lockyer's papers were read before the RS on 5 February 1891, but their publication was deferred.[38] In April, Lockyer sent an angry note to the RS complaining that his paper 'has been in the hands of the Society for upwards of four months & . . . Dr. Huggins has quite recently communicated another paper on the same subject by Mr. Keeler to which I have to reply.[39] The action of the Society then seems to be placing me at a disadvantage in a controversy which is none of my raising.'[40] In the end, neither paper read on 5 February ever appeared in print. When individual participants cannot and/or will not resolve their dispute, institutional forces can and do intervene in order to restore the community's public façade of comity and conformity.

13.4 President of the BAAS

This round of his all-too-public debate with Lockyer quieted in time for William to devote his attention to a whole new array of responsibilities. He had been asked to serve as President of the British Association for the Advancement of Science. He had resisted pressure to assume the role the previous year, but he told Young, 'the feeling is so strong that I ought to take the chair in 1891, that I have not been able to refuse'.[41] As a former President of the RAS (1876–7), Huggins was no stranger to the honour and challenges of such an office.[42] 'The difficulties of the position [President of the BAAS] are very great', he explained. 'I shall be expected to give judicially an account of the state of the astronomical side & results from spectrum analysis. I shall have to look to my friends, and especially for assistance and suggestions.'

Holden was elated by the news. 'I am more than pleased to know that you will accept the Presidency of the B.A.A.S. next year', he wrote in congratulation. 'I shall look with eagerness for one more of your luminous & suggestive aperçus of the State of Science. Certainly you cannot fail to examine M. Lockyer's positions. I am not a competent judge – but it seems to me that he has forced a theory to the very edge of breaking.'[43]

The tradition of launching the annual BAAS meeting with an address by the new President developed gradually over the first ten years of the organisation's existence.[44] By 1891 it had become a highlight of the programme. Although each address bore the stamp of the disciplinary and personal interests of its author, they all aimed to meet three principal goals: review the current state of science, applaud recent advances and rally the troops to meet the challenges looming on the horizon.

With this year's meeting slated for Cardiff in mid-August, Huggins got straight to work on preparing his speech. He had relied heavily on George Stokes's comments and suggestions while writing his Bakerian lecture in 1885. He turned once again to his old friend for counsel in April 1891. 'I am anxious to make the address suggestive, but at the same time not to speculate unwisely', he confided. He then turned to the subject of his real concern: 'Towards the middle (page 8) of the Part on the Nebulae, there is one sentence in which I mention

Professor Lockyer. I am uncertain whether I should put it in or not. What is your opinion on this point?' A stream of anxious questions flowed throughout the summer.

Meanwhile, Keeler, looking for professional advancement and more accommodating living arrangements for his new family, resigned his post at the Lick in June 1891 to return to Pittsburgh, Pennsylvania. There he assumed the directorship of the Allegheny Observatory, a small but notable research institution where he had launched his career a decade earlier as an assistant to Samuel Pierpont Langley. Holden appointed William Wallace Campbell to take over Keeler's key position on the Lick team of observers. Campbell must have expressed doubts that he was qualified for the task. Keeler offered both his congratulations and his encouragement:

You could count the men in the country who are familiar with astronomical spectroscopy on the fingers of one hand ... so that *somebody* would have to work the subject up, and I am sure that nobody could *do* it better or in less time than yourself.

I wrote to Dr. Huggins not long ago that I had resigned my position, but did not say who was to succeed me ... No doubt Dr. Huggins can give you some pointers. You know he is the founder of the science of astronomical spectroscopy.[45]

13.5 George Ellery Hale

Huggins first met George Ellery Hale in August 1891, just weeks before he was to deliver his BAAS address. The energetic young American was only twenty-three years old, but had already developed quite a reputation for himself as a man on the move. Born one month after Huggins read his watershed paper on stellar motion in the line of sight, Hale never knew a time when astronomy was practised without spectroscopy. Towards the end of his life, he recalled assembling a crude spectroscope as a child.[46] Putting it to use was exciting and frustrating at the same time. To be able to see with clarity the wonderful spectral lines he was reading about and – more importantly – to investigate them further on his own, Hale replaced his handmade spectroscope with one that had been professionally made. He found well-illustrated how-to books like Lockyer's *Studies in Spectrum Analysis* to be invaluable guides for launching a fruitful line of enquiry.[47]

In his second year of study at MIT, Hale volunteered to assist at the Harvard College Observatory. There he availed himself of the observatory's resources to conduct his own research on the solar spectrum. For his senior thesis he designed and built a prototype of an instrument he called a spectroheliograph.[48] With it, he aimed to capture monochromatic photographs of solar prominences without an eclipse.

Although the results of these early efforts showed more promise than real success, Hale was far from discouraged. He was, after all, the director of his own private astrophysical research facility back home in Chicago.[49] The Kenwood Physical Observatory had been built and equipped with financial assistance from his father.[50] Three years in the planning and execution, the observatory was dedicated with some fanfare in June 1891.[51] By then, Hale had improved the spectroheliograph's design and had identified the violet K line as the brightest and sharpest of spectral lines associated with solar prominences. Using the spectroheliograph to isolate that portion of the spectrum, he was finally able to obtain clear and dramatic photographs of solar prominences.[52]

In early July, he left the work of the Kenwood Observatory in the able hands of his well-trained teenage sister and brother, packed up his research notes and letters of introduction and set off with his young wife on a grand tour of the premier observatories in Europe. Hale wanted to introduce himself and his work to the world's astronomers. He hoped to have the opportunity to examine their instruments, learn their research methods and otherwise hobnob with these scientific worthies about their forthcoming projects. He also wanted to drum up support for his own plans to establish a new international journal devoted to astrophysics.

Hale penned a hasty article on his recent work for the *American Journal of Science* the day before he sailed from New York.[53] It was crucial to make his mark in this rapidly evolving line of investigation by getting some of his first photographic successes into print lest his efforts to photograph the solar prominences be forestalled during his long absence.[54]

The Hales began their trip in England, arriving in Southampton on 14 July.[55] On 1 August, Huggins invited them to visit Tulse Hill.[56] Having been burned by his experience with Draper, Huggins may have felt wary about allowing another young American enthusiast to explore his observatory. But Hale, he soon discovered, was different. The time they spent together at Tulse Hill opened up an entirely new kind of collegial relationship for the sixty-seven-year-old astronomer. Despite their age difference, the two men had a lot in common, from their shared boyhood fascination with microscopy to their dogged determination to force the Sun to yield its secrets. There was much to talk about.

They continued their conversation in a long and active correspondence that Margaret continued after William's death in 1910. They exchanged thoughts on solar physics and confided fledgling theories to explain both visible and invisible solar spectra phenomena. Huggins encouraged Hale's sunspot investigations. He provided him with counsel when asked and frank criticism when he felt it was needed. Hale rallied support for Huggins in the American astronomical community and helped smooth over disagreements that cropped up as Huggins's claims were challenged. He both sought Huggins's advice on spectroscopic matters and listened attentively to it.

Indeed, Hale was one of the few individuals – perhaps the only one – whom Huggins learned to trust and respect to the point of being able to accept critical comments and competitive efforts on his work. It is significant that Huggins never viewed Hale's attempts to photograph the solar corona out of eclipse as an intrusion even though the two had different ideas about how to go about accomplishing it. The childless Hugginses treated Hale like a son.[57]

Huggins was affected by his association with Hale in two important ways. On the one hand, it invigorated him to have someone who viewed him as a valued mentor and who seemed destined to do great things in astronomy. On the other, Hale's remarkable progress and successes reminded him just how quickly his own great contributions would be surpassed in the coming years.

Hale came away from his August meeting with Huggins sporting two valuable feathers in his cap. The first was Huggins's enthusiastic support for his plans for an international astrophysical journal. The endorsement he received from Huggins and other prominent

European astronomers assisted him in taking the first steps toward achieving that goal. Soon after his return to Chicago, he joined with William Wallace Payne (1837–1928), editor of *Sidereal Messenger*, to create *Astronomy and Astro-physics*.[58] Payne and Hale promised their readers that the new hybrid journal would devote equal space to articles on astrophysics and what they called 'general astronomy'. Huggins had been contributing articles to *Sidereal Messenger* since its first volume appeared in 1883. In October 1891, he sent Hale a letter of hearty congratulations, assuring him 'It will be a pleasure to us to send you any papers we may have, at as early a date as possible, for insertion in your new journal.' A few years later, he agreed to serve as an associate editor of Hale's new journal, *The Astrophysical Journal*: an internationally oriented younger sibling of *Astronomy and Astro-physics* devoted entirely to research in astrophysics.[59]

Hale's second plume was Huggins's encouragement to write up his recent work on photographing solar prominences and present it at the BAAS meeting in Cardiff.[60] As President, Huggins could and would make the necessary arrangements. Hale availed himself of Lockyer's laboratory facilities where he made 'use of a measuring machine' to determine the 'wave-length for the new prominence lines' from which he drew his conclusions.[61] He finished his paper on 13 August and delivered it a week later on the first day of the Mathematical and Physical Science section meetings.[62]

Afterward, he finally had time to write to his old college chum, Harry Manley Goodwin (1870–1955). 'I am treated like a *Grand-Duke!*' he boasted. 'Lina [Mrs. Hale] sits with Mrs. Huggins and Lord Bute's party [John Patrick Crichton-Stuart, 3rd Marquess of Bute (1847–1900)] at the evening lectures, &c, and altogether you would think we were celebrated people. Of course it is all Dr. Huggins' doings, for he, as President, is all powerful. He took a fancy to us and our work and hence the result.'[63]

13.6 The President's address

Huggins delivered his presidential address to the BAAS the evening before Hale's presentation. Rather than review the entire history of the development of 'spectroscopic astronomy', he aimed to concentrate on 'what we do know at present, as distinguished from what we do not know, of a few only of [astronomy's] more important problems'.[64] He urged '[v]ery great caution' in any 'attempt to reason by the aid of laboratory experiments to the temperature of the heavenly bodies from their radiation'.[65] It was reasonable, of course, to identify the spectral lines seen in a celestial body with a known terrestrial element, '[w]hen the coincidences are very numerous . . . or the lines are characteristically grouped'. But, he lamented, in the first of many thinly veiled expressions of disdain for the work of his unnamed nemesis, Lockyer, 'the progress of science has been greatly retarded by resting important conclusions upon the apparent coincidence of single lines, in spectroscopes of very small resolving power'.[66] By contrast, he declared, 'it is only by the laborious and slow process of most cautious observation that the foundations of the science of celestial physics can be surely laid'.[67] For those in the audience who were familiar with Huggins's feud with Lockyer over the meteoritic hypothesis, there was no mistaking the intended target of his remarks.

He turned from general complaints directed at Lockyer's method and analysis to more specific criticism of his claims regarding the causes of specific phenomena, all without once uttering Lockyer's name. Whether he took this approach because Stokes had advised him it was the wisest course, or because he concluded it was the only judicious way to deal publicly with his recent controversy with Lockyer, he would later pay a price for his choice.

'Recently', Huggins told his audience, 'the suggestion has been made [*tacit*: by Lockyer] that the Aurora is a phenomenon produced by the dust of meteors and falling stars, and that near positions of certain auroral lines to lines or flutings of manganese, lead, barium, thallium, iron, &c., are sufficient to justify us in regarding meteoric dust in the atmosphere as the origin of the auroral spectrum.'[68] But, he noted, careful and well-considered investigations by others – notably Liveing, Dewar and Schuster – had cast serious doubt on such a notion.

Although comets are undoubtedly composed of meteoritic material, Huggins continued, 'it would be perhaps unwise at present to attempt to define too precisely [*tacit*: as Lockyer does] the exact condition of the matter which forms the nucleus of the comet'.[69] Instead, he argued, '[w]e must look rather to disruptive electric discharges produced probably by processes of evaporation due to increased solar heat, which would be amply sufficient to set free portions of the occluded gases into the vacuum of space'.[70]

On the subject of stellar evolution, Huggins referred to a recent 'hypothesis [*tacit*: of Lockyer] that the nebulae and stars consist of colliding meteoric stones in different stages of condensation'.[71] He countered this notion with his own view, which he presented 'with reserve as the subject is so obscure'. In it, he suggested 'the white stars, which are most numerous, represent the early adult and most persistent stage of stellar life, the solar condition that of full maturity and of commencing age; while in the orange and red stars with banded spectra we see the setting in and advance of old age'.[72]

He admitted the error of his own first interpretations of nebular spectra. But, he explained, he had formulated those ideas nearly thirty years ago when 'our knowledge of stellar spectra was small'. In addition, he confessed, his writing at that time had 'probably' been subject to 'the undue influence of theological opinions then widely prevalent'. Since then, he had adopted a different, 'truer position'.[73]

These words became etched in Lockyer's brain. He interpreted them to mean his adversary had come to agree with his views of the nature and structure of nebulae. Outraged by Huggins's failure to attribute his conversion to the persuasive power of the meteoritic hypothesis, Lockyer became determined to have satisfaction.

Huggins unwittingly added further injury to the insult he had inflicted upon Lockyer when he turned to the topic of nebular spectra. Indeed, it was for him the most challenging part of his address. Determined to broach the subject of the meteoritic hypothesis directly without reference to Lockyer, he began by reviewing explanations for the origins of the solar system with a focus on the status of various modern iterations of Laplace's nebular hypothesis. He reacquainted his audience, first, with the work of French astronomer Hervé Faye (1814–1902) who tried to resolve mechanical difficulties in Laplace's hypothesis by proposing that the solar system's planets had evolved from meteorites, and, second, with George Darwin's study of the dynamics of swarms of meteorites. He reminded his listeners of the 'recent remarkable photograph' taken of the great nebula in Andromeda by

Isaac Roberts, which many had taken as evidence of 'cosmical evolution on a gigantic scale'. In it, the nebula's matter appears 'to be distributed ... in rings or spiral streams' suggesting 'a surge in a succession of evolutionary events not inconsistent with that which the nebular hypothesis requires'.[74]

Finally, Huggins introduced his audience to the work of Peter Guthrie Tait (1831–1901). In doing so, he came dangerously close to exposing the elephant that some in his audience knew had been in the room all along. After all, it was Tait's suggestion that the light of comets and nebulae is produced by collisions of meteorites which had sown the seeds of the meteoritic hypothesis in Lockyer's mind in the first place. One can well imagine an air of anxious anticipation as the assembly waited to hear Huggins opine in public on his adversary's views. But the would-be electric moment passed without so much as a whisper of Lockyer's name or his hypothesis. Instead, Huggins stated simply that spectroscopic study of the light from nebulae had uncovered no evidence for the elements associated with meteorites. He repeated his own view of the nebular spectrum as 'chiefly remarkable for a few brilliant lines, very narrow and defined'. Current research efforts had uncovered the chemical and physical causes of some of these lines, he said. He was confident they would all be interpreted correctly in time. Even 'those which have not been matched with a known terrestrial substance' may yet be found to be products of extreme temperature or identified with 'some of the unknown elements which are wanting in our terrestrial chemistry between hydrogen and lithium'.[75]

Huggins spent much of the remainder of his long address hailing the great successes made possible by the application of photography to astronomical research, including a nod to Hale's recent achievements with the spectroheliograph, and enumerating the many puzzles remaining to be solved. 'Happy is the lot of those who are still on the eastern side of life's meridian!' he declared wistfully.[76] His meeting with Hale clearly gave him hope for astronomy's future.

13.7 Nova Aurigae

On 23 January 1892, Scottish amateur astronomer Thomas David Anderson (1853–1932) discovered a new star in Auriga.[77] Astronomers around the world immediately scrutinised its light.[78] William and Margaret Huggins began their own exhaustive and exhausting study of its spectrum.[79]

There was really no clear understanding of the physical mechanisms at work in such events, but there was, by then, a firm belief that spectrum analysis would offer important clues. Of course, astronomers, physicists and chemists were still speculating on what causes spectral lines to appear in the first place, as well as what occasionally makes them disappear, widen, split or shift. Nevertheless, by 1892 the interpretive guidelines were richer. Since 1868, Huggins himself had linked the shift of spectral lines to relative motion of the light source along the line of sight.[80] The growing acceptance of photographic records of transient celestial events as scientific evidence had opened interpretive discussion of them to a global arena. An empirical formula had been derived by which the wavelengths of the four principal hydrogen lines could be generated.[81] Theoretical interest in this formula was enhanced when it was used to calculate the wavelength of a fifth hydrogen line, one that Huggins had observed in the ultraviolet two years earlier.[82]

Theories on what could cause material substances to emit such patterned radiant energy were being proposed by physicists and chemists alike.[83] At the heart of these theories lay many unanswered questions concerning the role of electricity, magnetism, temperature, density and the like in generating and sustaining what were surely well-ordered harmonic vibrations within the most fundamental units of matter. The lack of certain answers to these questions added much smoke and little light to the controversies that raged among astronomers over the interpretation of celestial spectra.

The gross features Huggins had noted over a quarter century earlier in the spectrum of T Coronae – namely the appearance of bright and dark lines – were universally observed in the spectrum of Nova Aurigae. But this time, the bright lines could be (and were) described by their many observers as greatly widened and shifted toward the red end of the spectrum. Some of these bright lines were accompanied by adjoining absorption lines which appeared to be shifted toward the blue. These shifts – toward the red or the blue – could be (and were) interpreted as indicating that the light being analysed was coming from at least two different sources – one a luminous gas or bright-line star moving away from Earth, the other a more Sun-like object moving towards it – each at tremendous speeds. Some of these dark and bright lines were described as doublets and possibly triplets, an observation which led to speculation that this event involved the explosive collision of perhaps as many as six bodies.

Huggins preferred to construct a scenario based on more familiar themes: perhaps the bright and dark spectral lines signify the onset of violent stellar eruptions in which hot gaseous material that had been trapped under the cooling crust of an ancient star was being spewed out into space. Such eruptions would be similar to, but more extreme than, those seen almost daily on the Sun. Alternatively, the lines could indicate the creation of multiple 'reversing layers' due to turbulence in the normal structure of the star's atmosphere caused, perhaps, by the tidal forces of a nearby companion or passing star. Such multiple reversals, Huggins argued, had been artificially induced in terrestrial laboratories and could signal the existence of similar dynamic processes within stars. He acknowledged that such extreme conditions would rarely arise in nature. Even so, he considered the explanation far more plausible than one involving six stars independently moving toward each other on a collision course.[84]

The length of time that the light in Nova Aurigae remained sufficiently bright to analyse gave astronomers around the world unprecedented opportunities to examine it carefully. In August 1892 – just nine months after its eruption – William Wallace Campbell at the Lick announced that recent photographs he had taken of the nova showed it to be distinctly nebular in appearance.[85] In fact, he claimed, subsequent examination of the nova's spectrum indicated that it had begun to display the distinctive spectral signature of a planetary nebula.[86]

Huggins found it difficult to imagine that such rapid change in the nova or its spectrum could realistically occur. He blamed the Lick's blurry photograph on a failure to focus the star's light properly for the range of wavelengths being emitted by the star during its outburst. And he forcefully questioned Campbell's interpretation of the nova's spectrum.

It is true that, in his earliest reports on T Coronae back in 1866, Huggins had noted the presence of 'a very faint nebulosity ... extending some little distance round the star, and gradually fading away at its outer boundary'.[87] After comparing T Coronae with neighbouring stars, he at first concluded that the 'nebulosity was due to the star itself'. He later rejected that notion after visual observations failed to confirm his initial suspicions, and scrutiny of its unusual spectrum failed to reveal any signs of telltale nebular lines.

Indeed, when Nova Aurigae returned to a favourable position in the evening sky the following February, Huggins examined its now feeble light specifically for signs of nebular characteristics. In March 1893, he boasted of having 'done more' with the nova than even the Lick.[88] He would know a planetary nebula spectrum if he saw one, and Nova Aurigae's spectrum, with its complex array of bright lines, did not display the features necessary to classify it as such.

Rather, he likened its spectrum to that of β Lyrae (Sheliak), a variable star with a spectrum of bright and dark lines that vary in brightness, width and structure in complex ways. It had only recently been observed that the emission lines in the star's spectrum were bordered by dark companions in an arrangement not too different from that in the nova spectrum. One popular explanation for β Lyrae's odd behaviour was that it was a dynamic binary system. With this as his model, Huggins conjectured in a letter to Hale that in Nova Aurigae there might also be two original stars that had

cooled very considerably – to be almost dark and to have an exterior formed crust – in all such conditions an unstable state of things would exist. Some time eruptions would take place in such a condition of sensitive equilibrium, a small tidal action from without would be adequate to 'pull the trigger' – to cause violent eruptions.

'The spectrum', he argued, 'points to eruptions similar to the sun's prominences.'[89] A few weeks later, feeling 'quite sure about the general accuracy of the main features as they were drawn quite independently by Mrs. Huggins and myself, without either knowing what the other saw', he informed Campbell that '[o]ne can only speak with great reserve about the Nova, but it seems to me very doubtful whether these groups have any physical connection whatever with the nebular lines'.[90]

As Huggins's relationship with Campbell developed an unpleasant adversarial edge, he came to miss the affable Keeler. In 1894, when Hale invited Campbell to join Huggins on the editorial board of his new *Astrophysical Journal*, he felt obliged to intervene lest unseemly discord threaten the publication's chances for success.[91] 'I am sure you will not take it amiss', Hale wrote to Campbell, 'if I ask you if, in case you become one of our Assistant Editors, you will moderate a little your tone in your replies to such men as Kayser [German spectroscopist Heinrich Kayser (1853–1940)] and Huggins. Even when there is not the slightest doubt of the correctness of your own position, it is Keeler's and my opinion that the age and long experience of such men entitle them to be treated with a very considerable degree of respect. Do not misunderstand me in this; I mention it because we are very anxious to have your cooperation, and at the same time can not afford to offend Kayser and Huggins in the smallest degree.'[92]

Campbell replied:

I agree that 'age and long experience' are entitled to a considerable degree of respect. Youth, of course, has no such claims to offer; but *conscientious work*, by whomsoever done, is entitled to consideration, even from those of age and experience. Now the truth is Kayser and Huggins rode rough-shod over some of my work. They did not even treat it *on its merits* . . .

Huggins was extremely unjust to me in his article on the present spectrum of *Nova Aurigae*, in that he misrepresented my description of the principal line. His recent note in A. & A.-P. is not at all candid. His ignoring Lockyer's February 1890 photographs of the Orion Nebula spectrum was outrageous. In my opinion *age* is no excuse for injustice: it only adds to the culpability.

I deeply regret the spirit of controversy which seems to be creeping into astronomical literature. On the other hand it is not right for any man, however young, who happens to collide with the 'authorities', to lie down and let them pass over him.[93]

Modern views support Campbell's contention that the nova's evolving spectrum was the result of a star's rapid transformation into a planetary nebula. But it is important to recognise that – based on the evidence available when the nova first appeared – it was not the only interpretation that was seriously considered at the time. Huggins's physical explanation for the events leading up to the nova's sudden outburst seems contrived to us today. But, from his perspective in 1893, it was Campbell's scheme that was hard to swallow if one played by the rules of scientific explanation. Legitimate evidentiary interpretation ought only to be based on the simplest known mechanisms and the least speculation. Extrapolating from laboratory findings to the field was a risky business requiring the utmost caution.

Nearly a decade later, when another nova, this time in the constellation Perseus, came into view in February 1901, observers were fortunate to spot it less than 28 hours after it erupted. It rapidly gained in brightness over the next day. For the first time, spectroscopists were able to examine a nova's light as it ascended to its peak magnitude. Needless to say, they tracked its rapid changes with great interest. But they had to look fast. Two days after it was first seen, the nova had already begun to fade.

Within months, observers launched into speculations on what appeared to be a nebulosity around the nova. In September, George Willis Ritchey (1864–1945) using the 24-in reflecting telescope at the Yerkes Observatory, obtained a photograph showing a 'spiral or annular nebula' around the central star.[94] In fact, the rapidity of the nebula's apparent expansion defied imagination, not to mention the laws of physics. 'It seems that from Ritchey's photo. that . . . [the star] is surrounded by a faint nebula, though itself practically stellar', Huggins wrote to Campbell in November. 'Its central position within the nebula does not suggest a run-a-way star rushing through it.'[95]

Huggins shared his own emerging interpretation of Nova Persei with Hale in October 1901. Steadfast in his conviction that nebulae are places for stars to form, not die, but unable this time to argue with the photographic evidence, Huggins sought theoretical guidance in a more commonplace example of filmy luminosity around a star – namely, the solar corona. His theory about the luminous structure surrounding the central star in Nova Persei betrays his long-standing belief that coronal features and eruptions in the Sun's atmosphere are the direct result of electrical forces within the Sun:

I do not yet know what to suggest as to surroundings of the Nova. I think that it is not well to use the term, *nebula*, as it rather assumes that we have to do with a true nebula; it may indeed be that we have, but it may be matter thrown out by the star, and rather coronal or cometary in its nature. Sidgreaves [Father Walter Sidgreaves (1837–1919)] finds the two strong cyanogen bands which I found in one of the early comets. Such matter under electrical repulsion after explosive projection would move rapidly. But is the motion of luminosity we see, that of true ponderable matter, or of energy through space sparsely filled with material particles?[96]

In January 1902, he wrote to Hale again:

The laws of physics forbid us to think of gases, even if shot out of the star by explosions, as moving with a velocity at all of the order of speed that the largest possible parallax demands. No doubt, however, that finely divided matter similar to that of the solar Corona, driven by electric repulsion might attain a very high velocity – for example, the particles in a Comet's tail. Could such matter continue luminous by emission or by reflection? . . .[97]

To find out would require more laboratory work, closer examination of the solar corona, more comparisons with photographs of the nova's changing spectrum, more trudging back and forth between the laboratory and the field.

13.8 The Yerkes telescope

In early 1893, Hale invited the Hugginses to visit Chicago. The invitation was one of many he sent out to astronomers he had met on his European tour. He hoped he could lure them to participate in an international meeting he wanted to hold during the World's Columbian Exposition. The fair, commemorating the 400th anniversary of Christopher Columbus's first voyage to the New World, was expected to attract millions of international visitors to Chicago from May through to October of that year.[98] The city had risen rapidly and creatively from the ashes of a destructive fire in 1871. Fair organisers touted the new Chicago as a modern hub of all the very best and latest in art, entertainment, science and technology. They relished the opportunity to show it off to the world. Hale and his father were both involved in the fair's planning and development.[99] Hosting an international gathering of the world's leading astronomers would add a certain celestial lustre to the celebration.

The Hugginses were honoured by Hale's invitation. The prospect of travel to Chicago at such a time 'is almost irresistible!' William replied, 'and yet, for several reasons, we must deny ourselves what would in many respects be a great treat to us'.[100] He wrote again in May to thank Hale and his wife once again 'for your very kind invitation to Chicago' and to reiterate 'our great regret that we can only just catch from afar the faint echoes of your grand doings. We are indeed in the *old world*!!'[101]

But others accepted. A delegation from Germany along with representatives from France, Italy, Russia, Holland, Switzerland and Hungary joined with astronomers from across the United States to convene the Congress on Mathematics, Astronomy and Astro-Physics from 21 to 26 August.[102] In addition to learning about and discussing their colleagues' latest research, attendees initiated the creation of a committee to establish a much-needed international standard unit of measure for wavelengths, they voiced support of the installation of a spectroheliograph at the Observatory at Pune in India, and they voted to confer the name 'Chicago' on a minor planet in honour of the meeting.

The highlight of the week was a trip to the Exposition's Manufactures and Liberal Arts Building on 23 August.[103] There the visiting astronomers and other dignitaries witnessed the unveiling of another of Hale's ambitious projects: the tube and mounting of a telescope which, when complete, would be the largest in the world. The previous October, Hale had charmed wealthy streetcar magnate Charles Tyson Yerkes into underwriting the costly effort.[104] By the time the telescope went on public display in August 1893, a site on Lake Geneva in neighbouring Wisconsin had been selected on which to build the observatory that would house it. Its 60-ft long tube and massive mounting had been completed in record time for the Exposition by Warner & Swasey, a machine-tool factory in Cleveland, Ohio.[105] Meanwhile, its gargantuan 40-in objective lenses were being polished under the supervision of Congress attendee Alvan Graham Clark (1832–97). Clark was the sole surviving member of the renowned optical family Clark & Sons. Less than a decade earlier, he had worked with his father, Alvan Clark, and older brother, George Bassett Clark (1827–91), to produce the Lick's 36-in lenses.[106] And so, on that warm Wednesday afternoon in August, the assembled guests watched as electric motors carried 'the mammoth instrument' through its 'evolutions and workings'.[107] Blind though it was, the Yerkes telescope would have been an impressive sight to see.

13.9 Photographing the corona without an eclipse

During the Congress in Chicago, Italian solar observer Pietro Tacchini (1838–1905) encouraged Hale to continue his efforts to photograph the solar corona without an eclipse at the observatory he had established on Mt Etna in Sicily. Hale had already made two unsuccessful attempts to achieve that elusive prize. In April 1893, he had employed the spectroheliograph at his Kenwood Observatory. He conceived an entirely new method of forcing the corona to reveal itself. To 'filter out' the light of the background sky, he set one slit of the spectroheliograph on a dark line in the sky's spectrum. Any additional light coming through at that wavelength, he reasoned, should be due to the corona.[108] 'I hope you may succeed', a truly optimistic Huggins wrote in February. 'I have little doubt that it is to be done.'[109] But the 'hopelessness of coronal photography beneath a Chicago sky' and 'shortcomings of the apparatus' led Hale to abandon his efforts at Kenwood.[110]

Huggins sent more encouragement along with words of wisdom based on his own years of wrestling with the challenge. 'I think your plan *promising*, but am not sure whether the increased glare from surfaces may not balance the advantages. I write at once in the hope *of persuading you to try also, my original simple plan*. The apparatus is easily put together. A concave lens would do, but diaphragms to stop all stray light & a long tube in front with diaphragms to keep out as far as possible all light but that of sun & corona.'[111] Two days later, he sent more detailed suggestions reminiscent of those he had shared with Gill years ago:

Much will depend upon *refinements*. Very *perfect polish* of the silvered mirror, or speculum. Suitably arranged diaphragms to get rid of all reflected light. Considerable light is *reflected back from* the photographic plate, which must be stopped as far as possible by diaphragms & *black velvet* (this is better than the ordinary blacking of the surface). Bromide paper is better than glass plates even if well-backed. A long diaphragmed tube in front is useful to keep all light but that also about the sun from entering. What I fear in your arrangement is that increased glare from scattering may perhaps over balance the real advantages . . .

I do hope you will succeed. I should be much interested to have my simple plan tried by such skilful hands as yours, at a good observatory and at a time when the corona is favourable.[112]

Hale teamed up with James Keeler for his second try at capturing the corona from mid-June to early July 1893. They used improved instrumentation assembled on the 14,147-ft summit of Pike's Peak in Colorado. But, even when the sky was free of clouds, they faced many other obstacles. Ubiquitous dust blew onto and scratched the mirror's surface, the low air pressure on the Peak hampered the smooth operation of the clepsydra that moved the instrument's slits during exposures, insects swarmed across their view of the Sun, and smoke and haze from frequent forest fires made photographing the corona a futile effort.

'I am so grieved that you have not succeeded with the corona', Huggins wrote to Hale after hearing of the failure at Pike's Peak. 'I thought you would, at least, be able to make a step in advance.'[113] Hale was disappointed, too. It had been a gruelling ordeal with little gain. As to the mountain's advantages for astronomical observations, he concluded, if they 'require a blue sky rather than good seeing, Pike's Peak (when not surrounded by forest fires) would seem to offer some important practical advantages over other mountains of equal altitude. But if good seeing is essential the Peak is not to be recommended.'[114]

And so Hale was happy to have Tacchini's invitation to tackle the corona a third time on Mt Etna. He and his wife were already planning a trip to Europe. After study and research in Germany, there would be time for solar work in Sicily. The Hales passed through London and called on the Hugginses both in November 1893 on their way to Germany and again the following July on their way home after a week of doing battle with both the corona and the great volcano.[115] Sulphurous fumes tarnished the mirror. Atmospheric conditions were often unsuitable for such a delicate photographic task: when smoke from the crater was not covering the Sun, bright haloes surrounded it. And, as on Pike's Peak, large swarms of insects interfered with the view. Having been 'assured by Professors Tacchini and Riccò [Annibale Riccò (1844–1919)] that the sky is frequently very good on Etna', Hale was unsure how to assess the experience. Perhaps 'the difficulties we encountered were exceptional', he concluded. 'If the wind had blown [the volcanic smoke] away from, instead of toward us, the sky would probably have been good, though I think by no means equal to the sky seen on Pike's Peak during the first part of our stay.'[116] Huggins wrote 'I am so sorry that you do no [*sic*] come crowned with the Corona! I hope the failure was due to some accident of bad weather & not clearly to the method itself.'[117] Hale concluded that the method was at least partly to blame for the failures. He devised yet another ingenious method for separating out the faint coronal light, this time using a bolometer to register what he believed would be its measurable amounts of heat radiation.[118]

Ultimately, the corona eluded its would-be captors until an as yet unborn French solar observer, Bernard Lyot (1897–1952), met with success in 1930.[119] It is worth noting that when Lyot addressed the RAS in 1939 upon receiving the Society's Gold Medal, he began with a brief history of previous attempts to photograph the solar corona without an eclipse.[120] He described the work of Henri-Alexandre Deslandres (1853–1948) in 1893 as well as the efforts of Hale and Riccò on Mt Etna in 1894, but made no mention of Huggins.

By omitting from 'The new astronomy' all references to his many years of fruitless struggle with the corona, Huggins succeeded in erasing his name from the collective memory of those who would chronicle the history of solar physics.[121]

13.10 The *Astrophysical Journal*

In September 1893, William Payne, Hale's *Astronomy and Astro-physics* co-editor, launched *Popular Astronomy*, a new journal aimed at teachers of astronomy in secondary schools.[122] Hale's partnership with Payne had been a marriage of convenience. The two were complementary personalities. Payne was tireless in his dedication to the cause of disseminating knowledge of astronomy and its latest discoveries to a broad spectrum of readers, but he was not a researcher and did not share Hale's enthusiasm for celestial spectroscopy. As astrophysicist Edward Arthur Fath (1880–1959) so lovingly noted in his late colleague's obituary, Payne's 'mind was not of the type which opens new paths'.[123]

In August 1894, after returning from Europe, Hale believed it was time for him to strike out on his own and create the international journal devoted solely to astrophysics that he had envisioned back in 1891. He recruited an impressive board of fifteen editors including five ranking European astronomers: Tacchini from Italy, Hermann Carl Vogel from Germany, Marie Alfred Cornu (1841–1902) from France, Nils Christoffer Dunér (1839–1914) from Sweden, and Huggins from England.

Recall Hale's attempt in October 1894 to bring Lick observer William Wallace Campbell into the group. Campbell had balked at the offer citing constraints on his time. 'I wish my work to be confined as closely as possible to *investigation*', he asserted. 'What little time I can grudgingly give to editorial work must be devoted wholly to *Publications* [of the Astronomical Society of the Pacific].' Hale persisted. 'I don't want to regard your refusal to join us as final', he told Campbell, 'for we are very anxious to have you with us.'[124] Hale knew he needed to assuage Campbell's concerns about confrontations with Huggins if he were to win him over.

I see that I must have failed to make my meaning in regard to Huggins, etc. quite clear. As to the value of a man's work being wholly independent of his age or reputation I most heartily agree with you, and I think you are quite right in replying to papers which criticise your own. All I mean to ask was that in your replies you make your case as strong as you please (the stronger the better), but word them in such a way that the facts, rather than the form of the language, shall indicate your position. Remember that much of Huggins' work was done at an early date (even 1888 is 'early' when photographs of nebular spectra are in question), and it is hard for a pioneer to see his work surpassed by a younger man.

As for Huggins' refusal to notice Lockyer's work you would wonder at it less if you knew the exact state of affairs in London. I could enter into details in regard to this subject, but hardly care to do so. It is sufficient to say that few men in Huggins' position would pay the least attention to Lockyer ...

Hale was a master of persuasion. Campbell signed on.

The first number of the new *Astrophysical Journal* appeared in January 1895 with articles on a wide array of subjects from the spectra of variable stars to observations of Mars.[125] It also contained a paper by Tacchini's colleague Annibale Riccò, a follow-up to Hale's review of his own work on the corona in the October number of *Astronomy and Astro-physics*.

Tacchini chronicled his unsuccessful efforts to photograph the solar corona without an eclipse using both a coronagraph constructed on Huggins's design and apparatus designed by Hale.[126]

Huggins avidly followed all reports concerning this new generation of attempts to photograph the corona. Soon after reading Hale's October review he wrote to his young friend to say he had found it 'full and interesting' and 'very admirable from every point of view'. Nevertheless, 'there is one statement ... which scarcely accords with my way of looking at the matter'.[127] He disputed Hale's assumption that longer exposure times were necessary to photograph the corona than to photograph solar prominences. 'I may be wrong,' Huggins wrote, 'but it certainly seems to me that the exposure necessary for photographing the true corona at times when there is no eclipse should be very greatly shorter than when the sun is eclipsed.' He explained his reasoning:

If, for argument's sake, we assume the intensity of the corona (C) during an eclipse to be = 1, and the time of exposure to be 1 *minute*; and if we assume further that the intensity of the air-glare, without an eclipse, (A) = 60, and time of exposure 1 *second*, then it seems to me what we have to photograph at ordinary times is not C requiring 1 minute, but A + C (= 61) needing only an exposure sixty times shorter. The shortness of the exposures, therefore, would not be an argument against, but really in favour of the possibility of the true corona having been caught. The great point is to be able to distinguish on the plate two different intensities differing by so small a proportion as $\frac{1}{60}$, or $\frac{1}{100}$, or $\frac{1}{200}$, –.

He repeated much of the same advice and cautions he had given to Gill and Woods in their work on the corona at the Cape. To bring out the barely perceptible differences between the light of the corona and that of the background sky, he recommended using sensitive plates. In addition, employing 'a very weak or slow acting developer' would make it easier to stop the chemical action at the moment the feeble image began to emerge. To work out the best materials and techniques in advance, he advocated laboratory experiments. 'If you prepared ... a card [of white paper partially painted with a nearly identical shade of white], it would be possible to find in the laboratory the photographic conditions most favourable for the discrimination of two illuminations differing very slightly from each other.' Above all, it was important to prevent the appearance of false effects by avoiding '(1) imperfect polish of surfaces [and] (2) reflection from edges, or diffraction produced by the shutter'.

Hale remained unconvinced and confused:

I can hardly induce myself to agree with you much as I would like to do so. It seems to me that if the exposure were made so short that no trace of the corona during an eclipse would be obtained with the same plate and apparatus, the interposition of no amount of atmospheric glare between the earth and the sun would serve to bring out a coronal image on the plate. Otherwise the prominences which are much brighter than the corona could be photographed directly without a spectro-heliograph. My difficulty is to see how the interposition of any luminous transparent object [the sky's light] between a photographic plate and the luminous object [the corona] can affect the time of exposure required by the latter object to produce a perceptible image on the plate.[128]

For guidance and support, Huggins turned to Stokes, always his trusted advisor on such knotty issues. 'There is a simple, yet fundamental question, in connection with attempts at photographing the corona, which I and Prof. Hale look at very differently', he explained. 'It seems to me that I am right, just as it seems to Prof. Hale that he is . . . I should be very greatly obliged if you will tell me in a word whether you would regard the point as I do, or as Hale does.'[129]

A few weeks later, Huggins sent along more background on the corona problem as well as further explanation of his reasoning behind his own views. They were, he reminded Stokes, solidly founded upon the advice of photographic expert William de Wiveslie Abney:

Abney has maintained from experiments that if an illuminated surface, some parts of which are brighter than the others in the proportion of 61 to 60, (or even 101 to 100) be photographed with an exposure just sufficient to give a medium density of deposit for the parts having an intensity of radiation 61, then the parts with a radiation of 60 only will show themselves by a sensibly less deposit . . .

It was on these premises, resting upon Abney's statements that it seemed to me to follow that if the corona during an eclipse requires *1 minute's* exposure, then at ordinary times with an air-glare 60 times as bright in front of it, the corona would show itself by a sensible difference of density of deposit on the parts of the sky-glare where it was, in *1 second* only, the exposure suitable for the air-glare + corona . . .[130]

Perhaps on advice from Stokes, Huggins sent virtually the same letter to Hale a week later. He enclosed a copy of a letter from Abney for good measure.[131] If Hale was not fully persuaded by all this, he nonetheless came to understand and appreciate the basis for Huggins's argument.

'I am very glad that you have come to see the *true principle* of the method I suggested for the Corona', Huggins cheered in April.[132]

I shall be grateful to you if you will in a note put quite clearly that this was my principle *from the first*.[133] Unless I had looked at the matter in this way I should never have made a single attempt. As the matter stands now I am represented as having put forward *an absurd – an impossible method*. If the Corona required as long as at an eclipse, of course the plate would be overdone by air glare long before any image could be obtained . . .

He sincerely hoped Hale would try the corona again. 'I feel very sure that it is to be done by photography.'

13.11 The Yerkes Observatory

In the spring of 1896, Hale began planning the dedication ceremony for the Yerkes Observatory. He had persuaded renowned astronomer Edward Emerson Barnard (1857–1923) to leave the Lick and accept a position at Yerkes.[134] The telescope's tube and mounting were long-since ready. The work on the giant lenses was complete.[135] They were to be shipped and installed as soon as the observatory building was finished. Besides, the Hales were expecting their first child in August. It would be good if everything were up and running before then.

He invited Huggins to deliver the dedicatory address and in March received an enthusiastic reply. '[I]t seems to me', Huggins wrote, '(1) That my Address should not exceed half an hour in time of delivery; (2) That it should not be of the nature of a paper, or a discussion of any scientific question; but rather of a general and popular character.'[136]

By July, however, Hale became resigned to the fact that the observatory would not be ready in time. 'I gather ... from your letter', Huggins wrote, 'that ... you and the President [of the University of Chicago] consider that it would be better to postpone the ceremony until next year.'[137] He assured Hale he did 'not see at present any reason why next summer should not suit me as well as now', adding, 'I hope that the date will be chosen to *suit the conveniences of the Observatory.*' He had been working on his dedicatory speech. 'I am glad that you think well of the lines of my address. I have already written it out in the rough.'

But that November, Huggins began to have second thoughts about making the trip to Chicago. 'I hope that it will not disappoint you too much, if I ask to be excused from making so long a journey for the Opening of the Yerks [*sic*]', he wrote to Hale.

I am suffering from my hands, and I am sorry to say that Mrs. Huggins has been very far from well for some time now. It is not one consideration alone, but quite a number of considerations which all pull one way, and that is against the American trip. I do not know whether you will care to have for any purpose, the notes I put down as a basis for my Address. They would require a little putting into shape according to the use you might think of making of them, even if you care to have them at all.[138]

He resisted Hale's efforts to change his mind. 'I see no alternative', he wrote the following March, 'but to ask you to accept my regret at not being able to give the Address.'[139] He offered several suggestions for replacement speakers. 'Certainly Vogel is the best man. After him, would come perhaps Dner [*sic*]; but I rather doubt, if Vogel fails you, whether it would be worth while to import an European Astronomer. I would *not* advise Janssen, and Cornu is not sufficiently an astronomer. Indeed, it would seem to me better if Vogel can not come, to fall back upon Young, who is more distinguished than Duner.'

In April, he declined Hale's invitation one last time. 'I fear that I must not alter my decision, in the summer there is the great heat which I should fear more even than the voyage. Failing Vogel, I think that you will do best to have Prof. Young.'[140] To offset that disappointing news, he announced excitedly, 'I have been persuaded to write an article on the rise of the new astronomy from the standpoint of my own work, in connection with the Queen's Jubilee.' The article would become his enduring retrospective essay.[141]

On 18 May, a headline in the Chicago *Tribune* announced the two giant lenses for the Yerkes telescope had been 'Packed with greatest care in soft material and mounted on springs' in a railway car for shipment to Lake Geneva.[142] The next day the lenses arrived safely in Chicago and were sent on to their final destination in Williams Bay, Wisconsin.[143] Everything was going smoothly until early Saturday morning, 29 May, when the movable floor of the observatory suddenly collapsed.[144] Eleven days later, Clark, the master optician who had overseen production of the lenses, suffered a fatal stroke.[145] Despite these major setbacks, all repairs were completed in record time. By 2 September, the *Tribune* reported that the floor was back in 'Working order'.[146]

The dedication, originally scheduled for June, was finally held in October 1897 accompanied by a week-long programme of special events.[147] A letter from Huggins, in which he described variations in the spectra of some stars and comparisons of the spectra of binary stars, was read at the first day's gathering on 18 October.[148] On the morning of 21 October, the Observatory was formally presented to the University of Chicago. Keeler delivered the dedicatory address.[149]

***

The Hugginses' alliances with American astronomers helped them cope with the distress of controversy with Lockyer over the chief nebula line. William was gratified by the honour of being chosen to serve as President of the BAAS. But they derived their greatest joy from what became a deep and lasting friendship with George Ellery Hale. It was a relationship that only strengthened over the years.

Huggins ended his 17 April 1895 letter to Hale with a small news item: 'I suppose you have seen that Ramsay thinks that he has got helium from cleavite [*sic*]. I have seen the line, but I have had no opportunity of comparing it with the sun. I believe there is only one tube.' If Ramsay's claim were proved true, the discovery promised to link terrestrial and celestial physics and chemistry in unimagined ways. He and Margaret were thrilled at the prospect. He knew his young American friend would be, too. They all joined in the global effort to resolve this unexpected mystery. It was, as we shall see in the next chapter, the first step of many the Hugginses would take into the new and unexplored world of wonders hidden in the basic building blocks of matter. They were never too old to learn new tricks.

Notes

1. Keeler to W. W. Campbell, 7 May 1891, MLSA, LO.
2. W. Huggins and M. L. Huggins, 'Note on the photographic spectra of Uranus and Saturn' (1889).
3. J. N. Lockyer, 'Notes on meteorites' (1888–9).
4. J. N. Lockyer, 'On the spectra of meteor-swarms (Group III)' (1889).
5. J. N. Lockyer, 'On the wave-length of the chief fluting seen in the spectrum of manganese' (1889).
6. J. N. Lockyer, 'On the cause of variability in condensing swarms of meteorites' (1889). Lockyer also published a report to the Solar Physics Committee in the same number: 'Further discussion of the sun-spot observations made at South Kensington. A report to the Solar Physics Committee', pp. 385–401.
7. J. N. Lockyer, 'Comparisons of the spectra of nebulae and stars of groups I and II with those of comets and aurorae' (1889).
8. J. N. Lockyer, 'The presence of bright carbon flutings in the spectra of celestial bodies' (1889), p. 39.
9. J. N. Lockyer, 'On the chief line in the spectrum of the nebulae' (1890).
10. *Ibid.*, pp. 190–7.
11. Lockyer excerpted statements from Huggins's papers 'On the spectrum of the Great Nebula in the sword-handle of Orion' (1865), p. 42, and 'On the spectra of some of the nebulae' (1864), p. 442.
12. 16 January 1890, 'Minutes of the Committee of Papers', CMB.90E, RSL.
13. For a discussion of the controversy from Lockyer's perspective, see Meadows, *Science and Controversy* (1972), ch. 7.
14. Liveing, 9 February 1890, Notebook 1, WCL/SC.
15. Liveing, 29 April 1890, Notebook 1, WCL/SC.
16. W. Huggins to Spottiswoode, 20 December 1874, MC 10.186, RSL.
17. W. Huggins to Holden, 3 October 1877, MLSA, LO.
18. W. Huggins to Holden, 22 January 1890, MLSA, LO.
19. W. Huggins to Holden, 15 March 1890, MLSA, LO.
20. Keeler authorised Huggins to make whatever use he wished of these general comments, but he was reluctant to publish his preliminary measures before having the opportunity to repeat them. Keeler to W. Huggins, 3 April 1890, Lick Observatory archives. Holden reassured Huggins 'it is a pleasure to all of us to endeavor (even) to aid you in the slightest way'. Holden to W. Huggins, 4 April 1890, MLSA, LO.
21. W. Huggins to Young, [April or May] 1890, CAY, DCL.

22. M. L. Huggins to Young, 8 May 1890, CAY, DCL. 'Truth hath the plague in his house' is a literary phrase of the day meaning 'Truth is carefully avoided'. Here Margaret jokingly accuses Nature of purposely making it difficult for observers to come into direct contact with Truth lest they become infected by it.

23. W. Huggins to Young, 10 May 1890, CAY, DCL.

24. In the end, it turned out to be a misunderstanding: Keeler suspected the source of error was in the Lick apparatus. See Keeler, 'On the motions of the planetary nebulae in the line of sight' (1890), p. 267.

25. Fowler, 'The elements of astronomy' (1890).

26. Lockyer's three papers were published together in number 293 of the *PRS*, **48** (1890): 'On the chief line in the spectrum of the nebulae'; 'Note on the spectrum of the nebula of Orion'; and 'Preliminary note on photographs of the spectrum of the nebula in Orion'. The Hugginses' two papers were published in number 294 of the same volume of the *PRS*: 'On a re-determination of the principal line in the spectrum of the nebula in Orion'; and 'Note on the photographic spectrum of the Great Nebula in Orion'. An additional paper by Lockyer, which was not involved in Liveing and Rücker's review, was also published in number 294: 'On the spectra of Comet *a* 1890 and the Nebula G. C. 4058 [NGC 5866 in Draco]'.

27. Liveing and Rücker, RR.10.360, RSL.

28. W. Huggins, 'On the spectrum of the nebulae', *The Times* of London, 16 June 1890, p. 10f.

29. W. Huggins to Holden, 30 June 1890, MLSA, LO.

30. J. N. Lockyer, *Meteoritic Hypothesis* (1890).

31. Christie, 14 January 1891, RR.11.63, RSL.

32. See Darwin, 'On the mechanical conditions of a swarm of meteorites and on theories of cosmogony' (1889).

33. Darwin, 17 January, 1891, RR.11.64, RSL.

34. J. N. Lockyer, 'On the causes which produce the phenomena of new stars' (1891).

35. Lockyer aimed his critique at the Hugginses' paper 'On a re-determination of the principal line in the spectrum of the nebula in Orion' (1890).

36. J. N. Lockyer, 'On the chief line in the spectra of the nebulae', AP.67.3, RSL.

37. W. Huggins, 'On the chief line in the spectrum of the nebulae: A reply', AP.67.8, RSL.

38. A notice that the papers were read and their publication deferred appeared in *PRS* **41** (1891), p. 136.

39. Keeler, 'On the chief line in the spectrum of the nebulae' (1891). This paper was received on 13 March 1891 and read on 19 March.

40. Lockyer, 24 April [1891], RR.11.66, RSL.

41. W. Huggins to Young, 10 May 1890, CAY, DCL.

42. Looking over Huggins's activity in the RAS up to the time he became president of the BAAS, we see he served in some official capacity every year beginning as a member of the RAS Council in 1864. He served as secretary from 1867 to 1871; foreign secretary from 1873 to 1875 and from 1883 until his death in 1910; vice president in 1872, 1878 to 1879 and 1881; and president in 1876 to 1877.

43. Holden to W. Huggins, 12 June 1890, MLSA, LO.

44. Initially, the president's role in the meetings was largely administrative. When there was an opening address, it was delivered by a distinguished member of the association. The Revd William Vernon Harcourt (1789–1871) was the first president to deliver the address (Birmingham meeting, 1839). The first in the modern series of presidents to do so was William Whewell who addressed the Plymouth meeting in 1841.

45. Keeler to Campbell, 7 May 1891, MLSA, LO.

46. It is risky, as Huggins's example amply illustrates, to rely on anecdotal accounts of events recalled many years after the fact by individuals who have a vested interest in shaping the story's historic legacy. Given the limited public access to Hale's mountain of personal papers and the fact that historians of science have focused largely on the arc and impact of Hale's professional career, Helen Wright's popular biography of Hale – a work commissioned by the National Academy of Sciences – remains the standard reference for details on his early life. She attempted to corroborate Hale's unpublished notes whenever possible with contemporary documents and personal interviews with surviving participants in the events she describes. Nevertheless, Hale's recollections, like Huggins's, must be read with a sceptical eye. Wright, *Explorer of the Universe* (1966), pp. 39–43.

47. J. N. Lockyer, *Studies in Spectrum Analysis* (1878). Hale described his early interest in celestial spectroscopy in autobiographical notes written in 1933. Box 92 [Reel 75] GEH, CIT.

48. Hale, 'Photography of the solar prominences' (1890); for Hale's description of his first spectroheliograph, see 'The spectroheliograph' (1893), pp. 247–9.

49. Hale to Goodwin, 9 June 1891, HM 28439, HLA.

50. Hale, 'The Kenwood Physical Observatory'(1891).

51. Young, 'Address at the dedication of the Kenwood Observatory' (1891).

52. Hale, 'Photography and the invisible solar prominences', *SM* **10** (1891), pp. 257–64.

53. Hale, 'Photographic investigation of solar prominences and their spectra' (1891). He datelined his article 'Brooklyn, July 6, 1891'. He and his wife set sail from New York to Southampton the next morning.

54. The *AJS* article was not the only one Hale had on his list of things to do before sailing to England. 'I have got to write two papers for the Sidereal Messenger, one for the Astronom. Nachrichten, one for the Am. Journ. of Sci., one for the Phi. Mag., and one (in French!) for the Comptes Rendus!' Hale to Goodwin, 21 June 1891, HM 28440, HLA. It was an overly ambitious goal. He completed only one article on the Kenwood Observatory for *Sidereal Messenger* and one on photographing solar prominences for *Astronomische Nachrichten*, both datelined 29 June.

55. 'The mails', *The Times* of London, 15 July 1891, p. 11b.

56. W. Huggins to Hale, 1 August 1891, Box 22 [Reel 19], GEH, CIT.

57. The Hales had two children: a daughter, Margaret, born in August 1896 and a son, William Ellery (named after Hale's father), born in November 1900. William and Margaret Huggins often inquired about the children in their correspondence. They surely delighted in their little 'namesakes'.

58. W. Huggins to Hale, 28 October 1891, published in 'The Astro-physical Journal', *AA* **11** (1892), pp. 17–22, p. 18.

59. Huggins eventually came to prefer publishing in the *Astrophysical Journal*. In 1898, he confessed to Hale, 'I am in some doubt about the way of publishing [recent photographs]. If I send them to the R.S., they will be probably several months . . . &, as you will have seen, the R.S. illustrations are not too well done. . . . My friends here wish me to publish in a British Society, the Royal, or the R. Astronomical, but on the other hand, the illustrations of the A.P.J., are usually well done, and I suppose you could arrange for them to appear in an *early* number of the J.' W. Huggins to Hale, 25 May 1898, Box 22 [Reel 19], GEH, CIT.

60. Wright, *Explorer of the Universe* (1966), p. 82; Hale to Goodwin, 21 August 1891, HM 28441, HLA.

61. Hale, 'The ultra-violet spectrum of the solar prominences' (1891). Hale described his encounter with Lockyer in a letter to his friend Harry Goodwin: 'Though I took the opposite side from Lockyer in several protracted discussions with him, he certainly treated me very well, and put his whole lab. at my disposal. It was there that I measured my photos, and ground out the material for the paper read yesterday.' Hale to Goodwin, 21 August 1891, HM 28441, HLA.

62. One of the papers preceding Hale's was a 'Report of the Committee . . . appointed to co-operate with Dr. C. Piazzi Smyth in his researches on the ultra-violet rays of the solar spectrum', *RBAAS, Cardiff* (1891), pp. 147–8.

63. Hale to Goodwin, 21 August 1891, HM 28441, HLA.

64. W. Huggins, 'Address of the president' (1891), p. 4.

65. *Ibid.*, p. 7.

66. *Ibid.*, p. 9.

67. *Ibid.*, p. 10.

68. *Ibid.*

69. *Ibid.*, p. 11.

70. *Ibid.*

71. *Ibid.*, p. 14.

72. *Ibid.*

73. *Ibid.*, p. 20.

74. *Ibid.*, p. 21.

75. *Ibid.*, p. 23.

76. *Ibid.*, p. 37.

77. W. Huggins, 'The new star in Auriga' (1892), p. 618.

78. The nova was observed by so many astronomers around the globe that it would be cumbersome to list them all. To see the findings of a small sample of British observers, see the following articles published in the *MNRAS* **52** (1892): Christie, 'On the photographic magnitude of Nova Aurigae, as determined at the Royal Observatory, Greenwich'; C. Pritchard, 'Preliminary note on the magnitude of the new star in Auriga'; Knott, 'The new star in Auriga'; Maunder, 'Note on the spectrum of Nova Aurigae'; and Roberts, 'Photograph of the region of Nova Aurigae'.

79. M. L. Huggins, 2 February 1892, Notebook 1, WCL/SC.

80. W. Huggins, 'Further observations on the spectra of some of the stars and nebulae' (1868).

81. Balmer, 'Notiz ber die Spektrallinien des Wasserstoff' (1885).

82. W. Huggins, 'On the spectrum of the flame of hydrogen' (1880).

83. McGucken, *Nineteenth-century Spectroscopy* (1969), ch. 3.

84. W. Huggins to Holden, 26 April 1893, ESH, MLSA, LO; W. Huggins and M. L. Huggins, 'On Nova Aurigae' (1892), p. 491.
85. W. W. Campbell, 'The spectrum of *Nova Aurigae'* (1892), p. 244.
86. *Ibid.*, p. 247
87. W. Huggins, 'On a new star' (1866), p. 275.
88. W. Huggins to Stokes, 23 March 1893, Add MS 7656.H1271, GSS, CUL.
89. W. Huggins to Hale, 5 May 1893, Box 22 [Reel 19], GEH, CIT.
90. W. Huggins to W. W. Campbell, 25 May 1893, MLSA, LO.
91. The first number of the *Astrophysical Journal* appeared in January 1895, with Hale and Keeler as editors, Campbell as one of five assistant editors, and Huggins as one of ten associate editors.
92. G. E. Hale to W. W. Campbell, 4 October 1894, MLSA, LO.
93. W. W. Campbell to Hale, 9 October 1894, MLSA, LO.
94. G. W. Ritchey, 'Nebulosity about *Nova Persei'*; 'Nebulosity about *Nova Persei*. Recent photographs' (1902).
95. W. Huggins to Campbell, 14 November 1901, MLSA, LO.
96. W. Huggins to Hale, 13 December 1901, Box 22 [Reel 19], GEH, CIT.
97. W. Huggins to Hale, 13 January 1902, Box 22 [Reel 19], GEH, CIT.
98. Chicago was selected to host the event over New York City and Washington, DC, American cities more familiar and accessible to overseas visitors. The choice put Chicago on the world map. Bolotin and Laing, *The Chicago World's Fair of 1893: The World's Columbian Exposition* (1992)
99. Wright, *Explorer of the Universe* (1966), p. 105.
100. W. Huggins to Hale, 11 February 1893, Box 22 [Reel 19], GEH, CIT.
101. W. Huggins to Hale, 5 May 1893, Box 22 [Reel 19], GEH, CIT. At age sixty-nine, Huggins may have deemed a trip to Chicago to be a daunting physical challenge given the rigours of contemporary trans-Atlantic and trans-continental travel.
102. Hale, 'The Congress on Mathematics, Astronomy and Astro-physics' (1893). 'Science has its day', Chicago *Tribune*, 22 August 1893, p. 9.
103. 'Will search the skies', *Daily Inter Ocean*, 23 August 1893, p. 7b.
104. 'Gives a telescope', Chicago *Tribune*, 12 October 1892, p. 1.
105. Descriptions of the telescope appeared with some frequency in Chicago's daily papers. Two articles offering details of its construction can be found in 'Work nearly done', Chicago *Tribune*, 20 April 1893, p. 9 and in 'Yerkes' costly gift', *Daily Inter Ocean*, 27 May 1893, p. 1.
106. Osterbrock *et al.*, *Eye on the Sky* (1988), p. 21 and pp. 50–1. The Lick's lenses were completed in autumn 1886. Alvan Graham Clark oversaw their placement in the telescope tube in December 1887, four months after the death of his father. *Ibid.*, p. 59.
107. A photograph of the mounted telescope tube taken by exposition planner, Daniel Hudson Burnham (1846–1912), was reproduced as the frontispiece for *Astronomy and Astro-physics*, October 1893.
108. Hale, 'On some attempts to photograph the solar corona without an eclipse' (1894), pp. 668–9.
109. W. Huggins to Hale, 11 February 1893, Box 22 [Reel 19], GEH, CIT.
110. Hale, 'On some attempts to photograph the solar corona without an eclipse' (1894), pp. 673–5.
111. W. Huggins to Hale, 3 May 1893, Box 22 [Reel 19], GEH, CIT.
112. W. Huggins to Hale, 5 May 1893, Box 22 [Reel 19], GEH, CIT.
113. W. Huggins to Hale, 2 August 1893, Box 22 [Reel 19], GEH, CIT.
114. Hale, 'On some attempts to photograph the solar corona without an eclipse' (1894), pp. 675–80.
115. W. Huggins to Hale, 5 November 1893, Box 22 [Reel 19], GEH, CIT.
116. Hale, 'On some attempts to photograph the solar corona without an eclipse' (1894), pp. 680–6.
117. W. Huggins to Hale, 25 July 1894, Box 22 [Reel 19], GEH, CIT.
118. Hale, 'On a new method of mapping the solar corona without an eclipse' (1895).
119. Lyot, 'La couronne solair étudiée en dehors des éclipses' (1930).
120. Lyot, 'A study of the solar corona and prominences without eclipses' (1939), p. 580.
121. As an example, see Hufbauer, *Exploring the Sun* (1991) especially pp. 90–4. In this excellent history of solar physics, Hufbauer begins his discussion of coronal photography with Lyot and credits him with having 'fashioned the first "coronagraph"', p. 92.
122. 'In retrospect', PA **40** (1932), pp. 593–5.
123. Fath, 'William Wallace Payne' (1928), p. 270.
124. Hale to Campbell, 27 October 1894, MLSA, LO.
125. See, for example, E. C. Pickering, 'Discovery of variable stars from their photographic spectra' (1895); and Ellery, 'Observations of Mars made in May and June, 1894, with the Melbourne Great Telescope'(1895).

126. Riccò, 'On some attempts to photograph the solar corona without an eclipse, made at the Mount Etna Observatory' (1895).

127. W. Huggins to Hale, 6 November 1894, Box 22 [Reel 19], GEH, CIT.

128. Hale to W. Huggins, n.d., Add MS 7656.H1276a, GGS, CUL. This item is, I believe, erroneously filed and numbered with Huggins's letter to Stokes dated 29 December 1894. It is clearly the 'extract from Prof Hale's reply to me' which Huggins tells Stokes he has enclosed with his letter dated 6 December 1894.

129. W. Huggins to Stokes, 6 December 1894, Add MS 7656.H1275, GGS, CUL.

130. W. Huggins to Stokes, 29 December 1894, Add MS 7656.H1276, GGS, CUL.

131. W. Huggins to Hale, 6 January 1895, Box 22 [Reel 19], GEH, CIT.

132. W. Huggins to Hale, 17 April 1895, Box 22 [Reel 19], GEH, CIT.

133. Hale published two notes in the *Astrophysical Journal* clarifying Huggins's views. Hale, 'Note on the exposure required in photographing the solar corona without an eclipse' (1895); 'Note on the Huggins method of photographing the solar corona without an eclipse' (1895).

134. 'Barnard coming to Chicago', *Daily Inter Ocean*, 31 May 1895, p. 3f; 'Comes to see stars', Chicago *Tribune*, 16 October 1895, p. 4.

135. 'The 40-inch lens for the great telescope given by Mr. Yerkes to the University of Chicago is finished', *Daily Inter Ocean*, 13 June 1895, p. 6g.

136. W. Huggins to Hale, 10 March 1896, Box 22 [Reel 19], GEH, CIT.

137. W. Huggins to Hale, 17 July 1896, Box 22 [Reel 19], GEH, CIT.

138. W. Huggins to Hale, 23 November 1896, Box 22 [Reel 19], GEH, CIT.

139. W. Huggins to Hale, 7 March 1896, Box 22 [Reel 19], GEH, CIT.

140. W. Huggins to Hale, 21 April 1897, Box 22 [Reel 19], GEH, CIT.

141. W. Huggins, 'The new astronomy' (1897).

142. 'Ship big Yerkes lens', Chicago *Tribune*, 18 May 1897, p. 12.

143. 'Yerkes lens in Chicago today', Chicago *Tribune*, 19 May 1897, p. 5; 'Yerkes lens reaches observatory', Chicago *Tribune*, 20 May 1895, p. 5.

144. Wright gives 28 May 1897, a Friday, as the date of the collapse, but the Chicago *Tribune* describes the incident as having taken place on a Saturday. See Wright, *Explorer of the Universe* (1966), pp. 130–1; 'Big telescope is safe', Chicago *Tribune*, 31 May 1897, p. 2; W. Huggins to Hale, 22 June 1897, Box 22 [Reel 19], GEH, CIT.

145. 'Alvan G. Clark dead', Chicago *Tribune*, 10 June 1897, p. 3.

146. 'Yerkes Observatory repairs', Chicago *Tribune*, 2 September 1897, p. 7.

147. 'To dedicate big lens', Chicago *Tribune*, 18 October 1897, p. 10.

148. 'Yerkes' big telescope', Chicago *Tribune*, 19 October 1897, p. 8.

149. Keeler, 'The importance of astrophysical research and the relation of astrophysics to other physical sciences' (1897).

14

The new astronomy

A pioneer rarely keeps the lead in a second generation as Sir William
Huggins has done.

– Agnes Mary Clerke[1]

The restless mix of ambition and curiosity that had spurred William Huggins down so
many unexplored paths in the past would not let him sit quietly upon the laurels he had
earned. The discovery of terrestrial helium and the puzzle of radium's spontaneous glow
brought him new brinks to teeter on, fresh risks to calculate, and undreamt-of wonders to
reveal. Because the published record of his research in these areas is scant and unrelated to
his more familiar astronomical investigations, Huggins's creative work, particularly on the
question of the nature of radium glow, has been ignored by historians of science. In this
chapter, that oversight will be rectified. The public record augmented by his unpublished
correspondence with fellow investigators bring to light how eagerly and ardently he and his
wife, Margaret, applied their spectroscopic and photographic expertise to these new and
tantalising problems.

The previous chapter introduced the young American astronomer George Ellery Hale,
and discussed the foundation of Huggins's alliance with him. In this chapter, we will
follow the growth and development of their close relationship as Hale launched his
career in astrophysics and pursued his ambitious plan to erect and direct the world's
largest refracting telescope. Huggins nurtured his friendship with Hale, and developed a
similar association with Irish mathematician and physicist Joseph Larmor. Over the years,
both Hale and Larmor became more like sons to the Hugginses than mere colleagues.
Their lively and collegial interchanges – a near continuous stream of creative give-and-
take on new and elusive puzzles – kept Huggins's investigative interests fresh and
enthusiastic.

During his eighth decade, Huggins began reaping the recognition of colleagues and the
nation for the fruits of his life's work. Knighthood and other honours were capped by
election as President of the Royal Society. Although he had no interest in retiring yet as
an active investigator, he nevertheless became increasingly nostalgic and wary of
encroachment upon his past accomplishments. In this important phase of his career,
he – with the invaluable assistance of his wife Margaret – began the challenging task
of carefully laying out the groundwork for what would become the foundations of his
historical image.

14.1 Helium

On 27 March 1895, three months after isolating a new component of Earth's atmosphere – an inert gas he called 'argon' – Scottish chemist William Ramsay (1852–1916) announced that he had found yet another in the emanation of the mineral clèveite, a 'uranate of lead containing rare earths'.[2] This second gas did not appear to be entirely new, however. When it was made luminous via electric spark, it produced a 'brilliant line in the yellow' which was close to, but not coincident with, the familiar sodium D lines. Ramsay believed it would prove to be the terrestrial source for the supposed solar element helium.[3]

Indeed, measurement revealed the line to be identical to the D_3 line that Lockyer had discovered nearly three decades before in the solar chromosphere.[4] Lockyer always suspected it was the signature of a new element.[5] His more conservative collaborator, chemist Edward Frankland (1825–99), argued persuasively against making any public claim of discovery to avoid the embarrassment of retraction if and when a match with a known substance was found. It now appeared likely that Lockyer had been right all along.

During the spring and summer of 1895, Huggins kept Hale apprised of the latest developments on Ramsay's discovery. On 28 April, he offered a glimpse of the exciting confusion that reigned:

Crookes uses only one prism, and his measures are uncertain within a tenthmeter or two, but Thalen [Swedish astronomer and physicist Tobias Robert Thalén (1827–1905)] has measured the line at Upsala, and I think that there is now little doubt that the new gas, if not helium, contains helium.

Apart from the helium line near D_1 great diversity exists as to the other lines. Ramsay gets one set, Crookes another, and Thalen others probably.

Lockyer has been trying the mineral in a vacuum. He professes to get the line & also those of hydrogen, but his lines are too faint to compare directly with the sun, so he says.

I saw the line first a fortnight ago, at the R. Institution. It appeared thin & well defined at the edges, but was rather faint. Last Thursday Ramsay showed a tube which gave a really brilliant yellow light due I suppose to the line but I did not see its spectrum ...

I think there is little doubt that helium has really been discovered, but probably mixed up with two or even more gases, possibly new ones.[6]

To corroborate Ramsay's discovery required having a tube of the new gas. As Huggins explained to Hale, 'If I had a tube I could compare it under resolving power sufficient to show the sun's rotation. Until Thursday the tubes appeared too faint and for this and some other personal reasons, I have not asked for a tube.'

Hale wanted to conduct his own experiments. He hoped Huggins might intercede on his behalf so that he could obtain a tube of the gas from Ramsay. But Ramsay was overwhelmed by the interest his new discovery had generated. 'Mr. Lockyer has the [tube]', he explained to Huggins. 'I can't take it out of his hands without consulting him. Moreover', Ramsay added, 'though I might be able to spare a vacuum tube of helium later, I am afraid that I have not one available at present. Each tube represents some particular stage of treatment and is numbered & labelled. I must keep them till the whole story becomes clear, or at least as clear as I can make it with Mr. Crookes' help.'[7] Under the circumstances, Huggins suggested that Hale might wish to 'get *some of the mineral* in *the States* & try for yourself? Several ores of uranite [*sic*] seem to contain it.'[8]

If Ramsay was unwilling or unable to ask Lockyer to share the tube with colleagues, Huggins was even less inclined. For one thing, to do so would require venturing into territory his nemesis had clearly staked out and claimed as his own. The tube of the new gas that Ramsay had lent Lockyer on 28 March was already so blackened from use that he was forced to try his hand at making his own from pieces of uraninite procured from a colleague. He presented the first photographs he obtained of the new gas's spectrum at the 25 April RS meeting.[9]

But there was more behind Huggins's reluctance to ask for a tube. In February, Lockyer had begun publishing in *Nature* transcriptions of notes from lectures on 'The Sun's place in Nature' that he had delivered the previous November and December at the Museum of Practical Geology.[10] The articles started out as a straightforward discourse on Lockyer's thinking about the Sun as a star. But things changed in April when, in his third instalment, Lockyer launched a lengthy and personal attack on Huggins which threatened to reignite the smouldering embers of their old hostilities over the nature of nebulae:

I am glad to say that among the first to accept the new evidence proving that nebulae are really early stages of evolution of stars was Dr. Huggins himself, the observer whose statement which I have quoted I had been fighting for years. That you see was a great victory. He says now not only that these bodies may represent early forms; places them in the line of evolution where I had placed them, but he even adduces the same evidence which I had brought forward in several of the arguments which I had employed. Dr. Huggins made a reference to this question as President of the British Association in the year 1891, and if any of you read that you will see that it is really an argument in favour of the views that I have been insisting upon since 1886, and his agreement seems all the more important since Dr. Huggins appears to have arrived at these conclusions quite independently. Not one word is said throughout the address of any arguments which I may have used, or any line of thought or observation on which I had founded the various statements which I had made; and therefore it would be charitable to suppose that he was unacquainted with my work when that address was given to the world. Of course you will acknowledge that there was a very extraordinary change of opinion, so extraordinary indeed that it is clear that Dr. Huggins felt it was of importance to himself that the change should be explained; and he confesses in the address, to which I refer, that the communication he made to the Royal Society in 1864 was not entirely founded on scientific evidence, but partly made under, to use his own words, 'the undue influence of theological opinions then widely prevalent.'

So after all I had been fighting partly an expression of theological opinion. If we had known that before, probably some trouble might have been saved.[11]

Huggins was shocked and dismayed. He naively believed he had brokered a truce through the disinterested tone of his BAAS lecture. Besides, he assumed Lockyer had given up on the meteoritic hypothesis to pursue his new interest in ancient astronomy.[12] But clearly the ceasefire was over and the gauntlet had been laid down once again.

What to say? What to do? On the one hand, responding to Lockyer's provocative remarks would only draw greater attention to them and bestow upon them an undeserved legitimacy. On the other, failing to respond could be interpreted as an admission that Lockyer's assertions were true. There was no obvious course of action that did not spell disaster. *Nature* was an international journal. Both William and Margaret recognised that Lockyer's words could damage their reputations far beyond their circle of English acquaintances.

Caught in uncertainty's powerful jaws, William added a confidential note to his already lengthy 28 April letter to Hale. He addressed it to both Hale and his wife 'as true friends' to whom he felt he could 'write a little freely'. '[T]hese attacks of Lockyer, *only just begun* I fear, are giving us much pain', he lamented. 'It is difficult to keep them from eating out of our life all joy of life.'[13]

The misrepresentations are so prolonged, so repeated, and so mixed at the same time as confounded, that it would not be possible to answer them without reprinting the whole papers with a copy in parallel columns, and it would lead to a flood of new abuse. I can hardly believe myself that the main accusation he is now bringing against me is that *I have stolen his ideas*!!

Before my [BAAS] Address he was constantly bringing up a sentence in one of my papers written more than 30 years ago, as representing my present views! So I thought it well to acknowledge the sentence in my Address, & show that *within two years* of writing that sentence I had taken different views. Now he again reprints the old sentence, & keeps back the explanation & makes out I had been recently converted by *his arguments*!!!

What could be done to rectify the situation? Lockyer 'is constantly at work against me by underground methods, such as dinners at a club to which he invites editors & literary people, and in this way, as well as in his lectures is doing all he can to "drown" us'. Huggins closed by asking the Hales to 'tell me quite frankly how the matter appears to you; you are out of it and see a wider horizon'. Unfortunately, we do not have Hale's reply. Whatever he wrote, his knack for soothing damaged egos surely eased the Hugginses' angst. In the end, they left Lockyer to flail away.

Their attention was soon pulled back to the spectrum of terrestrial helium by a startling announcement in *Nature*. In a letter dated 16 May, German mathematician and spectrum analyst Carl David Tolmé Runge (1856–1927) let it be known that he and his collaborator Louis Carl Heinrich Friedrich Paschen (1865–1947) had found the bright yellow line of Ramsay's new gas to be double.[14] 'We do not ... agree with the conclusion drawn by Mr. Crookes', Runge asserted, 'that the unknown element helium causing the line D_3 to appear in the solar spectrum is identical with the gas in cleveite, *unless D_3 is shown to be double*.'

As soon as they heard the news, William and Margaret sprang into action. They did not need a tube of the new gas to investigate Runge's claim. They had the Sun! Margaret recorded: 'W. & I on Saturday morning [8 June] made a number of observations on D_3, – using various powers.'[15]

We succeeded in getting a number of observations in all of which the line *seemed* single. The observations we thought best were some in which we got the Helium line very thin & well defined, when it ran up into a needle like point. It certainly seemed single then. I noticed however sometimes a varying behaviour about the sides which may have meant something. W. remarked – 'it is a single line unless it is that the sky is not clear enough to let us see a fainter line'.

William fired off a note to Hale. 'Runge has shown that Ramsay's gas line is *double*, so it is not true helium!' he exclaimed. 'If I had had a tube, I could have settled the matter at once in a couple of hours.'[16]

Margaret reported the results of their continuing efforts:

On the 12th we again observed & got some good observations all apparently confirming the singleness of D_3.

A private letter from Prof. Runge a few days later announced to us that he now found *all* the lines in Ramsay's gas to be *double*. He therefore thought that it must be Helium.

This was interesting. It did not follow of course that D_3 *must* be double. It might be a different condition in the Sun to what it is on Earth. But the probability that Helium had really been found had become very high indeed from Runge's observations showing that *a number of lines in the spectrum of Ramsay's gas were identical apparently with lines in the chromosphere*. Was it not probable that all these lines as well as D_3 belonged to Helium? And if this were granted – might not the solar lines *all be double really* – difficult doubles, – as the Ramsay gas lines were doubles?

Nothing but direct observation could settle the point & we felt that unless we had clearer skies we could do no more.

On 18 June, Huggins told Hale he had concluded that the 'new gas is *not solar helium*'.[17] A week later, he wrote, 'The Helium plot thickens. Runge, in a private letter, tells me that all the lines of the cleavite [*sic*] gas are doublets like the yellow one. Some fourteen doublets!'[18] Puzzled but undaunted by Runge's announcement, Huggins vowed 'We will try again.'

Finally, Margaret broke the good news:

It was the morning of July 10th that there seemed first for a long time a really blue looking sky. At breakfast time this morning oddly enough we had a letter from Prof. Runge telling us he had heard from Prof. Hale. He had seen D_3 double. We were inclined to put this success down to either clearer sky or – perhaps the use of another method.

We had looked most carefully all along – without any pre-convictions.

I was obliged to go to town so W. observed alone. It was as it had promised to be a much clearer sky and W. observing with the *4th* order of our Rowland grating (we could not use the 4th order well before) & a power of about 21 succeeded in picking up the fainter component. He was so delighted he thought of telegraphing to me in town! I wish he had![19]

William penned a hasty postcard to Hale: 'Since I wrote to you, with a bluer sky I have seen D_3 double in the sun, precisely similar to Runge's determination of the cleavite [*sic*] gas line.'[20] Margaret recorded: 'We observed again on the Saturday [13 July]. One has to wait for good moments, & be in good condition. It is a very delicate observation to make satisfactorily. Not of the highest difficulty, – but not an easy one. I got two or three good observations. D_3 is certainly double.'[21]

As William explained to Stokes, 'Our failure at first was due, I now believe to the thin cloud present, for with a bluer sky the duplicity can be certainly seen. At the limb both components are much expanded, and the interval between them very small indeed, but a little from the limb, and in prominences, the lines thin out, when the less refrangible component is much fainter.'[22]

The ability to generate terrestrial helium opened up greater opportunities for making direct laboratory/field spectral comparisons, particularly of eruptive features on the Sun and those stars sporting distinctive helium lines like the enigmatic β Lyrae.

14.2 Accolades and achievements

On the occasion of the Queen's Diamond Jubilee in June 1897, Huggins was created a Knight Commander of the Order of the Bath, thus becoming Sir William.[23] Margaret was thereafter addressed as Lady Huggins. That month's number of the popular magazine *The*

Nineteenth Century featured his essay – 'The new astronomy: A personal retrospect' – along with articles on 'British monarchy and modern democracy' and 'India under Queen Victoria'.[24] He sent copies to Hale and Yerkes. '[T]he article is wholly about the rise of the science in connection with my own work', he warned Hale. 'I do mention you once or twice, and at the end the Yerkes observatory. I trust that what I have said will be more than justified in the future by the work done there.'[25]

In January 1899, Huggins informed Hale he was planning 'to bring out our new spectra in a privately printed volume'.[26] By freeing themselves of editorial entanglements they reasoned they would 'be freer in many ways' to create the book exactly as they wished. 'We are working very hard to get the book out', he added, 'but it will be some weeks before all is ready.' Their principal motivation for publishing the book was 'to place in the hands of those interested in the subject representative spectra of the principal classes of stars through a long range of wave-length, together with scales attached, sufficient for the determination of the approximate positions of the stronger lines'.[27] They spared no effort in preparing the volume's twelve plates. To produce 'the most truthful representation of the photographs' required 'many trials and much consideration'.[28] In May Huggins complained to Hale 'I have lost much time over the plates.'[29] The Hugginses eventually settled for a scale of enlargement that sacrificed some detailed structure in order to reproduce correctly the tone of the originals.

Finally, the following February, William informed Hale 'Our book is now just ready. I have instructed [the publisher] to send a copy to you for the library of the Yerkes Observatory.'[30] The folio book, *An Atlas of Representative Stellar Spectra*, was identified on the title page as the first volume of the 'Publications of Sir William Huggins's Observatory'.[31] Clearly William and Margaret anticipated putting together a series of volumes on topics of historic and contemporary interest to practising astronomers. They were confident that this first volume would have special value to researchers because, as William told Hale:

I have gone into a theory of the order of stellar evolution, which I think has much in its favour, and which I trust will meet with the general acceptance of yourself, and Professor Frost, who I believe is now at the Yerkes Observatory.

I have not gone into stars of class IIIb, in which you have been working as none of these are included in our photographs.

We have succeeded in giving for the first time the *ultra-violet region* of stars of class IIIa, which leaves no doubt of their evolutional place.[32]

The Hugginses opened with a history of the Tulse Hill observatory. By including in it a lengthy excerpt from 'The new astronomy' essay, they made the retrospective's gripping narrative available to readers who had not seen it when it was first published, and refreshed the memories of those who had. If their run-ins with Lockyer had taught them anything, it was the importance of telling their story their way, and telling it often.

Reviewers admired the volume as 'magnificent', 'striking' and 'beautifully printed and bound'. All took great delight in the pen-and-ink illustrations contributed by Margaret Huggins. Thanks in large part to them, one reviewer quipped, the book was 'as far as possible, in appearance from the dull repulsiveness too commonly characteristic of observatory reports'.[33] In the *Edinburgh Review*, Agnes Clerke – as few others could – drew attention to Margaret's scientific contributions to the work:[34]

Since 1875 she has been, on equal terms, her husband's coadjutor, and while content to merge her initiative in his, she has known how to make its effect and influence tell as essential factors in the joint product of their labours. Apportionment of credit would be equally invidious and impossible. Suffice it to say that all belongs, in the truest and deepest sense, to each.[35]

Clerke instructed her readers on the range of current views on the subject of stellar evolution based on the book's plates of stellar spectra and the Hugginses' exposition of what those spectra reveal about the subtle 'life changes' in stars.

In his review, Hale acknowledged the inspirational value of the first chapter's riveting tale of astrophysics' origins for the general reader and specialist alike. But, in his opinion, the book's real contributions to modern astronomy were to be found in 'the photographs of the ultra-violet spectra of various types of stars [reproduced in the book's twelve plates], and the discussion of stellar evolution based upon them' in the sixth chapter. [36] The importance of the Hugginses' photographs 'will be recognized when it is remembered that they constitute practically our only source of information regarding the extreme ultra-violet region of stellar spectra'. Hale was particularly impressed by the 'excellent half-tone plates [which] bring out the details of the original negatives'. By using direct photographic reproductions, untouched by the engraver's tool' the Hugginses ensured 'the perfect reliability of the illustrations'. They wisely positioned each image on the plates so that 'common lines can be traced from star to star'. Furthermore, they included a wavelength scale, thus enabling interested readers to analyse the images themselves. 'It is indeed rather surprising', Hale remarked, 'that in the years which have elapsed since the first photographs of stellar spectra were obtained in this very region, practically no advances have been made except at the Tulse Hill Observatory'.

He found the Hugginses' ideas regarding stellar evolution both provocative and worthy of serious efforts to confirm or refute them. He was particularly intrigued by their view that 'some of the spectral changes observed in stars whose temperature is rising as a result of condensation, may be due primarily to increasing density and in much less degree to higher temperature'.[37] Heretofore, he noted, evolutionary schemes had predicted that the ultraviolet region of the spectra of stars passing from the first into the second class would show increasing degrees of absorption to the point of extinction. But the Hugginses' unparalleled photographs show that 'the position of maximum intensity in the spectrum advances toward the ultra-violet during this period of transition'.[38] Testing their claims, Hale noted, is a difficult but worthy challenge.[39] He concluded 'the work should be in the library of every spectroscopist'.[40]

No mention of the book was made in *Nature*. Huggins did not send a copy to the journal for review because – as he explained in confidence to Larmor when copies of his second volume, *The Scientific Papers*, were being sent out for review in October 1909 – he did 'not wish to risk it being put into the hands of one of L's assistants for review with disagreeable remarks suggested by himself; as was always the case with Miss Agnes Clerke's books'.[41]

In November 1900, William was elected to succeed Joseph Lister (1827–1912) as President of the Royal Society. Huggins was one of the few, and probably the last, self-taught individuals to hold the post. Although he held several honorary degrees from prestigious universities by the time of his election, he is the only President in the Society's long and varied history to come to the position with so little formal education.[42] Even Humphrey Davy (1770–1829) spent several years in grammar school before entering his apprenticeship.

As President, Huggins kept a busy social schedule. He represented the Society at banquets and other public events, addressed its annual November meetings, and served as host (accompanied by Lady Huggins, when warranted) of the Society's soirées each May and June.[43] Although he continued to do original research involving less physically demanding laboratory investigations, his career as an active astronomical observer was winding down. The last page of dated notes in Notebook 5, all recorded by Margaret in 1901, include comments on a photograph taken of the spectrum of α Leonis (Regulus) on 25 April, particulars of two photographs taken of the spectrum of α Lyrae (Vega) on 3 October and, finally, information on two more photographs taken of Vega on 1 November, a night she described as 'not very clear'.[44]

It was just at this time, that the Hugginses were reminded of the need to remain vigilant concerning the public record of William's contributions to the development of astrophysics. In November 1901, *The Observatory* reprinted the text of an address by David Gill on the occasion of the unveiling of an inscription stone for the Cape Observatory's new Victoria Telescope, a gift of Cape observer Frank McClean (1837–1904). Gill rejoiced that the telescope would make it possible at last for Cape-based astronomers to become full partners in the global effort to advance the new astronomy. He acknowledged William Huggins's many contributions to the rapidly developing field and drew special attention to his important work on stellar motion in the line of sight. It was an investigation, Gill noted, to which Huggins's 'attention was directed by Clerk Maxwell'.[45]

Huggins moved immediately to set the record straight. In point of fact, he asserted, 'the method suggested itself to me directly from Doppler's work, some time in 1862–3'.[46] It was true that he had included a letter from Maxwell in his first paper on stellar radial velocities, but, he emphasised, he intended its contents to serve as part of the historical context for his innovative research effort, not an explanation of its etiology.

Wishing to make the historical introduction to my paper as complete as possible, I asked my friend Clerk Maxwell, in 1867, to give me an account of some experiments which I had heard he had been making to detect the influence of motion on the refrangibility of light. His letter, which I did not receive until June 1867, appeared to me to be of so much interest that . . . I requested his permission to print it in full in my paper. Clerk Maxwell's reply [in] a letter dated March 23, 1868, shows clearly that my work had been independent, and not undertaken in consequence of a suggestion of his.

Concerned that Hale had also seen *The Observatory*'s reprint of Gill's address, Huggins sent a copy of his letter of clarification to his American colleague. 'Did you notice that [Gill] puts it that the Motion in the Line of Sight method was suggested to me by Clerk Maxwell!', he asked Hale. 'I can not imagine how he got this idea.'[47] It rankled Huggins that colleagues would think Maxwell had initiated or conceived of the plan to measure the radial velocity of luminous celestial bodies using stellar spectroscopy. 'This method being one of the most important of modern astronomical methods, there is now a desire on many sides to get some credit for it', he complained. In Maxwell's case it seemed that after the bright young physicist's untimely death in 1879 he and his work acquired a public reputation of near-mythic proportions: if a great discovery had been made and Maxwell had had anything to do with it, one could safely assume the process of discovery had been expedited by Maxwell's involvement. 'It seems to me a duty to keep the history of the subject clear', he forcefully declared to Hale. He then sought his American ally's assistance in repairing the damage. 'As no doubt Gill has sent out his Address very widely, I should be glad if you could see your way to put this reply, as *copied from the Observatory*, and *not sent by me*, in the "notes" of the Astrophysical Journal.'

Nevertheless, the Hugginses found Maxwell's letter more and more troublesome as time went by and William's 1868 *Philosophical Transactions* paper on stellar motion in the line of sight became the principal source from which knowledge of his groundbreaking work was revisited by their contemporaries, or encountered for the first time by new generations of researchers. They put themselves on guard to quash and rectify any and all future misunderstandings of the sort exemplified by Gill's address.

On 26 June 1902, Huggins became one of the first twelve individuals to be awarded the Order of Merit, a new honour founded by King Edward VII to acknowledge 'exceptionally meritorious service' in the military or 'towards the advancement of Art, Literature, Science' without regard for the individual's social standing.[48] Recipients were chosen privately by the royal family. The honorees included four scientists along with five major military figures, two noted political historians and one artist.[49] In addition to Huggins, the other scientists were Lords Rayleigh, Kelvin and Lister, all renowned for their contributions to British science. It is probable that Huggins was selected because he happened to be President of the Royal Society at the time. In what has become something of a tradition, the Order of Merit has been conferred on every President of the Royal Society, a distinction no other office enjoys.[50]

14.3 Radium

In the spring of 1903, the scientific world was abuzz with talk of radioactivity. Radium, the exotic new radioactive element, had been isolated the previous July by Polish chemist Marie Sklodowska Curie (1867–1934).[51] It seemed that ever since German physicist Wilhelm Röntgen (1845–1923) announced his serendipitous discovery of invisible penetrating '*X-Strahlen*' in December 1895,[52] there had been one mind-boggling revelation after another about the nature of the physical world.

At the Royal Academy banquet on 2 May, William Ramsay confided to Huggins that 'he expected to get a new gas from Radium'.[53] He hoped to have a paper on his discovery ready in time for the RS meeting at the end of the month. Huggins passed the privileged information along to his right-hand man, RS Physical Secretary Joseph Larmor. Since Ramsay's discovery of argon (December 1894) and helium (March 1895), he had isolated three other new gases in the same chemical family: krypton and neon (June 1898), and xenon (July 1898).[54] The possible discovery of another new gas was important news. Huggins authorised Larmor to push Ramsay's paper to the front of the queue as soon as it was received. In the meantime, he cautioned, 'the discovery should not be *mentioned to anyone* by you or me, as there are so many working on radium, and even a hint might enable someone to forestall him in publication'.[55]

Many in London *were* working on radium. William Crookes prepared an exhibit devoted to radium and its effects for the RS's 'black soirée' in May. He put 'auto-radiographs, photographs of radium emanations, luminous effects of radium emanations' on display alongside an instrument – the 'spinthariscope' – he had devised for rendering radioactive scintillations visible. He also showed off a 'permanent lamp' he had created by placing a solution of radium on a small plate.[56]

A month later, physicist Oliver Lodge (1851–1940) delivered the annual Romanes lecture at Oxford in which he espoused an electrical theory of matter and speculated on the nature of the spontaneous 'radio-activity' exhibited by such elements as uranium, thorium and radium.[57] Relying solely on known laws of mechanics and electricity, he likened radio-activity to 'the condensation or contraction of a nebula'. When 'the centrifugal force of the peripheral portions exceeds the gravitative pull of the central mass', Lodge explained, the particles left behind 'agglomerat[e] into a planet'. Meanwhile, 'the residue goes on shrinking and evolving fresh bodies and generating heat'. A radium atom behaves in a similar way, he declared. It 'has a great store of potential energy, immense in proportion to its mass, for it is controlled by electrical, not by gravitational forces'. Thus, the 'shrinkage of a few yards per century can account for all its tremendous emission'.[58]

The Curies travelled to London in June 1903. Pierre had been invited to talk about radium at the Royal Institution's Friday Evening Discourse on 19 June.[59] Although Marie was the one who had isolated the new element, measured its atomic weight and studied its properties, women were not permitted to address the Royal Institution at the time.

Earlier that evening, the Curies were received as guests by William and Margaret Huggins at the RS's June conversazione, its one annual event to which ladies were invited.[60] The grand array of exhibits included a mounted specimen of a newborn Indian elephant, fossilised remains of Eocene mammals from the lake beds of the Berket el Kerun and colour photographs of insects. There were also physical instruments and experiments to attract the Curies' attention. Oliver Lodge's exhibit on wireless telegraphy and Crookes's radium display from the May soirée were still on view. A new exhibit featured the recent work of Ernest Rutherford (1871–1937) from McGill University in Montreal and his collaborator Frederick Soddy (1877–1956) on the condensation of the radioactive emanations of radium and thorium.[61]

After the conversazione, Pierre Curie addressed the crowded hall at the Royal Institution. Speaking in French, he astonished his audience with demonstrations of radium's incredible energy. He showed that even the subfreezing temperature of liquefied air could not reduce the element's ability to generate heat. He made phosphorescent materials glow spontaneously simply by placing them near samples of radium. And he demonstrated that radium could expose covered photographic plates.[62]

The Curies' visit inspired William and Margaret to undertake their own investigations on radium. They believed the material's self-luminous property to be unlike either the phosphorescence or fluorescence exhibited by other substances. 'I have from the first been impressed by [radium's] sharply defined luminous surface', William later wrote to Larmor. 'The radium looks *like* a cleanly cut piece of solid light.'[63] Radium's spark and flame spectrum had been observed by others.[64] Always searching for some new and untried application for spectroscopy, the Hugginses were curious to see and study the spectrum of its spontaneous glow. They were the only ones to pursue this line of research. No one, then or since, has repeated their trials. It is difficult today to match their subjective descriptions of what they observed with modern expectations of radium's behaviour under similar circumstances.[65]

Unfortunately, the Wellesley notebooks contain no records of their investigations on radium. But William's correspondence with Larmor and others during this period allow us to flesh out the terse text of their published accounts and follow along as their thinking on the radium puzzle unfolded. 'I have been working very hard at the Ra problem', he told Larmor on 16 July. 'I have got 8 lines in the ultra-violet. No one of them appears to be a line of the spark spectrum of Ra. Several of them *do seem to agree with He lines*, but the spectrum on the whole, seems different.'[66]

The next day, he submitted a paper to the RS on the spectrum of the radium glow.[67] In it, he and Margaret described their fascination with radium's ability to glow 'spontaneously and without ceasing'. They suspected the continuous glow was generated 'by those more active molecules which are supposed, in consequence of a condition of internal instability, to be the source of all the phenomena of radioactivity'.[68] If their hunch was correct, this natural molecular agitation – just like that artificially induced in the radium by a flame or spark – should produce a distinctive spectrum.

They were encouraged in their thinking after preliminary visual observations of a 'thin fragment of some length of radium' revealed evidence of faint lines. They then attempted to record the spectrum photographically, beginning first with exposures of 24 hours. Because the spectrum was so faint, they were forced to make repeated trials, until finally, after exposing the plate for 72 hours they captured their first successful image.

The spectrum of the radium glow failed to match the pattern of lines seen in photographs of its spark spectrum. Perhaps, the Hugginses surmised, there are real physical differences after all between the light generated by natural rather than artificial excitation of the particles of radium. Still, there was always the chance that coincident lines were present but too faint to show up on their first set of photographs. Perhaps longer exposures would reveal them.

As William had already hinted to Larmor, he and Margaret believed they had detected signs of helium's characteristic signature. They were simply waiting until they replaced their spent helium tube with a fresh one to confirm their suspicion. Larmor was thrilled by the Hugginses' report and impatient to spread the news. With the Society on summer hiatus, he elected to publicise their pioneering efforts by means of an article in *The Times* of London.[69] It appeared with the eye-catching title 'Radium and Helium' and concluded with the suggestion that 'If ... the radium salt shines spontaneously in the dark largely by light belonging to the different element helium, another important step is gained in elucidating the nature of the instability of such chemical elements of high atomic weight and the radio-activity associated with it.'

Huggins was deeply fascinated by his radium work. Having focused his whole career exploring the universe at large, he was thrilled by the opportunity to delve into the mysteries of the realm of the very small. 'I do not know what view to take of electrons', he wrote to Larmor on 19 July. 'Are the electrons from H in any sense themselves H? Do they differ from electrons from other elements? Does the H molecule after it has lost electrons differ chemically from its original and normal state? Can Ra, after all, not be a true element, but a very stable compound of some substance, we may call X, with He? Say, when it contained 350 electrons instead of 700? If this is possible.'[70] The avalanche of questions tumbled freely from his typewriter like boulders set in motion by an all too curious child. '*On no accounct* [*sic*, read 'account'] *take the trouble to give any answer to any of these questions*', he admonished his conscientious secretary. He knew he had too many questions about things for which little was known at the time.

Despite their best efforts, the Hugginses' attempts to photograph the radium glow spectrum produced disappointing results. Although 'the lines do seem to agree with the He lines', William informed Larmor, 'the absence of the strongest lines of He ... has great weight against He being the source of the lines'.[71] He added, 'I think it well to put this clearly in the paper.'

Meanwhile, thanks to the *Times* article, news of the Hugginses' investigation of radium's spectrum reached a wide audience. Hale read about it in the journal *Science*.[72] 'This seems to me one of the most astonishing results obtained during many years', he wrote his old friend. 'I congratulate you most heartily on this renewed evidence of your genius as an explorer on open fields. In the light of this advance we must certainly readjust many of our ideas: it is of course far too soon to anticipate its full bearing on the future of spectroscopy.'[73]

But by 11 August William thought it best to admit defeat for now. Inconclusive photographs hampered his analysis of the radium glow spectrum. He confided in Larmor:

You will remember that I had misgivings when you were here. I have treated the matter in what appeared to me the most straightforward way, leaving as much as possible of the original paper and adding a dated addendum. There can be no doubt whatever of the correctness of the nitrogen interpretation.

May it not be occluded N? There is evidence of H, O, CO_2 and He, why not also N? Would not the massive Ra molecule attract and retain as satellites ... the lighter molecules of gaseous bodies? ...

I am anxious to see what comes out from the long exposure photograph now going on; I think that I shall develop it to-morrow when it will have had six days and nights.[74]

On 15 August, the RS published four short papers on radium including the Hugginses' report on the radium glow spectrum and Ramsay's investigation of the emanation obtained from radium. All four papers were received after the Society began its summer recess.

There are notable differences between the Hugginses' final published paper and the preliminary account of its contents given a month earlier in *The Times*. In their published paper, the Hugginses omitted any reference to the article's claim that a 'line, that of highest refrangibility, agrees with a line in the spark spectrum of radium itself, which, however, has not been recorded by other observers'. Instead they stated simply 'the two very strong characteristic rays of the spark spectrum of radium, in this part of the spectrum ... were not present on the plate'.[75] Also in an addendum, dated 5 August, they stepped back from their claim to have found evidence for helium in the radium spectra. Instead, they reported the spectrum of the radium glow most closely resembles that of nitrogen.

They tried to make it clear that identifying the chemical source of those spectral lines had never been the aim of their investigation. Rather, they emphasised, they were principally interested to see if natural radium glow could generate a bright-line spectrum. They believed they had demonstrated this important fact without question and concluded, 'in radium we have a body which at the ordinary temperature, sets up radiations which are similar to those which have hitherto only been obtained in connection with the electric discharge'.[76]

A few days after the special number of the *Proceedings* appeared in print, the Hugginses were horrified to find themselves the targets of an anonymous letter to the editor in *The Times* of London.[77] The letter's author, identified only as 'Inquirer', castigated the newspaper. The article, 'Radium and Helium', had heralded as 'a scientific discovery what now turns out to have been only an unaccountable mistake of observation'. It was a case of bait-and-switch, Inquirer complained. *The Times* had teased the public with the tantalising story that the Hugginses had found spectroscopic evidence for helium in the radium glow. 'This, if correct, was a very interesting and important contribution to the investigation of radium', Inquirer admitted, 'whether we suppose the helium to be due to the actual breaking-up of the radium molecule, or the radium to have the power of throwing accidentally associated helium into the state of vibration in which it produces its characteristic spectrum.' But, in fact, when the Hugginses' paper was made public, there really was nothing new to report – no helium as promised, only ordinary nitrogen. Inquirer demanded *The Times* make amends by giving 'similar publicity to the fact that the [Hugginses'] paper as printed completely abandons the position too hastily assumed in that portion of it which was described in your article'.

William was certain he knew Inquirer's true identity. The letter was 'evidently written by Dewar, whom we have known now for so many years', he informed Arthur Schuster.[78] James Dewar was the Royal Institution's Superintendent. Dewar had been in a feud with Ramsay since the discovery of argon in 1894. 'One reason doubtless that Dewar is angry', Huggins informed Schuster, 'is, that now in connection with Armstrong [chemist and Dewar confederate Henry Edward Armstrong (1848–1937)], he is seeking to discredit the new views introduced by Ra; and by Rutherford[,] Ramsay and Lodge, – precisely as he did some years ago, Argon & Rayleigh and Ramsay.' Huggins was referring to a vitriolic exchange played out in the pages of the *Chemical News* in 1898. Back then, in letter after letter, another anonymous writer, *Suum Cuique* [To Each His Own] held up Ramsay's claims of discovery to public ridicule.[79]

Huggins had no interest in becoming involved in such an affair. After all, he had had nothing to do with the *Times* article in question. 'When I sent the paper on July 17', he explained to Schuster, 'Larmor was so much interested in it that as the R.S. was no longer in session, he, *of his own initiative* sent a short notice of the paper to the Times in which doubtless he put more emphasis than we did especially in the title, on the *apparent* coincidence of four lines with lines of He.'

The '*Times* letter is especially disagreeable', he told Schuster in confidence, 'because of the animus shown in it, and the obvious desire to belittle as far as possible the results of the paper'. And, as he later explained to Larmor, the letter 'was so craftily worded as to give the general public the impression that we had tried to get the credit of a discovery for what was no discovery at all; and further, to do this, had manipulated the paper in some unusual and not quite straightforward way'.[80] It was upsetting to think that Dewar – 'whom we have known now for so many years' – might have written it.[81] But, for Huggins, the evidence in favour of that conclusion was hard to ignore. 'When I went to the RI to show Dewar the photographs I told him that the notice in Times [July] 20 was by Larmor, & that I was thinking of sending a note to the "Times" to say that the spectrum was *not* He but N. Dewar exclaimed "Oh, no! no! There is no occasion" No doubt, he was planning *his* letter!!'

Fortunately, the tempest was effectively quelled a few days later by an authoritative letter to *The Times* of London from Larmor's counterpart at the RS, Biological Secretary Michael Foster (1836–1907).[82] A few weeks later, on the eve of the upcoming BAAS meeting, Huggins gently chided Larmor: 'No doubt that it was a little unsatisfactory ... that your notice was entitled "Helium and Radium", as it made it easier to *Inquirer* to discredit the paper as of no importance, when the coincidence with He failed.' But he suggested a way Larmor could make amends: 'If an opportunity occurs at [the upcoming BAAS meeting in] Southport I hope that you will make it plain, as you say in your letter, that He does not occur in the title of the paper, and that *the inquiry was undertaken to see if Ra could not set up a spectrum of bright lines similar to that ordinarily only obtainable by an electric discharge, or in a flame.* The chemical significance of the spectrum was of *secondary* consequence.'[83]

Huggins need not have worried about his colleagues' response to his first report on the radium glow spectrum. At the Southport meeting, physicist Charles Vernon Boys (1855–1944) applauded the Hugginses' efforts in his presidential address to the Mathematical and Physical Science section. In particular, he expressed hope that their findings might help explain the luminosity of comets. 'It is possible that the internal motions set up by the separate parts [of a comet], each pursuing its individual orbit, may produce collisions numerous and violent enough to account for all the light that is seen, and for temperature sufficient to bring out the spectral lines that have been identified.'[84] Nevertheless, he was intrigued by the entirely new explanation opened up by the Hugginses' results, namely that 'radio-active bodies and their emanations can produce light independently of such action'. Indeed, the fact that radium glow generates a nitrogen spectrum raised important questions with far-reaching implications. 'Is it possible that the enveloping nitrogen has had its atoms so harried by the activity of the radium as to give a response hitherto only awakened by electric discharge?' he asked. If so, he conjectured, 'the hydrogen, the hydrocarbon, and possibly even the sodium or iron spectrum that has been observed [in the light of comets], may have come from cold atoms; and it is not even quite beyond the limits of imagination to picture, not from the comet matter itself, but from loose residual and highly attenuated matter through which the comet is passing'.[85] 'These are', he concluded, 'the days of rays.'

The Hugginses forged ahead with their pioneering observations of the spectra of the radium glow. In early September William reported to Larmor that the 'effect of the Ra emanation is to turn the quartz brown *right through its entire thickness*' and hinted that he had plans for another experiment.[86] Their progress was impeded by the length of exposure time required for each photograph. 'The photographic plate now cooking has had 7 days', he told Larmor, 'but I wish to give it at least 9 days.'[87] Another challenge was acquiring the radium necessary to pursue the investigations. 'I have purchased some Ra from a new source which professes to be of "absolute purity" and of very great activity. It is very greatly more luminous. The price is double that of the Ra hitherto procurable.'[88] Indeed, the demand for radium was rapidly outgrowing its availability.[89]

On 28 October 1903, he sent the RS a second paper on the radium glow spectrum.[90] The Hugginses had photographed the spectra of two specimens of radium bromide, one prepared by the German firm Buchler and Co., the other from the Société Centrale de Produits Chimiques in France. They used longer exposure times than those upon which they had based their first report. They walked readers through their efforts to answer questions raised by their first paper's findings and suggested that the so-called β rays (which they defined as synonymous with 'cathode corpuscles') might not be the agents responsible for the radium glow.[91] They remained puzzled as to why the principal spectrum observed was that of nitrogen, a common component of the atmosphere. Radium and bromine readily displayed their signature spectra when the radium samples were subjected to electrical excitation, but not under ordinary conditions, despite the Hugginses' painstaking efforts.[92]

The spontaneous radium glow never seemed to extend beyond the physical dimensions of the samples of radium bromide. The Hugginses would later retract this claim, but in autumn 1903 their conviction that the glow emanated from within these 'pieces of solid light' led them to suspect that the nitrogen generating the spectrum is in molecular proximity to the radium bromide surface. It also suggested to them that there were stringent physical limits on the power and effect of radioactivity's exciting mechanism.

William wondered why some samples of radium bromide exhibited visibly brighter glows than others, a phenomenon that appeared to be independent of nitrogen. 'I had a specimen offered me for sale which was so luminous as to be visible in day-light', he told Larmor in December 1903,

The specimen was sealed up in a glass tube. I took it out of the tube and gave an exposure of a a [*sic*] few hours ... I then gave an exposure of some days. I expected a very strong spectrum but nothing appeared on the plate! On examination I found the Ra had lost its brilliancy and was now less luminous and much less radio-active than my old specimens. My view was that some volatile phosphorescent substance had been added which had quickly evaporated when the Ra. was taken out of the sealed tube. On consulting Curie when he was here, he did not take this view but said that he found specimens to vary in brightness even when the chemical purity was apparently the same. Now the brightness of the N. glow is clearly proportional to the radioactivity of the specimen, and I am pretty sure that any greater brightness comes in from phosphorescence of some sort. My brilliant specimen was of a *very low* radioactivity, and contained a large proportion of barium. The two good specimens of Ra. B. which I have, show no trace of a continuous spectrum of phosphorescence. As I said, I am trying to get a photograph of the visible region. So far as I see at present, the glow of pure Ra. salts, consists entirely of N. luminescence; and the occasional greater brightness of specimens is altogether independent of their true radioactive power, and is due to some unexplained and adventitious phosphorescence which is set up no one knows how (at present).[93]

Pierre Curie had visited the Hugginses while he was in London in December 1903 to receive the RS's Davy Medal. Illness prevented Marie from accompanying him on this trip, but she sent her personal thanks to Margaret for their hospitality to her husband:

Mr Curie was completely envious to see you both, Mr Huggins and you in the very place where you spend your working life. There you have a truly beautiful existence, a worthy example. It is good to give all one's life in this way to the accomplishment of a beautiful work.[94]

Overwhelmed by questions, the Husginses continued their investigations, exposing plates up to fourteen days when necessary. And they planned a third series of experiments to monitor the radium glow spectrum when samples are removed from nitrogen-rich ambient air and immersed in an atmosphere of pure hydrogen for extended periods of time. Would the nitrogen spectrum continue to predominate? Or would the radium stimulate the production of a hydrogen spectrum?[95]

This round of investigations took a long time to complete.[96] Their first trial lasted only a few days. Soon after sealing up each of their two radium samples in hydrogen-filled tubes, they noticed the radium's visible glow begin to diminish. The samples did not regain their original brightness immediately after being returned to normal atmospheric conditions. Photographs of the glow's faint spectra revealed no signs of hydrogen, only nitrogen. Does nitrogen exist naturally as an occlusion in radium compounds? Or were traces of atmospheric air inadvertently left in the tubes?

To find out, they once again sealed the samples in tubes of hydrogen. After twenty-six days, they found even feebler nitrogen spectra! But, whatever the cause of radium's natural luminosity, it seemed unable to excite the enveloping hydrogen gas. This time, William and Margaret noted the samples' glow began to increase slightly a few days after removing them from the hydrogen tubes. Left to sit in the open air, they slowly and gradually regained their original brightness.

Ernest Rutherford encouraged them to pursue their experimental efforts in his book *Radio-activity*, which appeared in the spring of 1904. In it, he offered scientists and interested laymen the first 'complete and connected account . . . of the properties possessed by the naturally radio-active bodies' based on his own research and the numerous reports on the subject 'scattered throughout the various scientific journals'.[97] He proposed that atoms are complex entities and capable of change. It was a daring explanation for radioactive phenomena that broke new ground and old rules. Noted scientists, including the Curies and Lord Kelvin (William Thomson, 1824–1907), who viewed nature's building blocks as chemically distinct but immutable atoms, rejected Rutherford's proposal as heresy. Surely, once all the evidence was in, radioactivity would be explained using conventional laws of physics and chemistry.

Uncovering those natural laws presented special challenges. Radioactive materials seemed to emit energetic rays endlessly without suffering any measurable loss of size or mass. Subjecting the materials to the usual chemical tests yielded no discernable clues to their enigmatic radioactive properties. Investigators focused instead on observing the effects of the rays themselves: recording them photographically, observing their scintillations on a fluorescent screen, or measuring their ability to ionise the air using an electroscope.

For practical reasons, Rutherford preferred the electrical method. It enabled him to quantify and track small changes over time. But he commended the Hugginses' spectroscopic study of radium glow. To underscore the importance of their work, Rutherford reproduced the photographs published in their second paper on radium which compared the glow's spectrum with that of electrically excited nitrogen and radium.[98] 'Further experiments in this direction', he concluded, 'are much to be desired at the present time.'[99]

Rutherford delivered the Bakerian lecture to the RS on 19 May 1904.[100] The revolutionary new ideas of the up-and-coming young researcher would have been much on the

Hugginses' minds as they began the first of two extended trials on 24 June to locate once and for all the source of nitrogen responsible for the radium glow's spectrum. Is it, as they suspected, occluded in the radium compound? Or is it simply atmospheric nitrogen lingering in the tube despite all efforts to remove it?

Once again, they sealed one of their radium samples in a tube filled with hydrogen gas. And, as before, the radium glow all but vanished while the nitrogen spectrum became extremely feeble. But on 29 July, after the tube had been sealed for thirty-five days, William disclosed to Larmor '*In confidence*. At last I have got rid of the N. spectrum when the Ra is in hydrogen.'[101] It was exciting news, but there were still many uncertainties to resolve. 'Unfortunately the light is so feeble that it is difficult to get a sufficiently strong photograph. In a feeble one I suspect the *lines of H*.' Waiting for confirmatory results required tremendous patience. 'I have been nine months at it now', a suspense-filled Huggins reminded his secretary. 'As the exposures have to be nearly a fortnight, the work goes slowly.'

A few days later, he continued his on-going discussion with Larmor about the cause of the radium glow. 'Am I right', he asked, inserting three large question marks above the line, 'in supposing that the boundary of the solid radium would not be an impenetrable wall of radium molecules, but that there would be a very *narrow* neutral zone into which the atmospheric molecules would penetrate ???, if so, this would be probably the place where the N molecules are excited to luminosity.'[102]

Huggins used the occasion to provide an update on his hydrogen observations:

I *believe*, though I am not quite sure[,] that now I have got *luminous hydrogen*, that is, a spectrum in which the *bands of N are absent*, and are replaced by lines of H. I say *believe*, because two plates were too faint to be of use, the third shows a feeble spectrum, in which, we are pretty sure, there are lines falling at the positions of the hydrogen lines. Another photograph is now exposed which I intend to develop next Monday, and which I hope will be sufficiently strong to give certainty. Nearly two months have been taken up with the four photographs.

Perhaps radium can excite atoms of elements other than nitrogen after all! He announced plans to try a similar experiment immersing Ra in NO_2. Margaret added her own enthusiastic note regarding the anticipated confirmation of hydrogen in the spectrum:

I am delighted that we have got the result we have with H. Of course it is right to be even more than cautious; but I feel sure that in one of our photographs, though the image is faint as Sir William said, that with suitable care the H. spectrum can be made out. I certainly believe I see it, – and I really have had long experience in dealing with faint spectra.

Unfortunately, the following Monday their optimism turned to disappointment. 'I have just developed the photograph', William confided to Larmor, 'and to my astonishment, the N spectrum now reappears ... The Ra has turned brown, and is less luminous.'[103] The results were very disheartening. 'I must mount fresh Ra in H vacuum,' he complained, 'repeat all experiments: slow work.'

The Hugginses omitted any mention of the suspected hydrogen lines in their published account of these experiments. Instead, they focused on the changes in the sample's physical appearance during the nearly nine full months it was confined in the hydrogen tube, like the fact that it had turned from a 'yellowish-cream' colour to 'dark russet brown'. And, although its natural glow initially diminished as before, it gradually regained its full original brightness by the time the Hugginses were compelled to remove it from the hydrogen tube on 17 March 1905 in order to inspect its spectrum. No doubt they would have preferred keeping the sample in the tube longer, but it had become dislodged from its post inside the tube and could no longer be viewed *in situ*. Unfortunately, its glow faded away before they were able to subject it to spectroscopic examination that evening. This relatively sudden loss in brightness was both puzzling and frustrating. Tests with an electroscope demonstrated that the sample maintained a steady level of radioactivity throughout this trial despite changes in its colour or brightness, and regardless of whether it was immersed in hydrogen or in the open air.

By 9 May, over fifty days after being removed from the hydrogen tube, the sample had begun to glow slightly once again. Its dark brown colour was becoming less intense. By 13 August, an interval of nearly five months, the sample had regained its original colour and natural luminosity. Again, tests with the electroscope showed no change in its radioactivity.

Meanwhile, the Hugginses had sealed up their second radium sample back on 12 September 1904. It exhibited many of the same physical changes they had witnessed in the first long trial. This second sample remained properly positioned within the hydrogen tube, so they were able to photograph its spectrum up through 18 August 1905, when they submitted their third radium paper.

When the radium glow in the second tube had fallen to half its original brightness leaving the nitrogen spectrum quite faint, the Hugginses noticed a new green band, a spectral signature that they could not match with any known terrestrial element. The unexpected appearance of the unfamiliar band only deepened the radium glow mystery. To the naked eye, radium glow always looks the same. Had that fact masked important changes taking place in radium's chemical and physical makeup? William and Margaret were confident that in time the spectroscope, as always, would disclose what the eye could not see.

When radium is kept isolated from the atmosphere, does it exhaust its supply of nitrogen, and/or lose its power to excite the nitrogen responsible for its normal glow? Could there be a second type of natural glow triggered by a different radioactive process that only kicks in after radium has all but stopped exciting nitrogen? If so, that could explain why samples enclosed in hydrogen tubes for many months first faded and then regained their luminosity. Perhaps it is not a matter of the original nitrogen glow dying down and spontaneously resuscitating, but rather a complex sequence of events involving the production of a second, and as yet unexamined, form of radioactive glow.

That thought suggested a possible explanation for the rapid diminution in luminosity the Hugginses had observed after they removed their first sample from its hydrogen tube. Perhaps whatever it is that causes the secondary glow is not a constituent of the atmosphere at all, but is – like the nitrogen they deemed responsible for the primary radium glow – embedded somehow within the molecular structure of the radium itself. In the event, they took the absence of a hydrogen spectrum to mean that whatever the nature of the radioactive energy emitted by radium, it cannot do what flames, electric sparks or β rays can do: namely, cause hydrogen to generate its characteristic spectrum.

In their fourth and final paper on radium, the Hugginses addressed the question of the degree to which the radium glow can be seen beyond the surface of the radium itself.[104]

Previously, they had judged its extent by eye and with low magnification. The sharp and clean definition of the glow's edges led them to conclude that 'the glow appears to end with sudden abruptness at the boundary surface of the radium'. But after learning of recent studies by others who claimed to have found evidence of radioactive action as far as 2 cm from the radium's surface, they re-examined their old photographic plates of radium glow spectra.[105] They found that these photographs showed the radium glow extending 'some little distance outside the radium salt', a feature they confessed they had overlooked initially because their attention was focused on other glow phenomena.

They repeated the experiment just to be sure and exposed a new plate for fourteen days. Close comparison of the new plate with the old revealed the glow to be visible within, but not outside, the glass tube encasing the radium sample. The Hugginses concluded that β rays, which are impervious to glass, cannot be involved in producing the radium glow. Instead, they offered two possible explanations: perhaps 'the active cause is the α rays' which cannot penetrate glass; or alternatively, nitrogen molecules occluded in the radium compound may be 'broken up into ions, which are projected outwards'. The latter hypothesis involving nitrogen ions, he told Larmor, had been added 'as a *suggestion* only. I still cling to my first idea that the action on the N. is by direct encounters with active Ra molecules, and not at a distance by α particles. At *least, I wish to leave this view open for the present*. It fits in with some of my experiments.'[106] He closed the letter with an expression of relief and exultation upon concluding his term as RS President: 'I feel like an actor who has come down from the stage to the pit. I am now free to *applaud* – to *criticize*!'

William thanked Larmor for his interest in 'the little paper' when he finally submitted it on 12 December. '[W]hat is the state of the gaseous contents of the tube?', he wondered aloud, referring to their second tube of radium sealed in hydrogen. 'It has been closed for more than a year; the free space in the tube is *very small*, and into this the α rays and the emanations have been passing for months. What has become of them?'[107]

The Hugginses' work on the radium glow spectrum attracted the attention of physicist Robert John Strutt (1875–1947; son of Huggins's successor as RS President, John William Strutt, 3rd Baron Rayleigh, 1842–1919). In the second edition of his book *The Becquerel Rays and the Properties of Radium*, published in early 1906, Robert Strutt lauded the Hugginses' 'unrivalled experience in the photography of feeble spectra'.[108] He drew particular attention to the puzzling observations they reported in their third radium paper. He, too, expressed surprise that radium immersed in hydrogen gas did not generate any signs of a hydrogen spectrum. He shared their puzzlement over the mysterious green line's appearance, noting that 'This line has not been identified as forming part of any known spectrum.' Hoping, perhaps, to spur others to take up the challenge that these unusual results presented, Strutt concluded, 'It is difficult to conjecture any satisfactory explanation of these phenomena. They can only be described as an interesting and stimulating mystery.'[109]

Another book, published soon afterward, mentioned the Hugginses' radium glow work. This one, written by Rutherford, was titled *Radioactive Transformations*.[110] It contained updated versions of the ten Silliman Memorial Lectures that Rutherford had delivered in March 1905 at Yale University. Rutherford repeated the Hugginses' conclusion that the signature of nitrogen in radium glow is the result of the 'action of α particles, in free nitrogen close to the radium, or in nitrogen occluded within the radium compound'. He added, 'Such a result is of unusual interest, as it is the first example of a gas giving a spectrum when cold without the stimulus of a strong electric discharge.'[111]

That August, Larmor received another barrage of questions from Huggins, no doubt stimulated by the on-going debate on the nature of radium, radioactivity and the so-called transmutation of elements that began at the 1906 BAAS meeting at York and continued in the pages of *The Times* of London, *Chemical News* and *Nature*:[112]

What form of energy has the Ra by its greenhouse action appropriated and stored up? If ultraviolet light then there should be no renewal of emanation if the Ra was kept in the dark? Should it not be possible by some form of screening to stop any renewal of emanation after the existing amount is exhausted?

If Ra contains lead one would expect with an electric spark to get the known lead lines, but Ra gives an original spectrum of the same order as similar bodies we call elements.

I know that Rutherford suggests that the lead & helium are combined in some much closer way, than metals in alloys, or occluded gases in minerals. He expressly says a combination that neither chemical nor electrical (are not these at bottom the same?) forces can affect.

I wish I could see some new way of attack upon Ra. I fear that I am only multiplying words.[113]

Huggins was reading Rutherford's Yale lectures when he wrote to welcome the renowned physicist back to England in February 1907. The native New Zealander had spent nearly a decade in Canada after completing his graduate study at Cambridge's Cavendish Laboratory in 1898. Now he was ready to return to England and had accepted an offer from Arthur Schuster, the founder of the Manchester Laboratory, to take his place as its Director. Huggins looked forward to brushing elbows with Rutherford at the RS.[114] He was, as he told Rutherford, still 'working on photographing the spectrum of the radium light under different conditions' himself. He advertised the Society's supply of uranium and radium compounds available for loan to researchers.

In September he shared with Larmor his desire to repeat his spectrum experiments using actinium in place of radium. Unfortunately, he had not been able to acquire any samples containing the element. 'I suppose the α rays are the same', he mused, 'but possibly not identical in velocity. There seems a presumption that the Polonium (actinium?) rays are more powerful.'[115] In November he informed his friend he was using a sample of radium to turn a 'nearly white sapphire' a 'fine topaz colour' for Lady Huggins to wear in a ring.[116] But, over time, more and more of his correspondence turned back to astronomical topics.

***

Unfortunately for the Hugginses, the spectroscope did not prove to be as useful a tool for exposing and analysing substances' radioactive properties and behaviours as it had been for revealing their chemical and physical characteristics. 'For the detection of matter which possesses the radioactive property', Rutherford asserted in his Yale lectures, 'the electric method thus far transcends in delicacy the use of the spectroscope.'[117] Radioactive trans- formations occur 'at the rate of one atom per second', he explained. To sense and record such extremely tiny and rapid electrical fluctuations required special instruments. With the electroscope, he pointed out, 'we are able to detect the presence of an active substance like radium, when it exists in almost infinitesimal amount mixed with inactive matter, and also to determine with fair accuracy the amount present'.

Thanks to the discovery of radioactivity and all its attendant phenomena, the turn of the twentieth century was a time of transition for physicists and chemists from old aims, methods and theoretical underpinnings to new. As we shall see in the next chapter, astronomers, too, were challenged to accept a larger and more central role for astrophysics in routine astronomical investigation.

Notes

1. [Clerke], 'The evolution of the stars' (1900), p. 457.
2. 'Argon', *CN* **70** (1894), p. 296; Rayleigh and Ramsay, 'Argon, a new constituent of the atmosphere' (1895).
3. Ramsay, 'Discovery of helium' (1895); 'On a gas showing the spectrum of helium, the reputed cause of D_3, one of the lines in the coronal spectrum, preliminary note' (1895).
4. Crookes, 'The spectrum of the gas from clèveite' (1895).
5. For a discussion of the events surrounding Lockyer's discovery of the bright line D_3 associated with helium and its possible connection with the spectrum of Ramsay's unknown gas, see Meadows, *Science and Controversy* (1972), pp. 53–60. See also T. M. Lockyer and W. L. Lockyer, *The Life and Work of Sir Norman Lockyer* (1928), p. 42; and Jensen, 'Why helium ends in "-ium"' (2004). Ramsay himself objected to the name 'helium' and declared his intention to change it. 'I propose to change the name for "ium" is unsuitable for a Greek word; the substance is strongly analogous to Argon; and although helion is not Greek (helios being the sun), still it doesn't break through established usage to terminate in "*on*".' From 'Copy of *Ramsay's* reply', n.d. [filed after W. Huggins to Hale, 11 November 1899], Box 22 [Reel 19], GEH, CIT. The undated page titled 'Copy of *Ramsay's* reply' is, I believe, erroneously located after a letter from W. Huggins to Hale, dated 11 November 1899. It appears to be written on the same stationery as the 28 April 1895 letter cited above which is comprised of four numbered pages. The undated page is numbered 'five'. That evidence plus the fact that in the April 1895 letter, Huggins states that he hopes to be able to receive Ramsay's reply in time to enclose it – all seem to indicate that this page belongs with the 28 April 1895 letter.
6. W. Huggins to Hale, 28 April, 1895, Box 22 [Reel 19], GEH, CIT.
7. 'Copy of *Ramsay's* reply', n.d. [filed after W. Huggins to Hale, 11 November 1899], Box 22 [Reel 19], GEH, CIT.
8. W. Huggins to Hale, 28 April 1895, Box 22 [Reel 19], GEH, CIT.
9. Lockyer delivered five short reports to the RS on the new gas between 25 April and 13 June 1895: J. N. Lockyer, 'On the new gas obtained from uraninite' (1895).
10. The series appeared in eleven parts: J. N. Lockyer, 'The Sun's place in Nature' (1895). Lockyer later compiled these notes into a book: *The Sun's Place in Nature* (1897).
11. J. N. Lockyer, 'The Sun's place in Nature. III.' (1895), p. 566.
12. Lockyer had recently published a popular book on archaeo-astronomy: *Dawn of Astronomy: A Study of the Temple-worship and Mythology of the Ancient Egyptians*, (1894).
13. W. Huggins to Hale, 28 April 1895, Box 22 [Reel 19], GEH, CIT.
14. Runge, 'Terrestrial helium (?)' (1895).
15. M. L. Huggins, 8 June 1895, Notebook 5, WCL/SC. It is difficult to know with certainty when Margaret wrote her entries related to the Hugginses' efforts to confirm Runge's report. She clearly wrote them after the fact as summaries of their actions and findings.
16. W. Huggins to Hale, 8 June 1895, Box 22 [Reel 19], GEH, CIT.
17. W. Huggins to Hale, 18 June 1895, Box 22 [Reel 19], GEH, CIT.
18. W. Huggins to Hale, 25 June 1895, Box 22 [Reel 19], GEH, CIT.
19. M. L. Huggins, 10 July 1895, Notebook 5, WCL/SC.
20. W. Huggins to Hale, 10 July 1895, Box 22 [Reel 19], GEH, CIT.
21. M. L. Huggins, 13 July 1895, Notebook 5, WCL/SC.
22. W. Huggins to Stokes, Add MS 7656.H1277, GGS, CUL.
23. 'Diamond Jubilee honours', *Times* of London, 22 June 1897, p. 10e. Norman Lockyer and Edward Frankland were also created knights that day.
24. W. Huggins, 'The new astronomy' (1897); Lilly, 'British monarchy and modern democracy' (1897); and Lyall, 'India under Queen Victoria' (1897).
25. W. Huggins to Hale, June 1897, Box 22 [Reel 19], GEH, CIT.

26. W. Huggins to Hale, 17 January 1899, Box 22 [Reel 19], GEH, CIT.
27. W. Huggins and M. L. Huggins, *An Atlas of Representative Stellar Spectra* (1899), p. 132.
28. *Ibid.*, p. 133.
29. W. Huggins to Hale, 23 May 1899, Box 22 [Reel 19], GEH, CIT.
30. W. Huggins to Hale, 3 February 1899, Box 22 [Reel 19], GEH, CIT.
31. W. Huggins and M. L. Huggins, *An Atlas of Representative Stellar Spectra* (1899).
32. W. Huggins to Hale, 3 February 1899, Box 22 [Reel 19], GEH, CIT.
33. 'Publications of Sir William Huggins's Observatory, Vol. I.', *JBAA* **10** (1900), pp. 411–12; p. 411.
34. [Clerke], 'The evolution of the stars' (1900). All of Clerke's articles for the *Edinburgh Review* are listed in M. L. Huggins, *Agnes Mary Clerke and Ellen Mary Clerke* (1907), pp. 37–8.
35. [Clerke], 'The evolution of the stars' (1900), p. 458.
36. Hale, 'An Atlas of Representative Stellar Spectra' (1900), p. 293.
37. *Ibid.*, p. 294.
38. *Ibid.*, p. 295.
39. The review in the *Monthly Notices* notes 'The question is one that can be settled only by extreme care and with special instrumental appliances.' See 'Sir William and Lady Huggins's Atlas of Representative Spectra', *MNRAS* **60** (1900), pp. 392–4.
40. Hale, 'An Atlas of Representative Stellar Spectra' (1900), p. 297.
41. W. Huggins to Larmor, 25 October 1909, Lm.1016, JL, RSL.
42. Huggins received honorary degrees from Cambridge (1870), Oxford (1871), Edinburgh (1871), Dublin (1886), St Andrews (1893), and from universities in various foreign countries.
43. After William's term ended, William and Margaret prepared and published an annotated text of four of his presidential addresses (1902–5), each accompanied by one of her pen-and-ink illustrations: *The Royal Society: or, Science in the State and in the Schools* (1906).
44. M. L. Huggins, 25 April, 3 October and 1 November 1901, Notebook 5, WCL/SC.
45. D. Gill, 'Address on receipt of the McClean gift to the Cape Observatory' (1901), p. 400.
46. W. Huggins, 'Motion in the line of sight' (1901).
47. W. Huggins to Hale, 9 November 1901, Box 22 [Reel 19], GEH, CIT.
48. From a letter from the King's private secretary, Sir Arthur Davidson to Sir Arthur Ellis, 28 March 1902, quoted in Martin, *The Order of Merit* (2006), p. 24.
49. *Ibid.*, pp. 57–74.
50. *Ibid.*, p. 69. On William Huggins's death, the current RS President, William Crookes, was selected to take his place among the twelve.
51. P. Curie, Mme P. Curie and Bémont, 'Sur une nouvelle substance fortement radio-active, contenue dans la pechblende' (1898); M. Curie, 'Sur le poids atomique du radium' (1902).
52. Röntgen, 'Über eine neue Art von Strahlen' (1895).
53. 'The Royal Academy banquet', *The Times* of London, 4 May 1903, p. 4a.
54. Ramsay and Travers, 'On a new constituent of atmospheric air' (1898); 'On the companions of argon'; and 'On the extraction from air of the companions of argon and on neon' (1898).
55. W. Huggins to Larmor, Lm.838, JL, RSL.
56. 'Royal Society conversazione', *The Times* of London, 16 May 1903, p. 13a.
57. Lodge, *The Romanes Lecture 1903: Modern Views on Matter* (1903). For Lodge's discussion of radioactive elements, see pp. 16–27.
58. *Ibid.*, pp. 23–4.
59. P. Curie, 'Le radium' (1903).
60. 'The Royal Society's conversazione', *The Times* of London, 20 June 1903, p. 12a.
61. 'The Royal Society conversazione', *Nature* **68** (1903), p. 184. Rutherford had been elected to fellowship in the RS just a week earlier, on 11 June. However, he was not present at the conversazione.
62. 'Professor Curie on radium', *The Times* of London, 20 June 1893, p. 11f.
63. W. Huggins to Larmor, 28 October 1903, Lm.850, JL, RSL.
64. See, for example, Runge, 'On the spectrum of radium' (1900); Runge and Precht, 'On the flame spectrum of radium' (1903).
65. I wish to thank Arthur E. Champagne (W. C. Friday Professor of Physics, University of North Carolina) for sharing my interest in this problem and for his kind assistance in interpreting the Hugginses' radium glow observations.
66. W. Huggins to Larmor, 16 July 1903, Lm.839, JL, RSL.
67. W. Huggins and M. L. Huggins, 'On the spectrum of the spontaneous luminous radiation of radium at ordinary temperatures' (1903).

68. *Ibid.*, p. 196.
69. 'Radium and helium', *Times* of London, 20 July 1903, p. 10a.
70. W. Huggins to Larmor, 19 July 1903, Lm.840, JL, RSL.
71. W. Huggins to Larmor, 24 July 1903, Lm. 842, JL, RSL.
72. Hale to W. Huggins, 10 August 1903, Box 22 [Reel 19], GEH, CIT.
73. Hale published a version of the Hugginses' *Proceedings* paper in the *Astrophysical Journal*'s September number: W. Huggins and M. L. Huggins, 'On the spectrum of the spontaneous luminous radiation of radium at ordinary temperatures' (1903).
74. W. Huggins to Larmor, 11 August 1903, Lm.843, JL, RSL.
75. Huggins and Huggins, 'On the spectrum of the spontaneous radiation of radium at ordinary temperatures' (1903), pp. 197–8.
76. *Ibid.*, p. 199.
77. Inquirer, 'Radium and helium', *The Times* of London, 21 August 1903, p. 7f.
78. W. Huggins to Schuster, 23 August 1903, Sc.97, AS, RSL.
79. See, for example: *Suum Cuique*, 'Gas and gases' (1898); 'The new gases' (1898); and 'The gases of the atmosphere' (1898). For a brief discussion of *Suum Cuique*'s attack on Ramsay, see Brock, *William Crookes (1832–1919)* (2008), pp. 362–5.
80. W. Huggins to Larmor, 6 September 1903, Lm.844, JL, RSL.
81. W. Huggins to Schuster, 23 August 1903, Sc.97, AS, RSL.
82. Foster, 'The spectrum of radium', *The Times* of London, 25 August 1903, p. 6d.
83. W. Huggins to Larmor, 6 September 1903, Lm.844, JL, RSL.
84. Boys, 'Address' (1903), p. 531.
85. *Ibid.*, p. 532.
86. W. Huggins to Larmor, 8 September 1903, Lm.845, JL, RSL.
87. *Ibid.*
88. W. Huggins to Larmor, 17 September 1903, Lm.846, JL, RSL.
89. W. Huggins to Schuster, 23 August 1903, Sc.97, AS, RSL; and W. Huggins to Larmor, 2 May 1906, Lm.937, JL, RSL. The price of radium continued to rise. Huggins reported paying eight shillings for a milligram of radium bromide in August 1903. In May 1906, Huggins was shocked to learn the price of radium had quintupled to two pounds per milligram!
90. W. Huggins and M. L. Huggins, 'Further observations on the spectrum of the spontaneous luminous radiation of radium at ordinary temperatures' (1903).
91. *Ibid.*, p. 410.
92. W. Huggins to Larmor, 3 October 1903, Lm.851, JL, RSL.
93. W. Huggins to Larmor, 18 December 1903, Lm.855, JL, RSL.
94. Quinn, *Marie Curie: A Life* (1996), pp. 185–6. I am indebted to Ms. Quinn for providing me with the complete text of the original letter: M. Curie to M. L. Huggins, 6 December 1903, Curie papers, Bibliothèque Nationale. The translation provided here is my own.
95. W. Huggins to Larmor, 15 December 1903, Lm.854, JL, RSL.
96. W. Huggins and M. L. Huggins, 'On the spectrum of the spontaneous luminous radiation of radium. Part III. – Radiation in hydrogen' (1905).
97. Rutherford, *Radio-activity* (1904).
98. *Ibid.*, Fig. 33, opposite p. 169.
99. *Ibid.*, p. 170.
100. Rutherford, 'The succession of changes in radioactive bodies' (1905).
101. W. Huggins to Larmor, 29 July 1904, Lm.873, JL, RSL.
102. W. Huggins to Larmor, 3 August 1904, Lm.877, JL, RSL. Huggins inserted the large question marks above the text.
103. W. Huggins to Larmor, Monday morning [almost certainly 8 August 1904], Lm.1000, JL, RSL. William has written '*Confidential*' at the top of the page.
104. W. Huggins and M. L. Huggins, 'On the spectrum of the spontaneous luminous radiation of radium. Part IV. – Extension of the glow' (1905).
105. *Ibid.*, p. 131.
106. W. Huggins to Larmor, 10 December 1905, Lm.926, JL, RSL.
107. W. Huggins to Larmor, 12 December 1905, Lm.927, JL, RSL.
108. Strutt, *The Becquerel Rays and the Properties of Radium* (1906), p. 82.
109. *Ibid.*, p. 83.

110. Rutherford, *Radioactive Transformations* (1906).

111. *Ibid.*, p. 274.

112. Lankester, 'Address of the president' (1906), see especially, pp. 6–14; Soddy, 'The evolution of the elements' (1906); 'Physics at the British Association', *Nature* **74** (1906), pp. 453–5; 'The recent controversy on radium', *Nature* **74** (1906), pp. 517–18; Crookes, 'On radio-activity and radium' (1906); 'Radioactivity' (1906); 'Radium' (1906).

113. W. Huggins to Larmor, 18 August 1906, Lm.942, JL, RSL.

114. W. Huggins to Rutherford, 6 February 1907, Add MS 7653.H192, ERP, CUL.

115. W. Huggins to Larmor, 25 September 1907, Lm.959, JL, RSL.

116. W. Huggins to Larmor, 29 November 1907, Lm.960, JL, RSL.

117. Rutherford, *Radioactive Transformations* (1906), p. 36.

15

'One true mistress'

None of you know *how* hard we worked here just our two unaided selves.
– Margaret Lindsay Huggins[1]

In April 1892, three 'lady candidates' failed to receive sufficient votes to be elected to fellowship in the all-male RAS. Before the ballots were cast, the chair urged each Fellow to vote as he saw fit, with the caveat that admitting women as full members might violate the Society's Charter. One Fellow threatened to 'protest against the legality of the election in case the women should be elected'. Another cautioned that a vote for women was a vote for introducing a 'social element' into the RAS's normally 'dull meetings'. It would require 'a piano and a fiddle', and laying down a 'parquet flooring' so all could 'dance through most of the papers'.[2]

But change was afoot in the Society at the dawn of the twentieth century. On 8 May 1903, the RAS Council elected Margaret Huggins and her friend, historian of astronomy Agnes Clerke, as Honorary Members.[3] Only three other women had received such an honour: Caroline Lucretia Herschel (1750–1848) and Mary Somerville (1780–1872) in 1835, and Anne Sheepshanks (1789–1876) in 1862.[4] Over the next eleven years two more women, both Americans, joined their ranks: Scottish-born Williamina Paton Stevens Fleming (1857–1911) in May 1906 and Annie Jump Cannon (1863–1941) in March 1914.[5] The RAS finally amended its Charter to include women in February 1915.[6] That November, five women were among the nine individuals nominated for election.[7] In January 1916, all were elected.[8]

There were other signs of change at that May 1903 meeting. Society President Herbert Hall Turner addressed complaints he had overheard 'that at some of our recent meetings we have had rather a large number of papers of a somewhat technical character which did not possibly appeal to the majority'.[9] A review of the records of papers and discussions heard over the past two decades – roughly the duration of Turner's own Fellowship in the RAS – convinced him there *had* been a real change in the practice of astronomy.[10]

'Twenty years ago photographic work and spectroscopic work were largely in the qualitative stage', he noted. Photographers 'were seeing pictures for the first time and learning a great deal from a mere glance at them ... Since then we have had to settle down and measure millions of star-images on photographs ... and as a consequence our work has become quantitative.'

Spectroscopists were increasing their reliance on exacting measures of spectral line positions for determining motion in the line of sight. The New Astronomy had started out as a spindly child, born of, and nurtured by, the eclecticism and opportunism of independent research programmes like that of William Huggins. But, by the turn of the century, it had matured in the care of mathematically and scientifically trained astronomers into a technically and conceptually demanding enterprise that could be managed only by the few who possessed the instruments, skills and financial resources needed to stay competitive. In terms of individual practice, there was no denying the fact that increased reliance on precision measurement and quantitative analysis of spectroscopic data was excluding more and more of the Society's rank and file from active participation in this cutting-edge astronomical research.

Turner admitted that the resulting shift from 'qualitative to the quantitative stage has had an effect upon the papers read to us'.[11] But were such changes cause for complaint, or celebration? He pointed out, 'we should be very sorry if in this room the representative work of the time, whatever its character, did not come before us in all its freshness'. None present dared disagree. Fellows could not ignore the restless itch of change. Would they adapt more successfully by moulting like growing grasshoppers and breaking out of their too-tight exoskeletons as fresh replicas of their former selves? Or by metamorphosing like caterpillars and escaping their constraining cocoons to look, think and live in totally different ways? It was too soon to tell.

As we shall see in this chapter, Huggins himself was preparing to leave the field. He turned over his observing instruments to help train a new generation of aspiring astrophysicists. He fleshed out the bare bones of his retrospective essay. He and Margaret worked hard to establish this account as the authoritative version of events. Knowing he no longer possessed the tools and expertise and youth necessary to finish the race he had started, he made ready to pass the baton on to those who did: men who called themselves astrophysicists, like George Ellery Hale.[12]

15.1 Passing the baton

From the time of their first meeting in 1891, Huggins and Hale had enjoyed a symbiotic relationship with each giving as much as he gained to satisfy their united appetite for promoting the development and wider application of spectroscopic methods in astronomy. Now, in 1903, Hale was positioning himself to become a world leader in astrophysical research. The young director of the Yerkes Observatory had already been elected to the elite National Academy of Sciences (NAS) and invited to serve on the Advisory Committee on Astronomy of the Carnegie Institution that had been established recently to provide philanthropic support for basic scientific research.[13] Huggins eagerly served as mentor and midwife to his young colleague's ambitious plans.

Hale visited Huggins at Tulse Hill whenever he was in London. When they could not exchange their thoughts and ideas face to face, they posted them by mail. Although both corresponded with many others between 1903 and 1909, their letters to one another offer us a narrow but intimate glimpse at the complex developments in what was for each of them a dynamic phase in his career. In particular, they bring to light the comfort and ease with which the two men both supported and critiqued each other's work and ideas. Huggins routinely kept Hale apprised of ground-breaking developments announced in Continental journals. Hale informed Huggins of all his latest advances.

Hale strode into the new century full of bold ideas for new instruments, new observatory sites and new research programmes. For all its light-gathering power, the Yerkes Observatory's 40-in telescope simply could not satisfy the research needs of a new generation of astronomers like Hale, who were driven by the desire to gather, record and analyse the spectral signatures of *all* the light emitted by celestial bodies. Although Hale had been the prime mover behind the Yerkes project, he was already working hard to establish a research facility equipped with large reflectors, state-of-the-art photographic equipment, the latest laboratory facilities and specialised spectroscopic instruments, all to be operated by teams of trained investigators.

In January 1903, Hale alerted Huggins to his plans for 'a new astrophysical observatory'. He hoped to persuade the well-endowed Carnegie Institution to underwrite the project.[14] 'Suggestions of any kind will be most welcome', he added.[15] Huggins responded to Hale's plea with enthusiasm. '[I]t is of first importance', he wrote in support, 'to have a permanent observatory furnished with a large *reflecting* telescope and a complete equipment of auxiliary instruments for astrophysical research on some site with the most favorable conditions of atmosphere.'[16] The large reflector to which he referred was the one Hale hoped to create from a 60-in glass disc then being stored in the basement at the Yerkes Observatory.[17]

In August 1903, Hale confided to his friend, 'We found Wilson's Peak, near Pasadena, in southern California, to offer remarkable advantages.'[18] He had made a brief expeditionary trip in late June with William Wallace Campbell and William Joseph Hussey (1862–1926) of the Lick Observatory to check out the site's conditions.[19] 'The solar definition surpassed anything I have seen elsewhere, and the night conditions were equally good.'[20] The physical challenges of erecting an observatory on a mountaintop accessible only by a 2-foot wide, 9-mile long trail were daunting. The financial costs were prohibitive without a wealthy patron.

When Hale returned to Yerkes, he continued to pursue both observatory and laboratory work. He had improved both the optics and mechanics of his original spectroheliograph's design to better match the 40-in refractor's superior imaging capabilities.[21] The new Rumford spectroheliograph's monochromatic scans revealed details in the solar surface that had eluded earlier observers. New structures, which Hale called *flocculi*, appeared in areas covered by calcium and hydrogen vapour. He took advantage of their wavelength-specific nature to probe and photograph the solar atmosphere layer by layer.[22] The Sun is 'covered with *dark* hydrogen clouds', he announced to Huggins. These clouds 'frequently differ in the most marked way from the bright calcium clouds in the same region'.[23] He initiated a programme of photographing them every day in order to track changes in their appearance and position. 'If the new Carnegie observatory is established,' he told Huggins with obvious enthusiasm, 'there will be plenty of work for it to do in this direction.'[24]

The ever-cautious Huggins shared Hale's excitement, but he urged his young colleague to simulate solar conditions in laboratory experiments in order to compare the appearance of spectra of terrestrial elements with those on the Sun. Hale, as always, valued Huggins's comments and suggestions. He promised to try the experiments Huggins suggested, but cautioned that the project might have to await his return from another trip to California.[25]

Without waiting for confirmation of financial support from the Carnegie Institution, Hale left Yerkes for an extended stay in Pasadena on 18 December. He established a presence on the mountaintop, like a pioneer squatter, hauling instruments and equipment up the narrow, winding trail in order to conduct daily observations over a long period of time and test the site's suitability for routine solar work.

In May 1904, Hale wrote to Huggins with the good news 'that the Carnegie Institution has just given me a grant of $10,000, which will enable me to push forward my solar work on Mt. Wilson'. He was already planning the buildings and instruments that he would assemble on the site. One instrument in particular was a top priority. 'The Snow telescope will be brought out at once, and provided with powerful spectroscopes and spectroheliographs. I think we may count on further grants from the Carnegie Institution in the future.'[26] The Snow, a long-focus telescope already in use at the Yerkes Observatory, was designed to remain fixed in a horizontal position.[27] A coelostat directed the target object's light toward the telescope's primary mirror. Observers could easily examine the resulting image with any one of a battery of poised analytical instruments.[28] Hale was eager to move the telescope to Mt Wilson where he could put it to work on solar surface features.

Meanwhile, Hale was already at work on another project. Driven by his interest in establishing a coordinated collaborative research effort among solar observers around the globe, he had persuaded the NAS to sponsor a preliminary meeting at the International Congress of Science being planned for the 1904 World Exposition in St Louis, much like the successful Congress on Mathematics, Astronomy and Astro-Physics that Hale had organised during the Columbian Exposition in 1893.

Hale asked for Huggins's support in this grand endeavour. As President of the Royal Society, Huggins quickly assembled a joint RS/RAS committee to participate and helped convince other scientific societies to do the same. He also alerted Hale to possible conflicts with Lockyer, whose own interest in international solar and meteorological data collection might embolden him and his allies to seize control of Hale's agenda.[29]

Representatives from Britain, France, Germany, the Netherlands, Austria, Russia and Sweden joined their American colleagues at the first meeting of the International Union for Co-operation in Solar Research (IUCSR) in St Louis in September 1904. Together they began laying the ground rules for the organisation's structure and purpose. They outlined a provisional programme of observations and heard several papers on the problem of standardising wavelength measures.[30] And attendees agreed on the need for future meetings.[31]

Huggins avidly followed the progress at Mt Wilson, offering both congratulations and effusive praise for Hale's past, present and future successes with the instruments he had at his disposal. 'Congratulations that the 60″ mirror is finished', he wrote in November 1907.[32] 'I foresee grand results in the air of Mt. Wilson.' A few months later, in January, Hale reported 'our new laboratory in Pasadena is now under roof, the pit for the 30-foot spectrograph (similar to the one used with the new "tower" telescope on Mount Wilson) is completed, and the piers for the electric furnace and other apparatus are being built.'[33] News of the tower telescope led Huggins to exclaim, 'You are leading the world ... I am looking forward with intense interest to the triumphal progress of your results. It is in your power to turn over hitherto unread pages of Nature's secrets!'[34]

In early May 1908, Hale wrote to share news of his recent work on solar rotation using scans of the solar surface in Hα, the bright red line in the hydrogen spectrum. But he had even more exciting news to relate.

The most remarkable plate yet obtained is one made on April 29th, of which I will send you an enlarged print within a few days. This brings out, in a surprisingly clear way, the radial and spiral structure in regions surrounding sun-spots and other disturbed areas. As all of the spots on the plate show similar structure, I think we are undoubtedly observing, for the first time, great whirls in the sun's atmosphere which are naturally associated with spot formation.[35]

Hale had seen these patterns in earlier photographs. They recalled 'the appearance of iron filings in a magnetic field'. He hoped that examining the same features in wavelengths associated with calcium would show enough movement in the patterns 'to indicate the path of any currents which traverse it'. The currents might not be easy to detect 'if the whirls are in the form of our terrestrial tornadoes' for in that case, 'the diameter of the funnel-shaped structure is too small, where it pierces the calcium vapor, to appear in our photographs'.

The challenges were exciting given the real promise of solving the mysteries behind sunspot formation. 'I have always felt unprepared to attack the theory of sun-spots', he admitted to Huggins. '[T]he existing data seemed to me insufficient. If the new work can be followed up satisfactorily, however, I think we may soon be in a position to study this subject extensively.'

He was so excited over the evidence for swirling motion in sunspots that he sent a note to Lockyer for publication in *Nature*.[36] The 'photos are splendid', Lockyer replied with apparent enthusiasm.[37] He added that Hale's images appeared to show cool material descending into the sunspot, an observation that supported his own theories of sunspot formation. Lockyer handed off the task of reading and reviewing Hale's note to Thomas Francis Connolly, an assistant at the Solar Observatory. Connolly's commentary appeared in *Nature*'s 2 July number.[38] Meanwhile, Hale sent Lockyer a copy of his full report on the subject for publication, and promised to give serious thought to Lockyer's sunspot theory.[39]

After a long delay, a review of Hale's report finally appeared in *Nature*'s 20 August number.[40] Again, it was written by Connolly. Hale was perplexed and miffed. In October, he confided to Huggins, 'I do not understand why Lockyer did not publish the note I sent him.' The article that appeared 'did not represent my views at the time'.[41] Huggins, a veteran of unpleasant entanglements with Lockyer, understood the situation perfectly. 'It is easy to see why Lockyer substituted a statement for your own note', he replied.[42] 'He brings in two or three times *his own observations*.'

Huggins was in awe of both Hale's advances and the power of the instruments he had been able to assemble in his new mountaintop observatory. Hale's accomplishments served as bittersweet reminders to him that he was no longer able to keep up with the forefront of research. And they reinforced his decision to pass on his instruments to an institution that could put them to new and fresh use.

15.2 The Great Grubb telescope

Recall that the Great Grubb telescope, the centrepiece of the Hugginses' Tulse Hill observatory since 1870, was on loan from the Royal Society. In January 1908, convinced he could no longer put the instrument to full and proper use, Huggins began the process of relinquishing possession of it. But he did not wish to do so until he could be assured the instrument would have a good home to which it could be transferred and installed immediately without gathering dust in storage. In particular, he hoped to place it in the capable hands of Hugh Frank Newall, who had recently been elected to Cambridge University's new chair of astrophysics.

He spoke first to Newall 'in strict confidence' to learn if Cambridge would be interested in and able to receive the instrument. He then broached the subject informally with Larmor to test the RS's willingness to permit the transfer.[43] It seems that neither Newall nor Larmor was quick to respond.[44] In June Huggins explained to Larmor, who was but a young lad at the time the telescope's deed was drawn up, that it clearly stipulated that the instruments were to remain in his possession 'so long only as they are in constant use'.[45]

Finally, in October, having received Newall's personal agreement to take the instruments, he sent a formal letter to Larmor in his official capacity as RS Physical Secretary:

For some little time, for several reasons, especially on account of the weather conditions which make it all but impossible *now* to get work done by observing only during the early hours of the night, I have not been able to make such a use of the instruments as would justify me in retaining them.

I beg, therefore, now to be allowed to return them with the expression of my most sincere gratitude to the Society for the loan of them . . . .

The instruments are in perfect working order, and are much more available for research than they were in their original form. Originally only one telescope could be arranged for use at a time, balanced by a counterpoise. Some years ago Sir Howard Grubb carried out a suggestion of mine, by which, by means of a double declination axis, both telescopes could remain mounted together, and be equally available for use [see Figure 15.1]. Later additions were added by Troughton & Sims [*sic*] for the more stable attachment of spectroscopes. These main improvements were made by means of grants from the Donation Fund. A number of smaller desirable alterations and additions have also been made from time to time partly by the Society, and partly at my own expense.[46]

Huggins was still awaiting the University's formal approval. Indeed, in case the deal with Cambridge did not work out, he contemplated approaching Arthur Schuster at the University of Manchester. 'Schuster has spoken to me more than once of setting up a little observatory', he told Larmor on 2 November. 'If there is any hesitation [by Cambridge] on Thursday [5 November], I think it would be well for me to write *confidentially to Schuster at once*.'[47] He suggested that Larmor alert Newall – '[i]f you have an opportunity' – that he could have competition. Whether in consequence of a friendly nudge from Larmor, or – as is more likely – simply because Cambridge's bureaucratic processes had finally run their course, Huggins soon received the good news from Newall. The 'syndicate resolved unanimously to accept the telescope on the Society's conditions', he announced to Larmor on 8 November, adding, 'they see their way to provide a dome, &c.'[48]

Figure 15.1 Twin-equatorial telescope. The illustration is presumed to depict William Huggins's Great Grubb telescope. (Plate XVI, Chambers, *A Handbook of Descriptive and Practical Astronomy*, vol. 2, 4th edn (1889), opposite p. 96)

Huggins asked the telescope's maker, Howard Grubb, to dismount and 'pack the optical parts' of the instrument. Grubb later recalled the day he and Newall arrived at the Tulse Hill observatory to remove the telescope:

The Equatorial had been partially dismounted; all the numerous parts and attachments had been removed and were scattered over the floor, which was encumbered and littered with axes and various parts of the instrument, some of which had been already packed in packing cases; and in the midst of this litter, wrapped in a large cape and seated on a packing case, was Sir William himself, and his faithful collaboratrice who was flitting about watching the packing with keen interest and loving care.

It was a strange and interesting sight to see him sitting very quietly and patiently watching with regretful interest the dis-assembling of the very instrument which he had made so noble use of, and with which he had wrested secrets from nature by the labours and work of nearly half a century, but the pathetic side of the picture was presented to me when I proceeded to pack the great objective in the box.

Lady Huggins had asked me to let her know when I was ready to close the box, and when I intimated that I had it safely in the case, she took Sir William by the hand, and brought him across the room to have a last look at their very old friend, the object glass which had for so many years fulfilled its mission in bringing rays of light from a far distance to a focus, there to be submitted to the keen and searching analysis of the great scientist. They gazed long and sadly before I closed the lid.[49]

The day Grubb removed the instruments marked the instrument maker's last meeting with Huggins. His impression 'was that of a great man, gazing at what was apparently the ruins of his observatory and instruments, with a look in his eyes, at once sad and regretful, but at the same time, in a certain way, content and satisfied'.[50]

Margaret expressed her pleasure with the telescope's move to Cambridge in a letter to Sarah Frances Whiting, the Astronomy Professor at Wellesley College in Massachusetts, to whom, it will be recalled, Margaret later presented the notebooks and other items from the Tulse Hill observatory. The two women had met and developed a friendship during Whiting's visit to Europe in 1896.[51] 'The University', Margaret enthused, 'has gratified us by not only accepting the telescopes but by *recognizing* the *subject of our work, – Astrophysics.*'[52] Cambridge had 'created a new *Professorship* – viz. of *Astrophysics*' with Newall as the first to hold the position, and it had 'built a new Observatory for the instruments'. She was especially excited because 'the new Observatory will *bear our name*. I do rejoice for my Best & Dearest! who has worked – nay – *toiled* – so hard, for the spread of truth, with never a thought for himself.' She was relieved that despite the challenges of '*arranging* for the transfer' and the 'heavy work' required to complete the task, their careful preparation insured that 'all went well. Not the slightest mishap!'

She assured Whiting that relinquishing their instruments did not mean they were giving up on scientific investigation altogether. 'We retain our *Laboratories* where we have much going on . . . There is assuredly *much to* be done. We also retain our Observatory building so that we can *experiment* in *any new way* that occurs to us. There must be *new ways of working* – to be discovered.'

15.3 *Scientific Papers*

With the instruments' future set, the Hugginses turned to assembling a second volume of publications from the Tulse Hill observatory. As William announced to Hale in February 1909, they wanted to reprint 'my collected scientific papers, which as you know, contain the early pioneering work in nearly all the principal divisions of Astrophysics'.[53] It had been nearly ten years since the publication of their *Atlas*. Now they planned to pull everything together in one handy reference book. Hale was delighted. 'I need hardly say that in this form they will be of the greatest service to astronomers and spectroscopists.'[54] Margaret explained to Sarah Whiting 'The early papers of Sir William which are really the *foundations of Astrophysics*, were out of print, & students suffered much in not having them.' Their new book would fill that need.[55]

William thanked Larmor for contacting the press at Cambridge on his behalf. He was unfazed by their apparent lack of interest in the project. He was more interested in having the book printed under his own direction as he had the *Atlas*. RAS Assistant Secretary William

Wesley recommended another folio volume, but William was sure it 'will be more likely to be read if quarto' like the *Transactions*, with type similar in size to that used in the new *Proceedings*.[56]

There was to be little that was new in the volume other than occasional brief annotations deemed necessary to explain, clarify or update the text as originally published. Most of the book would be on astronomical topics, of course. Readers less familiar with the breadth of Huggins's interests were likely to be astonished to find articles on the chemical spectra of terrestrial elements, radium glow, microscopy, binocular vision and even the acoustics of violins!

William and Margaret decided to arrange the book topically, not chronologically, with sections devoted to categories such as the observatory and its instruments, stellar spectra, nebula spectra and motion in the line of sight. This meant splitting up some papers into smaller single-subject parts. The editing effort was worth it in order to enhance the accessibility of all the information the papers contained. As Margaret explained to Professor Whiting, 'We have tried in every way to make the volume useful, & easily useful, – to students.' She hoped the conscientious instructor would approve of their arrangement. 'We have reproduced all the original illustrations, and we have, wherever it seemed desirable, given *annotations* which we thought called for.'[57]

The influence of friends and colleagues reinforced their own desire to expand upon the outline of the history of astrophysics (and William's pioneering role in it) that he had sketched in his retrospective essay, 'The new astronomy'. They used excerpts from it to preface each section. Having the opportunity to retell this stirring narrative in his own voice without interruption or distraction from other points of view made it possible for Huggins to shape future generations' second-hand memory of the new discipline's origins. It still provides inspiration to astrophysicists who read it today.

'The book is progressing at the printers', William joyfully reported to Larmor in April.[58] He had decided on the image to use for the volume's frontispiece: the photograph showing him seated in his observatory near the eyepiece of the Great Grubb telescope which had been taken a few years ago to accompany an essay by Agnes Clerke on late-nineteenth-century developments in astronomy (see Figure 10.2).[59]

In May, the Hugginses were thrilled to have Hale in London once again. They had been looking forward to his visit since December when Hale first announced his plans for a European trip.[60] The trip was 'for the purpose of a change and rest', Hale told William in February.[61] Because he had been suffering 'a great deal . . . from nervousness', he explained, his physician had recommended he 'give up work altogether for a considerable period'. Instead, the irrepressible Hale planned a whirlwind tour abroad.

I find in actual practice that I gain much more than I lose through these trips, partly because of the rest, and also because of the new ideas which come from the stimulus of discussions with men in various departments of research. As the programme of research with the 60-inch reflector must be prepared with great care, I feel that such a discussion, with you and others, would be particularly valuable at the present time.

Besides, the Royal Society had elected Hale a Foreign Member on 25 March. He would have the opportunity to be formally admitted while he was in London. William and Margaret 'fully ... rejoiced, with all your other friends, at the recognition of your scientific work' by this 'limited and very select body'.[62] Hale would no doubt have been surprised to learn that he had won election to the RS despite Huggins's recommendation that 'according to usual custom' he 'might wait a little' since 'he is rather young'.[63]

Hale had also accepted, albeit reluctantly, a request to deliver an evening discourse at the Royal Institution during his stay. Exhausted though he was, he knew it would be politic to advertise the power and promise of his growing new astrophysical observatory on Mt Wilson. Thus, on 14 May 1909, he described the wonders of the solar atmosphere revealed by his improved spectroheliograph in a lecture on his recent discovery of vortical patterns and magnetically induced disturbances on the solar surface.[64]

In October, when *Scientific Papers* was finally ready, the Hugginses sent copies out for review. 'I am glad that you think well of the get-up of the book generally', William wrote to Larmor.[65] 'I shall *not* send [a review copy] to N[ature]; unless I have some assurance beforehand that the book will be put in the hands of a *competent* and *independent* reviewer.'

The volume was well-received at home and abroad. Father Aloysius Laurence Cortie (1859–1925), then Professor of Mathematics at Stonyhurst College, praised its topical structure which he was sure would make it useful to specialists and students alike.[66] The anonymous reviewer for the *Quarterly Review* described it, along with its companion first volume, as 'a monument of conscientious work carried out with the most painstaking assiduity for half a century'.[67] And Edwin Brant Frost (1866–1935), who had replaced Hale as Director of the Yerkes Observatory, hailed the volume as a record of the 'devotion of a lifetime to science'.[68] It was certain to help busy teachers with meagre resources impress their students with the 'marvel of the revelations of the spectroscope in its astronomical applications'.

15.4 'Life is work, and work is life'

Huggins suffered from bouts of neuralgia and occasional respiratory illness toward the end of his long life. Somehow he always managed to work around these minor setbacks giving those around him the impression that he was indomitable.[69] In 1905, he was featured in a popular magazine article on the secrets of longevity. Then eighty-one, Huggins reported that he 'eats with moderation, taking but very little meat, drinking coffee, but seldom using alcoholic liquors. He never smokes, and sleeps nine hours a day'.[70] A similar survey of elderly notables from all walks of life conducted in 1908 marvelled at his 'excellent health' which he maintained thanks to his moderate living habits enriched by recreations such as 'collecting of antique works of art, music, botany and fishing'.[71]

'Life is work, and work is life', he told an interviewer on the occasion of his eighty-sixth birthday. 'How could I have spent the day more happily than by dedicating it to the service of this thing which has been the one true mistress of my life?'[72]

As the years fly past, fresh knowledge becomes more and more difficult to grasp. It is not that I work less keenly as the days go by, but rather that as the science has developed, only details are left for discovery – but who knows? Every day I work for hours in my laboratory; and then on till nine o'clock of an evening. I am here in my study thinking and reading, reading and thinking. Every now and then some fresh detail is laid bare to me, and who is there to say that any day my experiments may not result in some new discovery altogether?[73]

On 5 May 1910, Huggins attended the second meeting of a special joint RS and RAS committee established to collect and reprint all of William Herschel's *Philosophical Transactions* papers. Scholars and astronomers alike had difficulty locating copies of the rare volumes that contained them.[74] Although a synopsis of these works had been compiled thirty years earlier by the Director of the Lick Observatory, Edward Holden, and Yale physicist Charles Sheldon Hastings (1848–1932), many, including Huggins, felt the reprinting of Herschel's papers, in full with all illustrations, to be a project that was long overdue.[75] He served on the committee with old friends and colleagues like RS Physical Secretary, Joseph Larmor, and David Gill, then RAS President.[76]

Huggins had come to this meeting 'full of his usual wisdom and common sense, and ready to aid, as ever, with his advice. His intellect was as clear as it ever was', Gill told RAS Fellows who had gathered on 13 May for what was already slated to be a solemn and abbreviated meeting devoted to mourning the sudden death of King Edward VII on 6 May. The news that Huggins had died on 12 May – just the day before! – took everyone by surprise and compounded the Fellows' grief.[77] '[N]one of us', Gill declared, 'expected that the end was so near.'

On Saturday, 14 May, following a private service at the Hugginses' Tulse Hill home, William's coffin was moved to Golder's Green Crematorium for the funeral service. The organist began with Chopin's 'Prelude No. 4' – the haunting and mournful piece played at the composer's own funeral. The coffin was brought into the chapel to the stirring strains of the 'Pilgrims' chorus' from Wagner's *Tannhäuser*. The service closed with 'O rest in the Lord' from Mendelssohn's oratorio, *Elijah*. Margaret's younger half-brothers – John Murray (1863–1943), a London surgeon, and his wife, and Colonel George Murray – were among chief mourners in attendance.[78] Representatives from the Royal Society included William Crookes and Frank Newall. David Gill, William Wesley and others from the RAS paid their last respects to their good friend and colleague. Other scientific notables also attended including Arthur Stanley Eddington (1882–1944), then Chief Assistant at the Greenwich Observatory, and renowned physics professor Sylvanus Phillips Thompson (1851–1916). Even James Dewar came with his wife. Thanks to Margaret's chance encounter with Dewar at the Royal Institution shortly before William's death, the old friends – estranged since the infamous letter to the editor from 'Inquirer' appeared in *The Times* of London in 1903 – had been happily reconciled.[79]

William's ashes were placed in a simple bronze urn. Margaret designed a lockable grille bearing the words 'In Thy Light, shall we see Light', which, she explained to Joseph Larmor, had 'always struck me as amazingly suitable as a motto for *spectroscopists*'.[80]

15.5 '... guardian of my Dearest's reputation'

'My sorrow is beyond all words', the anguished widow confided to Larmor just six weeks after her unspeakable loss. '[E]very day, I think it grows. It could not be otherwise, – we were so absolutely one.'[81] Over the months, in letter after letter she candidly and unabashedly sought from Larmor the solace only a son – even a surrogate son – could give for her 'awful aloneness', 'suffering' and 'exquisite pain'.

'*You* soon became dear to him', she told him on 17 October.[82] 'He was pleased when you spoke of yourself as *his* Secretary, – R.S.' She was personally grateful for Larmor's efforts to secure her £100 annual pension 'on proper grounds'. To select just the right '*personal remembrance*' for such a friend, Margaret had 'tried hard to choose as he [William] would wish; and also in a way that would please the recipient'. Thus, she promised Larmor a signet ring that her husband had '*valued* & liked much & often wore'.

But there is another side to this and the other poignant personal letters Margaret wrote to Larmor during this difficult time.[83] In extended excerpts from their correspondence we see that, while steeped in grief, she nevertheless played an active and powerful behind-the-scenes role as both architect and engineer in the important work of constructing and maintaining her husband's historic image.

Just five months after William's death, driven by her 'strong sense of Duty', she announced to Larmor, 'I think I see my way to write this winter, the *Life* which *should* be written, and which I believe only *I* can write. It shall not be too long!' Unlike the dry and recitative memoirs sure to be written by William's colleagues, Margaret assured him her account would 'be rather in the nature of *a vivid sketch*, & will deal at *most length* with my Dearest's early life which is in truth, not known at all'.[84]

As William's longtime collaborative partner, she had always followed closely and with great interest the progress of projects with which he was intimately involved. After his death, she felt a special obligation to pick up and run with some of the flags he had been forced to drop. Thus, she inserted herself into the work of the committee charged with reprinting William Herschel's papers. 'You may, I fear resent my intrusion into this matter', she wrote straightforwardly to Larmor, a member of the committee. 'Or, will you trust me? You really may.'[85]

Her wish to become involved, she explained to him, arose after discovering that, through no fault of his own, William Wesley was making poor reproductions of Herschel's illustrations. The originals upon which the RAS Assistant Secretary was relying to make the new plates were themselves in poor condition. Committee members David Gill and Herschel's great-grandson, Joseph Alfred Hardcastle (1868–1917), had been charged with overseeing Wesley's effort. Margaret was appalled to find that neither man possessed the requisite eye or experience to recognise the inferior quality of these originals. To ensure the project's success, she offered to loan William's own prized copies of Herschel's copper plate engravings on condition that '*proper care*' would be taken of them. 'I have taken the trouble I have, and risk vexing *you*', she explained to Larmor, 'because my Dearest was keenly interested in this Reprint; because I myself feel interest in it; & lastly, because I do feel for poor Mr. Wesley.'

She later protested the unwise decisions she felt were being made regarding the printing of the plates and strongly advocated that Wesley be made an active member of the committee so that he might better guide the project to a successful completion.[86] In January 1911, she wrote again 'about Mr. Wesley & the Herschel matters'. But this time her tone was more conciliatory, perhaps in response to a chastising hint from Larmor. 'I have an impression that you thought I had made needless trouble', she wrote almost apologetically. 'It is always difficult to act perfectly', she confessed. 'But *you* were away, and I do assure you that Mr. W[esley] had got into a *very rubbed-up state* through Mr. Hardcastle.' Besides, she added, 'I did succeed in smoothing down Mr. W & in getting him to be satisfied to wait until *you* returned & then to trust *your* knowledge & guidance. These are small things; but I think it only fair to myself to state them.'[87]

If her preserved correspondence with Larmor is any indication, Margaret *did* refrain from further interference with the work of the Herschel papers committee, and she expressed enthusiasm for the final product as it neared completion.[88] Perhaps Larmor had reined her in successfully. It is just as likely, however, that her attention was diverted to the daunting task of monitoring the obituaries and memorial essays being written by William's colleagues. She admitted to Larmor that she had been anxious 'from the first, about the articles written in our own Land, on my Dearest'.[89]

Indeed, an essay by Frank Newall – the lead article in the October number of *Science Progress*, a quarterly aimed at keeping scientists and interested laymen abreast of scientific advances being made in all fields – prompted her to seek her busy friend's immediate advice and support.[90] 'If you *have* read [Newall's essay], will you, *in strict confidence*, tell *me* what you think of it?', she implored on 29 October 1910. 'If you have *not* read the article, would you mind trying to make a little time to glance through it? and then tell me what you think of it?' She confessed 'I am *intensely* anxious about this Memoir. There are *so few* who have *the necessary* knowledge; & who have also *a right judgment*, – and the power of seeing things in just perspective. I do wish *you* might have written [it].'

In his obituary essay for *Science Progress*, Newall praised the judgement and caution with which Huggins harnessed the restless creativity that made possible his many pioneering discoveries in celestial and terrestrial spectroscopy. He extolled the *Atlas* and *Scientific Papers*, both valuable reference volumes. He lauded Huggins's many years of service to the nation's premier scientific societies. And he reminded readers of the collaborative assistance Huggins had received from his capable wife.

Many readers would rightly consider the work an unapologetic homage to a lost leader by a grateful disciple. But Margaret saw it otherwise. Newall's essay became the subject of eight long letters to Larmor over the next month. 'As regards Prof. Newall's article', she wrote to Larmor on 3 November, 'I feel in a most painful position'.

That his intentions were all that is good & kind, I do not doubt for a moment. He wrote to me *most kindly*, on his way home from America, & mentioned the article.

I was *dismayed* when I read the article, – at the serious *omissions*; and not only dismayed but *most deeply pained* by a statement in connection with a highly important piece of work, – a statement which is *absolutely untrue*, and which I cannot allow to pass unchallenged.[91]

In particular, Margaret was distraught over two statements she believed needed to be retracted. The first – that 'Huggins ... originally ascribed the brightest green [nebular] line to nitrogen' – was, in her view, erroneous.[92] The second – that her husband had received assistance 'on the theoretical aspect' of Doppler's principle 'from Clerk Maxwell' in his attempt to measure stellar motion in the line of sight – was misleading.[93]

It is unpardonable that a man in Prof. Newall's position should have made such a blunder – I am sorry for him; but it is my *duty* to see to it, that my husband is vindicated.

I have the facts before me, – as they were before Prof. Newall.

My position is sad enough without the pain of such an incident as this. But I am not in the least angry. Pray do not think this of me. Sorrow & mourning leave no room for poor, small feelings. But they do bring home to one the vital necessity of *Truth* . . .

My present feeling is to write frankly to Prof. Newall & then send short notes to both *The Astrophysical Journal* & *The Observatory*.[94]

'But', she inserted at the top of the page, 'we all know that once an untruth gets into print, it is *almost impossible to check it*.' Newall's article intensified her determination to 'lose no time in preparing my Dearest's *Life*; & as far as *Love* & *devoted Labour* can preserve his work for the future from such mistakes and misunderstanding as Prof. Newall's writings show, – he shall be safe.'

Larmor replied immediately hoping to ease Margaret's distress. He gently persuaded her to talk her concerns over with him rather than write to the journals. She trusted him to help set things right.

Meanwhile, Hale, the Hugginses' other 'son', was in London with his family. He had suffered a serious breakdown of his own earlier in the year. News of his mentor's death came about the same time he received word of a major setback in his ambitious plan to build a 100-in telescope for his astrophysical observatory at Mt Wilson. After the first disc had been declared unsuitable for figuring in December 1908, plans were made to cast another disc with a new furnace and annealing oven.[95] After a few unsuccessful attempts, a promising disc was cast in February 1910 and buried 'in a manure pile for annealing'.[96] But in May it was discovered that the disc had broken during that long and delicate process.[97]

Hale was unable to work productively for months afterward. The fourth meeting of the IUCSR was held as scheduled at the end of August 1910 on the site of the new Mt Wilson observatory in southern California. Larmor and Newall were among the Britons in attendance.[98] But Hale was so overwrought when the meeting convened he was forced to limit his participation.[99]

Hoping to restore his mental and physical health, he embarked with his family on a major European tour in October. While in London in early November, he visited Margaret at Tulse Hill. He shared her sense of loss over William's death and surely lent a sympathetic ear to all her woes. She gave him his own special keepsake – another ring that William had often worn – and confided in him her anxiety over the Newall essay.[100] Whenever danger threatened, it was always good to keep strong allies alert and ready to come to one's defence. In the event, after his visit, it was *she* who expressed worry about *his* health.[101]

On 7 November, Margaret thanked Larmor sincerely for the 'kind & quite frank letter' she had received from him.[102] She assured him she viewed such candour as a mark of friendship. 'I *need* Friends who can & will be so truly kind & loyal to me as always to speak plainly & tell me truly what they think.' By the same token, she expected Larmor to grant her the same privilege and 'allow *me* to speak freely'.

She sent him some rough notes she had compiled on Newall's essay with the explanation 'I felt that you thought I was not reasonable; & that I was making too much of errors – if such there are. Ah!' she exclaimed. 'But consider how I stand.'

Surely – while I live, – seeing I was *a partner* in our work, – I am bound to see that no wrong is done to my Dearest's reputation?

If the Magazine in question were of the order of the *Pall Mall*, or even of the *Contemporary*, errors would not matter so much.

But *Sci. Progress* is not a popular Maga. Not at all. It is meant for people interested in Science & with some knowledge; but even more, for *Scientific men* engaged with *one subject* but anxious to know what is going on in the subjects outside their own special one.

Also Prof. Newall is not a popular magazine writer. No; he is a Professor of no mean University; & he is being advertised widely as – 'the acknowledged master of the subject,' (Spectrum Analysis) in connection with his book – *The Spectroscope & Its Work*.[103]

She reminded Larmor of David Gill's 1901 address at the dedication of the Victoria Telescope in which he 'ascribed the *Line of Sight Method* boldly & fully to, – Clerk Maxwell'. That 'painful' experience had taught her the need to remain vigilant against the ever-present threat to historical truth of unchecked errors. 'Consulting together', she told Larmor, 'my Dearest & I, – I wrote to Sir David & insisted upon putting the truth to him. He *had* to give in, & thought to get off by withdrawing his false attribution, & giving Sir William the credit due to him, – in a *Cape newspaper*. That', she stated emphatically, 'was *not enough*.' To eradicate a widely disseminated error required broadcasting its correction even more widely. Thus, she told Larmor, 'we wrote a *note* to *The Observatory*; & reprinted that *Note* in our *Collected Papers*. See p. 230. How wise I was in begging my Dearest to reprint our *Observatory* Note, is now proved.'

She knew their actions then could not prevent the same error from being repeated, or entirely new errors from arising now or at some time in the future.

I thought it *very* likely that Sir D[avid]'s memory might again fail him! but – could not have believed that Prof. Newall would so *badly blunder*, – and, – not once, – but *twice*.

I assure you I am greatly harassed. Sorrow, & an awful aloneness are crushing enough. But, – from the quarter from which I *counted* on sufficient knowledge & support, – there comes, *an added sorrow* & a pain ...

Surely you cannot yourself think that these 2 false statements, – should be allowed to pass without protest?

Surely you cannot advise me to let them pass? or, do you? ...

You cannot think how much I feel these mistakes. And, – forgive me, – you cannot tell *how* I shrink from the idea even, of being disloyal to my Dearest by allowing to pass *what are serious injustices to him*, – to say the least.

But I am not quarrelsome, & I detest controversy.

But, I am the guardian of my Dearest's reputation while I live, – and I would rather lose every Friend I have – than fail in my duty to Truth, – *my duty* to my Dearest.

She closed her long letter by offering a possible remedy: 'It has occurred to me, that if he can be got to see that he is *in error & seriously*, Prof. Newall would be willing *himself* to write a short *Note* for the *next No.* of *Science Progress*. What do you think of this suggestion?'

Larmor tried his best to absorb, or at least deflect, some of Margaret's anger by shifting part of the blame for Newall's objectionable statements to himself. He referred her to *Aether and Matter*, his Adams Prize essay which had been published back in 1900.[104] In it, he confessed, he had linked Maxwell's theoretical views on the relative motion of the Earth and the ether with Huggins's 'fundamental memoir . . . on the spectroscopic determination of the velocity of movement of stars in the line of sight'. Perhaps, Larmor suggested, Newall had interpreted these remarks to mean that Huggins had built his method upon Maxwell's ideas. Margaret was not persuaded.

However much *you* may wish to take all the blame for these unhappy words this will not recall them. They are in print, & have gone forth now. But, the *responsibility* for them rests, to my mind, really on Mr. Newall. He was not *obliged* to accept your suggestion; & the subject [astrophysics] he was treating was *his*, – & not specially *your's*. [*sic*]

I have only one thing to consider. Do the words imply that Sir William was *helped* on the theoretical side of the Line of Sight Method, by Clerk Maxwell?

I consider that they *do*. Moreover I believe that *any* reader would take this meaning from the words.[105]

She wanted Larmor to appreciate fully the heavy responsibility she bore:

Well; I have been considering *most anxiously & painfully*, my duty. I feel it is a cruel position to have been placed in; – but *my duty must be done*.

I *deeply* regret to have such a duty to perform, – but I am wholly guiltless in the matter. No chance was given me of staving off what must now be done . . . .

I shall state my case to Prof. Newall, & I think *he* should write *a very plain Note* for the next No. of *Science Progress*, arranging that the Note shall be the *1st article* in the next No. so as to *ensure* as far as possible that those who read the erroneous statements, shall read their retractation [*sic*].

When I have Prof. Newall's reply to my letter, – I will consider whether I can then leave this matter.

It is more pain to me than I can tell you, – that the one article I had thought would give me unalloyed pleasure, is the *only one* that has given me pain & worry . . . .

It is my duty to consider all this; but I assure you that I have striven, & shall strive, to take large views & rise above the *merely personal*, part of things . . . .

Be *sure* I am grateful to you for hearing me out about that article; & for giving me so much of your time. And I truly value your frankness.

Larmor then took a different tack: trying to persuade Margaret to accept the situation as an unfortunate result of Newall's poor choice of words. Surely they did not reflect his true intentions. She remained steadfast in her demand for a retraction.

I have nothing to do with what may have been intended.

My position is clear & simple; & I have maintained it clearly & unwaveringly from the first; – I once more put it; – to anyone reading Prof. Newall's paper, the sentence occurring at the top of p. 189, –

'Basing his method (i.e. Sir W. Huggins) of attack on the correctness of Doppler's principle, *on the theoretical aspect of which he received assistance from Clerk Maxwell*,' – (The italicising is mine.) [Margaret's] plainly suggests that Sir William *was helped* in his *Line of Sight work*, by Clerk Maxwell.

Nothing you have said has altered my view; & I repeat, that I cannot allow it to pass uncontradicted. There must be absolute retraction . . . .

I shall look up the passage in your book to-morrow, when I am in town. But I shall keep to my plain position. I am not qualified, I well know, to argue with *you* on abstruse mathematical questions or points. Nor am I attempting to do so. On the contrary, if I *had* to deal with such questions, I should at once ask your help & *fully acknowledge it*. But, – the *wrong & misleading* statement which has been made in *Sci. Prog.* I *am* well able to pronounce upon. It *is* misleading & I call for retractation. It must be as plainly & fully retracted, as must be the wrong assertion about Nitrogen & the chief neb. line.

At this moment I do not feel inclined to be satisfied with anything short of a short note *retracting both statements as erroneous*, from Prof. Newall himself. Why not? He has made *two*, bad & injurious blunders. He can in *one way only*, do justice to the Great Dead, – and to his own best self, *i.e.* in being manly enough to tell the truth.

I *am delaying* my letter to him stating simply & kindly the sad & painful duty imposed on me by these mistakes, – until I hear what it is *you* would do, *supposing* (only – *supposing*) I authorized you to deal with the matter.

I will hear you calmly & fairly.

But, – I have *not* yielded one point of my position remember. I have *not* authorized you to arrange the matter. I simply am willing to hear what reparation you propose.

Why are you so fully of pity for Prof. Newall? with, it seems to me, – none for the victim.

Why should bad errors be covered up?

If I allow that false statement, – & false it is, – to pass, I know well that Sir D. Gill will say privately if not openly, – that he was not wrong after all . . .

I am *certain* that *many* will be led to take a wrong view by that unhappy sentence . . .

It does not matter about me. I may not be able to carry on very long the effort to live without the Life which was infinitely dearer to me than my own. I am quietly but *steadily* setting my house in order.

But, I shall strive to live long enough to do my duty in the present matters & some others, & to render to my husband such a tribute as few husbands have had.

I have written strongly; but my mind is clear, & not dominated by mere feeling.

You may therefore feel sure that all you may say – I will surely consider *most* carefully.[106]

On 15 November, she supplied Larmor with a sample letter for Newall to use that would satisfy all her objections:

Copy of Note to Prof. Newall

In my article on Sir William Huggins in *Science Progress* for October, I greatly regret to find I have made one erroneous statement, and another which is seriously misleading; and I desire at once to correct them.

(1) On p. 187 of *Sci. Progress*, I say that Huggins originally ascribed the brightest green line of the nebular spectrum, to nitrogen. My statement is wrong; he did *not* do so.

(2) On p. 189, of the same article, in speaking of Huggins' Line of Sight work, my words, – 'on the theoretical aspect of which he received assistance from Clerk Maxwell,' – are misleading as suggesting that Huggins was *helped* by Maxwell. Nothing could be further from the truth, for Huggins' work was done absolutely alone and was completed before he even heard that Clerk Maxwell had considered the subject.[107]

Larmor, in turn, drafted his own sample letter for Margaret's comments and, he hoped, approval. She was unimpressed. On 16 November, she replied:

On receiving your last letter & . . . the draft of the statement you proposed to send – 'perhaps to *Nature* & the *Observatory* at once, & certainly to *Science Progress*,' – you may feel sure I lost no time in giving your proposed statement my best attention.

After most careful reading & consideration, I decided that I *ought* not to accept it for some weighty reasons which are obvious; & for others which are also weighty . . .

First, let me consider your draft statement.

(1) It is *far too long* and too involved; & from beginning to end, anyone reading it must think of the acute old French proverb, – 'qui s'excuse, s'accuse' [he who excuses himself, accuses himself].

There can be no question that Prof. Newall made a wholly unwarrantable blunder about the nitrogen. He made it *shortly* and *plainly* in *nine words* – *a single line* of print.

You propose, instead of an equally short & plain retractation, to offer *a good half page* of apologetics – not too clear, – and, to throw discredit on the perfectly lucid & perfectly expressive English of Sir William's original paper, – by way of excusing your brother Professor . . .

(2) Your proposed statement is far too apologetic. If Prof. Newall had been asked to write that article by 'wireless' when on the sea, & had agreed, – there might be some reason in apologies & excuses about – no books &c.

But when he knew all he had undertaken to do before he started, – such excuses are out of place. Anyone so placed, should take what they need in books or notes, – or face the consequences.

No *woman* would have failed to take what was necessary. Men it seems are deficient in elementary common-sense in such matters.

(3) You conclude your *long* statement by a passage stating that the 'paper was not intended to be a complete judicial account of Sir W. Huggins' great work.'

No. That is obvious, I should have thought, to all. It was so to me; & that is one reason why I deprecated any complication about Clerk Maxwell being brought in. And, I maintain that what was said, was & is, *mischievously misleading*. And, Prof. Newall *ought* to have known & remembered what had happened about Sir D. Gill.

It is *deplorable* that a man who is now a Professor! of Astrophysics, should so fail in *general* knowledge of his own subject.

The subject is not *your's* [sic] *specially*, so I cannot regard you as to blame, at all as he is.

Bear with me now while I speak for a moment about Clerk Maxwell in relation to Sir William & his work.

You were good enough to refer me to your book *Aether & Matter* pp. 14–15 . . .

I devoted as much time as was necessary, & I did not leave that book until I could feel that I had grasped (as far as any one could grasp without mathematical training) the *gist* of those pages 14–15 of what I am quite aware is a *great* work.

I admit I feel a certain satisfaction of mind that the general conclusion I came to after close study of those pages . . . was virtually what you say at the end of your draft statement – namely *even now* those *theoretical* questions 'involve prominent topics of discussion.'

Just so. And so I consider it was most undesirable to introduce them into an article of a general & semi-popular character. And, I do not think it would be at all desirable or even fair, to use your apologetic statement for a long allusion to Clerk Maxwell's theoretical work – when the article is on Sir W. Huggins, & *not* on Clerk Maxwell.

Moreover, – and here is a point which has never yet been put forward – & I put it now to you as a point of some importance. It is that Clerk Maxwell's theoretical work on this subject has *gained immensely all these years* by having a place in the great Paper by Sir W. Huggins.

Had the Paper by C. M. been in the *Transactions* in the ordinary way, *few comparatively*, would have known of it. But, inserted into one of the most vital & potentially great papers of recent times, it has been easily seen & studied by *all* students of Astrophysics.

I do no injustice to Clerk Maxwell either, if I say that his own letter to Sir William shows in truth, – an *eagerness* to have his Paper included in Sir William Huggins' Paper. It was *no letter* he sent; but *a paper*.[108]

She turned to the subject of her planned biography of her husband. 'In the Life I am preparing of my husband, there will be place for some statement [on Maxwell's work] & I shall wish *there* for *your* help, – if you will give it to me?'

Were Clerk Maxwell living now, I should I can assure, have no doubt that *he* would *feel with me* about that *misleading* sentence in *Sci. Progress.*

I should have told him my pain & anxiety about it with perfect frankness, & I *know* he would have understood & *agreed* with me . . .

You need not, think for a moment that *I* would defraud him of any honour – any respect. I do think you might trust me a little . . .

Remember, those errors are *sure to have done harm.* In these days – mistakes get abroad – amazingly; & it is almost impossible to kill them though we *must* try to do so . . . .

You must not forget either, that you yourself saw no errors in the article until I drew your attention to them.

And most of the readers would be even less likely to see & mark them. Therefore, they must be *plainly* indicated, & *plainly* retracted.

15.6 'I now withdraw . . .'

In the end, Margaret decided to drop the matter, not because she finally won a retraction from Newall, or found she had made a wrongful accusation, but because, as she explained to Larmor in January 1911, she could not ask anyone to retract words they firmly believed to be true.

Prof. Newall after several letters at last said – what I wish he had said *at once*; – namely, that he could not conscientiously take any other view than that Sir William believed wholly & absolutely in the nebular brightest green line being due to Nitrogen from the first.

I cannot agree. I shall always say no! while I live. I *know* such a view is wrong.

But I will *never* wish anyone, or try to induce any one, to go against his or her conscience. So, I leave the matter. *I wish nothing to be said or written by Prof. Newall.* And the other point, I prefer too, to be left *unnoticed;* as most likely Prof. Newall *conscientiously* cannot change that view either.

Nevertheless, – I do not think that the word *'assist'* should have been used in connection with Clerk Maxwell, – especially after the Gill episode which I think Prof. Newall *ought* to have remembered.[109]

She felt fully justified in her response to Newall's essay and the actions they prompted her to take.

I think that it was *wholly natural* that I should in both cases feel acutely what was said in the article, and that I should feel it *a sacred duty* to protest.

Just think *how* recently I had been editing those Papers! Think how – almost in my ears – is my Dearest's voice, as we discussed the proofs sheet by sheet.

My share of the editing, I may tell *you* in confidence was *genuine work.* Every sheet, I went over *alone*, at least *three* times. & at least *twice* with Sir William.

Here she inserted an emphatic note in the margins: 'He [William] was equally careful; and he always consulted with me. *He* thought me worth hearing.'

And *you* at any rate, will believe me, when I say that all through the editing, the object most important to me was – I know my Husband's *own mind.* I tried in every way I could to *anticipate*, so that if I were *left*, – I might still be able to add useful notes in the future.

Well, – I did of course no more than my duty. Or rather, did no more than *try* to do my duty. Do we ever, any of use [*sic*, read 'us'] succeed wholly in doing it?

***

I do not regret my *protest*. And, while I now withdraw, & will have *nothing said or written* by Prof. Newall because his conscientious belief shall be respected by me, however mistaken I hold it to be; – I also believe that *his* view will not be the *general one* …

I understood you to agree with me *that* as regards the *Nitrogen*, Sir William while feeling bound to record *appearances*, from the first had his mind open & was *never satisfied* that the brightest green *nebular* line was due to Nitrogen.

Also, I understood you to at least feel, that the term '*assist*' was unfortunate in connection with the *Line of Sight Method*; especially considering the trouble there had been with Sir D. Gill (which Prof. Newall should have remembered.)

I have felt very thankful that I knew *your* mind, so far, on these points. It has been a comfort to me.

***

I still feel that it was, all things considered, – most natural that I should feel very strongly about the *Nitrogen* & about the term '*assist*'; & assuredly, suffering acutely as I was, – *only* an overwhelmingly strong sense of duty & loving loyalty *could have moved* me to remonstrate.

Intellectually I see nothing to blame myself for. I offered my reasons; I offered *full* proofs. And I did *not* make the points personal by pleading my intimate position. I do admit however that my language & tone were over-vehement; that indeed they amounted to an overbearingness which was not philo-sophic, & which unduly presumed upon Friendship.

Also, in rejecting your kindly effort to draft something to satisfy me, – I was not courteous.

In all this, I was wrong, – *very* wrong. I make no excuse for myself; I apologize to you *most sincerely.*

You, – never failed; & I deeply thank you for your chivalrous & true kindness, in not even reproaching me.

Be very sure I have not failed to much more than *reproach* myself; & that my apology to you is not a mere matter of words.

You will I think forgive me, & believe that I am sorry …

There could be no criticism at all, – if whenever objection was taken to statements, – *mean personal feeling* is to be read into purely intellectual matters treated *intellectually* …

Your kind, *warning* shall I call it? not to waste myself on ephemeral things like that article, – but to concentrate myself on my Dearest's *Life*, – I thank you for. It was kind & kindly meant; & I will tell you that that *Life* is in my thoughts every day; – & many a part of a night …

I am apologizing also to Mr. Newall …

God will help me to do my duty; – but – Oh! for the touch of *one* dear hand! The sound of *one* dear voice!

Newall redeemed himself by writing Huggins's obituary notice for the February 1911 *Monthly Notices*.[110] In it he carefully sidestepped the troublesome elements of his *Science Progress* article. In laying out his narrative of Huggins's life and work, he adhered to the template Huggins had laid out in his 'New astronomy' essay. He avoided suggesting that Huggins had identified the brightest line in the nebular spectrum with nitrogen. And he wisely excluded any mention of Maxwell in his discussion of Huggins's work on stellar motion in the line of sight.[111]

Margaret and Larmor each learned important lessons from their difficult encounter over the *Science Progress* article. Larmor came to recognise her as a force to be reckoned with in any future plans to memorialise William's life and work. She, meanwhile, made sure that an article on her husband would appear in the new supplement to the *Dictionary of National Biography* then being prepared.[112] And she asked that Larmor include her in the plans for William's obituary essay in the RS's *Proceedings* (see the Frontispiece), suggesting names of possible authors and offering '*several* portraits you could choose from – all of which are practically unknown'.[113]

Margaret promised not to meddle in the writing of the RS memoir. Nevertheless, she offered her assistance, '[i]f the writer really wished me to see the MS. before printing'.[114] In some respects, she was less concerned about this essay than she had been about the one in *Science Progress*. For one thing, the quarterly had a wider circulation than the *Proceedings*. For another, readers of *Science Progress* were unlikely to have followed her husband's work before encountering Newall's essay. Though scientifically trained, many would lack the specialised background and expertise to distinguish fact from fiction. On the other hand, the smaller community of *Proceedings* readers knew her husband personally and could themselves detect and correct any errors the essay might contain.

In the end, the RS tapped Frank Watson Dyson (1868–1939) to write Huggins's obituary essay.[115] Dyson was a wise choice. As Regius Professor of Astronomy at the University of Edinburgh, he held the prestigious title of Astronomer Royal for Scotland. He carried little of the baggage borne by the older and more familiar stakeholders in the London-based scientific community. His voice could rise with disinterested authority above the din of their philosophical and territorial squabbling. Whether by his own design or on the advice of Larmor, Dyson quoted Huggins's own accounts and interpretations of his work directly from his papers and thus avoided stoking Margaret's ire.

Larmor gave Margaret an opportunity to read the proof of Dyson's memoir. In October 1911, she sent him her corrections tempered with humour and embellished with personal anecdotes. 'All the changes … I have ventured to suggest', she was quick to assure him, 'are marked in with a *soft pencil*; you can put them in *properly with ink*, – or erase them as you may see fit.'[116] Overall, she found the essay 'weak on the *personal* side', but that was hardly surprising, she acknowledged, 'for Dr. Dyson did not know Sir William *intimately*'.

She was sure she had done both Larmor and Dyson a great service by reviewing the proof. 'I do not think you will regret having let me see it', she quipped, 'for at least I have prevented a number of undesirable errors. The date of death of my Dearest, for instance, was wrongly given.'

These positive errors speak for themselves so that I need not spend time commenting. Except perhaps in one case at foot of p. 1, where my Dearest & Prof. Miller are stated to have *walked* back to Tulse Hill together. The authority (XIX Cent.) Dr. Dyson imagines he quotes, does not state *how* they travelled. But it would have been a *very long* walk. – & in those days a *dark* one! It was not *likely* they would *walk*. As a matter of fact, I can state that they *drove*. …

p. 6. I venture to suggest that there should be – to be just – some slight modification regarding Mr. Maunder's Line-of-Sight observations. The truth will have to be told some day. I may tell it. Mr. Maunder was a very careful & competent observer.[117] Had those observations been made with a *proper spectroscope* there can be little doubt that the results would have been very different to what they were.

That detestable Half-Prism spectroscope which Mr. Christie (as he then was) designed & *would* not give up, – *we considered* brought discredit on poor Mr. Maunder, – & in some eyes, on the Line-of-Sight Method itself. Why – there were nights when the same star – *both* approached & receded. *Danced* in fact in the most flighty – shamelessly flighty – manner!

I was miserable over it; & Sir William felt it so much that he really *implored* Mr. Christie to change the spectroscope. But – no! nothing would move him. So the laughter went on at Potsdam & elsewhere – *They* were moved!

I do suggest some little modification. Will you you [*sic*] not ask Dr. Dyson to consider it?

p. 7. I do not wish to be troublesome, but I do venture to suggest that it is only fair & just to Sir William that some such few words as I have pencilled *re* the Prominences should be inserted.

p. 9. We never questioned that the *nebular lines* might appear, & *did appear* in Novae. Therefore I have suggested a *slight* verbal alteration. Sir William did not admit however that the presence of the *nebular lines* proved that a *Nova* had become a *nebula*. It is true what Dr. Dyson says that we could not examine *Nova Aurigae*, in its supposed nebular state, when Prof. Campbell declared it was a nebula. But it should be noted that he worked with an *achromatic*.

Later – some years – when there was another *Nova* – the later phase came of course & again Prof. Campbell said it became a *nebula*.

We, observing the alleged *nebula* with our *reflector* saw the object well and clearly as a *star*. We could see no difference between the *Nova* & other stars of similar magnitude.

Of course *all* this does not touch the proof; but I thought it might interest you.

p. 11. I thank you *deeply* for *your* little addition *apropos* of the Herschel Papers. He was intensely eager & interested about that work.

'Now', she paused as she calmly closed her comments and turned her attention back to the trusted friend who was reading them, 'you will not think I have been troublesome will you? Indeed, *indeed*, my only wish has been to *help* . . . that all may be as perfect as possible.' Without the original draft of Dyson's essay it is difficult to judge, but it seems from a careful reading of the published version that all her comments were taken into account. None of the few factual errors that remained were the sort which exercised her.[118]

15.7 The new Huggins Observatory

In July 1911, Margaret visited Cambridge to see 'the new Dome & its arrangements'.[119] She wanted to check on the new home for the Great Grubb telescope and the other instruments she and William had transferred to the University's care. 'I should like to go again for a *rather* longer visit', she told Larmor. 'Not only do I wish to see some of the work *at night* in the Observatories; I also wish to pay a quiet visit to *The Fitzwilliam Museum* with the idea in my mind of perhaps some of our collections ultimately finding a home there.'[120]

A few days later, she mustered the courage to speak more frankly to Larmor about her concerns after that visit to Cambridge. Although she insisted she was '*very satisfied*', that 'Prof. Newall has managed & arranged [the instruments] admirably', that '[g]ood work *ought* to be done with those telescopes' and that she had 'good hope that it will be done', she came away from the visit feeling troubled nonetheless.[121]

Just one thing disappointed me! May I tell you? Will you promise not to think it petty of me?

I risk – telling you. It disappointed – even grieved me – that there seems no *tablet*, – or *note* anywhere, of my Dearest's association with the Instruments & their new home.

I understood that it was to be a clear plain, feature, that the Observatory was in fact a Memorial of my Husband's long & splendid labours.

Surely there should be a suitable marble tablet in the walls, or on the walls, – somewhere? Prof. Newall said something about his wish to have a brassplate on the Telescopes. Yes; *this* there should be; but surely (?) there should be such a *tablet* as I have indicated.

If there is difficulty about money, may I be allowed to offer these things? I shrink from anything like asking Prof. Newall to be at expense. But I confess I am anxious about both these things & if you could promise me your help in *suggesting them*, I will *gladly* settle all expense.

I wouldn't interfere or be troublesome. You & Prof. Newall could settle what should be said. But, – I do hope *you* will agree with me that there should be a record in a stone or marble tablet (or brass) for the *Observatory*; & also (this is a minor matter) a *brass-plate* on the Duplex-Telescope.

A plaque was placed in the observatory announcing bearing the inscription: '1870–1908. These telescopes were used by Sir William Huggins & Lady Huggins in their Observatory at Tulse Hill in researches which formed the foundation of the science of astrophysics. Presented by the Royal Society to the University of Cambridge. 1908.'[122] The telescope remained in use until 1953, when it was dismantled to make way for a new Schmidt camera.[123]

In 1991, David Dewhirst, who joined the Observatory staff in 1950, recalled he 'was one of the last people to use the Huggins 15 inch refractor (the antique reflector was already dismounted and replaced by a balance weight), for photoelectric photometry.' Describing the Huggins telescope as having been 'dismantled', Dewhirst wrote, 'was a tactful way of saying *scrapped*'.

In retrospect I am sure the decision, not lightly taken, was a correct one. The telescope, after regular use for 80 years, was at the end of its useful life; the polar axis bearings worn out, the RA drive poor, and the whole mounting of design not suited to modern work. We preserved a number of parts: the fine Grubb 15 inch lens (in store here); the divided circles, 18 inch speculum metal mirror in its cell, and much of the early spectrographic equipment are now in the Whipple Museum of the History of Science.[124]

15.8 Wellesley College

Margaret's correspondence with Larmor touched on a wide range of subjects after the *Science Progress* episode: from passionate soliloquies on Japanese prints to Larmor's election to Parliament to books on Irish national history. She rejoiced over her renewed friendship with the Dewars. She shared the bittersweet satisfaction she derived from giving her beloved portrait of William to the National Gallery. She expressed her hope that Larmor might write a personal recollection of her husband to include in the biography she was preparing.

In late December 1912, she announced her plans to move from Tulse Hill to a small flat in Chelsea on the Embankment. She looked forward to enjoying a panoramic view of the Thames from her windows. Despite the ordeal of packing up and moving out of a house with so many memories, she believed 'much healing will come to me' in this new home located nearer to her friends and interests.[125]

She became focused on the difficult task of paring back her possessions. Her smaller quarters would not have room for everything. Besides she planned a simpler and less cluttered life for herself. Still, how could she part with the many cherished items she and William had collected over the years? Where could they be put to greatest use? In what venue would they continue to serve as a fitting tribute to her husband and his work? Who would be a worthy recipient?

After a visit from Sarah Whiting in July 1913, it seems her thoughts turned to Wellesley College. The women's college, established in a small town west of Boston in 1875, had a well-equipped teaching observatory thanks to Whiting's hard work and dedication and the generosity of trustee Sarah Elizabeth Whitin, whose financial support made the project possible. When the observatory building was dedicated in 1899, Margaret extended her hearty congratulations to Whiting and sincere thanks to the donor:

My heart goes out to your students! How well I remember my own early intense love of the heavens, & my early aspirations towards, & attempts to get, practical knowledge of astronomy. For me there was no adequate help; but for your students, – what prospects!

A beautiful & excellent Observatory fitted with all that the modern astronomer can desire, – a very Uraniburg which would have delighted the heart of Tycho Brahé, – given for the students' use by one stirred purely by the wish that they may enter one of the most fascinating realms of science, – surely such a gift so given must impress powerfully . . .

[Y]our students [will] think also . . . of the faithful teacher whose longing for them to get astrophysical training has led to the gift of the Observatory, & who proposes an admirable scheme for their instruction[. S]urely they will resolve to work with a devotion worthy of their teacher & of the splendid advantages now thrown open to them.

All departments of natural science are fascinating; but not one seems to me to be so perfectly so as that of Astrophysics, – or as it is well called, – the New Astronomy.

For, in the first place, it is many sided. It involves work not only in the observatory but in the Laboratory, and it asks from its students competent knowledge as observers, as measurers, as photographers, as chemists, as physicists. For training in inductive reasoning, – the great intellectual instrument in modern science, and I may add, modern life; – and for training of the senses Astrophysics offers from its very complexity, marvellous opportunities.

But it offers more. Imagination is as much to be desired in scientific research, and in life, as inductive reasoning, and the close & intimate dealing with many aspects of the vast heavenly hosts which Astrophysics renders possible, seems to me to offer most valuable training in that often neglected but most precious power . . .

I dare not say how important may be the work done in this Observatory in the training of those who directly & indirectly will guide the future of Astrophysics in America. But I do venture to say that no student working earnestly within its walls can fail to find much that is highest & best in her awakened and strengthened.[126]

The two women were nearly the same age, they shared a love of art and astronomy and they held similar aspirations for the future growth of girls' opportunities to learn how to satisfy their curiosity about the natural world. After they met in 1896, Margaret developed a deep affection and admiration for the dedicated American educator. Whiting created instructional exercises for spectroscopic study based in part on the plates of stellar spectra featured in the

Hugginses' *Atlas* and advocated enriching the teaching of introductory astronomy with materials and student projects drawn from the liberal arts.[127] After Whiting's London visit in 1913, Margaret wrote to describe the particular items she believed would be of aesthetic and/ or scientific interest to both Whiting and her students.[128]

The College purchased two telescopes, a sextant and a universal dial from Margaret Huggins.[129] But Margaret included many other items as gifts in the fourteen crates she packed and shipped off in 1914.[130] She particularly wanted Whiting to have '2 *panels* of stained glass for your Observatory or Observatory House'.[131] The panels form a pair on an astronomical theme. One depicts 'the group from the Bayeux Tapestry regarding the comet of 1066 – which you know was Halley's Comet'. On that panel, the tapestry's familiar caption begins with the Latin words '*Isti Mirant*' [these men wonder]. The caption is completed on the second panel (see cover illustration) with the word '*Stella*' [the star]. To complement the first panel pictorially, Margaret explained, the second panel 'contains the comet – towards which the group in Panel I looks. And besides the Comet, Panel II contains the Sun with Chromosphere, a Nebula, & other Figures. The Spectrum colours & the Planetary symbols are judiciously used.' Indeed, colourful representations of both the solar spectrum (crossed with Fraunhofer's principal dark lines) and the nebular spectrum (with its three characteristic lines) stretch dramatically across it.

Other gifts included a '*Diffraction Grating*' made by Lewis Rutherfurd. The grating 'performs well & we used it a good deal', she told Whiting. 'It is *his own work – ruled by himself with his own machine*. There are 13681 lines to the inch. This in the *early eighties* of the last century was a pretty *high* number.'[132] She also gave the College an astrolabe and a copy of her '*Monograph on the Astrolabe*' which she hoped 'some one of your clever students when familiar with this ancient Instrument, may devise a new form of it which would be accurate enough in use to be really valuable as well as superlatively convenient to travellers wishing to travel light'.[133]

Perhaps the most valuable gift Margaret gave Wellesley College was the six bound notebooks containing the records of William's and her own work at the Tulse Hill observatory.[134] It is quite remarkable that she sent these notebooks so far away, rather than include them in the items given to Cambridge for the use of students and researchers at the new Huggins observatory, or donate them to the Fitzwilliam Museum for historic preservation, or present them to the RAS or RS library where interested Fellows could access and study them.

In her last letter to the College, Margaret explained that she decided on Wellesley because of her 'deep interest in America and her great Mission' and her own fascination with 'the young women of America, and the intellectual justice now being granted to them'.[135] Such high-minded words mask the complex array of factors she undoubtedly weighed in making her decision. She certainly wanted to motivate young women to pursue the study of astronomy by giving them the opportunity to see firsthand the extent and importance of her own contributions to the work at Tulse Hill.[136] Perhaps, in light of the *Science Progress* episode, she feared that her husband's English colleagues would not – or could not – fully appreciate her husband's contributions. Perhaps she worried some would search the notebook record for errors or shortcomings in William's work. She could not burn the precious notebooks, as some widows did their deceased beloveds' papers, so she may have done the next best thing: entrusted them to friendly hands in the New World where they could remain safely out of his detractors' reach.

***

In January 1914, with her health in decline, Margaret sent William Wallace Campbell, then Director of the Lick Observatory, a reproduction of the portrait John Collier had painted of her husband. She regretted she had not written more frequently over the years. She wished she could write more now, but confessed that illness made the task too difficult. 'In a day or so I have to face an operation', she told Campbell, confessing she was worried despite the positive prognosis.[137] William, after all, had died unexpectedly in surgery. That same day, she sent another copy of Collier's painting to Hale for the Mt Wilson Observatory. 'I am facing Death calmly', she assured him after apprising him of her upcoming operation. 'Please give my love to your wife & to Margaret, and believe always that I have loved you all. In the event of my death, you will receive back the letters you lent me. I am calm.'

In February 1915, James Dewar wrote to American astronomer and instrument maker John Alfred Brashear (1840–1920) to express his great concern over the 'condition of our old friend Lady Huggins. We used to see her here [at the Royal Institution] two or three times a week, being an old member and intimate friend. Alas, she has been struggling for some time with that terrible disease cancer and is now given up by the doctors as beyond all hope. I doubt very much whether the Life of Sir William, on which she was engaged, can be at all far advanced. Her earnest hope was to live long enough to finish this Life.'[138]

Margaret died a month later on 24 March 1915.[139] Knowing she could not complete her husband's biography, she passed her notes and the documents she had gathered on to friends John Montefiore and his sister, Julia. *A Sketch of the Life of Sir William Huggins* was finally published in 1936 by Julia Montefiore's executor, Charles Mills, and his collaborator, C. F. Brooke.[140]

On 29 March 1917, two years after Margaret's death, a medallion honouring her husband's memory and her own was unveiled in the crypt of London's St Paul's Cathedral. Royal Society President, Joseph John Thomson (1856–1940), delivered the dedicatory address. Those in attendance included all the principal officers of the Royal Society, the President of the RAS, Dyson (now Astronomer Royal), Larmor, Newall, Maunder, Wesley and many others. Margaret had originally planned the memorial in William's honour. Thanks to Julia Montefiore's efforts, the funds Margaret had set aside for that purpose were used to design a medallion dedicated to both 'Sir William Huggins, astronomer 1824–1910' and 'Margaret Lindsay Huggins 1848–1915, his wife and fellow-worker'.[141]

Notes

1. M. L. Huggins to Larmor, 17 October 1910, Lm.790, JL, RSL.
2. 'Meeting of the Royal Astronomical Society. Friday, April 8, 1892', *OBS* 15 (1892), pp. 200–16; pp. 215–16. The three candidates were Elizabeth Brown (1830–99), Alice Everett (1865–1949) and Annie Russell (1868–1947). See Brück, *Women in Early British and Irish Astronomy* (2009), chs 10 and 13.
3. 'Meeting of the Royal Astronomical Society. Friday, 1903 May 8', *OBS* 26 (1903), pp. 229–45; pp. 229–30.
4. For discussions of the first two of these women and their astronomical contributions, see M. T. Brück, *Women in Early British and Irish Astronomy* (2009). Caroline Herschel is the subject of ch. 3, pp. 25–44, and Mary Somerville is the subject of ch. 6, pp. 67–89. Anne Sheepshanks, who was not a practising astronomer herself,

but rather a generous patron of the science, is memorialised in Dunkin, *Obituary Notices of Astronomers* (1879), pp. 237–40.

5. See Turner, 'Mrs. Fleming' (1912); 'Meeting of the Royal Astronomical Society. Friday, 1914 March 13', *OBS* **37** (1914), p. 148; and Merrill, 'Annie Jump Cannon' (1942). Fleming's election would have delighted Margaret who had 'followed her work with interest & admiration' for many years. See M. L. Huggins to S. F. Whiting, 22 September 1899, 'Souvenir Whitin Observatory: Memory Book compiled by Sarah Frances Whiting; including photographs and correspondence (1882–1913)', WCL/SC.

6. The vote was 59–3 in favour of the motion. 'Annual general meeting of the Royal Astronomical Society. Friday, 1915 February 12', *OBS* **38** (1915), pp. 119–23; pp. 120–22.

7. *MNRAS* **76** (1915), pp. 1–2. The candidates nominated were Mary Adela Blagg (1858–1944), Ella K. Church, Alice Grace Cook (d. 1958), Irene Elizabeth Toy Warner, and Mrs. Fiametta Wilson (1864–1920).

8. *MNRAS* **76** (1916), p. 195.

9. 'Meeting of the Royal Astronomical Society. Friday, 1903 May 8', *OBS* **26** (1903), pp. 229–45; p. 239.

10. Turner was elected to Fellowship on 9 January 1885. See 'Herbert Hall Turner', *MNRAS* **91** (1931), pp. 321–34.

11. 'Meeting of the Royal Astronomical Society. Friday, 1903 May 8', *OBS* **26** (1903), pp. 229–45; p. 239.

12. M. L. Huggins to Hale, 4 October 1905, Box 22 [Reel 19], GEH, CIT.

13. The Carnegie Institution was established in January 1902 with a $10 million endowment from the Scottish-born American industrialist, Andrew Carnegie (1835–1919).

14. Hale to W. Huggins, 28 January 1903, Box 22 [Reel 19], GEH, CIT.

15. On 14 March 1903, Hale thanked Huggins for his suggestions.

16. W. Huggins to Hale, 17 February 1903, in L. Boss, W. W. Campbell and G. E. Hale, *Report of Committee on Southern and Solar Observatories* (1903), pp. 146–7.

17. The glass disc had been purchased in 1894 by Hale's father. Hale's plans for the 60-in mirror and Huggins's support for them are discussed in Wright, *Explorer of the Universe* (1966), pp. 160–2. Huggins had already lent his weighty support to one of Hale's earlier unsuccessful attempts to secure funding for the project back in December 1901. See W. Huggins to Hale, 13 December 1901, Box 22 [Reel 19], GEH, CIT. The 60-in telescope was not completed until December 1908.

18. Hale to W. Huggins, 26 August 1903, Box 22 [Reel 19], GEH, CIT.

19. Hale, 'A study of the conditions for solar research at Mount Wilson, California' (1905), p. 125.

20. Hale to W. Huggins, 26 August 1903, Box 22 [Reel 19], GEH, CIT.

21. Hale and Ellerman, 'The Rumford spectroheliograph of the Yerkes Observatory' (1903).

22. Hale and Ellerman, 'Calcium and hydrogen flocculi' (1904).

23. Hale to W. Huggins, 10 August 1903, Box 22 [Reel 19], GEH, CIT.

24. Hale to W. Huggins, 26 August 1903, Box 22 [Reel 19], GEH, CIT.

25. Hale to W. Huggins, 16 December 1903, Box 22 [Reel 19], GEH, CIT.

26. Hale to W. Huggins, 21 May 1904, Box 22 [Reel 19], GEH, CIT.

27. For practical and financial reasons, the Snow telescope was erected to follow the slope of the terrain at Mt Wilson.

28. Hale, 'The solar observatory of the Carnegie Institution of Washington' (1905); 'Some tests of the Snow telescope' (1906).

29. W. Huggins to Hale, 17 June and 29 July 1904, Box 22 [Reel 19], GEH, CIT.

30. Hale and Perrine, 'Proceedings of the first Conference on Solar Research' (1904).

31. The next two meetings were held at Oxford in late September 1905 and at Meudon in May 1907. See 'International Union for Co-operation in Solar Research: Meeting at Oxford, September 27–29, 1905', *APJ* **22** (1905), pp. 276–80; and 'The meeting of the International Solar Union', *OBS* **30** (1907), pp. 243–5.

32. W. Huggins to Hale, 8 November 1907, Box 22 [Reel 19], GEH, CIT.

33. Hale to W. Huggins, 16 January 1908, Box 22 [Reel 19], GEH, CIT.

34. W. Huggins to Hale, 26 March 1908, Box 22 [Reel 19], GEH, CIT.

35. Hale to W. Huggins, 9 May 1908, Box 22 [Reel 19], GEH, CIT.

36. Hale to J. N. Lockyer, 20 May 1908, Box 26 [Reel 23], GEH, CIT.

37. J. N. Lockyer to Hale, 4 June 1908, Box 26 [Reel 23], GEH, CIT.

38. Connolly, 'Recent work with the spectroheliograph' (1908).

39. Hale to J. N. Lockyer, 22 June 1908, Box 26 [Reel 23], GEH, CIT.

40. Connolly, 'Solar vortices' (1908).

41. Hale to W. Huggins, 7 October 1908, Box 22 [Reel 19], GEH, CIT.

42. W. Huggins to Hale, 26 October 1908, Box 22 [Reel 19], GEH, CIT.

43. W. Huggins to Larmor, 5 January 1908, Lm.974, JL, RSL.

44. W. Huggins to Larmor, 22 May 1908, Lm.973, JL, RSL.

45. W. Huggins to Larmor, 7 June 1908, Lm.975, JL, RSL.

46. W. Huggins to the Secretary of the Royal Society, 17 October 1908, CD.145, RSL.

47. W. Huggins to Larmor, 2 November 1908, Lm.989, JL, RSL.

48. W. Huggins to Larmor, 8 November 1908, Lm.990, JL, RSL.

49. Mills and Brook, *Sketch of the Life* (1936), pp. 59–60.

50. *Ibid.*, p. 61.

51. Cannon, 'Sarah Frances Whiting' (1927), pp. 543–4.

52. M. L. Huggins to S. F. Whiting, 21 January 1910, 3L. Correspondence . . . regarding Huggins gifts to Wellesley College, WCL/SC.

53. W. Huggins to Hale, 10 February 1909, Box 22 [Reel 19], GEH, CIT.

54. Hale to W. Huggins, 24 February 1909, Box 22 [Reel 19], GEH, CIT.

55. M. L. Huggins to S. F. Whiting, 21 January 1910, 3L. Correspondence . . . regarding Huggins gifts to Wellesley College, WCL/SC.

56. W. Huggins to Larmor, 1 February 1909, Lm.1002, JL, RSL.

57. M. L. Huggins to S. F. Whiting, 21 January 1910, 3L. Correspondence . . . regarding Huggins gifts to Wellesley College, WCL/SC.

58. W. Huggins to Larmor, 15 April 1909, Lm.1009, JL, RSL.

59. Clerke, 'Astronomy, 1846–1885' (1904). The photograph appeared on p. 691.

60. Hale to W. Huggins, 21 December 1908, Box 22 [Reel 19], GEH, CIT.

61. Hale to W. Huggins, 24 February 1909, Box 22 [Reel 19], GEH, CIT.

62. W. Huggins to Hale, 4 May 1909, Box 22 [Reel 19], GEH, CIT. This letter is, if not the last Huggins wrote to Hale, then the last that is preserved in the Hale papers.

63. W. Huggins to Larmor, 28 February 1909, Lm.1006, JL, RSL. After his official admission to the RS, Hale wrote to his wife, 'When I was proposed by Larmor as a Foreign Associate of the R.S., there was some opposition in the Council (Larmor says only on account of my age), and some of the members urged that a well-known biological be chosen instead. Dyson told me that at this point Gill sailed in, and promptly annihilated the objector! Turner thinks the election an unusual honor as the last American chosen 2 years ago (Pickering) was also an astronomer, and it would be customary to take a man in some other subject. He also thought they would not elect another American in any subject so soon.' Hale to Evelina Hale, 4 May 1909, Box 80 [Reel 68], GEH, CIT.

64. Hale, 'Solar vortices and magnetic fields' (1909).

65. W. Huggins to Larmor, 25 October 1909, Lm.1016, JL, RSL.

66. Cortie, 'The foundations of astrophysics' (1909).

67. 'The new astronomy', *The Quarterly Review* **212** (1910), pp. 439–55.

68. Frost, 'The scientific papers of Sir William Huggins' (1910).

69. 'Sir William Huggins', *The Times* of London, 13 May 1910, p. 11c; T. J. J. See, 'Tribute to the memory of Sir William Huggins' (1910).

70. Originally published in the February number of *Grand Magazine*, and reviewed in 'How to live long – By some of those who have done it', *The American Monthly Review of Reviews* **31** (1905), pp. 366–7; p. 366.

71. 'What to eat, drink, and avoid. The experience of experts in the art of living', *The Review of Reviews* **37** (1908), pp. 136–46; p. 138. Some of the others surveyed included William Crookes (aged 76), Alfred Russel Wallace (aged 85), and the relative youngster, George Bernard Shaw (aged 52).

72. 'Sir W. Huggins's 60 years watching the stars', *Grey River Argus*, 28 April 1910, p. 1.

73. Cited in L. Whiting, *Life Transfigured* (1910), pp. 110–11. This quote, as well as the previous one appear to have been excerpted from the same source: a longer published interview Huggins gave on or about 7 February 1910. I have been unable to identify the original source.

74. T. J. J. See, 'Need of the collected works of Sir Wm. Herschel' (1909).

75. Holden and Hastings, *A Synopsis of the Scientific Writings of Sir William Herschel* (1881).

76. 'The Collected Scientific Works of Sir William Herschel', *APJ* **35** (1912), pp. 296–7.

77. Some obituaries erroneously reported that Huggins died at a nursing home, but, in fact, he died following a minor surgical procedure at the home of Dr Dudley Bishop in Clapham Common. See, for example, one of the most authoritative notices: Dyson, 'Sir William Huggins' (1910), p. xix.

78. 'Funerals. Sir William Huggins, O. M.', *The Times* of London, 16 May 1910, p. 11b. A Mrs Green is also listed as a chief mourner. Margaret had a half-sister (Jane Evelyn) in addition to her two half-brothers (John and George). These three were born of her widowed father's marriage (c. 1860) to Elizabeth Pott (1827?–1922).

Thus far, research has yet to uncover biographical information on George or Jane Evelyn, thus it is not possible to identify the latter as Mrs Green. Margaret appears not to have maintained close ties with her family. As she later told Joseph Larmor (2 January 1913, Lm.823, JL, RSL), 'my relatives are few & hardly count'.

79. M. L. Huggins to Larmor, 18 January 1911, Lm.807, JL, RSL.

80. The phrase is from Psalm 36:9. M. L. Huggins to Larmor, 17 October 1910, Lm.790, JL, RSL.

81. M. L. Huggins to Larmor, 26 June 1910, Lm.786, JL, RSL.

82. M. L. Huggins to Larmor, 17 October 1910, Lm.790, JL, RSL.

83. Since May 1990, when I first examined and transcribed Margaret Huggins's letters to Joseph Larmor (RSL), excerpts from this important correspondence have been published independently by McKenna-Lawlor in *Whatever Shines Should be Observed* (1998), pp. 97–123. However, she included no dates or other locating information for the excerpts she cited. I hope readers may find that lack remedied here.

84. M. L. Huggins to Larmor, 17 October 1910, Lm.790, JL, RSL.

85. M. L. Huggins to Larmor, 16 October 1910, Lm.789, JL, RSL.

86. M. L. Huggins to Larmor, n.d. Lm.792, JL, RSL.

87. M. L. Huggins to Larmor, 12 January 1911, Lm.808, JL, RSL.

88. M. L. Huggins to Larmor, 17 August 1911, Lm.814, JL, RSL.

89. M. L. Huggins to Larmor, 29 October 1910, Lm.795, JL, RSL.

90. Newall, 'Sir William Huggins' (1910–11).

91. M. L. Huggins to Larmor, 3 November 1910, Lm.796, JL, RSL.

92. Newall, 'Sir William Huggins' (1910–11), p. 187.

93. *Ibid.*, p. 189.

94. M. L. Huggins to Larmor, 3 November 1910, Lm.796, JL, RSL.

95. Wright, *Explorer of the Universe* (1966), p. 254–7.

96. *Ibid.*, p. 258.

97. Wright states 'second 100-in disk broken in annealing oven, May 10, 1910', *Explorer of the Universe* (1966), note 28, p. 451. Unfortunately, she does not provide a source for that information. Factual errors elsewhere in the book urge caution in accepting such statements without confirmation. Lucien Jouvaud, the New York agent who handled Mt Wilson's orders for the Saint Gobain glassworks, notified George Ritchey on 12 May that he 'had no further news as yet regarding the large disk' but that 'my people will keep me posted and I will let you know whenever I have any advice.' L. Jouvaud to G. W. Ritchey, 12 May 1910, Box 155 [Reel 87], GEH, CIT. On 20 May, he telegraphed Ritchey that 'Breaking of last disk in annealing was accidental. New disk being cast and all possible will be done to make it a success this time.' Jouvaud to Ritchey, 20 May 1910, Box 155 [Reel 87], GEH, CIT. From an independent source we learn that Ritchey expressed 'sorrow' that the disc had broken during the annealing process in a letter (16 May 1910) to Lucien Delloye (1856–1938), directeur générale du service des Glaceries at the Saint Gobain glassworks. Ritchey's letter is described in 'Note on the Manufacture of the Large 100-inch disk of the Mount Wilson Observatory', Service Technique des Glaceries, 12 October 1934, translated by Jérôme Remy. I should like to thank Robin Mason (Mason Productions, Inc.) for making 'Note on the Manufacture ...' available to me.

98. Other Britons attending the meeting included Fowler, from South Kensington, Schuster from Manchester, Turner from Oxford, A. L. Cortie from Stonyhurst College, and F. W. Dyson, Astronomer Royal of Scotland. See 'The International Union for Co-operation in Solar Research', *APJ* **32** (1910), pp. 258–63; pp. 261–3.

99. Wright, *Explorer of the Universe* (1966), pp. 260–2. The Hales hosted a garden party for conference attendees in Pasadena on 29 August. Hale greeted the assembled delegates on Mt Wilson on Wednesday, 31 August, and described the observatory's Tower Telescope. He and his wife hosted a dinner in Pasadena at the close of the meeting on Saturday, 3 September. See Chant, 'The Mount Wilson Conference of the Solar Union' (1910).

100. M. L. Huggins to Hale, 4 November 1910, Box 22 [Reel 19], GEH, CIT. Hale noted this gift in his personal diary: 'Lady Huggins gave me Sir William's ring.' 'Friday, Nov. 4, 1910', Box 93 [Reel 76], GEH, CIT.

101. 'I am *very* anxious about George Hale. I do not like his look at all.' M. L. Huggins to Larmor, 7 November 1910, Lm.797, JL, RSL.

102. *Ibid.*

103. Newall, *The Spectroscope and its Work* (1910).

104. Larmor, *Aether and Matter* (1900), pp. 14–15.

105. M. L. Huggins to Larmor, 9 November 1910, Lm.798, JL, RSL.

106. M. L. Huggins to Larmor, 11 November 1910, Lm.799, JL, RSL.

107. M. L. Huggins to Larmor, 15 November 1910, Lm.800, JL, RSL.

108. M. L. Huggins to Larmor, 16 November 1910, Lm.801, JL, RSL.

109. M. L. Huggins to Larmor, 12 January 1911, Lm.808, JL, RSL.

110. Newall, 'William Huggins' (1911).

111. *Ibid.*, p. 266.

112. M. L. Huggins to Larmor, 12 January 1911, Lm.808, JL, RSL.

113. M. L. Huggins to Larmor, 9 November 1910, Lm.796, JL, RSL.

114. *Ibid.*

115. Dyson was appointed to succeed Christie as Astronomer Royal upon the latter's retirement from that post on 1 October 1910.

116. M. L. Huggins to Larmor, 11 October 1911, Lm.819, JL, RSL.

117. Margaret's comments about Maunder are quite interesting. Maunder was nominated for Fellowship in the Royal Society several years in a row, but William strongly and successfully opposed his election each time. In a letter marked '*Strictly confidential*', he told Larmor in February 1909: 'The astronomical candidates are all weak. I am astonished that Darwin & Gill should use the argument: – "that he [Maunder] has been left so often"! Some have yet to learn that persistency in candidature is to be taken as a makeweight for a weak record!! After the third time, each passing over by the Council should be reckoned against him. Surely the passing over of Maunder by some 5 or 6 new Councils in succession should be taken that he has not been considered a strong enough candidate.' W. Huggins to Larmor, 17 February 1909, Lm.1005, JL, RSL. Assuming that Margaret's positive view of Maunder reflects her husband's as well, William clearly separated technical expertise from scientific ability.

118. Huggins was knighted in 1897 (not 1887), and he died at the operating physician's residence (not at a nursing home).

119. M. L. Huggins to Larmor, 15 August 1911, Lm.813, JL, RSL.

120. The Fitzwilliam Museum, founded in 1816, is Cambridge University's art and antiquities museum.

121. M. L. Huggins to Larmor, 17 August 1911, Lm.814, JL, RSL.

122. The plaque (HIN444) is now in the Library Rare Book Room of the Institute of Astronomy at the University of Cambridge. Photographs of the plaque are displayed on their website: www.dspace.cam.ac.uk/handle/1810/224260. An illustration of the plaque can be found in M. Stratton, 'The history of the Cambridge observatories' (1949), Plate V(b).

123. Hurn, *Sir William Huggins* (2008).

124. D. W. Dewhirst to the author, 14 November 1991. Twenty years later, Dewhirst, now eighty-four, relates: 'I was fortunate to have worked with Huggins's telescope. If two astronomers work with the same telescope and bang their heads in the dark on awkward corners in the mounting, they develop a rather unusual bond even if they have never met . . .'. Dewhirst to the author, 22 February 2010.

125. M. L. Huggins to Larmor, 28 December 1912, Lm.822, JL, RSL.

126. M. L. Huggins to S. F. Whiting, 22 September 1899, Souvenir Whitin Observatory: Memory Book compiled by S. F. Whiting; including photographs and correspondence (1882–1913), WCL/SC.

127. Whiting wrote a number of articles for *Popular Astronomy* on ideas for improving methods of teaching astronomy at the college level. See, for example, S. F. Whiting, 'Spectroscopic work for classes in astronomy' (1905); 'A pedagogical suggestion for teachers of astronomy' (1912).

128. M. L. Huggins to S. F. Whiting, 15 July 1913, Correspondence: S. F. Whiting and Lady Margaret Huggins regarding Huggins gifts to Wellesley College (undated, 1910–1914), WCL/SC. For a description of many of the items Margaret gave to Wellesley, see S. F. Whiting, 'Priceless accessions' (1914).

129. M. L. Huggins to S. F. Whiting, 9 January 1914, Correspondence: S. F. Whiting and Lady Margaret Huggins regarding Huggins gifts to Wellesley College (undated, 1910–1914), WCL/SC.

130. S. F. Whiting, 'Priceless accessions' (1914); 'An international gift' (1915).

131. M. L. Huggins to S. F. Whiting, 28 November 1913, Correspondence: S. F. Whiting and Lady Margaret Huggins regarding Huggins gifts to Wellesley College (undated, 1910–1914), WCL/SC.

132. M. L. Huggins to S. F. Whiting, n.d., Correspondence: S. F. Whiting and Lady Margaret Huggins regarding Huggins gifts to Wellesley College (undated, 1910–1914), WCL/SC.

133. M. L. Huggins, 'The astrolabe' (1894).

134. S. F. Whiting, 'Diaries' (1917); and Morgan, 'Huggins archives' (1980).

135. S. F. Whiting, 'An international gift', p. 699.

136. In her obituary of Margaret Huggins, renowned astronomer Annie Jump Cannon, herself a Wellesley graduate and former student of Sarah Whiting, wrote 'Her last letters were filled with the joy and pleasure of sending the gifts to the Wellesley College Observatory. "Happy it makes me to think these things will help on and inspire young life in the Astronomy of the future. Women in the Astronomy of the future must play a great part, I am sure."' Cannon, 'Lady Huggins' (1915).

137. M. L. Huggins to W. W. Campbell, 17 January 1914, MLSA, LO.

138. Dewar to J. A. Brashear, 20 February 1915, WCL/SC.
139. 'Lady Huggins: Her work in astronomy', *The Times* of London, 25 March 1915, p. 10
140. Mills and Brooke, *A Sketch of the Life* (1936).
141. 'Memorial to Sir William and Lady Huggins', *Nature* **99** (1917), pp. 153–4; 'General notes', *PA* **25** (1917), pp. 416–17; M. T. Brück and I. Elliott, 'The family background of Lady Huggins (Margaret Lindsay Huggins)' (1992), p. 211. A photograph of the medallion can be seen in M. T. Brück, *Women in Early British and Irish Astronomy* (2009), p. 180.

16

Conclusion

May you enter and make it possible for others to enter many a Promised Land.

– Margaret Lindsay Huggins[1]

In 1897 *The Nineteenth Century*, a popular magazine, published an essay by William Huggins on 'The new astronomy' and his role in its development. In it, Huggins imposed an artificial order and rationale upon his programmatic decisions by enumerating a neat sequence of pioneering projects that began with his first efforts in stellar spectroscopy and ended with his design of a spectroscopic method to determine stellar motion in the line of sight. He fleshed out the story with vivid descriptions of the career risks he took, the instrumental and methodological challenges he faced, as well as the rewards he gained throughout his long and illustrious career in consequence of his decision to devote his observing programme to the spectroscopic study of celestial bodies. The captivating eye-witness account carried readers behind the scenes of scientific discovery in a very personal and dramatic way.

Huggins and his wife, Margaret, later reprinted extended excerpts from it in both their *Atlas* (1899) and *Scientific Papers* (1909) thus making its passages readily available to interested scientists and laymen alike. Indeed, after his death in 1910, 'The new astronomy' was the reference of choice for obituarists and biographers who lacked the time and energy required to read Huggins's many published papers, let alone the access necessary to examine his unpublished correspondence and notebook records. Margaret's vigilant watch over the content of everything written about her husband's work may well have pushed authors to adhere even more strictly to Huggins's own published words.

Even today, it is tempting to place this essay in a different category from Huggins's other published work: to see it as truer for its candour, as more accurate for its detail, as closer to the way things actually happened than any of his formal scientific papers. But it is not a deposition. It is a synthetic account composed of specially selected events recalled some thirty-five years after the fact. Huggins was not a disinterested observer of those events. He was an active participant in them with much to gain from characterising his own career development as symbiotic with the wider growth of interest in subjecting the light of celestial bodies to prismatic analysis.

Reading 'The new astronomy' is like tracing the path of the string out of a labyrinth once the puzzle has been solved. The multitude of choices Huggins faced at every crossroads no longer clutter the way. The investigative dead ends he reached have fallen victim to the narrator's desire for brevity and clarity. Huggins omitted mention of projects – his attempts to measure the heat of stars, his search for evidence of change in a lunar crater's appearance, his innovative efforts to photograph the solar corona, to name a few – that, in his view, had either failed or did not fit his narrative of discovery. He avoided any reference to the controversies and priority disputes that caused him so much anxiety over the years. He left out reminders of his colleagues' objections to his pursuit of personal rather than public research interests with the telescope that the Royal Society had placed in his custody. The absence of these episodes from 'The new astronomy' is hardly surprising. But their absence from others' subsequent renditions of the story demonstrates the control the Hugginses gained over what eventually became the authoritative account of the origins of astronomical physics.

My aim in this work has been to restore some of the lost complexity and noisy confusion that energised the bustling and diverse scientific community of which Huggins was a part and made the formation of a new species of scientific investigation possible. To that end, I have looked well beyond Huggins's version of events to locate less-filtered accounts available in contemporaneous records.

The unpublished record shows that his rise to prominence as a serious amateur astronomer was the result of a conscious career strategy pursued on a number of fronts by a bright and ambitious man. He may have been a risk-taker, but in keeping with his entrepreneurial background his risks were calculated to maximise success. He could not tame his entrepreneurial research style to fit the selfless and disinterested stereotype of a man of science in the late nineteenth century, but he could and did construct a more conforming public account of himself and his work. Contrary to the impression given in his reminiscent account, Huggins's research programme did not so much undergo a radical transformation as a gradual and fitful shift that was contemporaneous with the growth of interest in subjecting the light emitted by a variety of celestial bodies to prismatic analysis by an admittedly eclectic group of independent and self-directed observers in the international astronomical community.

The Tulse Hill observatory notebooks open a unique, though imperfect, window on the Hugginses' research efforts. Many notebook entries were recorded after-the-fact rather than in real time. And, perhaps inevitably for such a lifelong effort, the log they purport to keep is incomplete. Huggins illustrated his early telescopic observations of comets and planetary surfaces, for example, but he left no record of his first look at stellar or nebular spectra. He described his pioneering efforts at measuring the heat of stars, but made no notes on his first attempts to photograph their spectra. Neither William nor Margaret made any notebook entries from June 1882 to December 1886.

Despite these lapses, the notebooks are of inestimable value in putting meat on the bones, so to speak, of Huggins's published accounts of his research. Indeed, comparing his notebook description of an observation against the published account of that same observation exposes his acumen in presenting his work to the community of researchers of which he considered himself a part. The notebooks bring to light his considerable skill at manipulating balky or temperamental instruments, the wide range of his observing interests, and his ability to transform tentative observational notes into confident published reports.

They also reveal for the first time the breadth and depth of Margaret's important contributions to the work of the Tulse Hill observatory. In published accounts, she successfully cloaked herself in the invisible garb of the proper Victorian lady, taking care that her collaborative assistance did not contradict or interfere with the image she had helped to create of her husband as the innovator and principal observer in the team. In contrast to that image, her entries in the notebooks make it clear that her presence and expertise not only strengthened, but also shaped the research agenda at Tulse Hill. From them we learn that she figured prominently in introducing photography into their methodological toolkit, in designing and modifying instruments and procedures, in selecting subjects for examination, and in communicating the results of their joint efforts in scientific journals.

The notebook record is supplemented by Huggins's extensive personal correspondence. It is in his letters, not in his published papers, that Huggins aired his methodological concerns, his reactions to controversy and his anxieties about the accuracy of his measurements. And it is through his correspondence that we can identify the individuals whose advice and counsel he valued at various stages of his long career.

The papers of George Gabriel Stokes, for example, are a treasure trove of correspondence from all manner of science professionals and serious amateurs. To all who wrote to him, Stokes was a mentor, interlocutor, sounding board and much more. It was a trial to read the nearly illegible scrawl of his replies, but all, including Huggins, agreed that was a small price to pay for the quiet satisfaction of having merited Stokes's attention. Huggins relied on Stokes's expertise when perplexed by questions regarding experimental method or physical theory. He routinely sought Stokes's counsel and occasional intervention when he became embroiled in controversy. He called on Stokes for editorial guidance when crafting major addresses such as his Bakerian Lecture in 1885 and his presidential address to the BAAS in 1891.

Huggins's lengthy and often intense correspondence with David Gill on photographing the solar corona without an eclipse has lain buried in the archives for over a century. It contains tutorial epistles from an anxious yet hopeful Huggins which are bound tightly with the blurry book press ghosts of Gill's vain pleas for Huggins to come to the Cape and oversee the delicate process himself. These often poignant letters expose the motivations and frustrations that drive the work of cutting-edge science.

The twenty-year correspondence Huggins shared with American astronomer George Ellery Hale reveals – as no published paper can – the intimate dynamic of their close personal relationship. Huggins was charmed by the young man's magnetic mix of wit and ambition. He was also fascinated by the powerful instruments, institutions and publications that Hale was able to assemble and use to advance his research agenda and the cause of astrophysics. The mutual trust and admiration the two built up over the years enabled Huggins to engage in what was for him a rarity, namely candid exchanges of ideas and critical remarks with a colleague. Their correspondence helps us understand why, when Huggins sensed he was losing his place on the cutting edge of research in the new astronomy, he anointed Hale as his successor.

Margaret Huggins's handwriting appears often in her husband's correspondence. The thoughts, suggestions and criticisms she added to his letters to Hale and Gill, for example, reinforce the testimony of the notebook record regarding the integral role she played in many of the investigative efforts at Tulse Hill. Her heartfelt correspondence with Joseph Larmor after her husband's death reveals her important role in developing and preserving her husband's historical legacy.

Once evidence from Huggins's notebooks and correspondence is mapped onto the chronological grid of his published papers, the private Huggins that emerges turns out to be both more interesting and certainly more complex than the cautious, focused and methodical individual portrayed in the public record. We discover that, by circumstance or design, he was introduced to the method of prismatic analysis at a point in his life when he had not yet found an investigative focus to channel his astronomical interests. The recently retired entrepreneur thrived in the methodologically and programmatically diverse community of late-nineteenth-century amateur astronomers. His curiosity and ambition intersected in dynamic and generally positive ways with his nagging personal insecurities to fuel his successful efforts to become one of the elite in scientific London.

Free to select the objects he wished to observe, Huggins pursued an eclectic and opportunistic observing programme. Rather than exhaust a single line of investigation, he routinely looked for novel ways to make his mark. He tinkered with new gadgets, embraced new methods, then rummaged through old intractable problems in search of new puzzles to solve.

He sought advice on observing technique and problem design from those more experienced than he. He learned the ropes from skilled mentors like the Revd William Rutter Dawes. He adapted select elements of others' methods and research agendas to suit his own specialised purposes. He encouraged and nurtured a cooperative investigative venture with his neighbour, chemist William Allen Miller, a high-ranking officer in the Royal Society. His partnership with Miller offered him an entrée to that elite scientific circle. Once inside, he developed strong collegial ties with Stokes and other prominent fellows like Thomas Romney Robinson and James Clerk Maxwell to advance his own research plans.

Perhaps Huggins's greatest asset was his inclination, born of his entrepreneurial background, to treat each and every innovation and discovery as a commodity to be packaged and sold like an exotic bolt of cloth. He deftly advertised his 'wares', taking care to avoid becoming identified as a charlatan or mountebank. He treated his colleagues in the scientific community like discriminating and sophisticated clientele, making every effort to instill in each individual a desire – even a need – to 'buy' and use his new-fangled methods. He executed the moves of this delicate dance with surefooted ease.

In the case of measures of stellar motion in the line of sight, for example, Huggins worked hard to persuade his colleagues of the method's theoretical authenticity, its practical utility and its reproducibility. He bolstered his plan's theoretical underpinnings with references to the pioneering contributions of Continental researchers. He invoked the name of James Clerk Maxwell to endorse its practicality. To demonstrate that he possessed the tools and the expertise to accomplish what he claimed, he took care to emphasise his reluctance to speculate, his caution in conducting his observations and his concern for accuracy.

Although Huggins was not always successful in achieving the aims of his research projects, he had a knack for diminishing the negative impact of his failures. He worked hard to control the public account of his work at every step. He personally corrected or retracted statements he later deemed to be erroneous rather than leave them for others to discover. He acted quickly to point out and rectify others' mistaken interpretations of his work. Indeed, he made certain that whenever others discussed his work they did so in conformity with his own account of it.

Like any savvy businessman, he remained alert to threats from competitors. As others were attracted to the new field of astronomical spectroscopy, he took steps to preserve his place in the forefront of this rapidly evolving research niche. He vigorously protected his claims for credit and priority for discoveries. He established strong alliances with influential astronomers, particularly American astronomers like Edward Holden and George Ellery Hale who had access to the world's newest and largest telescopes. He trusted them and relied upon them to accept, verify and support his observational and interpretational claims.

In the second half of the nineteenth century, chemists, physicists and astronomers alike came to appreciate the growing intimacy and necessary interdependency of their areas of expertise. The opportunity to observe spectra produced by celestial bodies changed the practice of qualitiative astronomers. No longer limited to the descriptive research methods of natural historians and curiosity collectors, astronomers who had sought to acquire certain knowledge of the chemical constituents and physical structure of celestial bodies now found the analytical and mensurational tools of laboratory science available to them. Interpretations of celestial spectra fuelled a fertile dialectic between laboratory and field – that is, the terrestrial and celestial realms – and nourished the very human desire to find intelligible patterns in them. But without the ability to modify artificially the conditions under which celestial bodies produce their light, observers like Huggins could only extrapolate from the laboratory to the field and back again by working fast and loose within an as yet incomplete and uncertain system of physical theory. It was a risky business.

The symbiosis of theory and practice allowed a wide range of interpretive schemes to exist simultaneously and to interact fruitfully. The lack of clear explanatory guidelines in this fledgling discipline proved to be a help rather than a hindrance. It fostered productive interdependence of laboratory and field observations that provoked and sustained a level of discussion, comparison, criticism and controversy necessary to develop confidence in the interpretive power of spectrum analysis as applied to the light of celestial bodies.

By the dawn of the twentieth century, astronomers found the questions they could ask about the bodies they observed and the methods deemed appropriate to examine them had changed in ways their predecessors could not have imagined. The expertise required to tackle those questions became increasingly specialised. Huggins and amateurs like him found their places in the vanguard of the new astronomy taken over by young recruits with university training in astronomy, mathematics and physics. Contemporaneous developments in atomic theory enhanced the explanatory power of all spectral signatures. A generation later, individuals choosing careers in astronomy could look forward to learning the nature of novae; the substance of the corona; the structure of the Milky Way; the dynamics of stellar motions; and even the source of stellar luminosity itself.

William Huggins participated with vigour and vision in the events that shaped the new scientific specialty. Coupling the spectroscope to the telescope suggested new questions which, in turn, generated new mensurational tasks and ultimately altered the familiar boundaries of acceptable research within the astronomical community. The choices that Huggins made as he moved from the periphery of scientific London toward its inner circle expose the dynamic and often uncertain process by which the boundaries of acceptable research were redefined in astronomy during his lifetime. One individual's incremental career choices do not determine the direction of a developing research agenda. Nevertheless, as concrete instances of personal effort to establish a foothold in the community at large, they make visible otherwise tacit acts of negotiation and manoeuvring strategies. When a novice like Huggins succeeds despite his lack of access to the proper channels, the historian can hope to find in the unpublished record the telltale signs of the tunnel that was dug to undermine the walls.

Note

1. M. L. Huggins to Hale, 4 October 1905, Box 22 [Reel 19], GEH, CIT.

Appendix

The new astronomy: a personal retrospect[1]

[907] While progress in all branches of knowledge has been rapid beyond precedent during the past sixty years, in at least two directions this knowledge has been so unexpected and novel in character that two new sciences may be said to have arisen: the new medicine, with which the names of Lister and of Pasteur will remain associated; and the new astronomy, of the birth and early growth of which I have now to speak.

The new astronomy, unlike the old astronomy to which we are indebted for skill in the navigation of the seas, the calculation of the tides, and the daily regulation of time, can lay no claim to afford us material help in the routine of daily life. Her sphere lies outside the earth. Is she less fair? Shall we pay her less court because it is to mental culture in its highest form, to our purely intellectual joys that she contributes? For surely in no part of Nature are the noblest and most profound conceptions of the human spirit more directly called forth than in the study of the heavens and the host thereof.

> That with the glorie of so goodly sight
> The hearts of men . . .
> . . . may lift themselves up hyer.[2]

May we not rather greet her in the words of Horace: 'O matre pulchra filia pulchrior'[3]?

As it fell to my lot to have some part in the early development of this new science, it has been suggested to me that the present Jubilee year of retrospect would be a suitable occasion to give some account of its history from the standpoint of my own work.

Before I begin the narrative of my personal observations, it is desirable that I should give a short statement of the circumstances which led up to the birth of the new science in 1859, and also say a few words of the state of scientific opinion about the matters of which it treats, just before that time.

It is not easy for men of the present generation, familiar with the knowledge which the new methods of research of which I am [908] about to speak have revealed to us, to put themselves back a generation, into the position of the scientific thought which existed on those subjects in the early years of the Queen's reign. At that time any knowledge of the chemical nature and of the physics of the heavenly bodies was regarded as not only impossible of attainment by any methods of direct observation, but as, indeed, lying altogether outside the limitations imposed upon man by his senses, and by the fixity of his position upon the earth.

It could never be, it was confidently thought, more than a matter of presumption, whether even the matter of the sun, and much less that of the stars, were of the same nature as that of the

328

earth, and the unceasing energy radiated from it due to such matter at a high temperature. The nebular hypothesis of Laplace at the end of the last century required, indeed, that matter similar to that of the earth should exist throughout the solar system; but then this hypothesis itself needed for its full confirmation the independent and direct observation that the solar matter was terrestrial in its nature. This theoretical probability in the case of the sun vanished almost into thin air when the attempt was made to extend it to the stellar hosts; for it might well be urged that in those immensely distant regions an original difference of the primordial stuff as well as other conditions of condensation were present, giving rise to groups of substances which have but little analogy with those of our earthly chemistry.

About the time of the Queen's accession to the throne the French philosopher Comte put very clearly in his *Cours de Philosophie Positive* the views then held, of the impossibility of direct observations of the chemical nature of the heavenly bodies. He says:

On conçoit en effet, que nous puissions conjecturer, avec quelque espoir de succès, sur la formation du système solaire dont nous faisons partie, car il nous présente de nombreux phénomènes parfaitement connus, susceptibles peut-être de porter un témoignage décisif de sa véritable origine immédiate. Mais quelle pourrait être, au contraire, la base rationnelle de nos conjectures sur la formation des soleils eux-mêmes? Comment confirmer ou infirmer à ce sujet, d'àpres [*sic*] les phénomènes, aucune hypothèse cosmogonique, lorsqu'il n'existe vraiment en ce genre aucun phénomène exploré, ni même, sans doute, EXPLORABLE?[4] [The capitals are mine.]

We could never know for certain, it seemed, whether the matter and the forces with which we are familiar are peculiar to the earth, or are common with it to the midnight sky,

> All sow'd with glistering stars more thicke than grasse,
> Whereof each other doth in brightnesse passe.[5]

For how could we extend the methods of the laboratory to bodies at distances so great that even the imagination fails to realise them?

The only communication from them which reaches us across the [909] gulf of space is the light which tells us of their existence. Fortunately this light is not so simple in its nature as it seems to be to the unaided eye. In reality it is very complex; like a cable of many strands, it is made up of light rays of many kinds. Let this light-cable pass from air obliquely through a piece of glass, and its separate strand-rays all go astray, each turning its own way, and then go on apart. Make the glass into the shape of a wedge or prism, and the rays are twice widely scattered.

> First the flaming red
> Sprung vivid forth: the tawny orange next;
> And next delicious yellow; by whose side
> Fell the kind beams of all-refreshing green.
> Then the pure blue, that swells autumnal skies,
> Ethereal played; and then, of sadder hue,
> Emerged the deepened indigo, as when
> The heavy-skirted evening droops with frost;
> While the last gleamings of refracted light
> Died in the fainting violet away.[6]

Within this unravelled starlight exists a strange cryptography. Some of the rays may be blotted out, others may be enhanced in brilliancy. These differences, countless in variety, form a code of signals, in which is conveyed to us, when once we have made out the cipher in which it is written, information of the chemical nature of the celestial gases by which the different light rays have been blotted out, or by which they have been enhanced. In the hands of the astronomer a prism has now become more potent in revealing the unknown than even was said to be 'Agrippa's magic glass.'[7]

It was the discovery of this code of signals, and of its interpretation, which made possible the rise of the new astronomy. We must glance, but very briefly, at some of the chief steps in the progress of events which slowly led up to this discovery.

Newton, in his classical work upon the solar spectrum, failed, through some strange fatality, to discover the narrow gaps wanting in light, which, as dark lines, cross the colours of the spectrum and constitute the code of symbols. His failure is often put down to his using a round hole in place of a narrow slit, through the overlapping of the images of which the dark lines failed to show themselves. Though Newton did use a round hole, he states distinctly in his *Optics* that later he adopted a narrow opening in the form of a long parallelogram – that is, a true slit – at first one-tenth of an inch in width, then only one-twentieth of an inch, and at last still narrower. These conditions under which Newton worked were such as should have shown him the dark lines upon his screen. Professor Johnson has recently repeated Newton's experiments under strictly similar conditions, with the result that the chief dark lines were well seen. For some reason Newton failed to discover them. A possible cause [910] may have been the bad annealing of his prism, though he says that it was made of good glass and free from bubbles.

The dark lines were described first by Wollaston in 1792, who strangely associated them with the boundaries of the spectral colours, and so turned contemporary thought away from the direction in which lay their true significance. It was left to Fraunhofer in 1815, by whose name the dark lines are still known, not only to map some 600 of them, but also to discover similar lines, but differently arranged in several stars. Further, he found that a pair of dark lines in the solar spectrum appeared to correspond in their position in the spectrum, and in their distance from each other, to a pair of bright lines which were nearly always present in terrestrial flames. This last observation contained the key to the interpretation of the dark lines as a code of symbols: but Fraunhofer failed to use it; and the birth of astrophysics was delayed. An observation by Forbes [Scottish physicist James David Forbes (1809–68)] at the eclipse of 1836 led thought away from the suggestive experiments of Fraunhofer; so that in the very year of the Queen's accession the knowledge of the time had to be summed up by Mrs. Somerville in the negation: 'We are still ignorant of the cause of these rayless bands.'[8]

Later on, the revelation came more or less fully to many minds. Foucault, Balfour Stewart, Ångstrom prepared the way. Prophetic guesses were made by Stokes and by Lord Kelvin. But it was Kirchhoff who, in 1859, first fully developed the true significance of the dark lines; and by his joint work with Bunsen on the solar spectrum proved beyond all question that the dark lines in the spectrum of the sun are produced by the absorption of the vapours of the same substances, which when suitably heated give out corresponding bright lines; and, further, that many of the solar absorbing vapours are those of substances found upon the earth. The new astronomy was born.

At the time that I purchased my present house, Tulse Hill was much more than now in the country and away from the smoke of London. It was after a little hesitation that I decided to give my chief attention to observational astronomy, for I was strongly under the spell of the rapid discoveries then taking place in microscopical research in connection with physiology.

In 1856 I built a convenient observatory opening by a passage from the house, and raised so as to command an uninterrupted view of the sky except on the north side. It consisted of a dome twelve feet in diameter, and a transit room. There was erected in it an equatorially mounted telescope by Dolland of five inches aperture, at that time looked upon as a large rather than a small instrument. I commenced work on the usual lines, taking transits, observing and making drawings of planets. Some of Jupiter now lying before me, I venture to think, would not compare [911] unfavourably with drawings made with the larger instruments of the present day.

About that time Mr. Alvan Clark, the founder of the American firm famous for the construction of the great object-glasses of the Lick and the Yerkes Observatories, then a portrait-painter by profession, began, as an amateur, to make object-glasses of large size for that time, and of very great merit. Specimens of his earliest work came into the hands of my friend Mr. Dawes and received the high approval of that distinguished judge. In 1858 I purchased from Mr. Dawes an object-glass by Alvan Clark of eight inches diameter, which he parted with to make room for a lens of larger diameter by a quarter of an inch, which Mr. Clark had undertaken to make for him. I paid the price it had cost Mr. Dawes – namely, 200*l.* This telescope was mounted for me equatorially and provided with a clock motion by Mr. Cooke of York.

I soon became a little dissatisfied with the routine character of ordinary astronomical work, and in a vague way sought about in my mind for the possibility of research upon the heavens in a new direction or by new methods. It was just at this time, when a vague longing after newer methods of observation for attacking many of the problems of the heavenly bodies filled my mind, that the news reached me of Kirchhoff's great discovery of the true nature and the chemical constitution of the sun from his interpretation of the Fraunhofer lines.

This news was to me like the coming upon a spring of water in a dry and thirsty land. Here at last presented itself the very order of work for which in an indefinite way I was looking – namely, to extend his novel methods of research upon the sun to the other heavenly bodies. A feeling as of inspiration seized me: I felt as if I had it now in my power to lift a veil which had never before been lifted; as if a key had been put into my hands which would unlock a door which had been regarded as for ever closed to man – the veil and door behind which lay the unknown mystery of the true nature of the heavenly bodies. This was especially work for which I was to a great extent prepared, from being already familiar with the chief methods of chemical and physical research.

It was just at this time that I happened to meet at a soirée of the Pharmaceutical Society, where spectroscopes were shown, my friend and neighbour, Dr. W. Allen Miller, Professor of Chemistry at King's College, who had already worked much on chemical spectroscopy. A sudden impulse seized me to suggest to him that we should return home together. On our way home I told him of what was in my mind, and asked him to join me in the attempt I was about to make, to apply Kirchhoff's methods to the stars. At first, from considerations of the great relative faintness of the stars, and the great delicacy of the work from the earth's motion, [912] even with the aid of a clockwork, he hesitated as to the probability of our success. Finally he agreed to come to my observatory on the first fine evening, for some preliminary experiments as to what we might expect to do upon the stars.

At that time a star spectroscope was an instrument unknown to the optician. I remember that for our first trials we had one of the hollow prisms filled with bisulphide of carbon so much in use then, and which in consequence of a small leak smelt abominably. To this day this pungent odour reminds me of star spectra!

Let us look at the problem which lay before us. It is difficult for any one, who has now only to give an order for a star spectroscope, to understand in any true degree the difficulties which we met with in attempting to make such observations for the first time. From the sun with which the Heidelberg professors had to do – which, even bright as it is, for some parts of the spectrum has no light to spare – to the brightest stars is a very far cry. The light received at the earth from a first magnitude star, as Vega, is only about the one forty thousand millionth part of that received from the sun.

Fortunately, as the stars are too far off to show a true disk, it is possible to concentrate all the light received from the star upon a large mirror or object-glass, into the telescopic image, and so increase its brightness.

We could not make use of the easy method adopted by Fraunhofer of placing a prism before the object-glass, for we needed a terrestrial spectrum, taken under the same conditions, for the interpretation, by a simultaneous comparison with it of the star's spectrum. Kirchhoff's method required that the image of a star should be thrown upon a narrow slit simultaneously with the light from a flame or from an electric spark.

These conditions made it necessary to attach a spectroscope to the eye-end of the telescope, so that it would be carried with it, with its slit in the focal plane. Then, by means of a small reflecting prism placed before one half of the slit, light from a terrestrial source at the side of the telescope could be sent into the instrument together with the star's light, and so form a spectrum by the side of the stellar spectrum, for convenient comparison with it.

This was not all. As the telescopic image of a star is a point, its spectrum will be a narrow line of light without appreciable breadth. Now for the observation of either dark or of bright lines across the spectrum a certain breadth is absolutely needful. To get breadth, the pointlike image of the star must be broadened out. As light is of first importance, it was desirable to broaden the star's image only in the one direction necessary to give breadth to the spectrum; or, in other words, to convert the stellar point into a short line of light. Such an enlargement in one direction only could be given by the device, first employed by Fraunhofer himself, of a lens convex or [913] concave in one direction only, and flat; and so having no action on the light, in a direction at right angles to the former one.

When I went to the distinguished optician, Mr. Andrew Ross, to ask for such a lens, he told me that no such lenses were made in England, but that the spectacle lenses then very occasionally required to correct astigmatism – first used, I believe, by the then Astronomer Royal, the late Sir George Airy – were ground in Berlin. He procured for me from Germany several lenses; but not long after, a cylindrical lens was ground for me by Browning. By means of such a lens, placed within the focus of the telescope, in front of the slit the pointlike image of a star could be widened in one direction so as to become a very fine line of light, just so long as, but no longer than, was necessary to give to the spectrum a breadth sufficient for distinguishing any lines by which it may be crossed.

It is scarcely possible at the present day, when all these points are so familiar as household words, for any astronomer to realise the large amount of time and labour which had to be devoted to the successful construction of the first star spectroscope. Especially was it difficult to provide for the satisfactory introduction of the light for the comparison spectrum. We soon found, to our dismay, how easily the comparison lines might become instrumentally shifted, and so be no longer strictly fiducial. As a test we used the solar lines as reflected to us from the moon – a test of more than sufficient delicacy with the resolving power at our command.

Then it was that an astronomical observatory began, for the first time, to take on the appearance of a laboratory. Primary batteries, giving forth noxious gases were arranged outside one of the windows; a large induction coil stood mounted on a stand on wheels so as to follow the positions of the eye-end of the telescope, together with a battery of several Leyden jars; shelves with Bunsen burners, vacuum tubes, and bottles of chemicals, especially of specimens of pure metals, lined its walls.

The observatory became a meeting place where terrestrial chemistry was brought into direct touch with celestial chemistry. The characteristic light-rays from earthly hydrogen shone side by side with the corresponding radiations from starry hydrogen, or else fell upon the dark lines due to the absorption of the hydrogen in Sirius or in Vega. Iron from our mines was line-matched, light for dark, with stellar iron from opposite parts of the celestial sphere. Sodium, which upon the earth is always present with us, was found to be widely diffused through the celestial spaces.

This time was, indeed, one of strained expectation and of scientific exaltation for the astronomer, almost without parallel; for nearly every observation revealed a new fact, and almost every night's work was red-lettered by some discovery. And yet, notwithstanding, we had to record 'that the inquiry in which we had been engaged has been [914] more than usually toilsome; indeed it has demanded a sacrifice of time very great when compared with the amount of information which we have been able to obtain.'

Soon after the close of 1862 we sent a preliminary note to the Royal Society, 'On the Lines of some of the Fixed Stars,' in which we gave diagrams of the spectra of Sirius, Betelgeux, and Aldebaran, with the statement that we had observed the spectra of some forty stars, and also the spectra of the planets Jupiter and Mars. It was a little remarkable that on the same day on which our paper was to be read, but some little time after it had been sent in, news arrived there from America that similar observations on some of the stars had been made by Mr. Rutherfurd. A very little later similar work on the spectra of the stars was undertaken in Rome by Secchi, and in Germany by Vogel.

In February 1863 the strictly astronomical character of the observatory was further encroached upon by the erection, in one corner, of a small photographic tent furnished with baths and other appliances for the wet collodion process. We obtained photographs, indeed, of the spectra of Sirius and Capella; but from want of steadiness and more perfect adjustment of the instruments, the spectra, though defined at the edges, did not show the dark lines as we expected. The dry collodion plates then available were not rapid enough; and the wet process was so inconvenient for long exposures, from irregular drying, and draining back from the positions in which the plates had often to be put, that we did not persevere in our attempts to photograph the stellar spectra. I resumed them with success in 1875, as we shall see further on.

At that time no convenient maps of the spectra of the chemical elements, which were then but imperfectly known, were available for comparison with the spectra of the stars. Kirchhoff's maps were confined to a few elements, and were laid down on an arbitrary scale, relatively to the solar spectrum. It was not always easy, since our work had to be done at night when the solar spectrum could not be seen, to recognise with certainty even the lines included in Kirchhoff's maps. To meet this want, I devoted a great part of 1863 to mapping, with a train of six prisms, the spectra of twenty-six of the elements; using as a standard scale the spark-spectrum of common air, which would be always at hand. The lines of air were first carefully referred to those of purified oxygen and nitrogen. The spectra were obtained by the discharge of a large induction coil furnished with a condenser of several Leyden jars. I was much assisted by specimens of pure metals furnished to me by Dr. W. A. Miller and Dr. Matthiessen. My paper on this subject, and its accompanying maps, appeared in the volume of the Transactions of the Royal Society for 1864.

During the same time, whenever the nights were fine, our work [915] on the spectra of the stars went on, and the results were communicated to the Royal Society in April 1864; after which Dr. Miller had not sufficient leisure to continue working with me. The general accuracy of our work, so far as it was possible with the instruments at our disposal is shown by the good agreement of the spectra of Aldebaran and Betelgeux with the observations of the same stars made later in Germany by Vogel.

It is obviously unsafe to claim for spectrum comparisons a greater degree of accuracy than is justified by the resolving power employed. When the apparent coincidences of the lines of the same substance are numerous, as in the case of iron; or the lines are characteristically grouped, as are those of hydrogen, of sodium, and of magnesium, there is no room for doubt that the same substances are really in the stars. Coincidence with a single line may be little better than trusting to a bruised reed; for the stellar line may, under greater resolving power, break up into two or more lines, and then the coincidence may disappear. As we shall see presently, the apparent position of the star-line may not be its true one, in consequence of the earth's or the star's motion in the line of sight. Our work, however, was amply sufficient to give a certain reply to the wonder that had so long asked in vain of what the stars were made. The chemistry of the solar system was shown to prevail, essentially at least, wherever a star twinkles. The stars were undoubtedly suns after the order of our sun, though not all at the same evolutional stage, older or younger it may be, in the life history of bodies of which the vitality is heat. Further, elements which play a chief *rôle* in terrestrial physics, as iron, hydrogen, sodium, magnesium, calcium, were found to be the first and the most easily recognised of the earthly substances in the stars.

Soon after the completion of the joint work of Dr. Miller and myself, and then working alone, I was fortunate in the early autumn of the same year, 1864, to begin some observations in a region hitherto unexplored; and which, to this day, remain associated in my memory with the profound awe which I felt on looking for the first time at that which no eye of man had seen, and which even the scientific imagination could not foreshow.

The attempt seemed almost hopeless. For not only are the nebulae very faintly luminous – as Marius [German astronomer Johann Simon Mayr (1573–1624)] put it, 'like a rush-light shining through a horn' – but their feeble shining cannot be increased in brightness, as can be that of the stars, neither to the eye nor in the spectroscope, by any optic tube, however great.

Shortly after making the observations of which I am about to speak, I dined at Greenwich, Otto Struve being also a guest, when, on telling of my recent work on the nebulae, Sir George Airy said: 'It seems to me a case of "Eyes and No Eyes."'[9] Such work indeed it was, as we shall see, on certain of the nebulae.

[916] The nature of these mysterious bodies was still an unread riddle. Towards the end of the last century the elder Herschel, from his observations at Slough, came very near suggesting what is doubtless the true nature, and place in the Cosmos, of the nebulae. I will let him speak in his own words: –

A shining fluid of a nature unknown to us.

What a field of novelty is here opened to our conceptions!... We may now explain that very extensive nebulosity, expanded over more than sixty degrees of the heavens, about the constellation of Orion; a luminous matter accounting much better for it than clustering stars at a distance ...

If this matter is self luminous, it seems more fit to produce a star by its condensation, than to depend on the star for its existence.[10]

This view of the nebulae as parts of a fiery mist out of which the heavens had been slowly fashioned, began, a little before the middle of the present century, at least in many minds, to give way before the revelations of the giant telescopes which had come into use, and especially of the telescope, six feet in diameter, constructed by the late Earl of Rosse at a cost of not less than 12,000*l*.

Nebula after nebula yielded, being resolved apparently into innumerable stars, as the optical power was increased; and so the opinion began to gain ground that all nebulae may be capable of resolution into stars. According to this view, nebulae would have to be regarded, not as early stages of an evolutionary progress, but rather as stellar galaxies already formed, external to our system – cosmical 'sandheaps' too remote to be separated into their component stars. Lord Rosse himself was careful to point out that it would be unsafe from his observations to conclude that all nebulosity is but the glare of stars too remote to be resolved by our instruments. In 1858 Herbert Spencer showed clearly that, notwithstanding the Parsonstown revelations, the evidence from the observation of nebulae up to that time was really in favour of their being early stages of an evolutionary progression.

On the evening of the 29th of August, 1864, I directed the telescope for the first time to a planetary nebula in Draco. The reader may now be able to picture to himself to some extent the feeling of excited suspense, mingled with a degree of awe, with which, after a few moments of hesitation, I put my eye to the spectroscope. Was I not about to look into a secret place of creation?

I looked into the spectroscope. No spectrum such as I expected! A single bright line only! At first, I suspected some displacement of the prism, and that I was looking at a reflection of the illuminated slit from one of its faces. This thought was scarcely more than momentary; then the true interpretation flashed upon me. The light of the nebula was monochromatic, and so, unlike any other light I had as yet subjected to prismatic examination, could not be extended [917] out to form a complete spectrum. After passing through the two prisms it remained concentrated into a single bright line, having a width corresponding to the width of the slit, and occupying in the instrument a position at that part of the spectrum to which its light belongs in refrangibility. A little closer looking showed two other bright lines on the side towards the blue, all the three lines being separated by intervals relatively dark.

The riddle of the nebulae was solved. The answer, which had come to us in the light itself, read: Not an aggregation of stars, but a luminous gas. Stars after the order of our own sun, and of the brighter stars, would give a different spectrum; the light of this nebula had clearly been emitted by a luminous gas. With an excess of caution, at the moment I did not venture to go further than to point out that we had to do with bodies of an order quite different from that of the stars. Further observations soon convinced me that, though the short span of human life is far too minute relatively to cosmical events for us to expect to see in succession any distinct steps in so august a process, the probability is indeed over-whelming in favour of an evolution in the past, and still going on, of the heavenly hosts. A time surely existed when the matter now condensed into the sun and planets filled the whole space occupied by the solar system, in the condition of gas which then appeared as a glowing nebula, after the order, it may be, of some now existing in the heavens. There remained no room for doubt that the nebulae, which our telescopes reveal to us, are the early stages of long processions of cosmical events, which correspond broadly to those required by the nebular hypothesis in one or other of its forms.

Not indeed that the philosophical astronomer would venture to dogmatise in matters of detail, or profess to be able to tell you pat off by heart exactly how everything has taken place in the universe, with the flippant tongue of a Lady Constance after reading The Revelations of Chaos –

'It shows you exactly how a star is formed; nothing could be so pretty. A cluster of vapour – the cream of the Milky Way; a sort of celestial cheese churned into light.'[11]

It is necessary to bear distinctly in mind that the old view which made the matter of the nebulae to consist of an original fiery mist – in the words of the poet:

> . . . a tumultuous cloud
> Instinct with fire and nitre –[12]

could no longer hold its place after Helmholtz had shown, in 1854, that such an originally fiery condition of the nebulous stuff was quite unnecessary, since in the mutual gravitation of widely separated [918] matter we have a store of potential energy sufficient to generate the high temperature of the sun and stars.

The solution of the primary riddle of the nebulae left pending some secondary ques-tions. What chemical substances are represented by the newly found bright lines? Is solar matter common to the nebulae as well as to the stars? What are the physical conditions of the nebulous matter?

Further observations showed two lines of hydrogen; and recent observations have shown associated with it the new element recently discovered by Professor Ramsay, occluded in certain minerals, and of which a brilliant yellow line in the sun had long been looked upon as the badge of an element as yet unknown. The principal line of these nebulae suggests probably another substance which has not yet been unearthed from its hiding place in terrestrial rocks by the cunning of the chemist.

Are the nebulae very hot, or comparatively cool? The spectroscope indicates a high temperature: that is to say, that the individual molecules or atoms, which by their encounters are luminous, have motions corresponding to a very high temperature, and in this sense are very hot. On account of the great extent of the nebulae, however, a comparatively small number of luminous molecules might be sufficient to make them as bright as they appear to us; taking this view, their mean temperature, if they can be said to have one, might be low, and so correspond with what we might expect to find in gaseous masses at an early stage of condensation.

In the nebulae I had as yet examined, the condensation of nearly all the light into a few bright lines made the observations of their spectra less difficult than I feared would be the case. It became, indeed, a case of 'Eyes and No Eyes' when a few days later I turned the telescope to the Great Nebula in Andromeda. Its light was distributed throughout the spectrum, and consequently extremely faint. The brighter middle part only could be seen, though I have since proved, as I at first suggested might be the case, that the blue and the red ends are really not absent, but are not seen on account of their feebler effect upon the eye. Though continuous, the spectrum did not look uniform in brightness, but its extreme feebleness made it uncertain whether the irregularities were due to certain parts being enhanced by bright lines, or the other parts enfeebled by dark lines.

Out of sixty of the brighter nebulae and clusters, I found about one-third, including the planetary nebulae and that of Orion, to give the bright-line spectrum. It would be altogether out of place here to follow the results of my further observations along the same lines of research, which occupied the two years immediately succeeding.

I pass at once to a primary spectroscopic observation of one of [919] those rare and strange sights of the heavens of which only about nineteen have been recorded in as many centuries:

> ... those far stars that come in sight
> Once in a century.[13]

On the 18th of May, 1866, at 5 P. M. a letter came with the address 'Tuam, from an unknown correspondent, one John Birmingham.' Mr. Birmingham afterwards became well known by his observations of variable stars, and especially by his valuable catalogue of Red Stars in 1877. The letter ran: –

I beg to direct your attention to a new star which I observed last Saturday night, and which must be a most interesting object for spectrum analysis. It is situated in Cor. Bor.; and is very brilliant, of about the second magnitude. I sent an account of it to the *Times* yesterday, but as that journal is not likely to publish communications from this part of the world, I scarcely think that it will find a place for mine.

Fortunately the evening was fine, and as soon as it was dusk I looked, with not a little scepticism, I freely confess, at the place of the sky named in the letter. To my great joy, there shone a bright new star, giving a new aspect to the Northern Crown; of the order doubtless of the splendid temporary star of 1572, which Tycho supposed to be generated from the ethereal substance of the Milky Way, and afterwards dissipated by the sun, or dissolved from some internal cause.

I sent a messenger for my friend Dr. Miller; and an hour later we directed the telescope, with spectroscope attached, to the blazing star. Later in the evening a letter arrived from Mr. Baxendale [*sic*], who had independently discovered the star on the 15th.

By this evening, the 18th, the star had already fallen in brightness below the third magnitude. The view in the spectroscope was strange, and up to that time unprecedented. Upon a spectrum of the solar order, with its numberless dark lines, shone out brilliantly a few very bright lines. There was little doubt that at least two of these lines belonged to hydrogen. The great brilliancy of these lines as compared with the parts of the continuous spectrum upon which they fell, suggested a temperature for the gas emitting them higher than that of the star's photosphere.

A few of days, as indeed had been its forbears appearing at long intervals, the new star waned with a rapidity little less remarkable than was the suddenness of its outburst, without visible descent, all armed in a full panoply of light from the moment of its birth. A few hours only before Birmingham saw it blazing with second-magnitude splendour, Schmidt, observing at Athens, could testify that no outburst had taken place. Rapid was the decline of its light, falling in twelve days from the second down to the eighth magnitude.

[920] It was obvious to us that no very considerable mass of matter could cool down from the high temperature indicated by the bright lines in so short a time. At the same time it was not less clear that the extent of the mass of the fervid gas must be on a very grand scale indeed, for a star at its undoubted distance from us, to take on so great a splendour. These considerations led us to suggest some sudden and vast convulsion, which had taken place in a star so far cooled down as to give but little light, or even to be partially crusted over; by volcanic forces, or by the disturbing approach or partial collision of another dark star. The essential character of the explanation lay in the suggestion of a possible chemical combination of some of the escaping highly heated gases from within, when cooled by the sudden expansion, which might give rise to an outburst of flame at once very brilliant and of very short duration.

The more precise statement of what occurred during our observations, as made afterwards from the pulpit of one of our cathedrals – 'That from afar astronomers had seen a world on fire go out in smoke and ashes' – must be put down to an excess of the theological imagination.

From the beginning of our work upon the spectra of the stars, I saw in vision the application of the new knowledge to the creation of a great method of astronomical observation which could not fail in future to have a powerful influence on the progress of astronomy; indeed, in some respects greater than the more direct one of the investigation of the chemical nature and the relative physical conditions of the stars.

It was the opprobrium of the older astronomy – though indeed one which involved no disgrace, for *à l'impossible nul n'est tenu* – that only that part of the motions of the stars which is across the line of sight could be seen and directly measured. The direct observation of the other component in the line of sight, since it caused no change of place and, from the great distance of the stars, no appreciable change of size or of brightness within an observer's lifetime, seemed to lie hopelessly quite outside the limits of man's powers.

Still, it was only too clear that, so long as we were unable to ascertain directly those components of the stars' motions which lie in the line of sight, the speed and direction of the solar motion in space, and many of the great problems of the constitution of the heavens, must remain more or less imperfectly known.

Now as the colour of a given kind of light, and the exact position it would take up in a spectrum, depends directly upon the length of the waves, or, to put it differently, upon the number of waves which would pass into the eye in a second of time, it seemed more than probable that motion between the source of the light and the observer must change the apparent length of the waves [921] to him, and the number reaching his eye in a second. To a swimmer striking out from the shore each wave is shorter, and the number he goes through in a given time is greater than would be the case if he had stood still in the water. Such a change of wave-length would transform any given kind of light, so that it would take a new place in the spectrum, and from the amount of this change to a higher or to a lower place, we could determine the velocity per second of the relative motion between the star and the earth.

The notion that the propagation of light is not instantaneous, though rapid far beyond the appreciation of our senses, is due, not as is sometimes stated to Francis, but to Roger Bacon, 'Relinquitur ergo,' he says, in his *Opus Majus*, 'quod lux multiplicatur in tempore ... sed tamen non in tempore sensibili et perceptibili a visu, sed insensibili ...'[14] The discovery of its actual velocity was made by Roemer in 1675, from observations of the satellites of Jupiter. Now though the effect of motion in the line of sight upon the apparent velocity of light underlies Roemer's determinations, the idea of a change of colour in light from motion between the source of light and the observer was announced for the first time by Doppler in 1841. Later, various experiments were made in connection with this view by Ballot, Sestini, Klinkerfues, Clerk Maxwell, and Fizeau. But no attempts had been made, nor were indeed possible, to discover by this principle the motions of the heavenly bodies in the line of sight. For, to learn whether any change in the light had taken place from motion in the line of sight, it was clearly necessary to know the original wave-length of the light before it left the star.

As soon as our observations had shown that certain earthly substances were present in the stars, the original wave-lengths of their lines became known, and any small want of coincidence of the stellar lines with the same lines produced upon the earth might safely be interpreted as revealing the velocity of approach or of recession between the star and the earth.

These considerations were present to my mind from the first, and helped me to bear up under many toilsome disappointments: 'Studio fallente laborem.'[15] It was not until 1866 that I found time to construct a spectroscope of greater power for this research. It would be scarcely possible, even with greater space, to convey to the reader any true conception of the difficulties which presented themselves in this work, from various instrumental causes, and of the extreme care and caution which were needful to distinguish spurious instrumental shifts of a line from a true shift due to the star's motion.

At last, in 1868, I felt able to announce in a paper printed in the Transactions of the Royal Society for that year, the foundation of this new method of research, which, transcending the wildest dreams of an earlier time, enables the astronomer to measure off directly in [922] terrestrial units the invisible motions in the line of sight of the heavenly bodies.

To pure astronomers the method came before its time, since they were then unfamiliar with Spectrum Analysis, which lay completely outside the routine work of an observatory. It would be easy to mention the names of men well known, to whom I was 'as a very lovely song of one that hath a pleasant voice.'[16] They heard my words, but for a time were very slow to avail themselves of this new power of research. My observations were, however, shortly afterwards confirmed by Vogel in Germany; and by others the principle was soon applied to solar phenomena. By making use of improved methods of photography, Vogel has recently determined the motions of approach and of recession of some fifty stars, with an accuracy of about an English mile a second. In the hands of Young, Dunèr, Keeler, and others, the method has been successfully applied to a determination of the rotation of the sun, of Saturn and his rings, and of Jupiter.

It has become fruitful in another direction, for it puts into our hands the power of separating double stars which are beyond the resolving power of any telescope that can ever be constructed. Pickering and Vogel have independently discovered by this method an entirely new class of double stars.

Double stars too close to be separately visible unite in giving a compound spectrum. Now, if the stars are in motion about a common centre of gravity, the lines of one star will shift periodically relatively to similar lines of the other star, in the spectrum common to both; and such lines will consequently, at those times, appear double. Even if one of the stars is too dark to give a spectrum which can be seen upon that of the other star, as is actually the case with Algol and Spica, the whirling of the stars about each other may be discovered from the periodical shifting of the lines of the brighter star relatively to terrestrial lines of the same substance. It is clear that as the stars revolve about their common centre of gravity, the bright star would be sometimes advancing, and at others receding, relatively to an observer on the earth, except it should so happen that the stars' orbit were perpendicular to the line of sight.

It would be scarcely possible, without the appearance of great exaggeration, to attempt to sketch out even in broad outline the many glorious achievements which doubtless lie before this method of research in the immediate future.

Comets in the olden time were looked upon as the portents of all kinds of woe:

> There with long bloody haire, a blazing star
> Threatens the World with Famin, Plague, and War.[17]

Though they were no longer, at the time of which I am speaking, a [923] terror to mankind, they were a great mystery. Perhaps of no other phenomenon of nature had so many guesses at truth been made on different, and even on opposing principles of explanation. It was about this time that a beam of light was thrown in, for the first time, upon the night of mystery in which they moved and had their being, by the researches of Newton of Yale College, by Adams, and by Schiaparelli. The unexpected fact came out of the close relationship of the orbits of certain comets with those of periodic meteor-swarms. Only a year before the observations of which I am about to speak were made, Odling [English chemist William Odling (1829–1921)] had lighted up the theatre of the Royal Institution with gas brought by a meteorite from celestial space. Two years earlier, Donati showed the light of a small comet to be in part self-emitted, and so not wholly reflected sunshine.

I had myself, in the case of three faint comets, in 1866, in 1867, and January 1868, discovered that part of their light was peculiar to them, and that light of the last one consisted mainly of three bright flutings. Intense, therefore, was the great expectancy with which I directed the telescope with its attached spectroscope to the much brighter comet which appeared in June 1868.

The comet's light was resolved into a spectrum of three bright bands or flutings, each alike falling off in brightness on the more refrangible side. On the evening of the 22nd, I measured the positions in the spectrum of the brighter beginnings of the flutings on the red side. I was not a little surprised the next morning to find that the three cometary flutings agreed in position with three similar flutings in the brightest part of the spectrum of carbon. Some time before, I had mapped down the spectrum of carbon, from different sources, chiefly from different hydrocarbons. In some of these spectra, the separate lines of which the flutings are built up are individually more distinct than in others. The comet bands, as I had seen them on the previous evening, appeared to be identical in character in this respect, as well as in position in the spectrum, with the flutings as they appeared when I took the spark in a current of olefiant gas. I immediately filled a small holder with this gas, arranged an apparatus in such a manner that the gas could be attached to the end of the telescope, and its spectrum, when a spark was taken in it, seen side by side with that of the comet.

Fortunately the evening was fine; and on account of the exceptional interest of confronting for the first time the spectrum of an earthly gas with that of a comet's light, I invited Dr. Miller to come and make the crucial observation with me. The expectation which I had formed from my measures was fully confirmed. The comet's spectrum when seen together with that from the gas agreed in all respects precisely with it. The comet, though 'subtle as Sphinx,'[18] had at last yielded up its secret. The principal part of its light was emitted by luminous vapour of carbon.

[924] The result was in harmony with the nature of the gas found occluded in meteorites. Odling had found carbonic oxide as well as hydrogen in his meteorite. Wright [American physicist Arthur Williams Wright (1836–1915)], experimenting with another type of meteorite, found that carbon dioxide was chiefly given off. Many meteorites contain a large percentage of hydrocarbons; from one of such sky-stones a little later I observed a spectrum similar to that of the comet. The three bands may be seen in the base of a candle flame.

Since these early observations the spectra of many comets have been examined by many observers. The close general agreement as to the three bright flutings which form the main feature of the cometary spectrum, confirms beyond doubt the view that the greater part of the light of comets is due to the fluted spectrum of carbon. Some additional knowledge of the spectra of comets, obtained by means of photography, will have its proper place later on.

About this time I devoted some attention to spectroscopic observations of the sun, and especially to the modifications of the spectrum which take place under the influence of the solar spots.

The aerial ocean around and above us, in which finely divided matter is always more or less floating, becomes itself illuminated, and a source of light, when the sun shines upon it, and so conceals, like a luminous veil, any object less brilliant than itself in the heavens beyond. From this cause the stars are invisible at midday. This curtain of light above us, at all ordinary times shuts out from our view the magnificent spectacle of red flames flashing upon a coronal glory of bright beams and streamers, which suddenly bursts upon the sight, for a few minutes only, when at rare intervals the light-curtain is lifted by the screening of the sun's light by the moon, at a total eclipse.

As yet the spectrum of the red flames had not been seen. If as seemed probable, it should be found to be that of a gas, consisting of bright lines only, it was conceivable that the spectroscope might enable us so to weaken by dispersion the air-glare, relatively to the bright lines which would remain undispersed, that the bright lines of the flames might become visible through the atmospheric glare.

The historic sequence of events is as follows. In November 1866 Mr. Lockyer asked the question: 'May not the spectroscope afford us evidence of the existence of the red flames, which total eclipses have revealed to us in the sun's atmosphere; though they escape all other methods of observation at other times?'[19]

In the Report of the Council of the Royal Astronomical Society, read in February 1868, occurs the following statement, furnished by me, in which the explanation is fully given of the principle on which I had been working to obtain the spectrum of the red flames without an eclipse:

[925] During the last two years Mr. Huggins has made numerous observations for the purpose of obtaining a view, if possible, of the red prominences seen during an eclipse. The invisibility of these objects at ordinary times is supposed to arise from the illumination of our atmosphere. If these bodies are gaseous, their spectra would consist of bright lines. With a powerful spectroscope the light reflected from our atmosphere near the sun's limb edge would be greatly reduced in intensity by the dispersion of the prisms, while the bright lines of the prominences, if such be present, would remain but little diminished in brilliancy. This principle has been carried out by various forms of prismatic apparatus, and also by other contrivances, but hitherto without success.[20]

At the total eclipse of the sun, August 18, 1868, several observers saw the light of the red flames to be resolved in their spectroscopes into bright lines, among which lines of hydrogen were recognised. The distinguished astronomer, Janssen, one of the observers in India, saw some of the bright lines again the next day, by means of the principle described above, when there was no eclipse.

On October 29th, Mr. Lockyer sent a note to the Royal Society to say that on that day he had succeeded in observing three bright lines, of a fine prominence.

About the time that the news of the discovery of the bright lines at the eclipse reached this country, in September, I was altogether incapacitated for work for some little time through the death of my beloved mother. We had been all in all to each other for many years. The first day I was sufficiently recovered to resume work, December 19, on looking at the sun's limb with the same spectroscope I had often used before, now that I knew exactly at what part of the spectrum to search for the lines, I saw them at the first moment of putting my eye to the instrument.

As yet, by all observers the lines only of the prominences had been seen, and therefore to learn their forms, it was necessary to combine in one design the lengths of the lines as they varied, when the slit was made to pass over a prominence. In February of the following year, it occurred to me that by widening the opening of the slit, the form of a prominence, and not its lines only, might be directly observed. This method of using a wide slit has been since universally employed.

It does not fall within the scope of this article to describe an ingenious photographic method by which Hale has been able to take daily records of the constantly varying phenomena of the red flames and the bright faculae, upon and around the solar disk.

The purpose of this article is to sketch in very broad outline only, the principal events, in the order of their succession in time, *quorum pars magna fui*,[21] which contributed in an important degree to the rise of the new astronomy. As a science advances it follows naturally that its further progress will consist more and more in matters of [926] detail, and in points which are of technical, rather than of general interest.

It would, therefore, be altogether out of place here, to carry on in detail the narrative of the work of my observatory, when, as was inevitable, it began to take on the character of a development only, along lines of which I have already spoken: namely, the observation of more stars, and of other nebulae, and other comets. I pass on, at once, therefore, to the year 1876, in which by the aid of the new dry plates, with gelatine films, introduced by Mr. Kennett, I was able to take up again, and this time with success, the photography of the spectra of the stars, of my early attempts at which I have already spoken.

I was now better prepared for work. My observatory had been enlarged from a dome of 12 feet in diameter, to a drum having a diameter of 18 feet. This alteration had been made for the reception of a larger telescope made by Sir Howard Grubb, at the expense of a legacy to the Royal Society, and which was placed in my hands on loan by that society. This instrument was furnished with two telescopes: an achromatic of 15 inches aperture, and a Cassegrain of 18 inches aperture, with mirrors of speculum metal. At this time, one only of these telescopes could be in use at a time. Later on, in 1882, by a device which occurred to me, of giving each telescope an independent polar axis, the one working within the other, both telescopes could remain together on the equatorial mounting, and be equally ready for use.

By this time I had the great happiness of having secured an able and enthusiastic assistant, by my marriage in 1875.

The great and notable advances in astronomical methods and discoveries by means of photography since 1875, are due almost entirely to the great advantages which the gelatine dry plate possesses for use in the observatory, over the process of Daguerre, and even over that of wet collodion. The silver-bromide gelatine plate, which I was the first, I believe, to use for photographing the spectra of stars, except for its grained texture, meets the need of the astronomer at all points. This plate possesses extreme sensitiveness; it is always ready for use; it can be placed in any position; it can be exposed for hours; lastly, immediate development is not necessary, and for this reason, as I soon found to be necessary in this climate, it can be exposed again to the same object on succeeding nights; and so make up by successive instalments, as the weather may permit, the total long exposure which may be needful.

The power of the eye falls off as the spectrum extends beyond the blue, and soon fails altogether. There is therefore no drawback to the use of glass for the prisms and lenses of a visual spectroscope. But while the sensitiveness of a photographic plate is not similarly limited, glass like the eye is imperfectly transparent, and soon becomes [927] opaque, to the parts of the spectrum at a short distance beyond the limit of the visible spectrum. To obtain, therefore, upon the plate a spectrum complete at the blue end of stellar light, it was necessary to avoid glass, and to employ instead Iceland spar and rock crystal which are transparent up to the limit of the ultra-violet light which can reach us through our atmosphere. Such a spectroscope was constructed and fixed with its slit at the focus of the great speculum of the Cassegrain telescope.

How was the image of a star to be easily brought, and then kept, for an hour or even for many hours, precisely at one place on a slit so narrow as about the one two-hundredth of an inch? For this purpose the very convenient device was adopted of making the slit-plates of highly polished metal, so as to form a divided mirror, in which the reflected image of a star could be observed from the eye-end of the telescope by means of a small telescope fixed within the central hole of the great mirror. A photograph of the spectrum of α Lyrae, taken with this instrument, was shown at the Royal Society in 1876.

In the spectra of such stars as Sirius and Vega, there came out in the ultra-violet region, which up to that time had remained unexplored, the completion of a grand rhythmical group of strong dark lines, of which the well-known hydrogen lines in the visible region form the lower members. Terrestrial chemistry became enriched with a more complete knowledge of the spectrum of hydrogen from the stars. Shortly afterwards, Cornu succeeded in photographing a similar spectrum in his laboratory from earthly hydrogen.

I presented in 1879 a paper, with maps, to the Royal Society, on the photographic spectra of the stars, which was printed in their Transactions for 1880. In this paper, besides descriptions of the photographs, and tables of the measures of the positions of the lines, I made a first attempt to arrange the stars in a possible evolutional series from the relative behaviour of the hydrogen and the metallic lines. In this series, Sirius and Vega are placed at the hotter and earlier end; Capella and the sun, at about the same evolutional stage, somewhere in the middle of the series; while at the most advanced and oldest stage of the stars which I had then photographed, came Betelgeux, in the spectrum of which the ultra-violet region, though not wanting, is very greatly enfeebled.

Shortly afterwards, I directed the photographic arrangement of combined spectroscope and telescope to the nebula in Orion, and obtained for the first time information of the nature of its spectrum beyond the visible region. One line a little distance on in the ultra-violet region came out very strongly on the plate. If this kind of light came within the range of our vision, it would no doubt give the dominant colour to the nebula, in place of its present blue-greenish hue. Other lines of the hydrogen series, as might be expected, were seen in the photograph, together with a number of other bright lines.

[928] In 1881, for the first time since the spectroscope and also suitable photographic plates had been in the hands of astronomers, the coming of a bright comet made it possible to extend the examination of its light into the invisible region of the spectrum at the blue end. On the 22nd of June, by leaving very early a banquet at the Mansion House, I was able, after my return home, to obtain with an exposure of one hour, a good photograph of the head of the comet. I was under a great tension of expectancy that the plate was developed, so that I might be able to look for the first time into a virgin region of nature, as yet unexplored by the eye of man.

The plate contained an extension and confirmation of my earlier observations by eye. There were the combined spectra of two kinds of light – a faint continuous spectrum, crossed by Fraunhofer lines which showed it to be reflected solar light. Upon this was seen a second spectrum of the original light emitted by the comet itself. This spectrum consisted mainly of two groups of bright lines, characteristic of the spectra of certain compounds of carbon. It will be remembered that my earlier observations revealed the three principal flutings of carbon as the main feature of a comet's spectrum in the visible region. The photograph brought a new fact to light. Liveing and Dewar had shown that one of these bands consisted of lines belonging to a nitrogen compound of carbon. We gained the new knowledge that nitrogen, as well as carbon and hydrogen, exists in comets. Now, nitrogen is present in the gas found occluded in some meteorites. At a later date, Dr. Flight [English mineralogist Walter Flight (1841–85)] showed that nitrogen formed as much as 17 per cent. of the occluded gas from the meteorite of Cranbourne, Australia.

I have now advanced to the extreme limit of time within which the rise of the new astronomy can be regarded as taking place. At this time, in respect of the broad lines of its methods, and the wide scope of the directions in which it was already applied, it had become well established. Already it possessed a literature of its own, and many observatories were becoming, in part at least, devoted to its methods.

In my own observatory work has gone on whenever our unfavourable climate has permitted observations to be made. At the present moment more than one research is in progress. It would be altogether beyond the intention, and limited scope, of the present article to follow this later work.

We found the new astronomy newly born in a laboratory at Heidelberg; to astronomers she was

> . . . a stranger,
> Born out of their dominions.[22]

We take leave of her in the full beauty of a vigorous youth, receiving homage in nearly all the observatories of the world, some of [929] which indeed are devoted wholly to her cult. So powerful is the magic of her charms that gifts have poured in from all sides to do her honour. It has been by some free gifts that Pickering, at Cambridge, United States, and in the southern hemisphere, has been able to give her so devoted a service. In this country, where from almost the hour of her birth she won hearts, enthusiastic worshippers have not been wanting. By the liberality of the late Mr. Newall, and the disinterested devotion of his son, a well-equipped observatory is now wholly given up to her worship at Cambridge. This Jubilee year is red-lettered at Greenwich by the inauguration of a magnificent double telescope, laid at her feet by Sir Henry Thompson. Next year, the Royal Observatory at the Cape will be able to add to its devotion to the old astronomy a homage not less sincere and enthusiastic to the new astronomy, by means of the splendid instruments which Mr. McClean, who personally serves under her colours, has presented to that Observatory. In Germany, the first National Observatory dedicated to the new astronomy in 1874, under the direction of the distinguished astrophysicist, Professor Vogel, is about to be furnished by the Government with new and larger instruments in her honour.

In America, many have done liberally, but Mr. Yerkes has excelled them all. This summer will be celebrated the opening of a palatial institution on the shore of Lake Geneva, founded by Mr. Yerkes and dedicated to our fair lady, the new astronomy. This observatory, in respect of the great size of its telescope, of forty inches in aperture, the largest yet constructed, its armoury of instruments for spectroscopic attack upon the heavens, and the completeness of its laboratories and its workshops, will represent the most advanced state of instrument making; and at the same time render possible, under the most favourable conditions, the latest and the most perfect methods of research of the new astronomy. Above all, the needful men will not be wanting. A knightly band, who have shown their knighthood by prowess in discovery, led by Professor Hale in chivalrous quest of Truth will surely make this palace of the new astronomy worthy to be regarded as the Uraniborg of the end of the nineteenth century, as the Danish Observatory, under Tycho and his astronomers, represented the highest development of astronomy at the close of the sixteenth.

Notes

1. W. Huggins, *Nineteenth Century: A Monthly Review*, **41** (1897), pp. 907–29. Huggins offered no citations for the literary excerpts in the essay.
2. Edmund Spenser (1552–99), 'An hymne of heavenly beautie' (1596), lines 15–19.
3. 'O fairer daughter of a fair mother!' Horace (65–8 BCE), *Odes*, Book I, Ode xvi, line 1.
4. Auguste Comte, *Cours de Philosophie Positive* (Paris: Bachelier, 1835), vol. 2, pp. 364–5.
5. Spenser, 'An hymne of heavenly beautie', lines 53–4.
6. James Thomson (1700–48), 'A poem sacred to the memory of Sir Isaac Newton' (1727), lines 102–11.
7. A fabulous mirror attributed to alchemist Cornelius Agrippa (1486–1535) reputed to display images of the observer's thoughts and memories.
8. Mary Somerville, *On the Connexion of the Physical Sciences* (London: John Murray, 1849), p. 178.
9. A reference to a popular children's story by John Aiken (1747–1822) titled 'Eyes and no eyes; or, the art of seeing', which relates the very different reports offered by two boys – one dull, the other curious – after each had walked through the same countryside. The same title was used by William Schwenck Gilbert (1836–1911) for a one act musical play he wrote in 1875 with composer Thomas German Reed (1817–88). However, Gilbert based his libretto loosely on another tale of personal differences in observation: 'The emperor's new clothes' by Hans Christian Anderson (1805–75).
10. William Herschel, 'On nebulous stars, properly so-called', *PTRSL* **81** (1791), pp. 71–88; pp. 83–5.
11. Benjamin Disraeli, *Tancred, or the New Crusade* (Leipzig: Bernhard Tauchnitz, 1847), vol. 1, p. 123.
12. John Milton, *Paradise Lost*, Book II, lines 936–7.
13. James Russell Lowell (1819–91), 'Incident in a railroad car' (1842), lines 75–6.
14. Roger Bacon, *Opus Majus*, V, 'Perspectivae' I, IX, 3.
15. From Horace, *Satires* ii, 2, line 12: 'molliter austerum studio fallente laborem' [in which the interest in the game makes the player enjoy the exercise, forgetting how severe it is]. See Edward P. Morris (ed.), *Horace: The Satires and Epistles* (New York: American Book Company, 1911), pp. 157–8.
16. Ezekiel 33:32.
17. From Joshua Sylvester (1563–1618), *Du Bartas: His Divine Weekes And Workes with A Compleate Collectio [sic] of all the other most delight-full Workes* (London: Humfray Lownes, 1621), p. 33. This is Sylvester's translation of 'La Sepmaine; ou Creation du monde' (1578), by French poet, Guillaume de Saluste du Bartas (1544–90).
18. William Shakespeare, *Love's Labour's Lost*, IV, iii, line 337.
19. J.N. Lockyer, 'Spectroscopic observations of the Sun', *PRS* **15** (1866), pp. 256–8; p. 258.
20. W. Huggins, 'Mr. Huggins' Observatory', *MNRAS* **28** (1868), pp. 86–8; p. 88.
21. From Vergil (70–19 BCE), *Aeneid*, Book II, line 6. This fragment, meaning 'in which I played a great part', is excerpted out of context from a longer phrase in a speech by Aeneas to the Carthaginians: 'quaeque ipse miserrima vidi, et quorum pars magna fui' [and those terrible things I saw, and in which I played a great part].
22. Based on words spoken by Catherine, Queen of England, in Shakespeare's *Henry VIII*, II, iv, lines 16–17: 'I am a most poor woman, and a stranger, Born out of your dominions . . .'.

Bibliography

Abbott, Francis 'Notes on η Argus', *MNRAS* **24** (1863), pp. 2–7.
    'Observations on η *Argus*', *MNRAS* **24** (1864), p. 110.
    'On the variability of η Argus and surrounding nebula', *MNRAS* **28** (1868), pp. 266–8.
    'Reply [to John F. W. Herschel]', *MNRAS* **31** (1871), p. 230.
Ackland, William, 'The collodio-albumen process', *Photographic Journal* **3** (1856), pp. 106–9 and pp. 122–5.
    'The difficulties of the dry processes', *Photographic Journal* **6** (1860), pp. 99–103.
Airy, George Biddell, 'President's address', *MNRAS* **3** (1836), pp. 167–74.
    'Observations of the total solar eclipse of 1842, July 7', *MNRAS* **5** (1842), pp. 214–21.
    'Address of the president on presenting the Gold Medal', *MNRAS* **15** (1855), p. 148.
    'Apparatus for the observation of the spectra of stars', *MNRAS* **23** (1863), pp. 188–91.
    'Response to Col. Strange', *AR* **10** (1872), pp. 114–15, pp. 118–19.
    'Proposed devotion of an observatory to observation of the phenomena of *Jupiter's* satellites', *MNRAS* **32** (1872), p. 80.
Allen, David Elliston, *The Naturalist in Britain: A Social History* (London: Allen Lane, 1976).
Alter, Peter, *The Reluctant Patron: Science and the State in Britain 1850–1920*, trans. Angela Davies (Oxford: Berg, 1987).
Ångström, Anders Jonas, *Recherches sur le Spectre Solaire* (Upsala: W. Schultz, 1868).
    Spectre Normal du Soleil (Berlin: F. Dümmler, 1869).
Anstis, Frank Harold, 'The scientific work of Sir William Huggins', M.Sc. thesis (University College, London, 1961).
Armstrong, William George, 'Address', *RBAAS, Newcastle-upon-Tyne* (1863), pp. li–lxiv.
Baedeker, Karl, *Switzerland, and the Adjacent Portions of Italy, Savoy, and the Tyrol. Hand book for Travellers*, 10th edn (London: Dulau and Co., 1883).
Baily, Francis, 'On a remarkable phenomenon that occurs in total and annular eclipses of the Sun', *MNRAS*, **4** (1836), pp. 15–19.
    'Some remarks on the total eclipse of the Sun, on July 8th, 1842', *MNRAS* **5** (1839–1843), pp. 208–14.
Ball, Robert Stawell, 'Thomas Romney Robinson', *MNRAS* **43** (1883), pp. 181–3.
Ball, William Valentine (ed.), *Reminiscences and Letters of Sir Robert Ball* (London: Cassell & Co., 1915).
Balmer, Johann Jakob, 'Notiz über die Spektrallinien des Wasserstoff', *APC* **25** (1885), pp. 80–5.
Bartholomew, C. F., 'The discovery of the solar granulation', *QJRAS* **17** (1976), pp. 263–89.
Barton, Ruth, '"An influential set of chaps": The X-Club and Royal Society politics 1864–85', *BJHS* **23** (1990), pp. 53–81.
Bazerman, Charles, *Shaping Written Knowledge: The Genre and Activity of the Experimental Article in Science* (Madison: University of Wisconsin Press, 1988).
Becker, Barbara J., Ecclecticism, opportunism, and the evolution of a new research agenda: William and Margaret Huggins and the origins of astrophysics, PhD thesis, The Johns Hopkins University (1993).

'Margaret and William Huggins at work in the Tulse Hill Observatory', in H. Pycior, N. Slack and P. G. Abir-Am (eds), *Creative Couples in Science: Multidisciplinary Perspectives* (New Brunswick: Rutgers University Press, 1996), pp. 98–111.

'David Gill', 'Edmond Halley', 'Thomas Hariot', 'William Huggins', and 'Astronomers Royal', in John Lankford (ed.), *History of Astronomy: An Encyclopedia* (New York: Garland Press, 1997).

'Priority, persuasion, and the virtue of perseverance: William Huggins's efforts to photograph the solar corona without an eclipse', *JHA* **31** (2000), pp. 223–43.

'Visionary memories: William Huggins and the origins of astrophysics', *JHA* **32** (2001), pp. 43–62.

'Celestial spectroscopy: Making reality fit the myth', *Science* **301** (2003), pp. 1332–3. Reprinted in Dava Sobel (ed.), *The Best American Science Writing: 2004* (New York: HarperCollins Publishers, 2004).

'La spettroscopia e la nascita dell'astrofisica [Spectroscopy and the rise of astrophysics]', in S. Petruccioli (ed.), *La Storia della Scienza, Vol. VII: L'Ottocento* (Istituto della Enciclopedia Italiana, 2003), pp. 265–81.

'William Huggins' and 'Margaret Lindsay Huggins', in Bernard Lightman (ed.), *Dictionary of Nineteenth-century British Scientists*, vol. 2 (Bristol: Thoemmes Press, 2004), pp. 1021–9.

'William Huggins' and 'Margaret Lindsay Huggins' in *The New DNB* (Oxford: Oxford University Press, 2004).

'From dilettante to serious amateur: William Huggins's move into the inner circle', *JAHH* **13** (2010), pp. 112–19.

Becker, Bernard H., *Scientific London* (New York: Appleton & Co., 1875).

Bede, Cuthbert [Edward Bradley], *Photographic Pleasures: Popularly Portrayed with Pen & Pencil* (London: T. McLean, 1855).

Beer, Gillian, '"The death of the Sun": Victorian solar physics and solar myth', in J. B. Bullen (ed.), *The Sun is God: Painting, Literature and Mythology in the Nineteenth Century* (Oxford: Clarendon Press, 1989), pp. 159–80.

Beeson, Gilbert W., Influences on the identification of wives with the Air Force Organization: An examination of the two-person career, PhD thesis, University of North Carolina (1986).

Bennett, James A., *Church, State and Astronomy in Ireland: 200 Years of Armagh Observatory* (Armagh: Armagh Observatory, 1990).

Bessel, Friedrich Wilhelm, 'Über den gegenwärtigen Standpunkt der Astronomie', in *Populäre Vorlesungen über wissenschaftliche Gegenstände* (Hamburg: Perthes-Besser & Mauke, 1848), pp. 5–6.

Birmingham, John, 'The new variable near η Coronae', *MNRAS* **26** (1866), p. 310.

Birt, William Radcliffe, 'On the appearance of Saturn's ring, 1862', *MNRAS* **22** (1862), p. 295.

'On the Crater Linné', *MNRAS* **28** (1868), p. 220.

'On the Crater Linné', *MNRAS* **29** (1869), p. 63.

Bolotin, N. and C. Laing, *The Chicago World's Fair of 1893: The World's Columbian Exposition* (Washington, DC: Preservation Press, 1992).

Bolzano, Bernard, 'Ein Paar Bemerkungen über die neue Theorie in Herrn Professor Chr. Doppler's Schrif: "Über das farbige Licht der Doppel-sterne und einiger anderer Gestirne des Himmels"', *APC* **60** (1843), pp. 83–8.

Bond, George Phillips, 'On the spiral structure of the Great Nebula of Orion', *MNRAS* **21** (1861), pp. 203–7.

'Remarks upon the statements of Messrs. Stone and Carpenter relating to Sir John Herschel's figure of the Sinus Magnus in the Nebula of Orion', *MNRAS* **24** (1864), pp. 177–81.

Truman Henry Safford (ed.), *Observations upon the Great Nebula of Orion* (Cambridge, MA: Riverside Press, 1867).

Bond, William Cranch, 'Discovery of a new satellite of Saturn', *MNRAS* **9** (1848), pp. 1–2.

Booth, Charles, *Life and Labour of the People in London*, vol. 7 (London: Macmillan & Co., 1896).

Boss, Lewis, William Wallace Campbell and George Ellery Hale, *Report of Committee on Southern and Solar Observatories* (Washington, DC: Carnegie Institution, 1903).

Bowen, Ira, 'The origin of the nebulium spectrum', *Nature* **120** (1927), p. 473.

Boyle, Robert, *Experiments and Considerations Touching Colours* (London: Henry Herringman, 1664).

Boys, Charles Vernon, 'Address', *RBAAS, Southport* (1903), pp. 525–35.

Bradley, James, 'A letter . . . giving an account of a new discovered motion of the fix'd stars', *PTRSL* **35** (1727), pp. 637–61.

Brashear, Ronald S., 'Radiant heat in astronomy, 1830–1880', Unpublished manuscript.

Brewster, David, *A Treatise on New Philosophical Instruments, for Various Purposes in the Arts and Sciences* (Edinburgh: John Murray and William Blackwood, 1813)
  'Description of a monochromatic lamp for microscopical purposes, &c. with remarks on the absorption of the prismatic rays by coloured media', *TRSE* **9** (1823), pp. 433–44.
  'Report on the recent progress of optics', *RBAAS, Oxford* (1832), pp. 318–22.

Briggs, Asa, *Victorian People: A Reassessment of Persons and Themes 1851–67* (Chicago, IL: University of Chicago, 1973).

Brock, William H., *William Crookes (1832–1919) and the Commercialization of Science* (Aldershot: Ashgate, 2008).

Brodie, Frederick, 'Note on the telescopic appearance of Mars', *MNRAS* **16** (1856), pp. 204–5.

Brown, Lieut. [A. B.], 'Report of observations of the solar eclipse', *MNRAS* **31** (1871), pp. 52–9.

Browning, John, 'On a spectroscope in which the prisms are automatically adjusted to the minimum angle of deviation for the particular ray under examination', *MNRAS* **30** (1870), pp. 198–202.

Brück, Hermann Alexander and Maire Theresa Brück, *The Peripatetic Astronomer: The Life of Charles Piazzi Smyth* (Bristol: Adam Hilger, 1988).

Brück, Mary Theresa, 'Companions in astronomy: Margaret Lindsay Huggins and Agnes Mary Clerke', *IAJ* **20** (1991), pp. 70–7
  'An astronomical love affair', in M. Mulvihill and P. Deevy (eds), *Stars, Shells and Bluebells* (Dublin: Women in Technology & Science, 1997), pp. 76–83.
  Women in Early British and Irish Astronomy: Stars and Satellites (London: Springer, 2009).

Brück, Maire Theresa and Ian Elliott, 'The family background of Lady Huggins (Margaret Lindsay Huggins)', *IAJ* **20** (1992), pp. 210–11.

Buijs-Ballot, Christoph Hendrik Diedrik, 'Akustische Versuche auf der Niederländischen Eisenbahn nebst gelegentlichen Bemerkungen zur Theorie des Hrn. Prof. Doppler', *APC* **66** (1845), pp. 321–51.

Burnett, John E. and Alison D. Morrison-Low, *'Vulgar and Mechanick': The Scientific Instrument Trade in Ireland 1650–1921* (Dublin: Royal Dublin Society, 1989).

Burnham, Robert, *Burnham's Celestial Handbook*, 3 vols (New York: Dover, 1978).

Burnham, Sherburne Wesley, 'Note on Hind's variable nebula in Taurus', *MNRAS* **51** (1890), pp. 94–6.

Burr, Thomas William, 'Spectrum analysis applied to the stars', *AR* **2** (1864), pp. 73–4.
  'Spectrum analysis of the stars and nebulae', *AR* **2** (1864), pp. 253–6.

Burstyn, Joan N., *Victorian Education and the Ideal of Womanhood* (New Brunswick, NJ: Rutgers University Press, 1984).

Burton, Charles Edward, 'On the present dimensions of the white spot Linné', *MNRAS* **34** (1874), pp. 107–8.

Campani, Giuseppe, 'Observations made in Italy, confirming the former, and withall fixing the period of the revolution of Mars', *PTRSL* **1** (1666), pp. 242–5.

Campbell, Lewis and William Garnett, *The Life of James Clerk Maxwell* (New York: Johnson Reprint Corporation, 1969).

Campbell, W. [M.], 'Lieut. Campbell's report', *PRS* **17** (1868), pp. 120–4.

Campbell, William Wallace, 'The spectrum of Nova Aurigae', *PASP* **4** (1892), pp. 231–47.
  'Sir William Huggins', *Annual Report of the Smithsonian Institution* (1910), pp. 307–17.

A Brief Account of the Lick Observatory of the University of California, 7th edn (Berkeley, CA: University of California Press, 1927).

Cannon, Annie Jump, 'Lady Huggins', *OBS* **38** (1915), pp. 323–4.

'Sarah Frances Whiting', *PA* **35** (1927), pp. 539–45.

Capron, John Rand, *Photographed Spectra* (London: E. & F. N. Spon, 1877).

Cardwell, Donald Stephen Lowell, *The Organisation of Science in England* (London: Heinemann, 1972).

Carpenter, William Benjamin, 'Spectrum analysis', *Good Words* **14** (1873), pp. 356–64 and pp. 528–31.

The Microscope and its Revelations (Philadelphia, PA: Lindsay & Blakiston, 1876).

Carrington, Richard Christopher, *Observations of Spots on the Sun, from November 9th 1853 to March 24th 1861, Made at Redhill* (London: Williams and Norgate, 1863).

Carroll, Lewis [Charles Lutwidge Dodgson], *Through the Looking-glass and What Alice Found There* (New York: Random House, 1946).

Chacornac, Jean, 'On the missing nebula in Coma Berenices', *MNRAS* **22** (1862), p. 277.

Chambers, George Frederick, *A Handbook of Descriptive Astronomy*, 3rd edn (Oxford: Clarendon Press, 1877).

4th edn (Oxford: Clarendon Press, 1890).

Chandler, Seth Carlo, 'Catalogue of variable stars', *AJ* **8** (1888), pp. 81–94.

'Second catalogue of variable stars', *AJ* **13** (1893), pp. 89–110.

Chant, Clarence Augustus, 'The Mount Wilson conference of the Solar Union', *JRASC* **4** (1910), pp. 356–72.

Chapman, Allan, *The Victorian Amateur Astronomer: Independent Astronomical Research in Britain 1820–1920* (Chichester: Praxis Publishing Ltd, 1998).

Chapman, D. C., 'Astronomical photography', *BJP* **22** (1875), pp. 630–1.

Christie, William Henry Mahoney, 'Dr. Huggins' method of photographing the solar corona without an eclipse', *MNRAS* **43** (1883), pp. 231–2.

'On the photographic magnitude of Nova Aurigae, as determined at the Royal Observatory, Greenwich', *MNRAS* **52** (1892), pp. 357–66.

Clarke, Thomas, 'The beer and albumen process', *BJP* **22** (1875), pp. 221–2.

Clerke, Agnes Mary, *A Popular History of Astronomy during the Nineteenth Century* (Edinburgh: Adam & Charles Black, 1885).

2nd edn (New York: Macmillan & Co., 1887).

3rd edn (London: Adam & Charles Black, 1893).

'William Allen Miller', in Sidney Lee (ed.), *DNB*, vol. 37 (New York: Macmillan & Co., 1894), pp. 429–30.

'William Huggins', *Encyclopedia Britannica*, 11th edn, vol. 13 (New York: Encyclopedia Britannica, 1910), pp. 856–7.

'The evolution of the stars', *ER* **191** (1900), pp. 455–77.

'Astronomy, 1846–1885', in Henry Duff Traill and James Suamarez Mann (eds), *Social England: A Record of the Progress of the People . . . from the Earliest Times to the Present Day*, vol. 6 (London: Cassell and Company, Limited, 1904), pp. 686–96.

Comte, Auguste, *Cours de Philosophie Positive*, vol. 2 (Paris: Bacheliere, Imprimeur-Libraire, 1833). English translation: Harriet Martineau, *The Positive Philosophy of August Comte* (London: George Bell, 1896).

Connolly, Thomas Francis, 'Recent work with the spectroheliograph', *Nature* **78** (1908), pp. 200–1.

'Solar vortices', *Nature* **78** (1908), pp. 368–9.

Cortie, Aloysius Laurence, 'The foundations of astrophysics', *OBS* **32** (1909), pp. 465–8.

Crary, Jonathan, *Techniques of the Observer: On Vision and Modernity in the Nineteenth Century* (Cambridge, MA: MIT Press, 1990).

Crookes, William, 'Early researches on the spectra of artificial light from different sources', *CN* **3** (1861), pp. 184–5; pp. 261–3; pp. 303–7.

'On the existence of a new element probably of the sulphur group', *CN* **3** (1861), pp. 193–5.

'The total solar eclipse of August last', *QJS* **7** (1870), pp. 28–43.

'The spectrum of the gas from clèveite', *CN* **71** (1895), p. 151.

'On radio-activity and radium', *CN* **94** (1906), p. 125.

'Radioactivity', *CN* **94** (1906), p. 125.

'Radium', *CN* **94** (1906), pp. 125–7; pp. 144–5; and pp. 153–4.

Croswell, Ken, 'Will the lion roar again?', *Astronomy* **19** (November, 1991), pp. 44–9.

Crowther, James G., *Statesmen of Science* (Bristol: Dufour, 1966).

Curie, Marie, 'Sur le poids atomique du radium', *CR* **135** (1902), pp. 161–3.

Curie, Pierre, 'Le radium', *PRI* **17** (1903), pp. 389–402.

Curie, Pierre, Mme. P. Curie and Gustave Bémont, 'Sur une nouvelle substance fortement radio-active, contenue dans la pechblende', *CR* **127** (1898), pp. 1215–17.

D'Albe, Edmund Edward Fournier, *The Life of Sir William Crookes* (London: T. Fisher Unwin, 1923).

Darwin, George Howard, 'On the mechanical conditions of a swarm of meteorites and on theories of cosmogony', *PTRSL* **180** (1889), pp. 1–69.

Dawes, Revd William Rutter, 'On the appearance of round bright spots on one of the belts of Jupiter', *MNRAS* **18** (1857), pp. 6–7 and pp. 49–50.

'Saturnian phenomena', *MNRAS* **22** (1862), pp. 297–9.

Dawson, George, *A Manual of Photography founded on Hardwich's Photographic Chemistry*, 8th edn (London: J. A. Churchill, 1873).

De la Rue, Warren, 'On the markings on the scales of Amathusia Horsfieldii', *TRMSL* **1** (1844), pp. 36–40.

'Mars in opposition in 1852', *MNRAS* **12** (1852), pp. 173–4.

'Report on the present state of celestial photography in England', *RBAAS, Aberdeen* (1859), pp. 130–53.

'The Bakerian Lecture: On the total solar eclipse of July 18th, 1860', *PTRSL* **152** (1862), pp. 333–416.

'The president's address', *MNRAS* **25** (1864), pp. 1–17.

'On some attempts to render the luminous prominences of the Sun visible without the use of the spectroscope', *MNRAS* **30** (1869), pp. 22–4.

Descartes, René, *Les Météores* (Leyde: I. Maire, 1637).

DeVorkin, David H., An astronomical symbiosis: stellar evolution and spectral classification (1860–1910), PhD thesis, University of Leicester (1978).

'W. W. Campbell's spectroscopic study of the martian atmosphere', *QJRAS* **18** (1977), pp. 37–53.

'Community and spectral classification in astrophysics: The acceptance of E. C. Pickering's system in 1910', *Isis* **72** (1981), pp. 29–49.

Dewhirst, David, 'The Greenwich–Cambridge axis', *Vistas in Astronomy* **20** (1976), pp. 109–11.

Dingle, Herbert, 'A hundred years of spectroscopy', *BJHS* **1** (1967), pp. 199–216.

Donati, Giovanni Battista, 'Memorie astronomiche', *MNRAS* **23** (1863), pp. 100–7.

Donkin, Alice E., 'Margaret Lindsay Huggins', *The Englishwoman* (May, 1915), pp. 152–9.

Doppler, Christian, 'Über das farbige Licht der Doppelsterne und einiger anderer Gestirne des Himmels', *Abhandlungen der Königlich Böhmischen Gesellschaft der Wissenschaften*, 5th Series **2** (1843), pp. 467–82.

Douglas-Smith, A. E., *The City of London School* (Oxford: Blackwell, 1965).

Draper, Henry, 'Photographs of the spectra of Venus and α Lyrae', *AJS*, 3rd Series **13** (1877), p. 95. Reprinted in *PM*, 5th Series **3** (1877), p. 238.

'On photographing the spectra of the stars and planets', *AJS*, 3rd Series **18** (1879), pp. 419–25. Reprinted in *Nature*, **21** (1879), pp. 83–5.

'On photographs of the spectrum of the nebula in Orion', *AJS*, 3rd Series **23** (1882), pp. 339–41.

Dreyer, John Louis Emil, Herbert Hall Turner, *et al.*, *History of the Royal Astronomical Society 1820–1920* (Oxford: Blackwell Scientific Publications, 1987; London: Wheldon & Wesley, 1923).

Dunkin, Edward, 'Lieutenant-Colonel Alexander Strange', *MNRAS* **37** (1877), pp. 154–9.

Obituary Notices of Astronomers: Fellows and Associates of the Royal Astronomical Society (London: Williams & Norgate, 1879).

'President's address on presenting the Gold Medal of the Society to William Huggins', *MNRAS* **45** (1885), pp. 277–93.

Dupré, Friedrich Wilhelm and August, 'On the existence of a fourth member of the calcium group of metals', *CN* **3** (1861), pp. 116–17.

Dyson, Frank Watson, 'Sir William Huggins', *PRS* **86A** (1910), pp. i–xix.

Edge, David O. and Michael J. Mulkay, *Astronomy Transformed: The Emergence of Radio Astronomy in Britain* (New York: John Wiley & Sons, 1976).

Ellery, Robert Lewis John, 'Observations of Mars made in May and June, 1894, with the Melbourne Great Telescope', *APJ* **1** (1895), pp. 47–8 and Plate IV.

Elliott, Ian, 'The Huggins' sesquicentenary', *IAJ* **26** (1999), pp. 65–8.

Ennis, Jacob, 'The influence of the Earth's atmosphere on the color of the STARS', *PANSP* **16** (1864), pp. 161–5.

The Origin of the Stars, and the Causes of their Motions and their Light (New York: D. Appleton and Company, 1867), pp. 102–163.

Fath, Edward Arthur, 'William Wallace Payne', *PA* **36** (1928), pp. 267–70.

Fizeau, Armand-Hippolyte-Louis, 'Acoustique et Optique', *Société Philomatique de Paris; Extraits des Procès-verbaux des Séances pendant l'Année 1848* (Paris: Imprimerie de Cosson, 1848), pp. 81–3.

'Des effets du mouvement sur le ton des vibrations sonores et sur la longueur d'onde des rayons de lumière', *ACP*, 4th Series **19** (1870), pp. 211–21.

Flammarion, Camille, *La Planète Mars et ses Conditions D'Habilité* (Paris: Gauthier-Villars et Fils, 1892).

Fleck, Ludwik, *Genesis and Development of a Scientific Fact* (Chicago, IL: University of Chicago Press, 1970). Originally published as *Entstehung und Entwicklung einer wissenschaftlichen Tatsache: Einführung in die Lehre vom Denkstil und Denkkollektiv* (Basel: Benno Schwabe & Co., 1935).

Foucault, Michel, *The Order of Things: An Archaeology of the Human Sciences* (New York: Pantheon, 1970). Originally published as *Les Mots et Les Choses: Une Archéologie des Sciences Humaines* (Paris: Gallimard, 1966).

Fowler, Alfred, 'The elements of astronomy', *Nature* **41** (1890), p. 485.

'Sir Norman Lockyer', *PRS* **104** (1920), pp. i–xiv.

Fraunhofer, Joseph, 'Bestimmung des Brechungs- und Farbenzerstreuungs-Vermögens verschiedener Glasarten, in Bezug auf die Vervollkommung achromatischer Fernröhre', *Denkschriften der Königlichen Akademie der Wissenschaften zu München für die Jahre 1814 und 1815* **5** (1817), pp. 193–226. English translation: 'Determination of the refractive and the dispersive power of different kinds of glass, with reference to the perfecting of achromatic telescopes', *EPJ* **9** (1823), pp. 288–99, and **10** (1824), pp. 26–40.

'Neue Modifikation des Lichtes durch gegenseitige Einwirkung und Beugung der Strahlen, und Gesetze derselben', *Denkschriften der königlichen Akademie der Wissenschaften zu München* **8** (1821), pp. 1–76. English translation: 'New modification of light by the mutual influence and the diffraction of the rays, and the laws of this modification', in J. S. Ames (ed.), *Prismatic and Diffraction Spectra: Memoirs by Joseph von Fraunhofer* (New York: Harper & Brothers, 1898), pp. 13–38.

'Kurzer Bericht von den Resultaten neuerer Versuche über die Gesetz des Lichtes, und die Theorie derselben', *AP*, 1st Series **74** (1823), pp. 337–78. English translation: 'A short account of the results of recent experiments upon the laws of light and its theory', *EJS* **7** (1827), pp. 101–13, 251–62, and **8** (1828), pp. 7–10.

Fresnel, Augustin-Jean, 'Mémoire sur la diffraction de la lumière, où l'on examine particulièrement le phénomène des franges colorées que présentent les ombres des corps éclairé par un point lumineux', *AC* **1** (1816), pp. 239–81.

'Mémoire sur la diffraction de la lumière', *AC* **11** (1819), pp. 246–96, 337–78.

Frew, David J., 'The historical record of η Carinae. I. The visual light curve, 1595–2000', *JAD* **10** (2004), pp. 1–76.

Frost, Edwin Brant, 'The Scientific Papers of Sir William Huggins', *APJ* **32** (1910), pp. 323–5.

G. J. W., 'Solar spectrum', *AR* **8** (1870), p. 64.

Gassiot, John P., 'On spectrum analysis . . .', *PRS* **12** (1863), pp. 536–8.

Gieryn, Thomas F. and Richard F. Hirsh, 'Marginality and innovation in science', *SSS* **13** (1983), pp. 87–106.

Gilbert, Ludwig Wilhelm, [Note to Readers], *AP*, New Series **1** (1809), p. 225.

Gill, David, 'Address on receipt of the McClean gift to the Cape Observatory', *OBS* **24** (1901), pp. 397–403.

Gill, Mrs [Isobel Sarah], *Six Months in Ascension: An Unscientific Account of a Scientific Expedition* (London: John Murray, 1880).

Giltay, K. M., 'On spectrum analysis', *CN* **4** (1861), pp. 328–9.

Gingerich, Owen, 'The first photograph of a nebula', *ST* **60** (1980), pp. 364–7.

(ed.), *Astrophysics and Twentieth Century Astronomy to 1950: Part A* (Cambridge: Cambridge University Press, 1984).

Gladstone, John H., 'Notes on the atmospheric lines of the solar spectrum, and on certain spectra of gases', *CN* **4** (1861), pp. 140–2.

Glass, Ian S., *Victorian Telescope Makers: The Lives and Letters of Thomas and Howard Grubb* (Bristol: Institute of Physics Publishing, 1997).

Revolutionaries of the Cosmos: The Astrophysicists (Oxford: Oxford University Press, 2006).

Goldberg, Leo and Lawrence H. Aller, *Atoms, Stars and Nebulae* (Philadelphia, PA: Blakiston Co., 1943).

Grandeau, Louis, *Instruction Pratique sur L'Analyse Spectrale* (Paris: Mallet-Bachelier, 1863).

Grant, Robert, *History of Physical Astronomy from the Earliest Ages to the Middle of the Nineteenth Century* (London: Robert Baldwin, 1852).

[Gregory, Richard A.], 'Sir William Huggins', *Nature* **83** (1910), pp. 342–3.

Grubb, Howard, 'Automatic spectroscope for Dr. Huggins' sun observations', *MNRAS* **31** (1870), pp. 36–8.

Hagstrom, Warren O., *The Scientific Community* (Carbondale, IL: University of Southern Illinois Press, 1965).

Hale, George Ellery, 'Photography of the solar prominences', *Technology Quarterly* **3** (1890), pp. 310–16.

'Photography and the invisible solar prominences', *SM* **10** (1891), pp. 257–64.

'The Kenwood Physical Observatory', *SM* **10** (1891), pp. 321–3.

'Photographic investigation of solar prominences and their spectra', *AJS*, 3rd Series **42** (1891), pp. 160–6 and Plate III.

'The ultra-violet spectrum of the solar prominences', *RBAAS, Cardiff* (1891), pp. 557–8. Full text in *AA* **11** (1892), pp. 50–9.

'The Congress on Mathematics, Astronomy and Astro-physics', *AA* **12** (1893), pp. 743–9.

'The spectroheliograph', *AA* **13** (1893), pp. 241–57.

'On some attempts to photograph the solar corona without an eclipse', *AA* **13** (1894), pp. 662–87.

'On a new method of mapping the solar corona without an eclipse', *APJ* **1** (1895), pp. 318–34.

'Note on the exposure required in photographing the solar corona without an eclipse', *APJ* **1** (1895), pp. 438–9.

'Note on the Huggins method of photographing the solar corona without an eclipse', *APJ* **2** (1895), p. 77.

'An Atlas of Representative Stellar Spectra', *APJ* **12** (1900), pp. 291–7.

'A study of the conditions for solar research at Mount Wilson, California', *APJ* **21** (1905), pp. 124–50.

'The Solar Observatory of the Carnegie Institution of Washington', *Contributions from the Solar Observatory of the Carnegie Institution of Washington*, No. 2 (Washington, DC: Carnegie Institution, 1905), pp. 7–17.

'Some tests of the Snow Telescope', *APJ* **23** (1906), pp. 6–10.

The Study of Stellar Evolution: An Account of some Recent Methods of Astrophysical Research (Chicago, IL: University of Chicago Press, 1908).

'Solar vortices and magnetic fields', *Nature* **82** (1909), pp. 20–3 and pp. 50–3.

'The work of Sir William Huggins', *APJ* **37** (1913), pp. 145–53.

Hale, George Ellery and Ferdinand Ellerman, 'The Rumford Spectroheliograph of the Yerkes Observatory' in *Publications of the Yerkes Observatory*, vol. 3 (Chicago, IL: University of Chicago Press, 1903).

'Calcium and hydrogen flocculi', *APJ* **19** (1904), pp. 41–51, with twelve plates.

Hale, George Ellery and Charles Dillon Perrine, 'Proceedings of the first Conference on Solar Research, held at St. Louis, September 23, 1904', *APJ* **20** (1904), pp. 301–6.

Hall, Angelo, *An Astronomer's Wife: The Biography of Angeline Hall* (Baltimore, MD: Nunn & Co., 1908).

Hall, Marie Boas, *All Scientists Now: The Royal Society in the Nineteenth Century* (Cambridge: Cambridge University Press, 1984).

Halley, Edmund, [Edmundus Halleius] *Catalogus Stellarum Australium sive Supplementum Catalogi Tychonici* (London: Thomae James, 1679).

'Considerations on the change of latitudes of some of the principal fixt stars', *PTRSL* **30** (1718), pp. 736–8.

Hard, Arnold Henry, *The Story of Rayon and other Synthetic Textiles* (London: United Trade Press Ltd, 1939).

Hargens, Lowell L., Nicholas C. Mullins and Pamela K. Hecht, 'Research areas and stratification processes in science', *SSS* **10** (1980), pp. 55–74.

Harman, Peter (ed.), *The Scientific Letters and Papers of James Clerk Maxwell*, 2 vols (Cambridge: Cambridge University Press, 1990).

The Natural Philosophy of James Clerk Maxwell (Cambridge: Cambridge University Press, 1998).

Harrison, James Park, 'Inductive proof of the Moon's insolation', *MNRAS* **28** (1868), p. 29.

Harrison, John F. C., *The Early Victorians 1832–1851* (New York: Praeger, 1971).

Hearnshaw, John B., 'Doppler and Vogel – Two notable anniversaries in stellar astronomy', *Vistas in Astronomy* **35** (1992), pp. 157–77.

Heathcote, Bernard V. and Pauline F., 'The feminine influence: Aspects of the role of women in the evolution of photography in the British Isles', *History of Photography* **12** (1988), pp. 259–73.

Hentschel, Klaus, *Mapping the Spectrum: Techniques of Visual Representation in Research and Teaching* (Oxford: Oxford University Press, 2002).

Herrmann, Dieter, *The History of Astronomy from Herschel to Hertzsprung* (Cambridge: Cambridge University Press, 1984).

Herschel, John Fredrick William, 'Presidential address on presenting the Gold Medal to Francis Baily', *MNRAS* **1** (1827), pp. 14–20.

'Observations of nebulae and clusters of stars, made at Slough, with a twenty-feet reflector, between the years 1825 and 1833', *PTRSL* **123** (1833), pp. 359–509.

'Extract of a letter from Sir John Herschel to the President . . .', *MNRAS* **4** (1838), pp. 121–2.

'Auszug eines Schreibens von Sir John Herschel . . .', *AN* **15** (1838), pp. 311–12.

Results of Astronomical Observations Made during the Years 1834, 5, 6, 7, 8, at the Cape of Good Hope; being the Completion of a Telescopic Survey of the Whole Surface of the Visible Heavens, Commenced in 1825 (London: Smith, Elder and Co., 1847).

Outlines of Astronomy (London: Longman, Brown, Green, Longmans, and Roberts, 1849); 4th edn (Philadelphia, PA: Blanchard & Lea, 1861); 7th edn (London: Longman, Green, Roberts, & Green, 1864); 11th edn (London: Longman, Brown, Green, Longmans, and Roberts, 1871)

'Letter . . . to Mr. Hind on the disappearance of a nebula in Coma Berenices', *MNRAS* **22** (1862), pp. 248–50.

'Variability of nebulae', *CM* **7** (1863), pp. 143–4.

'Catalogue of nebulae and clusters of stars', *PTRSL* **154** (1864), pp. 1–137.

'On light', *Good Words* **6** (1865), pp. 358–64.

'On a possible method of viewing the red flames without an eclipse', *MNRAS* **29** (1868), pp. 5–6.

'On the variable star η Argus and its surrounding nebula', *MNRAS* **28** (1868), pp. 225–9.

'Remarks on Mr. Abbott's paper on η Argus', *MNRAS* **31** (1871), p. 228.

Herschel, John Fredrick William and Lieut John Herschel, 'The Great Nebula round η Argus', *MNRAS* **29** (1869), p. 82.

Herschel, Lieut John, 'The great eclipse of the Sun. I', *The Engineer* **6** (1868), pp. 345–6.

'Account of the solar eclipse of 1868, as seen at Jamkhandi in the Bombay Presidency', *PRS* **17** (1868), pp. 104–20

Herschel, William, 'On the remarkable appearances at the polar regions of the planet Mars, the inclination of its axis, the position of its poles . . .', *PTRSL* **74** (1784), pp. 233–73.

'Catalogue of one thousand new nebulae and clusters of stars', *PTRSL* **76** (1786), pp. 457–99.

'Catalogue of a second thousand of new nebulae and clusters of stars; with a few introductory remarks on the construction of the heavens', *PTRSL* **79** (1789), pp. 212–55.

'On nebulous stars, properly so called', *PTRSL* **81** (1791), pp. 71–88.

'Catalogue of 500 new nebulae, *PTRSL* **92** (1802), pp. 477–528.

'Astronomical observations relating to the construction of the heavens', *PTRSL* **101** (1811), pp. 269–336.

Hetherington, Norriss S., *Science and Objectivity: Episodes in the History of Astronomy* (Ames, IA: Iowa State University Press, 1988).

Heysinger, Isaac Winter, *The Source and Mode of Solar Energy throughout the Universe* (Philadelphia, PA: Lippincott, 1895)

Hind, John, '*Auszug aus einem Schreiben des Herrn Hind an die Redaction*', *AN* **35** (1853), pp. 371–2.

'Note on the variable nebula in *Taurus*', *MNRAS* **24** (1864), p. 110.

Hirsh, Richard F., 'The riddle of the gaseous nebulae', *Isis* **70** (1979), pp. 197–212.

Hodgkins, Louise Manning, 'Lady Huggins: Astronomer', *The Christian Advocate* (21 October 1915), pp. 1417–18.

Hofmann, J. G., 'Sur un nouveau modèle de prisme pour spectroscope à vision direct', *CR* **79** (1874), p. 581.

Hogg, Helen Sawyer, 'Variable stars' in Owen Gingerich (ed.), *Astrophysics and Twentieth-century Astronomy to 1950: Part A* (Cambridge: Cambridge University Press, 1984), pp. 73–89.

Holberg, Jay B., *Sirius: Brightest Diamond in the Sky* (New York: Springer, 2007), pp. 90–2.

Holden, Edward Singleton and C. S. Hastings, *A Synopsis of the Scientific Writings of Sir William Herschel* (Washington, DC: Government Printing Office, 1881).

Holmes, Frederic L., 'Scientific writing and scientific discovery', *Isis* **78** (1987), pp. 220–35.

Hoskin, Michael A., *William Herschel and the Construction of the Heavens* (New York: W. W. Norton & Co., 1963).

'Apparatus and ideas in mid-nineteenth-century cosmology', in A. Beer (ed.), *New Aspects in the History and Philosophy of Astronomy* (Oxford: Pergamon Press, 1967), pp. 79–85.

Hufbauer, Karl, *Exploring the Sun: Solar Science since Galileo* (Baltimore: The Johns Hopkins University Press, 1991).

'Amateurs and the rise of astrophysics 1840–1910', *Berichte zu Wissenschaftsgeschichte* **9** (1986), pp. 183–90.

Huggins, Margaret Lindsay, 'The astrolabe', *PA* **2** (1894), pp. 199–202 and pp. 261–6.

'. . . Teach me how to name the . . . light', *APJ* **8** (1898), p. 54.

Agnes Mary Clerke and Ellen Mary Clerke: An Appreciation (Printed for Private Circulation, 1907).

Hughes, Jabez, 'Photography as an industrial occupation for women', *BJP* **20** (1873), pp. 222–3.

Hunt, Robert, *A Manual of Photography* (London: John Joseph Griffin & Co., 1853).

Hurn, Mark, *Sir William Huggins: The Cambridge Connection*, (Cambridge: Institute of Astronomy, 2008).

Hutchins, Roger, *British University Observatories 1772–1939* (Aldershot: Ashgate, 2008).

Hyde, Walter Lewis, 'The calamity of the Great Melbourne Telescope', *PASA* **7** (1987), pp. 227–30.

Inkster, Ian and Jack Morrell, *Metropolis and Province: Science in British Culture, 1780–1850* (London: Hutchinson, 1983).

Jackson, Myles W., *Spectrum of Belief: Joseph von Fraunhofer and the Craft of Precision Optics* (Cambridge, MA: The MIT Press, 2000).

James, Frank A. J. L., 'Lockyer as astronomer', *Nature* **225** (1970), pp. 230–2.

    The early development of spectroscopy and astrophysics, PhD thesis, Imperial College of Science and Technology (1981).

    'Thermodynamics and sources of solar heat, 1846–1862', *BJHS* **15** (1982), pp. 155–81.

    'The conservation of energy, theories of absorption and resonating molecules, 1851–1854: G. G. Stokes, A. J. Angstrom and W. Thomson', *NRRSL* **38** (1983), pp. 79–107.

    'The debate on the nature of the absorption of light, 1830–1835', *HS* **21** (1983), pp. 335–68.

    'The establishment of spectro-chemical analysis as a practical method of qualitative analysis, 1854–1861', *Ambix* **30** (1983), pp. 30–53.

    'The study of spark spectra, 1835–1859', *Ambix* **30** (1983), pp. 137–62.

    'The creation of a Victorian myth: The historiography of spectroscopy', *HS* **23** (1985), pp. 1–24.

    'The discovery of line spectra', *Ambix* **32** (1985), pp. 53–70.

    'Spectro-chemistry and myth: A rejoinder', *HS* **24** (1986), pp. 433–7.

    'The tales of Benjamin Abbott: A source for the early life of Michael Faraday', *BJHS* **25** (1992), pp. 229–40.

Janssen, Pierre Jules, 'On the solar protuberances', *PRS* **17** (1868), pp. 276–7.

    'Summary of some of the results obtained during the total solar eclipse of August 1868, Parts I and II', *AR* **7** (1869), pp. 107–10 and pp. 131–4.

Jensen, William B., 'Why helium ends in "-ium"', *Journal of Chemical Education* **81** (2004), p. 944.

Jones, K. G., 'S Andromedae, 1885: An analysis of contemporary reports and a reconstruction', *JHA* **7** (1976), pp. 27–40.

Kargon, Robert Hugh, 'Arthur Schuster', *DSB*, vol. 12 (New York: Scribner & Son, 1970), pp. 237–9.

    Science in Victorian Manchester: Enterprise and Expertise (Manchester: Manchester University Press, 1977).

Keeler, James Edward, 'On the motions of the planetary nebulae in the line of sight', *PASP* **2** (1890), pp. 265–80.

    'On the chief line in the spectrum of the nebulae' [communicated by W. Huggins], *PRS* **49** (1891), pp. 399–403.

    'Spectroscopic observations of nebulae made at Mount Hamilton, California', *Publications of the Lick Observatory* **3** (1894), pp. 161–229.

    'The importance of astrophysical research and the relation of astrophysics to other physical sciences', *APJ* **6** (1897), pp. 271–88.

Kincaid, Sidney Bolton, 'On the estimation of star colours', *MNRAS* **27** (1867), pp. 264–6.

King, Henry Charles, *The History of the Telescope* (London: Charles Griffin & Company Limited, 1955).

Kirchhoff, Gustav, 'Über die Fraunhofer'schen Linien', *Monatsberichte der Königlichen Preussische Akademie der Wissenschaften zu Berlin aus dem Jahre 1859* (1860), pp. 662–5. Reprinted in *AP* **109** (1860), pp. 148–51. English translation: G. Stokes, 'On the simultaneous emission and absorption of rays of the same definite refrangibility', *PM*, 4th Series **20** (1860), pp. 195–6.

'Contributions towards the history of spectrum analysis and of the analysis of the solar atmosphere', *PM*, 4th Series **25** (1863), pp. 250–62.

Kirchhoff, Gustav, and Robert Bunsen, 'Chemische Analyse durch Spectralbeobachtungen', *APC* **110** (1860), pp. 161–89. English translation: 'Chemical analysis by spectrum-observations', *PM*, 4th Series **20** (1860), pp. 89–109.

'Chemical analysis by spectrum-observations – Second memoir', *PM*, 4th Series **22** (1861), pp. 329–49 and pp. 498–510, and Plate VI.

Klinkerfues, Ernst Friedrich Wilhelm, 'Fernere Mittheilungen über den Einfluss der Bewegung der Lichtquelle auf die Brechbarkeit eines Strahls', *Nachrichten von der Königliche Gesellschaft der Wissenschaften zu Göttingen* No. 4 (1866), pp. 33–60.

Knight, David M., 'The vital flame', *Ambix* **23** (1976), pp. 5–15.

Knobel, Edward Ball, 'Warren De La Rue', *MNRAS* **50** (1890), pp. 155–64.

Knott, George, 'The new star in Auriga', *MNRAS* **52** (1892), pp. 367–8.

Kohler, Robert, *From Medical Chemistry to Biochemistry: The Making of a Biomedical Chemistry* (Cambridge: Cambridge University Press, 1982).

Krisciunas, Kevin, *Astronomical Centers of the World* (Cambridge: Cambridge University Press, 1988).

Kuhn, Thomas, *The Structure of Scientific Revolutions*, 2nd edn (Chicago, IL: University of Chicago Press, 1970).

Lacaille, Nicolas Louis de, 'Sur les étoiles nébuleuses du ciel Austral', *Histoire de L'Académie Royale des Sciences, 1775* (Paris: l'Imprimerie Royale, 1756), pp. 194–9.

Lankester, Edwin Ray, 'Address of the president', *RBAAS, York* (1906), pp. 3–42.

Lankford, John, 'Amateur versus professional: The transatlantic debate over the measurement of jovian longitude', *JBAA* **89** (1979), pp. 574–82.

'A note on T. J. J. See's observations of craters on Mercury', *JHA* **11** (1980), pp. 129–32.

'Amateurs and astrophysics: A neglected aspect in the development of a scientific specialty', *SSS* **11** (1981), pp. 275–303.

'Amateurs versus professionals: The controversy over telescope size in late Victorian science', *Isis* **72** (1981), pp. 11–28.

'The impact of photography on astronomy', in Owen Gingerich (ed.), *Astrophysics and Twentieth Century Astronomy to 1950* (Cambridge: Cambridge University Press, 1984), pp. 16–39.

Laplace, Pierre Simon de, *Exposition du système du monde*, 2 vols (Paris: Cercle Social, 1796).

Larmor, Joseph, *Aether and Matter: A Development of the Dynamical Relations of the Aether to Material Systems on the Basis of the Atomic Constitution of Matter* (Cambridge: Cambridge University Press, 1900).

(ed.), *Memoir and Scientific Correspondence of the Late Sir George Gabriel Stokes*, 2 vols (Cambridge University Press: Cambridge, 1971; 1907).

Lassell, William, 'Discovery of a new satellite of Saturn', *MNRAS* **8** (1848), pp. 195–7.

'On a satellite of Neptune', *MNRAS* **11** (1850), pp. 61–2.

'In a letter dated . . .', *MNRAS* **11** (1851), p. 201.

Lee, John, 'Address delivered by the president, on presenting the Gold Medal of the Society to Mr. Warren De la Rue', *MNRAS* **22** (1862), pp. 131–40.

Lemaine, Gérard, Roy McLeod, *et al.* (eds), *Perspectives on the Emergence of Scientific Disciplines* (The Hague: Mouton, 1976).

Lightman, Bernard (ed.), *Victorian Science in Context* (Chicago, IL: University of Chicago Press, 1997).

'Celestial objects for common readers', in Janet and Mark Robinson (eds), *The Stargazer of Hardwicke: The Life and Work of Thomas William Webb* (Leominster: Gracewing, 2006), pp. 215–34.

Lilly, W. S., 'British monarchy and modern democracy', *The Nineteenth Century* **41** (1897), pp. 853–64.

Lockyer, Joseph Norman, 'Note on communication of a note on the "Lines in the spectra of some of the fixed stars"', *MNRAS* **23** (1863), pp. 179–80.

'Science', *The Reader* **1** (1863), p. 20 and p. 47.

'The paper on stellar spectra', *The Reader* **1** (1864), p. 248. Reprinted (nearly verbatim) in 'Stellar spectra', *AR* **1** (1863), p. 54.

'Stellar physics', *The Reader* **2** (1864), pp. 14–16.

'Variable nebulae', *The Reader* **3** (1864), p. 109.

'Celestial analysis', *The Reader* **4** (1864), pp. 577–9.

'What is a nebula?', *The Reader* **5** (1865), pp. 631–2.

'Spectroscopic observations of the Sun', *PRS* **15** (1866), pp. 256–8.

'Notice of an observation of the spectrum of a solar prominence', *PRS* **17** (1868), pp. 91–2.

'Spectroscopic observation of the "red prominences" without an eclipse of the Sun', *MNRAS* **29** (1869), pp. 162–3.

'Note on a paper by Mr. Huggins', *AR* **7** (1869), p. 42.

The Spectroscope and Its Applications (London: Macmillan & Co, 1873).

'On spectrum photography', *Nature* **10** (1874), pp. 109–12 and pp. 254–6.

Contributions to Solar Physics (London: Macmillan & Co., 1874).

Studies in Spectrum Analysis (New York: Appleton & Co., 1878).

Chemistry of the Sun (London: Macmillan & Co., 1887).

'Researches on meteorites', *Nature* **37** (1887), pp. 55–61 and pp. 80–7.

'Researches on the spectra of meteorites. A report to the Solar Physics Committee', *PRS* **43** (1887), pp. 117–56.

'Suggestions on the classification of the various species of heavenly bodies – The Bakerian Lecture', *PRS* **44** (1888), pp. 1–93.

'Appendix to Bakerian Lecture: Suggestions on the classification of the various species of heavenly bodies', *PRS* **45** (1889), pp. 157–262.

'On the spectra of meteor-swarms (Group III)', *PRS* **45** (1889), pp. 380–92.

'On the wave-length of the chief fluting seen in the spectrum of manganese', *PRS* **46** (1889), pp. 35–40.

'On the cause of variability in condensing swarms of meteorites', *PRS* **46** (1889), pp. 401–23.

'Comparisons of the spectra of nebulae and stars of groups I and II with those of comets and aurorae', *PRS* **47** (1889), pp. 28–39.

'The presence of bright carbon flutings in the spectra of celestial bodies', *PRS* **47** (1889), pp. 39–40.

'On the chief line in the spectrum of the nebulae', *PRS* **48** (1890), pp. 167–98.

'Note on the spectrum of the nebula of Orion', *PRS* **48** (1890), pp. 198–9.

'Preliminary note on photographs of the spectrum of the nebula in Orion', *PRS* **48** (1890), pp. 199–201.

'On the spectra of Comet *a* 1890 and the nebula G. C. 4058 [NGC 5866 in Draco]', *PRS* **48** (1890), pp. 217–20.

'Notes on meteorites', *Nature* **38** (1888), pp. 424–8, pp. 456–8, pp. 530–3, pp. 556–9 and pp. 602–5; *Nature* **39** (1889), pp. 139–42, pp. 233–6 and pp. 400–2; and *Nature* **40** (1889), pp. 136–9.

The Meteoritic Hypothesis: A Statement of the Results of a Spectroscopic Inquiry into the Origin of Cosmical Systems (London: Macmillan & Co., 1890).

'On the causes which produce the phenomena of new stars', *PTRSL* **182** (1891), pp. 397–448. Abstract in *PRS* **49** (1891), pp. 443–6.

Dawn of Astronomy: A Study of the Temple-worship and Mythology of the Ancient Egyptians (London: Macmillan and Co., 1894).

'On the new gas obtained from uraninite', *PRS* **58** (1895), pp. 67–70; pp. 113–16; pp. 116–19; p. 192; pp. 193–5.

'The Sun's place in nature', *Nature* **51** (1895), pp. 374–7, pp. 396–9, pp. 565–7, and pp. 590–2; *Nature* **52** (1895), pp. 12–14, pp. 156–8, pp. 204–7, pp. 253–5, pp. 327–9, pp. 422–5, and pp. 446–50.

The Sun's Place in Nature (London: Macmillan and Co., 1897).

Recent and Coming Eclipses: Being Notes on the Total Solar Eclipses of 1893, 1896, and 1898 (London: Macmillan & Co., 1897).

Inorganic Evolution: As Studied by Spectrum Analysis (London: Macmillan & Co., 1900).

Lockyer, Thomazina Mary and Winifred Lucas Lockyer, *The Life and Work of Sir Norman Lockyer* (London: Macmillan, 1928).

Lodge, Oliver, *The Romanes Lecture 1903: Modern Views on Matter* (Oxford: The Clarendon Press, 1903).

Lohne, Johannes A., 'Thomas Harriott (1560–1621): The Tycho Brahe of optics', *Centaurus* **6** (1959), pp. 113–21.

'The fair fame of Thomas Harriott: Rigaud *versus* Baron von Zach', *Centaurus* **8** (1963), pp. 69–84.

Love, E. F. J., 'On a method of discriminating real from accidental coincidences between lines of different spectra', *PM*, 5th Series **25** (1888), pp. 1–6.

Lyall, Alfred, 'India under Queen Victoria', *The Nineteenth Century* **41** (1897), pp. 865–882.

Lynn, W. T., 'William John Burchell and η Argus', *OBS* **30** (1907), pp. 138–40.

Lyons, Henry G., *The Royal Society 1660–1940: A History of its Administration under its Charters* (New York: Greenwood Press, 1968).

Lyot, Bernard, 'La couronne solair étudiée en dehors des éclipses', *CR* **191** (1930), pp. 834–7.

'A study of the solar corona and prominences without eclipses', *MNRAS* **99** (1939), pp. 580–94.

Mack, Edward C., *Public Schools and British Opinion 1780 to 1860* (London: Methuen & Co., 1936).

Public Schools and British Opinion since 1860 (New York: Columbia University Press, 1941).

MacLeod, Roy M., 'The Royal Society and the Government Grant: Notes on the administration of scientific research, 1849–1914', *Historical Journal* **14** (1971), pp. 323–58.

'Of medals and men: A reward system in Victorian science 1826–1914', *NRRSL* **26** (1971), pp. 81–105.

'Resources of science in Victorian England: The endowment of science movement, 1868–1900', in Peter Mathias (ed.), *Science and Society 1600–1900* (Cambridge: Cambridge University Press, 1972), pp. 111–66.

Mädler, Johann Heinrich von, 'On the Crater Linné', *MNRAS* **27** (1867), p. 303.

Main, Revd Robert, 'Observations of Comet II, 1861', *MNRAS* **22** (1862), pp. 55–7.

'President's address on presenting the Gold Medal to Mr. De la Rue', *MNRAS* **22** (1862), pp. 131–40.

'Extract of a letter from Mr. A. Auwers . . .', *MNRAS* **22** (1862), pp. 148–50.

Malmgreen, Gail, *Silk Town: Industry and Culture in Macclesfield, 1750–1835* (Hull: Hull University Press, 1985).

Martin, Stanley, *The Order of Merit: One Hundred Years of Matchless Honour* (London: I. B. Tauris, 2006).

Mason, Joan, 'Hertha Ayrton (1854–1923) and the admission of women to the Royal Society of London', *NRRSL* **45** (1991), pp. 201–20.

'Women in science: Breaking out of the circle', *NRRSL* **46** (1992), pp. 177–82.

Mathias, Peter (ed.), *Science and Society 1600–1900* (Cambridge: Cambridge University Press, 1972).

Maunder, Edward Walter, 'The motions of stars in the line of sight', *OBS* **8** (1885), pp. 117–22.

'Note on the spectrum of Nova Aurigae', *MNRAS* **52** (1892), pp. 369–71.

Sir William Huggins and Spectroscopic Astronomy (London: T. C. and E. C. Jack, 1913).

May, Trevor, *An Economic and Social History of Britain 1760–1970* (New York: Longman, 1987).

McCarthy, M. F., S. J., 'Fr. Secchi and stellar spectra', *PA* **58** (1950), pp. 153–69.

McGucken, William, *Nineteenth Century Spectroscopy: Development of the Understanding of Spectra 1802–1897* (Baltimore, MD: The Johns Hopkins University Press, 1969).

McKenna-Lawlor, Susan, *Whatever Shines Should be Observed* (Dublin: Samton Limited, 1998).

McRae, Robert James, The origin of the conception of the continuous spectrum of heat and light, PhD thesis, University of Wisconson (1969).

Meadows, Arthur Jack, *Early Solar Physics* (Oxford: Pergamon Press, 1970).
  Science and Controversy: A Biography of Sir Norman Lockyer (Cambridge, MA: MIT Press, 1972).
  Greenwich Observatory, Volume 2: Recent History (1836–1975) (London: Taylor & Francis, 1975).
  'The Airy Era', *Vistas in Astronomy* **20** (1976), pp. 197–201.
  'The origins of astrophysics', in Owen Gingerich (ed.), *Astrophysics and Twentieth Century Astronomy to 1950: Part A* (Cambridge: Cambridge University Press, 1984), pp. 3–15.
  'The new astronomy', in Owen Gingerich (ed.), *Astrophysics and Twentieth Century Astronomy to 1950: Part A* (Cambridge: Cambridge University Press, 1984), pp. 59–72.
Meadows, Arthur Jack and J. E. Kennedy, 'The origin of solar-terrestrial studies', *Vistas in Astronomy*, **25** (1981), pp. 419–26.
Meinel, Aden and Margorie Meinel, *Sunsets, Twilights, and Evening Skies* (Cambridge: Cambridge University Press, 1983).
Merrill, Paul, 'Annie Jump Cannon', *MNRAS* **102** (1942), pp. 72–6.
Messier, Charles, 'Catalogue des nébuleuses et des amas d'étoiles', *Connaissance des Temps, ou Mouvement des Astres, pour l'Année Commune 1784* (Paris: l'Imprimerie Royale, 1781), pp. 227–72.
Miller, William Allen, 'On action of gases on the prismatic spectrum', *The Chemist* **6** (1845), p. 404.
  'On the action of gases on the prismatic spectrum' [abstr.], *RBAAS, Cambridge* (1845), pp. 28–9.
  Full text: 'Experiments and observations on some cases of lines in the prismatic spectrum produced by the passage of light through coloured vapours and gases, and from certain coloured flames', *PM*, 3rd Series **27** (1845), pp. 81–91. Reprinted in *CN* **3** (1861), pp. 304–7.
  Elements of Chemistry: Theoretical and Practical, 2nd edn (London: John W. Parker & Son, 1860).
  'Address to the Chemistry Section', *RBAAS, Manchester* (1861), pp. 75–6. Summary in *CN* **4** (1861), pp. 159–61.
  'On spectrum analysis', *JPS*, 2nd Series **3** (1862), pp. 399–412. Reprinted in *CN* **5** (1862), pp. 201–3; pp. 214–18.
  'On the photographic transparency of various bodies, and on the photographic effects of metallic and other spectra obtained by means of the electric spark', *PTRSL* **152** (1862), pp. 861–87.
Mills, Charles E. and C. F. Brooke, *A Sketch of the Life of Sir William Huggins* (Richmond: Times Printing Works, 1936).
Mitchell, Sally (ed.), *Victorian Britain: An Encyclopedia* (New York: Garland, 1988).
Moigno, L'Abbé François-Napoléon Marie, *Répertoire d'Optique Moderne, ou Analyse Complète des Travaux Modernes Relatifs aux Phénomènes de la Lumière*, 3 vols (Paris: A. Franck, 1850).
  Le Révérend Père Secchi, sa Vie (Paris: Gauthier-Villars, 1879).
Morgan, Julie, 'The Huggins Archives at Wellesley College', *JHA* **11** (1980), p. 147.
Morren, M., 'On spectrum analysis' (Extract from a letter to the Abbé Moigno, Editor of 'Cosmos'), *CN* **4** (1861), pp. 302–3.
Morus, Iwan Rhys, 'Currents from the underworld: Electricity and the technology of display in early Victorian England', *Isis* **84** (1993), pp. 50–69.
  Frankenstein's Children: Electricity, Exhibition, and Experiment in Early-Nineteenth-century London (Princeton, NJ: Princeton University Press, 1998).
Mulkay, Michael and David Edge, *Astronomy Transformed: The Emergence of Radio Astronomy in Britain* (New York: John Wiley & Sons, 1976).
Narrien, John, *An Historical Account of the Origin and Progress of Astronomy* (London: Baldwin & Cradock, 1833).
Nasmyth, James, 'On the structure of the luminous envelope of the Sun', *MLPSM*, 3rd Series **1** (1862), pp. 407–11.
Newall, Hugh Frank, *The Spectroscope and its Work* (London: Society for Promoting Christian Knowledge, 1910).
  'Sir William Huggins, K.C.B., O. M.', *SP* **5** (1910–11), pp. 173–90.
  'William Huggins', *MNRAS* **71** (1911), pp. 261–70.

'Dame Margaret Lindsay Huggins', *MNRAS* **76** (1916), pp. 278–82.

'The decade 1860–1870', in J. L. E. Dreyer and H. H. Turner (eds), *History of the Royal Astronomical Society 1820–1920* (Oxford: Blackwell Scientific Publications, 1987; London: Wheldon & Wesley, 1923), pp. 129–66.

Newcomb, Simon, 'Reminiscences of an astronomer. I', *The Atlantic Monthly* **82** (1898), pp. 244–53.

The Reminiscences of an Astronomer (Boston, MA: Houghton, Mifflin and Company, 1903).

Newhall, Beaumont, *The History of Photography from 1839 to the Present Day* (New York: Museum of Modern Art, 1964).

Newton, Hubert Anson, 'The original accounts of the displays in former times of the November star-shower; together with a determination of the length of its cycle, its annual period, and the probable orbit of the group of bodies round the Sun', *AJS*, 2nd Series **37** (1864), pp. 377–89, and **38** (1864), pp. 53–61.

Newton, Isaac, 'A letter of Mr. Isaac Newton . . . containing his new theory about light and colours', *PTRSL* **80** (1671/72), pp. 3075–87.

Opticks, or A Treatise of the Reflections, Refractions, Inflections & Colours of Light, based on the 1730 edn (New York: Dover Publications, 1979).

Nichol, John Pringle, *Views of the Architecture of the Heavens* (Edinburgh: William Tait, 1837).

Norberg, Arthur L., 'Simon Newcomb's early astronomical career', *Isis* **69** (1978), pp. 209–25.

Norman, Daniel, 'The development of astronomical photography', *Osiris* **5** (1938), pp. 560–94.

North, Sheryl J., 'The telescope widow syndrome', *ST* **80** (1990), p. 228.

Ogilvie, Marilyn Bailey, 'Marital collaboration: An approach to science', in P. Abir-Am and D. Outram (eds), *Uneasy Careers and Intimate Lives: Women in Science 1789–1979* (New Brunswick, NJ: Rutgers University Press, 1987), pp. 104–25.

'Obligatory amateurs: Annie Maunder (1868–1947) and British women astronomers at the dawn of professional astronomy', *BJHS* **33** (2000), pp. 67–84.

Olson, Roberta J. M., *Fire and Ice: A History of Comets in Art* (New York: Walker & Co., 1985).

Osterbrock, Donald Edward, *James E. Keeler: Pioneer American Astrophysicist and the Early Development of American Astrophysics* (Cambridge: Cambridge University Press, 1984).

'The rise and fall of Edward S. Holden. Parts I and II', *JHA* **15** (1984), pp. 81–127 and 151–76.

'Failure and success: Two early experiments with concave gratings in stellar spectroscopy', *JHA* **17** (1986), pp. 119–29.

Osterbrock, Donald Edward, John R. Gustafson and W. J. Shiloh Unruh, *Eye on the Sky: Lick Observatory's First Century* (Berkeley, CA: University of California Press, 1988).

Ottewell, Guy, *The Under-standing of Eclipses* (Greenville, SC: Astronomical Workshop, 1991).

Pang, Alex Soojung-Kim, Spheres of interest: Imperialism, culture, and practice in British solar eclipse expeditions 1860–1914, PhD thesis, University of Pennsylvania (1991).

'The social event of the season: Solar eclipse expeditions and Victorian culture', *Isis* **84** (1993), pp. 252–77.

'Victorian observing practices, printing technology, and representations of the solar corona, (1): The 1860s and 1870s', *JHA* **25** (1994), pp. 249–74.

'Gender, culture, and astrophysical fieldwork: Elizabeth Campbell and the Lick Observatory-Crocker eclipse expeditions', *Osiris*, 2nd Series **11** (1996), pp. 15–43.

Empire and the Sun: Victorian Solar Eclipse Expeditions (Stanford, CA: Stanford University Press, 2002).

Papanek, Hanna, 'Men, women, and work: Reflections on the two-person career', *AJS* **78** (1973), pp. 852–72.

Parsons, Lawrence, 4th Earl of Rosse, 'On the radiation of heat from the Moon', *PRS* **17** (1869), pp. 436–44.

'Note on the construction of thermopiles', *PRS* **18** (1870), pp. 553–6.

'On the radiation of heat from the Moon. Part II', *PRS* **19** (1870), pp. 9–14.

'On the radiation of heat from the Moon, the law of its absorption by our atmosphere, and its variation in amount with her phases, The Bakerian Lecture', *PRS* **21** (1872), pp. 241–2; and *PTRSL* **163** (1873), pp. 587–627.

Parsons, William, 3rd Earl of Rosse, 'Observations on some of the nebulae', *PTRSL* **134** (1844), pp. 321–4.

'Observations on the nebulae', *PTRSL* **140** (1850), pp. 499–514.

Penlake, Richard [Percy R. Salmon], 'The question of backed plates', *Wilson's Photographic Magazine* **48** (1911), pp. 349–51.

Pickering, Edward Charles, 'Discovery of variable stars from their photographic spectra', *APJ* **1** (1895), pp. 27–8 and Plate III.

Pickering, William Hayward, 'An attempt to photograph the corona', *Science* **5** (1885), pp. 266–7.

'An attempt to photograph the corona without an eclipse', *Science* **6** (1885), pp. 131–3.

Pigott, Edward, 'On the periodical changes of brightness of two fixed stars', *PTRSL* **87** (1797), pp. 133–41.

Plotkin, Howard Neil, Henry Draper: A scientific biography, PhD thesis, The Johns Hopkins University (1972).

'Henry Draper, the discovery of oxygen in the Sun and the dilemma of interpreting the solar spectrum', *JHA* **8** (1977), pp. 44–51.

'Harvard College Observatory', in Owen Gingerich (ed.), *Astrophysics and Twentieth Century Astronomy to 1950* (Cambridge: Cambridge University Press, 1984), pp. 122–4.

Porter, Theodore, *The Rise of Statistical Thinking* (Princeton, NJ: Princeton University Press, 1986).

Powell, Baden, 'Article XIII. – 1. Prof. Schumacker's *Astronomische Abhandlungen*, Altona, 1823. A memoir on refractive and dispersive powers, by M. Frauenhofer. 2. *Transactions of the Royal Society of Edinburgh*, Vol. IX. On a monochromatic lamp, &c. by Dr. Brewster. – On the absorption of light by coloured media, by J. F. W. Herschel. 3. Some account of the late M. Guinand and his improvements in the manufacture of glass', *British Critic*, New Series **23** (1825), pp. 263–74.

Powell, Eyre Burton, 'Variations in the light of η Argus, observed at Madras from 1853 to 1861', *MNRAS* **22** (1861), pp. 47–8.

'Notes on α Centauri and other southern binaries, and on the nebula about η Argus', *MNRAS* **24** (1864), pp. 170–2.

Powell, Tristram (ed.), *Victorian Photographs of Famous Men & Fair Women* (Boston, MA: David R. Godine, 1973).

Price, Derek John de Solla, *Little Science, Big Science* (New York: Columbia University Press, 1963).

Pritchard, Ada, *Charles Pritchard: Memoirs of his Life* (London: Seeley and Co., 1897).

Pritchard, Charles, 'On the telescope, its modern form, and the difficulties of its construction', *PRI* **4** (1866), pp. 641–4.

'Remarks of the president', *AR* **4** (1866), pp. 301–2.

'President's address on presenting the Gold Medal to Mr. Huggins and Prof. Miller', *MNRAS* **27** (1867), pp. 146–65.

'A true story of the atmosphere of a world on fire', *Good Words* **8** (1867), pp. 249–56.

'Perceiving without seeing', *Good Words* **10** (1869), pp. 45–53.

'Stars and lights; or, the structure of the sidereal heavens – V. The arrival of Herschel's faithful assistant', *Good Words* **10** (1869), pp. 609–14.

'Thomas Cooke', *MNRAS* **29** (1869), pp. 130–5.

'Historical sketch of solar eclipses', *Good Words* **12** (1871), pp. 628–37.

'John F. W. Herschel', *MNRAS* **32** (1872), pp. 122–42.

'Preliminary note on the magnitude of the new star in Auriga', *MNRAS* **52** (1892), pp. 366–7.

Proctor, Richard Anthony, 'Theoretical considerations respecting the corona. Parts I and II', *MNRAS* **31** (1871), pp. 184–94 and pp. 254–62.

'A giant Sun', *Good Words* **13** (1872), pp. 98–104.

Pycior, Helena Mary, Nancy Slack and Pnina Geraldine Abir-Am (eds), *Creative Couples in Science: Multidisciplinary Perspectives* (New Brunswick, NJ: Rutgers University Press, 1996).

Quinn, Susan, *Marie Curie: A Life* (New York: Simon & Schuster, 1996).

Ramsay, William, 'Discovery of helium', *CN* **71** (1895), p. 151.

'On a gas showing the spectrum of helium, the reputed cause of D₃, one of the lines in the coronal spectrum, preliminary note', *PRS* **58** (1895), pp. 65–7.

Ramsay, William and Morris William Travers, 'On a new constituent of atmospheric air', *PRS* **63** (1898), pp. 405–8.

'On the companions of argon', *PRS* **63** (1898), pp. 437–40.

'On the extraction from air of the companions of argon and on neon', *RBAAS, Bristol* (1898), pp. 828–30.

Ranyard, Arthur Cowper, 'Note on Dr. Henry Draper's photograph of the nebula in Orion', *Science* **2** (1881), pp. 82–3.

'On the connection between photographic action, the brightness of the luminous object and the time of exposure, as applied to celestial photography', *MNRAS* **46** (1886), pp. 305–9.

Rayleigh, Lord and William Ramsay, 'Argon, a new constituent of the atmosphere', *PRS* **57** (1895), pp. 265–87.

Rayner-Canham, Marlene F. and Geoffrey William Rayner-Canham, 'Pioneer women in nuclear science', *AJP* **58** (1990), pp. 1036–43.

Reynolds, Osborne, 'The tails of comets, the solar corona, and the aurora, considered as electric phenomena. Parts I and II', *MLPSM*, 3rd Series **6** (1870), pp. 44–52 and pp. 53–6.

Riccò, Annibale 'On some attempts to photograph the solar corona without an eclipse, made at the Mount Etna Observatory', *APJ* **1** (1895), pp. 18–26.

Rimmer, Gordon, 'Francis Abbott', *Australian Dictionary of Biography*, vol. 3 (Melbourne: Melbourne University Press, 1969), pp. 2–3.

Ritchey, George Willis, 'Nebulosity about Nova Persei', *APJ* **14** (1901), pp. 167–8, Plate VII.

'Nebulosity about Nova Persei. Recent photographs', *APJ* **15** (1902), pp. 129–31, Plates VI–X.

Roberts, Isaac, 'Photograph of the region of Nova Aurigae', *MNRAS* **52** (1892), pp. 371–2.

Robinson, Thomas Romney, 'On spectra of electric light, as modified by the nature of the electrodes and the media of discharge', *PTRSL* **152** (1862), pp. 939–86.

Robinson, Thomas Romney and Thomas Grubb, 'Description of the Great Melbourne Telescope', *PTRSL* **159** (1869), pp. 127–61.

Rohr, Moritz von, 'Fraunhofer's work and its present-day significance', *TOS* **27** (1925–6), pp. 277–94.

Röntgen, Wilhelm Conrad, 'Über eine neue Art von Strahlen', *Sitzungsberichte der Würzburger Physikalischen-Medicinischen Gesselschaft* (December, 1895), pp. 3–12. English translation by G. F. Barker, 'On a new kind of rays', in G. F. Barker (ed.), *Röntgen Rays: Memoirs by Röntgen, Stokes and J. J. Thompson*, (New York: Harper & Brothers Publishers, 1899), pp. 3–13.

Roscoe, Henry Enfield, 'On Bunsen and Kirchhoff's spectrum observations', *PRI* **3** (1861), pp. 323–8. Reprinted in *CN* **3** (1861), pp. 153–5 and pp. 170–2.

'On the application of the induction coil to Steinheil's apparatus for spectrum analysis', read before the Chemical Society, 20 June 1861, *CN* **4** (1861), pp. 118–22 and pp. 130–3.

'A course of three lectures on spectrum analysis', *CN* **5** (1862), pp. 218–22, pp. 261–5 and pp. 287–93.

Spectrum Analysis: Six Lectures Delivered in 1868, before the Society of Apothecaries of London (New York: Appleton & Co., 1869).

Rossiter, Margaret W., '"Women's work" in science, 1880–1910', *Isis* **71** (1980), pp. 381–98.

Rothenberg, Marc, The educational and intellectual background of American astronomers, 1825–1875, PhD thesis, Bryn Mawr College (1974). 'Organization and control: Professionals and amateurs in American astronomy', *SSS* **11** (1981), pp. 305–25.

Rothermel, Holly, 'Images of the Sun: Warren De la Rue, George Biddell Airy and celestial photography', *BJHS* **26** (1993), pp. 137–69.

Royal Commission on Scientific Instruction and the Advancement of Science, *Minutes of Evidence, Appendices, and Analyses of Evidence*, Vol. 2 (London: Eyre & Spottiswoode, 1874).

Royal Society, *Correspondence Concerning the Great Melbourne Telescope. In three parts: 1852–1870* (London: Taylor & Francis, 1871).

  The Eruption of Krakatoa and subsequent phenomena (London: Royal Society, 1888).

  The Record of the Royal Society of London for the Promotion of Natural Knowledge, 4th edn (Edinburgh: Morrison & Gibb Ltd., 1940).

Runge, Carl, 'Terrestrial helium (?)', *Nature* **52** (1895), p. 128.

  'On the spectrum of radium', *APJ* **12** (1900), pp. 1–3.

Runge, Carl and Julius Precht, 'On the flame spectrum of radium', *APJ* **17** (1903), pp. 147–9.

Ruskin, John, 'Of Queens' Gardens', in *Sesame and Lilies* (London: George Allen, 1905).

Russell, John Scott, 'On certain effects produced on sound by the rapid motion of the observer', *RBAAS, Swansea* (1849), pp. 37–8.

Rutherford, Ernest, *Radio-activity* (Cambridge: Cambridge University Press, 1904).

  'The succession of changes in radioactive bodies', *PTRSL* **204A** (1905), pp. 169–219.

  Radioactive Transformations (New Haven, CT: Yale University Press, 1906).

Rutherfurd, Lewis Morris, 'Astronomical observations with the spectroscope', *AJS* **35** (1863), pp. 71–7.

Sabine, Edward, 'President's address', *PRS* **13** (1864), pp. 499–502.

  'President's address', *PRS* **15** (1866), pp. 270–85.

  'President's address', *PRS* **17** (1868), pp. 135–50.

  'President's address', *PRS* **18** (1869), pp. 102–12.

Sabra, Abdelhamid I., *Theories of Light from Descartes to Newton* (Cambridge: Cambridge University Press, 1981).

Sagan, Carl and Ann Druyan, *Comet* (New York: Random House, 1985).

Salter, Frank R., *Dissenters and Public Affairs in Mid-Victorian England* (London: Dr Williams's Trust, 1967).

Schaffer, Simon, '"The great laboratories of the universe": William Herschel on matter theory and planetary life', *JHA* **11** (1980), pp. 81–111.

  'Herschel in Bedlam: Natural history and stellar astronomy', *BJHS* **13** (1980), pp. 211–39.

  'Astronomers mark time: Discipline and the personal equation', *Science in Context* **2** (1988), pp. 115–45.

  'The nebular hypothesis and the science of progress', in James R. Moore (ed.), *History, Humanity and Evolution: Essays for John C. Greene*, (Cambridge: Cambridge University Press, 1989), pp. 101–31.

  'Experimenting with objectives: Table-top trials in Victorian astronomy', unpublished manuscript, 1990.

Schellen, Heinrich, *Die Spectralanalyse in Ihrer Anwendung auf die Stoffe der Erde und die Natur der Himmelskörper* (Braunschweig: Druk und Verlag von George Westermann, 1870). English translation: Jane and Caroline Lassell, *Spectrum Analysis* (London: Longmans, Green & Co., 1872).

Schmidt, Johann Friedrich Julius, 'Über den Mondcrater Linné', *AN* **68** (1867), pp. 365–6.

Seater, Barbara B., 'Two person career: The pastor and his wife', *Free Inquiry in Creative Sociology* **10** (1982), pp. 75–9.

Secchi, Father Angelo, 'Note sur les spectres prismatiques des corps célestes', *CR* **57** (1863), p. 71.

  'Étoile Rouge Singulaire', *Les Mondes: Revue Hebdomadaire des Sciences* **8** (1865), pp. 141–2.

  'On the spectrum of the nebula of Orion', *MNRAS* **25** (1865), pp. 153–5.

  'Schreiben des Herrn Prof. Secchi . . .', *AN* **68** (1866), pp. 63–4.

  'Communications relative à l'analyse spectrale de la lumière de quelques étoiles', *CR* **63** (1866), p. 324.

'Stellar spectrometry', *Bulletino Meteorologico* (31 August 1866). Reprinted in *AR* **4** (1866), pp. 253–5.

'Spectrometric studies', *Bulletino Meteorologico* (1866). Reprinted in *AR* **4** (1866), pp. 280–5.

'Sur les spectres stellaires', *CR* **66** (1868), p. 398.

See, Thomas Jefferson Jackson, 'Need of the collected works of Sir Wm. Herschel', *OBS* **32** (1909), p. 473.

'Tribute to the memory of Sir William Huggins', *PA* **18** (1910), pp. 387–401.

Seitz, Adolf, *Joseph Fraunhofer und sein optisches Institut* (Berlin: Julius Springer, 1926).

Sellers, Ian, *Nineteenth-Century Nonconformity* (New York: Holmes & Meier, 1977).

Sestini, Benedetto, 'On the colors of stars', *AJ* **1** (1850), pp. 88–91.

Shapin, Steven, 'The invisible technician', *American Scientist* **77** (1989), pp. 555–63.

Sharp, Evelyn, *Hertha Ayrton 1854–1923: A Memoir* (London: Edward Arnold & Co., 1926).

Shirley, John William, 'An early experimental determination of Snell's Law', *AJP* **19** (1951), pp. 507 8.

Siegel, Daniel M., 'Balfour Stewart and Gustav Robert Kirchhoff: Two independent approaches to "Kirchhoff's Radiation Law"', *Isis* **67** (1976), pp. 565–600.

Sigsworth, Eric M., *In Search of Victorian Values: Aspects of Nineteenth-century Thought and Society* (Manchester: Manchester University Press, 1988).

Skeat, Walter William (ed.), *An Etymological Dictionary of the English Language*, 2nd edn (Oxford: Clarendon Press, 1893).

Smiles, Samuel, *Thrift, or How to Get On in the World* (Chicago, IL: Belford, Clarke & Co., 1881).

Self-Help (Chicago, IL: Belford, Clarke, & Co., 1884).

Smith, J. W. Ashley, *The Birth of Modern Education: The Contribution of the Dissenting Academies, 1660–1800* (London: Independent Press Ltd, 1954).

Smith, Robert William, 'The heavens recorded: Warren De La Rue and the 1860 eclipse', paper given at the 16th International Congress of the History of Science (Bucharest, 1981).

The Expanding Universe: Astronomy's 'Great Debate' 1900–1931 (Cambridge: Cambridge University Press, 1982).

'William Lassell and the discovery of Neptune', *JHA* **14** (1983), pp. 30–32.

'The Cambridge network in action: The discovery of Neptune', *Isis* **80** (1989), pp. 395–422.

'A national observatory transformed: Greenwich in the nineteenth century', *JHA* **22** (1991), pp. 1–14.

Smith, Robert William, and Richard Baum, 'William Lassell and the ring of Neptune: A case-study in instrumental failure', *JHA* **15** (1984), pp. 1–17.

Smyth, Charles Piazzi, 'Color in practical astronomy spectroscopically examined', *PRS* **29** (1879), pp. 779–849.

Micrometrical Measures of Gaseous Spectra under High Dispersion (Edinburgh: Neill & Co., 1886).

Smyth, William Henry, *A Cycle of Celestial Objects, for the Use of Naval, Military, and Private Astronomers* (London: John W. Parker, 1844).

Aedes Hartwellianae, or Notices of the Manor and Mansion of Hartwell (London: John Bowyer Nichol and Son, 1851).

The Cycle of Celestial Objects continued at the Hartwell Observatory to 1859, with a Notice of Recent Discoveries, including Details from the Aedes Hartwellianae (London: John Bowyer Nichols and Sons, 1860).

Sidereal Chromatics; Being a Re-print, with Additions, from the 'Bedford Cycle of Celestial Objects', and its 'Hartwell Continuation', on the Colours of Multiple Stars (London: John Bowyer Nichols and Sons, 1864).

Soddy, Frederick, 'The evolution of the elements', *RBAAS, York* (1906), pp. 122–31.

Stewart, Balfour, 'Reply to Kirchhoff on history of spectrum analysis', *PM*, 4th Series **25** (1863), p. 354.

'Some points on the history of spectrum analysis', *Nature* **21** (1880), p. 35.

Stewart, Balfour and Joseph Norman Lockyer, 'The Sun as a type of the material universe. Parts I & II', *Macmillan's Magazine* **18** (1868), pp. 246–57 and pp. 319–27.

Stokes, George Gabriel, 'On the long spectrum of electric light', *PRS* **12** (1862), pp. 166–8.

'Solar physics. Parts I and II', *Nature* **24** (1881), pp. 593–8 and pp. 613–18.

Stone, Edward James, 'Approximate determinations of the heating-powers of Arcturus and α Lyrae', *PRS* **18** (1870), pp. 159–65.

Stonequist, Everett V., *The Marginal Man* (New York: Scribner's, 1937).

Strange, Alexander, 'On the necessity for state intervention to secure the progress of physical science', *RBAAS, Norwich* (1868), pp. 6–8.

'On the insufficiency of existing national observatories', *MNRAS* **32** (1872), pp. 238–41.

'On the insufficiency of existing national observatories', *AR* **10** (1872), pp. 113–20.

Stratton, Frederick John Marrian, *Astronomical Physics* (London: Methuen & Co., 1925).

'The history of the Cambridge Observatories', in *Annals of the Solar Physics Observatory, Cambridge*, vol. 1 (London: Cambridge University Press, 1949), Plate V(b).

Strutt, Robert John, *The Becquerel Rays and the Properties of Radium*, 2nd edn (London: Edward Arnold, 1906).

Struve, Otto von, 'On the missing nebula in Taurus . . .', *MNRAS* **22** (1862), pp. 242–4.

Sutton, Michael A., Spectroscopy and the structure of matter: A study in the development of physical chemistry, PhD thesis, University of Oxford (1972).

'Sir John Herschel and the development of spectroscopy in Britain', *BJHS* **7** (1974), pp. 42–60.

'Spectroscopy and the chemists: A neglected opportunity?', *Ambix* **23** (1976), pp. 16–26.

'Spectroscopy, historiography and myth: The Victorians vindicated', *HS* **24** (1986), pp. 425–32.

Suum Cuique, 'Gas and gases', *CN* **77** (1898), p. 284.

'The new gases', *CN* **78** (1898), p. 11, p. 34 and p. 46.

'The gases of the atmosphere', *CN* **78** (1898), pp. 290–1.

Swan, William, 'On the prismatic spectra of the flames of compounds of carbon and hydrogen', *TRSE* **21** (1857), pp. 411–29 and pl. XXI. German translation: 'Über die prismatischen Spectra der Flammen von Kohlenwasserstoffverbindungen', *AP*, 2nd Series **100** (1857), pp. 306–35 and Plate 1.

Taylor, Shephard Thomas, *The Diary of a Medical Student during the Mid-Victorian Period 1860–1864* (Norwich: Jarrold & Sons, 1927).

Tebbutt, John, 'On the variability of η Argus', *MNRAS* **28** (1868), pp. 266–8.

Tempel, Wilhelm, 'Schreiben des Herrn Wilh. Tempel . . .', *AN* **54** (1861), pp. 285–6.

Tennant, Maj. James Francis, 'On the solar eclipse of 1868, August 17', *MNRAS* **27** (1867), p. 79.

'On the eclipse of August 1868', *MNRAS* **27** (1867), pp. 173–7.

'Report of the observations of the total solar eclipse of August 1868', *AR* **7** (1869), pp. 35–9.

Thompson, Silvanus Phillips, *Elementary Lessons in Electricity and Magnetism* (London: Macmillan and Co., 1881).

Todd, David Peck, *A New Astronomy* (New York: American Book Company, 1897).

Trotter, Coutts, 'William Allen Miller', *PRS* **19** (1870), pp. xix–xxvi.

'Lieut.-Col. Alexander Strange', in Sidney Lee (ed.), *DNB*, vol. 55, (New York: Macmillan, 1898), pp. 20–1.

Turnbull, J. M., 'A few words on the beer and albumen process', *BJP* **21** (1874), pp. 495–6.

Turner, Herbert Hall, 'George Biddell Airy', *MNRAS* **52** (1892), pp. 212–29.

'Lewis Morris Rutherfurd', *MNRAS* **53** (1893), pp. 229–31.

'Mrs. Fleming', *MNRAS* **72** (1912), pp. 261–4.

'The decade 1820–1830', in J. L. E. Dreyer and H. H. Turner (eds), *History of the Royal Astronomical Society 1820–1920* (Oxford: Blackwell Scientific Publications, 1987; London: Wheldon & Wesley, 1923), pp. 1–49.

Tyndall, John, 'Voyage to Algeria to observe the eclipse. 1870', in *Fragments of Science: A Series of Detached Essays, Addresses, and Reviews*, vol. 1 (New York: D. Appleton and Company, 1897), pp. 142–74.

Vaucouleurs, Gérard de, 'Discovering M31's spiral shape', *ST* **74** (1987), pp. 595–8.

Vico, Francescoe de, 'Schrieben des Herrn Professors De-Vico ...', *AN* **29** (1849), pp. 187–92.

Vogel, Hermann Carl, 'Über die Bestimmung der Bewegung von Sternen im Visionradius durch Spectrographische Beobachtung', *Sitzungsberichte der Königlich Preussischen Akademie der Wissenschaften zu Berlin* **26** (1888), pp. 397–401 and Plate II.

'On the progress made in the last decade in the determination of stellar motions in the line of sight', *APJ* **11** (1900), pp. 373–92.

Vogel, Hermann Wilhelm, *The Chemistry of Light and Photography*, 4th edn (London: Kegan Paul, Trench & Co., 1883).

W. P., 'Spectroscope construction', *AR* **8** (1870), p. 76.

Wardle, David, *English Popular Education: 1780–1975* (Cambridge: Cambridge University Press, 1976).

Warner, Deborah Jean, 'Lewis M. Rutherfurd: Pioneer astronomical photographer and spectroscopist', *TC* **12** (1971), pp. 190–216.

Watts, William Marshall, *An Introduction to the Study of Spectrum Analysis* (London: Longmans, Green, & Co., 1904).

Webb, Thomas William, 'Note on the telescopic appearance of the planet Mars', *MNRAS* **16** (1856), p. 188.

Celestial Objects for Common Telescopes (London: Longman, Green, Longman, and Roberts, 1859).

'Nebula in the Pleiades', *The Reader* **2** (1863), pp. 633–4.

'Clusters of stars and nebulae ...', *IO* **4** (August 1863), pp. 56–62.

'Clusters and nebulae ...', *IO* **4** (November 1863), pp. 257–66.

'Clusters, nebulae, and occultations', *IO* **4** (December 1863), pp. 346–52.

'Clusters and nebulae ...', *IO* **4** (January 1864), pp. 448–52.

'Clusters and nebulae ...', *IO* **5** (February 1864), pp. 54–60.

Webb, William Larkin, *Brief Biography and Popular Account of the Unparalleled Discoveries of T. J. J. See* (Lynn, MA: Thos. P. Nichols & Son, 1913).

Weinberg, Alvin Martin, *Reflections on Big Science* (Cambridge, MA: MIT Press, 1967).

'Scientific teams and scientific laboratories', *Daedalus* **99** (1970), pp. 1056–75.

Weinreb, Ben and Christopher Hibbert (eds), *The London Encyclopedia* (London: Macmillan, 1983).

Weiss, Edmund, 'Presentation of results of 1868 solar eclipse expedition', *AR* **7** (1869), pp. 100–3.

Whewell, William, 'Address to the General Assembly', *RBAAS, Cambridge* (1833), pp. xi–xxvi.

Whiting, Lilian, *Life Transfigured* (Boston: Little, Brown, and Company, 1910).

Whiting, Sarah Frances, 'Spectroscopic work for classes in astronomy', *PA* **13** (1905), pp. 387–91

'A pedagogical suggestion for teachers of astronomy', *PA* **20** (1912), pp. 156–60.

'Priceless accessions to Whitin Observatory, Wellesley College', *PA* **22** (1914), pp. 487–92.

'An international gift', *PA* **23** (1915), pp. 698–9.

'The Tulse Hill Observatory diaries', *PA* **25** (1917), pp. 158–63.

'Lady Huggins', *Science* **51** (1915), pp. 853–5.

'Margaret Lindsay Huggins', *APJ* **42** (1915), pp. 1–3.

Whitney, Charles A., 'Henry Draper', *DSB*, vol. 4 (New York: Scribner & Son, 1970), pp. 178–81.

Wildt, Rupert, 'The chemistry of the cosmos', in Harlow Shapley (ed.), *Source Book in Astronomy 1900–1950* (1960). Originally in *Scientia* **67** (1940), pp. 85–90.

Williams, Henry Smith, 'Astronomical progress of the century', *Harper's New Monthly Magazine* **94** (1897), pp. 140–56.

The Great Astronomers (New York: Newton Publishing Co., 1932).

Williams, Mari E. W., 'Astronomy in London: 1860–1900', *QJRAS* **28** (1987), pp. 10–26.

Wollaston, William Hyde, 'A method of examining refractive and dispersive powers, by prismatic
    reflection', *PTRSL* **92** (1802), pp. 365–80.
    'Neue Methode, die brechenden und zerstreuenden Kräfte der Körper vermittelst prismatischer
    Reflexion zu erforschen', *AP*, New Series **1** (1809), pp. 235–51 and 398–416.
Woods, Charles Ray, 'Photo-astronomy at the Riffel', in five parts, *PN* **28** (1884), pp. 490–2,
    pp. 523–4, pp, 582–3, pp. 611–13, and pp. 771–2.
    'Photographing the solar corona', *OBS* **7** (1884), pp. 376–8.
Woolf, Harry, 'The beginnings of astronomical spectroscopy', in *Melanges Alexandre Koyré*, vol. 1
    (Paris: Hermann, 1964), pp. 619–34.
    'Astrophysics in the early nineteenth century', *Actes du XIe Congres International d'Histoire des
    Sciences, 1965* **3** (1968), pp. 127–35.
Wright, Helen, *Explorer of the Universe: A Biography of George Ellery Hale* (New York: E. P.
    Dutton & Co., 1966).
Young, Charles Augustus, 'An attempt to photograph the corona', *Science* **5** (1885), p. 307.
    The Sun, (New York: D. Appleton and Company, 1881).
    'Address at the dedication of the Kenwood Observatory', *SM* **10** (1891), pp. 312–21.

Published papers of William Huggins

1856

'Description of an observatory erected at Upper Tulse Hill', *MNRAS* **16**, pp. 175–7.
'Observation of the occultation of Jupiter by the Moon, November 8, 1856', *MNRAS* **17**, p. 2.
'Note accompanying drawings of Jupiter, Mars, &c.', *MNRAS* **17**, p. 23.

1857

'Occultation of Spica [α] Virginis. 1857, May 6', *MNRAS* **17**, p. 203 and p. 246.

1862

'On the periodical changes in the belts and surface of Jupiter', *MNRAS* **22**, p. 294.
'On some phenomena attending the disappearance of Saturn's ring, May 19th, 1862', *MNRAS* **22**,
    pp. 295–6.

1863

'Note on the lines in the spectra of some of the fixed stars' [with W. A. Miller], *PRS* **12**, pp. 444–5.
'On the spectra of some of the chemical elements' [with W. A. Miller], *PRS* **13**, pp. 43–4; *PTRSL* **154**
    (1863), pp. 139–60. Abstract in *PM*, 4th Series **27** (1863), pp. 541–2.
'Spectrum analysis applied to the stars; or, the stars, what are they?', *IO* **3** (June 1863), pp. 338–49.

1864

'On the spectra of some of the nebulae' [communicated by W. A. Miller], *PRS* **13**, pp. 492–3; *PTRSL*
    154, pp. 437–44.
'On the spectra of some of the fixed stars' [with W. A. Miller], *PRS* **13**, pp. 242–4; *PTRSL* **154**,
    pp. 413–35.

1865

'On the disappearance of the spectrum of ε Piscium at its occultation of January 4th, 1865', *MNRAS* **25**,
    pp. 60–2.
'In the observatory of Mr. Huggins . . .', *MNRAS* **25**, pp. 107–9.

'Response to a letter of Father Secchi', *MNRAS* **25**, pp. 154–5.
'On the spectrum of the Great Nebula in the sword-handle of Orion' [communicated by W. A. Miller], *PRS* **14**, pp. 38–42.
'On the physical and chemical constitution of the fixed stars and nebulae', *PRI* **4**, pp. 441–9.
'Note on the prismatic examination of microscopic objects', *TRMSL*, New Series **13** (1865), pp. 85–7.

1866

'On the stars within the Trapezium of the nebula of Orion', *MNRAS* **26**, pp. 71–3.
'Mr. Huggins's observatory', *MNRAS* **26**, pp. 144–5.
'Note on the spectrum of the variable star α Orionis, with some remarks on the letter of the Rev. Father Secchi', *MNRAS* **26**, pp. 215–17.
'Results of some observations on the bright granules of the solar surface, with remarks on the nature of these bodies', *MNRAS* **26**, pp. 260–5
'On a new star', *MNRAS* **26**, pp. 275–7.
'Diagram of the spectrum of absorption and the spectrum of bright sines [*sic*, read 'lines'] forming the compound spectrum of the temporarily bright star near ε Corona Borealis', *MNRAS* **26**, p. 297.
'Binocular vision', *AR* **4**, pp. 47–9 and pp. 102–4.
'Schreiben des Herrn Huggins . . .', *AN* **67**, pp. 29–32.
'Further observations on the star in Corona', *AN* **67**, pp. 125–6.
'On the spectrum of Comet 1, 1866', *PRS* **15**, pp. 5–7.
'Further observations on the spectra of some of the nebulae, with a mode of determining the brightness of these bodies', *PRS* **15**, pp. 17–19; *PTRSL* **156**, pp. 381–97.
'On the spectrum of a new star in Corona Borealis' [with W. A. Miller], *PRS* **15**, pp. 146–9.
'On a temporary outburst of light in a star in Corona Borealis', *QJS* **3**, pp. 376–82.
On the Results of Spectrum Analysis Applied to the Heavenly Bodies: A Discourse Delivered at Nottingham, before the British Association, August 24, 1866 (London: W. Ladd).

1867

'Mr. Huggins' observatory', *MNRAS* **27**, p. 129.
'On the spectrum of Mars, with some remarks on the colour of that planet', *MNRAS* **27**, pp. 178–81.
'Note on the spectrum of Comet II. 1867', *MNRAS* **27**, p. 288.
'Note on the lunar crater Linné', *MNRAS* **27**, pp. 296–8.
'Stellar spectrometry', *AR* **5**, p. 34.
'The Trapezium of Orion', *AR* **5**, pp. 54–5.

1868

'Mr. Huggins' observatory', *MNRAS* **28**, pp. 86–8.
'On a possible method of viewing the red flames without an eclipse', *MNRAS* **29**, pp. 4–5.
'On the appearance of Mercury at its transit, November 5, 1868', *MNRAS* **29**, pp. 25–8.
'Spectrum analysis of Comet II. 1868', *AR* **6**, pp. 169–70.
'Aus einem Schreiben des Herrn William Huggins . . .', *AN* **71**, pp. 381–4.
'Description of a hand spectrum-telescope', *PRS* **16**, pp. 241–3.
'Further observations on the spectra of the Sun, and of some of the stars and nebulae, with an attempt to determine therefrom whether these bodies are moving towards or from the Earth', *PRS* **16**, pp. 382–6.
'On the spectrum of Brorsen's Comet, 1868', *PRS* **16**, pp. 386–9.
'On the spectrum of Comet II., 1868', *PRS* **16**, pp. 481–2.
'Further observations on the spectra of some of the stars and nebulae, with an attempt to determine therefrom whether these bodies are moving towards or from the Earth, also observations on the spectra of the Sun and of Comet II', *PTRSL* **158**, pp. 529–64.

1869

'Mr. Huggins' observatory', *MNRAS* **29**, pp. 142–3.
'Motion of Sirius in the line of sight', *MNRAS* **29**, p. 164.
'Note on Mr. De La Rue's paper "On some attempts to render the luminous prominences visible without the use of the spectroscope"', *MNRAS* **30**, pp. 36–7.
'Note on a method of viewing the solar prominences without an eclipse', *PRS* **17**, pp. 302–3.
'Note on the heat of the stars', *PRS* **17**, pp. 309–12.
'On some further results of spectrum analysis as applied to the heavenly bodies', *PRI* **5**, pp. 475–9.
'On some spectrum observations of comets', *PM*, 4th Series **37**, pp. 456–60.

1870

'Mr. Huggins' observatory', *MNRAS* **30**, p. 100.
'Note on the spectra of erbia and some other earths', *PRS* **18**, pp. 546–53.

1871

'Mr. Huggins' observatory', *MNRAS* **31**, pp. 110–11.
'On a registering spectroscope', *PRS* **19**, pp. 317–8.
'Note on the spectrum of Uranus and the spectrum of Comet I, 1871', *PRS* **19**, pp. 488–91.
'Note on the spectrum of Encke's Comet', *PRS* **20**, pp. 45–7.
'Note on the telescopic appearance of Encke's Comet', *PRS* **20**, pp. 87–9.
'Über das Spectrum des *Encke*'schen Cometen (III. 1871)', *AN* **78**, pp. 357–8.

1872

'Mr. Huggins' observatory', *MNRAS* **32**, pp. 156–7.
'Dr. Huggins's spectroscopic observations of stars and nebulae', *MNRAS* **32**, pp. 359–62.
'On the spectrum of the Great Nebula in Orion, and on the motions of some stars towards or from the Earth', *PRS* **20**, pp. 379–94.

1873

'Motions of stars in the line of sight', *MNRAS* **33**, pp. 238–9.
'Linné', *AR* **12**, pp. 142–3.
'Note on the wide-slit method of viewing the solar prominences', *PRS* **21**, pp. 127–8.
'Note on the proper motions of nebulae', *RBAAS, Bradford*, pp. 34–5; and *AR* **11**, pp. 269–71.

1874

'Note on the lunar crater Linné', *MNRAS* **34**, pp. 108–11.
'Mr. Huggins' observatory, Upper Tulse Hill', *MNRAS* **34**, p. 170.
'On the motions of some of the nebulae towards or from the Earth', *PRS* **22**, pp. 251–4.

1875

'Mr. Huggins' observatory, Upper Tulse Hill', *MNRAS* **35**, pp. 192–3.
'On the spectrum of Coggia's Comet', *PRS* **23**, pp. 154–9.

1876

'Note on the photographic spectra of stars', *PRS* **25**, pp. 445–6.

'Letter in reply to Father Secchi's letter on the displacement of stellar lines', *PM*, 5th Series **2** (1876), pp. 72–4.

1877

'Mr. Huggins' observatory', *MNRAS* **37**, pp. 174.
'On the inferences to be drawn from the appearance of bright lines in the spectra of irresolvable nebulae', *PRS* **26**, pp. 179–81.
'The photographic spectra of stars', *OBS* **1**, pp. 4–7.

1878

'On a cyclonic arrangement of the solar granules', *MNRAS* **38**, pp. 101–2.
'Mr. Huggins' observatory', *MNRAS* **38**, pp. 183–4.
'Address delivered by the president, Mr. Huggins, on presenting the Gold Medal of the Society to Baron Dembowski', *MNRAS* **38**, pp. 249–53.
'Heat of the stars', *AR* **16**, p. 309.

1879

'Mr. Huggins' observatory', *MNRAS* **39**, p. 252.
'On the photographic spectra of stars', *PRS* **30**, pp. 20–1.

1880

'Mr. Huggins' observatory, Upper Tulse Hill', *MNRAS* **40**, pp. 227–8.
'On the spectrum of the flame of hydrogen', *PRS* **30**, pp. 576–81.
'On the photographic spectra of the stars', *PTRSL* **171**, pp. 669–90.

1881

'William Lassell', *MNRAS* **41**, pp. 188–91.
'Mr. Huggins' observatory, Upper Tulse Hill', *MNRAS* **41**, pp. 208–9.
'Photographic spectrum of Comet *b* 1881', *AN* **100**, pp. 143–4.
'Preliminary note on the photographic spectrum of Comet *b*, 1881', *PRS* **33**, pp. 1–3; and *OBS* **4**, pp. 233–4.

1882

'Mr. Huggins' observatory, Upper Tulse Hill', *MNRAS* **42**, pp. 166–7.
'The photographic spectrum of the Great Nebula of Orion', *OBS* **5**, pp. 106–7.
'Note on the photographic spectrum of the Great Nebula in Orion', *PRS* **33**, pp. 425–8.
'On the photographic spectrum of Comet (Wells) I, 1882', *PRS* **34**, pp. 148–50.
'On a method of photographing the solar corona without an eclipse', *PRS* **34**, pp. 409–14.
'On comets', *PRI* **10**, pp. 1–11.

1883

'Mr. Huggins' observatory, Upper Tulse Hill', *MNRAS* **43**, pp. 199–200.
'On a method of photographing the solar corona without an eclipse', *AN* **104**, pp. 113–8; and *AR* **21**, pp. 38–9.
'Schreiben des Herrn Dr. W. Huggins', *AN* **104**, pp. 315–16.
'On the function of the sound-post, and on the proportional thickness of the strings of the violin', *PRS* **35**, pp. 241–8.

'On some results of photographing the solar corona without an eclipse', *RBAAS, Southport*, pp. 346–51.

1884

'Mr. Huggins' observatory, Upper Tulse Hill', *MNRAS* **44**, pp. 171–2.

1885

'The solar corona', *OBS* **8**, pp. 153–9.
'Photographing the corona', *OBS* **8**, pp. 376–7.
'On the corona of the Sun – The Bakerian lecture', *PRS* **39** (1885), pp. 108–35.
'On the solar corona', *PRI* **11**, pp. 202–14.
'The new star in Andromeda', *Nature* **32** (1885), p. 465.
'On the spectrum of the Stella Nova visible on the Great Nebula in Andromeda', *RBAAS, Aberdeen* (1885), p. 935.
'The Sun's corona', *The Nineteenth Century* **17**, pp. 676–89.
'An attempt to photograph the solar corona', *Science* **5** (1885), pp. 397–8.

1886

'Photography of the solar corona', *AN* **115**, pp. 191–2.
'Photography of the solar corona', *Nature* **34** (1886), pp. 469–70.

1888

'Mr. Huggins' observatory', *MNRAS* **48**, pp. 195–6.

1889

'Mr. Huggins' observatory', *MNRAS* **49**, p. 202.
'The photographic spectrum of the nebula of Orion', *MNRAS* **49**, pp. 403–4.
'The spectrum of Uranus; the spectrum of Saturn', *MNRAS* **49**, pp. 404–5.
'On the spectrum of Uranus', *AN* **121**, pp. 369–70.
'On the wave-length of the principal line in the spectrum of the aurora', *PRS* **45**, pp. 430–6.
'On the spectrum, visible and photographic, of the Great Nebula in Orion' [with M. L. Huggins], *PRS* **46**, pp. 40–60.
'On the limit of solar and stellar light in the ultra-violet part of the spectrum', *PRS* **46**, pp. 133–5.
'Note on the photographic spectra of Uranus and Saturn' [with M. L. Huggins], *PRS* **46**, pp. 231–3.

1890

'Dr. Huggins' observatory', *MNRAS* **50**, pp. 207–9.
'On a re-determination of the principal line in the spectrum of the nebula in Orion, and on the character of the line' [with M. L. Huggins], *PRS* **48**, pp. 202–13.
'Note on the photographic spectrum of the Great Nebula in Orion' [with M. L. Huggins], *PRS* **48**, pp. 213–16.
'On a new group of lines in the photographic spectrum of Sirius' [with M. L. Huggins], *PRS* **48**, pp. 216–17.
'On Wolf and Rayet's bright-line stars in Cygnus' [with M. L. Huggins], *PRS* **49**, pp. 33–46.

1891

'Dr. Huggins' observatory', *MNRAS* **51**, pp. 228–9.
'Address of the president', *RBAAS, Cardiff*, pp. 3–37.

1892

'On Nova Aurigae', *AA* **11**, pp. 571–81.
'Preliminary note on Nova Aurigae' [with M. L. Huggins], *PRS* **50**, pp. 465–9.
'On Nova Aurigae' [with M. L. Huggins], *PRS* **51**, pp. 486–95.
'The new star in Auriga', *PRI* **13**, pp. 615–24.

1893

'Note on the spectrum of Nova Aurigae', *AN* **132**, pp. 143–4.
'Note on the spectrum of Nova Aurigae', *AA* **12**, pp. 349–50.
'On the bright bands in the present spectrum of Nova Aurigae' [with M. L. Huggins], *PRS* **54**, pp. 30–6; and *AA* **12**, pp. 609–15.
'The Tulse Hill spectroscope', *AA* **12**, pp. 615–19.

1894

'On the visual appearance of Nova (T) Aurigae', *AN* **134**, pp. 309–10; and *OBS* **17**, pp. 108–9.
'Note on the spectrum of Mars', *OBS* **17**, pp. 353–4.

1895

'On the duplicity of the solar line D$_3$', *AN* **138**, pp. 229–30.
'On the duplicity of the solar line D$_3$ (with editorial comment)', *OBS* **18**, pp. 295–7.
'Note on the atmospheric bands in the spectrum of Mars', *APJ* **1**, pp. 193–5.
'The Tulse Hill ultra-violet spectroscope', *APJ* **1**, pp. 359–65.

1897

'Dr. Huggins's observatory, Upper Tulse Hill', *MNRAS* **57**, p. 260.
'On the relative behaviour of the H and K lines of the spectrum of calcium' [with M. L. Huggins], *PRS* **61**, pp. 433–41.
'On an automatic arrangement for giving breadth to stellar spectra on a photographic plate', *APJ* **5**, pp. 8–10.
'On the mode of printing maps of spectra and tables of wave-lengths', *APJ* **6**, pp. 55–6.
'On the spectra of the stars in the Trapezium of the Great Nebula of Orion' (read at the dedication of the Yerkes Observatory), *APJ* **6**, pp. 322–7.
'The new astronomy: A personal retrospect', *The Nineteenth Century* **41**, pp. 907–29.

1898

'Sir William Huggins' observatory, Upper Tulse Hill', *MNRAS* **58**, pp. 185–6.

1899

'Sir William Huggins's observatory', *MNRAS* **59**, p. 263.
'Oxygen in helium stars', *AN* **149**, pp. 231–2.
'Nitrogen in some helium stars', *AN* **150**, pp. 109–10.
An Atlas of Representative Stellar Spectra from λ 4870 to λ 3300 [with M. L. Huggins] (London: William Wesley and Son).

1900

'Sir William Huggins's observatory, Tulse Hill', *MNRAS* **60**, pp. 356–7.

'A suggested explanation of the solar corona', *APJ* **12**, pp. 279–80.

1901

'Motion in the line of sight', *OBS* **24** (1901), p. 459; and *APJ* **14**, p. 369.

1902

'The scientific life', *OBS* **25**, pp. 88–90.

1903

'Sir William Huggins's observatory, Upper Tulse Hill', *MNRAS* **63**, p. 229.
'On the spectrum of the spontaneous luminous radiation of radium at ordinary temperatures' [with M. L. Huggins], *PRS* **72**, pp. 196–9. Reprinted in *APJ* **18** (1903), pp. 151–5 and Plate IV.
'Further observations on the spectrum of the spontaneous luminous radiation of radium at ordinary temperatures' [with M. L. Huggins], *PRS* **72**, pp. 409–13.

1904

'Sir William Huggins's observatory, Upper Tulse Hill', *MNRAS* **64**, p. 328s.
'Address delivered by the president, Sir William Huggins', *PRS* **76**, pp. 1–29.

1905

'Sir William Huggins's observatory, Upper Tulse Hill', *MNRAS* **65**, p. 371.
'On the spectrum of the spontaneous luminous radiation of radium. Part III – Radiation in hydrogen' [with M. L. Huggins], *PRS* **76**, pp. 488–92.
'Address delivered by the president, Sir William Huggins', *PRS* **77**, pp. 100–21.
'On the spectrum of the spontaneous luminous radiation of radium. Part IV – Extension of the glow' [with M. L. Huggins], *PRS* **77**, pp. 132–5.

1906

'Sir William Huggins's observatory, Upper Tulse Hill', *MNRAS* **66**, p. 263.
The Royal Society, or Science in the State and in the Schools (London: Methuen & Co.).

1907

'Sir William Huggins's observatory, Upper Tulse Hill', *MNRAS* **67**, p. 208.

1908

'Sir William Huggins's observatory, Upper Tulse Hill', *MNRAS* **68**, p. 278.

1909

'Sir William Huggins's observatory, Upper Tulse Hill', *MNRAS* **69**, p. 283.
The Scientific Papers of Sir William Huggins [with M. L. Huggins] (London: William Wesley and Son).

1910

'Sir William Huggins's observatory, Upper Tulse Hill', *MNRAS* **70**, p. 331.

Index

Abbott, Francis (1799–1883), 67–9
Abney, William de Wiveleslie (1843–1920), 182, 185, 196, 198, 199, 202, 208, 211, 213–14, 260
Acland, Henry Wentworth (1815–1900), 9, 89
Adelaide Gallery, 29
Airy, George Biddell (1801–92), 8, 53–4, 91, 120–2, 132, 135, 137, 138–9, 140–2, 158–9, 162, 208, 222, 332, 335
  testimony, Devonshire Commission, 140–1
amateur astronomers, 12–13
Anderson, Thomas David (1853–1932), 251
Archer, Frederick Scott (1813–57), 31
Astronomical Register, 15, 52, 144
Astronomy and Astrophysics, 249, 258
Astrophysical Journal, 249, 253
astrophysics
  early development, 11–12
  Huggins on, 1897, 345–6
  turn of the twentieth century, 291–2, 326
Auwers, Arthur von (1838–1915), 66
Ayrton, Hertha (1854–1923), 235

BAAS, presidential addresses, 246
Baily, Francis (1774–1844), 91
Baker, Henry (1698–1774), 205
Ball, Robert Stawell (1840–1913), 94
Barnard, Edward Emerson (1857–1923), 260
Baxendell, Joseph (1815–87), 87–9, 90, 96, 338
Beer, Wilhelm Wolff (1797–1850), 95
Bessel, Friedrich Wilhelm (1784–1846), 13
Birmingham, John (1816–84), 86, 337–8
Birt, William Radcliff (1804–81), 95
Bishop, Sereno Edwards (1827–1909), 202
Bolzano, Bernard (1781–1848), 107
Bond, George Phillips (1825–65), 51, 70
Bond, William Cranch (1789–1859), 31, 38
Bontemps, Georges (1801–82), 127
Bowen, Ira Sprague (1898–1973), 226
Boyle, Robert (1627–91), 15, 16
Boys, Charles Vernon (1855–1944), 280
Bradley, Edward [Cuthbert Bede] (1827–89), 182
Bradley, James (1693–1762), 107
Brashear, John Alfred (1840–1920), 316

Brewster, David (1781–1868), 17, 20, 21, 107, 110
Brodie, Frederick (1823–96), 36
Brothers, Alfred (1826–1912), 159, 165
Browning, John (1835–1925), 114, 154, 155, 159, 171, 332
Buchler and Co., 281
Buijs-Ballot, Christoph Hendrik Diedrik (1817–90), 108–9, 111
Bunsen, Robert (1811–99), 15, 20–2, 187, 330, 333
Burchell, William John (1781–1863), 67

Cameron, Julia Margaret (1815–79), 182
Campbell, William Maxwell, 152
Campbell, William Wallace (1862–1938), 9, 247, 252–4, 258, 293, 312, 316
Cannon, Annie Jump (1863–1941), 291
Carnegie Institution, 292–4
Carrington, Richard Christopher (1826–75), 70, 85
Cassini, Giovanni Domenico (1625–1712), 35–6
Cavendish, William, 7th Duke of Devonshire (1808–91), 137, 139
Cayley, Arthur (1821–95), 15, 142, 144
Chacornac, Jean (1823–73), 67
Chevallier, Temple (1794–1873), 91
Christie, William Henry Mahoney (1845–1922), 122, 208, 222, 245, 312
City of London School, 30
Clark, Alvan (1804–87), 129, 331
Clark, Alvan Graham (1832–97), 256
Clark, George Bassett (1827–91), 256
Clerke, Agnes Mary (1842–1907), 4, 11, 49, 94, 95, 155, 229, 273, 274, 291, 299
Columbian Exposition, 1893, 255
comets
  1843, 32
  1858, Donati's, 39
  b 1881, 224
Comte, Auguste (1798–1857), 14, 329
Congress on Mathematics, Astronomy and Astro-Physics, 1893, 255–6, 294
Connolly, Thomas Francis, 295
Cooke and Sons, 129–30
Cooke, Thomas (1807–68), 39, 127–8, 154, 331

Cornu, Marie Alfred (1841–1902), 258, 261, 344
Cortie, Aloysius Laurence (1859–1925), 300
crater Linné, 95
Crookes, William (1832–1919), 8, 22, 48, 144, 158,
    162–4, 268, 270, 276, 301
  electrical apparatus, 205
  spinthariscope, 276
Curie, Marie Sklodowska (1867–1934), 275–81
Curie, Pierre (1859–1906), 276, 281

Dallmeyer, John Henry (1830–83), 159
Darwin, George Howard (1845–1912), 245, 250
Darwin, Leonard (1850–1943), 212
Davy, Humphrey (1770–1829), 274
Dawes, William Rutter (1799–1868), 113, 325, 331
  mentor to William Huggins, 41
  Royal Astronomical Society, Gold Medal, 38
  willow leaves controversy, 85–6
De la Rue, Warren (1815–89), 8, 9, 22, 31, 77, 85, 86,
    91, 96, 126, 130–1, 134, 138, 150, 153, 155–6,
    159–60, 161, 182, 208
  Mars, 36–7
Descartes, René (1596–1650), 16
Deslandres, Henri-Alexandre (1853–1948), 257
Devonshire Commission, 136–7
Dewar, James (1842–1923), 187, 250, 279–80, 301,
    316, 345
Dewhirst, David, 313
diffraction grating, 19–20
Donati, Giovanni Battista (1826–73), 4, 23, 39, 53,
    58, 340
Doppler, Christian Andreas (1803–53), 6, 23, 99,
    107–9, 274, 339
Doppler's principle, 107–8, 109, 111–14, 120–2,
    303, 306
  tests of, 108–9
Draper, Anna Palmer (1839–1914), 224–5
Draper, Henry (1837–81), 9, 179, 180, 193, 224–5,
    242, 248
  Memorial, 224–5
Draper, John William (1811–82), 9
Dunér, Nils Christoffer (1839–1914), 258, 261, 340
Dunkin, Edward (1821–98), 204
Dupré, August (1835–1907), 73
Dupré, Friedrich Wilhelm (c. 1835–1908), 73
Dyson, Frank Watson (1868–1939), 28, 316
  William Huggins obituary, 311–12

Eddington, Arthur Stanley (1882–1944), 301
Edward VII, King of the United Kingdom and Emperor
    of India (1841–1910), 24, 275, 301
Eginitis, Demetrius (1865–1934), 215
Ennis, Jacob (1807–90), 111
Euler, Leonhard (1707–83), 107

Faraday, Michael (1791–1867), 22
Fath, Edward Arthur (1880–1959), 258
Faye, Hervé (1814–1902), 250
Fizeau, Armand-Hippolyte-Louis (1819–96), 108–9,
    111, 113, 339

Flammarion, Camille (1842–1925), 36
Fleming, Williamina Paton Stevens (1857–1911), 291
Floyd, Richard Samuel (1843–90), 180
Foster, Michael (1836–1907), 280
Fowler, Alfred (1868–1940), 244
Frankland, Edward (1825–99), 268
Fraunhofer lines, 46, 53, 109, 163–4, 331
  British response to, 20–1
  discovery of, 19
  Kirchhoff and Bunsen interpretation of, 21
Fraunhofer, Joseph von (1786–1826), 4, 18–20, 23, 49,
    58, 110, 127, 154, 229, 330–2, 345
  celestial spectra, 19, 47, 49
Fresnel, Augustin Jean (1788–1827), 19
Frost, Edwin Brant (1866–1935), 272, 300

Gassiot, John Peter (1797–1877), 98
Geissler tube, 83
Gill, David (1843–1914), 8, 186, 203–4, 205–6,
    208–14, 222, 227, 256, 259, 274–5, 301, 302, 305,
    307, 308, 309, 310, 324
Gladstone, John Hall (1827–1902), 159
Gladstone, William Ewart (1809–98), 161
Goldschmidt, Herman (1802–66), 88
Good Words, 176–8
Goodwin, Harry Manley (1870–1955), 249
Gopal, Ram Chandra Rao (1833–97), 152
Gould, Benjamin Apthorp (1824–96), 203
Grant, Robert (1814–92), 14
Great Exhibition, 1851, 31, 134–5
Great Grubb Equatorial, 171
  Cambridge University, 296
  removal from Tulse Hill, 297–8
Great Melbourne telescope, 83, 126–7
Greenwich Observatory
  Jupiter's satellites, 137
  motion in the line of sight, 121–2, 222, 311–12
  spectrum analysis, 53–4
  timekeeping, 137, 138
Grubb and Son, 129–30, 131–2
Grubb, Howard (1844–1931), 126–30, 134, 206, 213,
    296–8, 343
Grubb, Thomas (1800–78), 83, 126, 127–8, 130
Guinand, Pierre Louis (1748–1824), 18, 114, 127

HMS *Captain*, 160, 162
Haig, Charles Thomas (1834–1907), 153
Hale, George Ellery (1868–1938), 9, 28, 215, 247–9,
    251, 253, 254–5, 262, 267, 268, 270–1, 272–3,
    274, 278, 292–5, 298–300, 304, 316, 324–6,
    343, 346
  100-in telescope, 304
  Astrophysical Journal, 258–9
  flocculi, 293
  Royal Society of London, Foreign Member, 299
  Rumford spectroheliograph, 293
  Snow telescope, 294
  solar corona out of eclipse, 256–7, 259–60
  solar rotation, 295
  spectroheliograph, 247–8, 256

sunspots, 295
tower telescope, 294–5
Yerkes Observatory, 260–2
Yerkes telescope, 255–6
Halley, Edmond (1656–1742), 67, 104, 114, 315
Hardcastle, Joseph Alfred (1868–1917), 302–3
Harkness, William (1837–1903), 157, 164
Harriot, Thomas (1560–1621), 16
Harrison, James Park, 96
Hartwig, Carl Ernst (1851–1923), 222
Hastings, Charles Sheldon (1848–1932), 301
Hawarden, Clementina Elphinstone, Countess of Rosse (1822–65), 182
helium
    discovery of, 268–9
    spectral lines, 270–1
Herschel, John (1837–1921), 84, 151–2
Herschel, John Frederick William (1792–1871), 8, 13, 22, 65, 66, 87, 150, 151
    Good Words, 177
Herschel, William (1738–1822), 11, 13, 35, 37, 67, 72, 88, 106, 335
    scientific papers, 301, 302–1, 302–3
Hind, John (1823–95), 66, 87, 154
Holden, Edward Singleton (1846–1914), 9, 180, 193, 199, 203, 207, 214, 224–5, 242–3, 245, 246–7, 301, 326
Huggins Observatory, Cambridge University, 298, 312–13
    plaque, 312–13
Huggins, Margaret Lindsay (1848–1915), 29, 170–1, 343
    crafting William Huggins's historical image, 302–13
    death of, 316
    early life, 176
    first notebook entry, 183
    initiative, 184–6
    interest in astronomy, 176
    Larmor, correspondence with, 301–14
    move from Tulse Hill, 313–14
    photography, 182, 183–6
    Wellesley College gift, 9–10, 315
Huggins, William (1824–1910), 24–4, 150
    aurorae, 233–4
    BAAS, President, 246, 249
    BAAS, presidential address, 246–7, 249–51
    Bakerian lecture, 205–6, 209
    chief nebular line, 303, 306–7, 308, 309, 310
    comets, 340–1
    compound spectroscope, 171
    coronagraph, Cape, 206, 208
    coronagraph, prototype, 196–7
    crater Linné, 95
    death of, 300–1
    double-image micrometer, 113
    early life, 29
    education, 29–30
    entrepreneurship, 31, 40–1, 60–1, 82, 98–9, 149, 221, 240, 267, 325–6
    funeral of, 301

Great Grubb Equatorial, 134, 149, 343
Hale, relationship with, 248
high-dispersion spectroscope, 114
historical image, 28, 46, 64–5, 105, 149, 176, 180–1, 215, 221, 236, 267, 292, 316
Jupiter, 39
knighthood, 271
Lockyer, ally, 75–6
Lockyer, competitor, 92
Lockyer, criticism from, 233–4, 269–70
Lockyer, response to, 269–70
Mars, 34–5, 36–8
Maxwell, letter from, 112, 113, 274–5, 303, 305, 306–7, 308–9, 310
meteoritic hypothesis, 250–1
meteors, 93–4
Miller, collaboration with, 50–2, 57–8, 98
motion in the line of sight, 120–1
motion in the line of sight, others' response to, 222–3, 338–40
move to Tulse Hill, 33–4
nebular spectra, 334–7
Nova Aurigae, interpretation, 252
Oliveira bequest, 129
Order of Merit, 275
photography, 1863, 59, 179, 180–1, 333–4
priority concerns, 193–4
Royal Astronomical Society, Fellow, 33
Royal Astronomical Society, Gold Medal, 98, 204
Royal Astronomical Society, President, 246
Royal Microscopical Society, Fellow, 31
Royal Society of London, Fellow, 78
Royal Society of London, President, 267, 273–4
Royal Society of London, Royal Medal, 98
Saturn, 40, 50–1
self-registering spectroscope, 163–4
solar corona, theory of, 205
solar prominences, 92–3, 154–6, 342–3
spectra, maps of, 56–7, 334
spectra, terrestrial metals, 54
star colour, 110
stellar evolution, 250
T Coronae, 337–8
telescope, 1842, 32
telescope, 1853, 33
telescope, 1858, 39, 127, 132
thermometrics, 96–8
thermopile, 97–8
Yerkes Observatory, 260–1
Huggins, William and Margaret
    Atlas of Representative Spectra, 7, 272, 298
    chief nebular line, 226, 229–32
    collaboration, 184, 186–7, 235
    Comet *b* 1881, 228–9
    correspondence, 8–9, 324–5
    first co-authored paper, 232–3
    marriage, 178–9
    meeting, 178
    memorial, St Paul's Cathedral, 316
    photography, 180, 181, 343–5

Huggins, William and Margaret (cont.)
  photography, Orion nebula, 193
  photography, solar corona out of eclipse, 194–8, 259–60
  radium glow experiments, 276–86
  S Andromedae, 223
  Scientific Papers of Sir William Huggins, 298–300
Hussey, William Joseph (1862–1926), 293
Huygens, Christian (1629–95), 35, 107

Inquirer, 279–80
International Union for Co-operation in Solar Research, 1904, 294
International Union for Co-operation in Solar Research, 1910, 304

Janssen, Pierre Jules (1824–1907), 153, 154–6, 170, 261, 342

Kapteyn, Jacobus Cornelius (1851–1922), 214
Kayser, Heinrich (1853–1940), 253–4
Keeler, James Edward (1857–1900), 9, 243–6, 247, 257, 262, 340
Kenwood Physical Observatory, 247
Kew Gardens, solar observations, 138
Kincaid, Sidney Bolton (1849–98), 89, 90, 112–13
  Metrochrome, 112
Kirchhoff and Bunsen
  'Chemical analysis by spectrum-observations', 48
  apparatus, 15, 55
Kirchhoff, Gustav (1824–87), 4, 6, 15, 46, 47, 55, 158, 187, 330–2, 334
  radiation law, British response to, 21–2
Klinkerfues, Friedrich Wilhelm (1827–84), 111, 339

Ladd, William (1815–85), 171
Langley, Samuel Pierpont (1834–1906), 247
Larmor, Joseph (1857–1942), 8, 235, 236, 267, 273, 275, 276–86, 296, 298–9, 300, 301–13, 316, 325
Lassell, William (1799–1880), 9, 51, 126, 132, 158–9, 196, 372
Lawrance, H. A., 197–8
LeSueur, Albert Adolphus Adalbert (1849–1906), 126
Leverrier, Urbain Jean Joseph (1811–77), 66
Lick Observatory, 203, 242–3, 245, 247, 252–3, 260, 293, 316, 331
Lick, James (1796–1876), 180, 203, 240, 242
Linné, Carl von (1707–78), 95
Lister, Joseph (1827–1912), 273, 275, 328
Liveing, George Downing (1827–1924), 187, 241–4, 250, 345
Lockyer, Joseph Norman (1836–1920), 7, 8–9, 53, 69, 70, 75, 99, 110–11, 143–5, 149, 159, 163, 170, 179, 197, 230, 240–1, 243–7, 249–51, 254, 258, 262, 272, 295, 342, 360
  aurorae, 233
  Bakerian lecture, 227
  chief nebular line, 232–4
  early solar observations, 92–3
  helium, 268–9

meteoritic hypothesis, 226–9, 240–1
  solar eclipse expedition 1870, 160–1, 165
  solar eclipse expedition 1871, 165
  solar eclipse expedition 1882, 192–3
  solar prominences, 154–6
Lodge, Oliver (1851–1940), 276, 279
Loewy, Benjamin (1831–92), 138
Lohrmann, Wilhelm Gotthelf (1796–1840), 95
Lyot, Bernard (1897–1952), 257

Mädler, Johann Heinrich (1794–1874), 67, 95
Mars
  nineteenth-century observations, 36
  early telescopic observations, 35
Maunder, Edward Walter (1851–1928), 121, 222–3, 311–12, 316
Maxwell, James Clerk (1831–79), 109–10, 111, 112, 115, 120, 274–5, 305, 306, 310, 325, 339
  motion in the line of sight, 113, 305–6, 308–9
McClean, Frank (1837–1904), 274, 345
meteor showers
  1832, 32
  1866, 93–4
Miller, William Allen (1817–70), 4, 8–9, 20, 21, 29, 47–51, 52, 53, 54, 58, 61, 72, 75, 84, 90, 94, 105–6, 109–11, 113, 115, 130, 170, 177, 224, 227, 311, 325, 331–4, 338, 341
  BAAS address, 1845, 48
  BAAS address, 1861, 47, 49
  death of, 171
  on history of spectrum analysis, 15–19
  Pharmaceutical Society soirée, 47, 49
  photography, 180–1
  Royal Astronomical Society, Gold Medal, 98
  T Coronae, 88–9
Moigno, François-Napoléon Marie (1804–84), 109
Molyneux, Samuel (1689–1728), 107
Montefiore, John, 29, 316
Montefiore, Julia, 29, 182, 316
Mt Etna, 256, 257
Mt Krakatoa, 201–2
Mt Wilson Observatory, 293–5, 316
Murray, George, 301
Murray, John (1863–1943), 301

Narrien, John (1782–1860), 13
Nasmyth, James (1808–90), 85
National Academy of Sciences, 292, 294
Nature, 7, 161, 240, 269–70, 273, 286, 295, 307
nebulae
  'Cat's Eye' (NGC 6543), 72, 75
  Andromeda (M31), 70, 73
  chief nebular line, 226
  illustrating observations of, 70–1
  Orion (M42), 66, 69, 70, 82
nebulae, variable
  Hind's, 66, 70
  Merope, 69, 70
  η Argus [Carinae], 67–9
nebulium, 229

Newall, Hugh Frank (1857–1944), 9, 53, 296–8, 301, 312–13, 316, 345
  William Huggins obituary, 303–11
Newcomb, Mary Caroline Hassler (1840–1921), 162
Newcomb, Simon (1835–1909), 161, 180, 194
Newton, Hubert Anson (1830–96), 93–4, 340
Newton, Isaac (1642–1727), 11, 13, 15, 16, 17, 330
Noble, William (1828–1904), 164, 233
novae, 87
  Nova Aurigae, 251–4, 312
  Nova Persei, 254–5
  S Andromedae, 222, 245
  T Coronae, 87, 89–90, 245, 252, 253

observatory assistants, invisibility of, 175
Order of Merit, 275

Paris Exhibition, 1867, 135
Parsons, Lawrence, 4th Earl of Rosse (1840–1908), 9, 96
Parsons, William, 3rd Earl of Rosse (1800–67), 14, 65, 72, 94, 126, 335
Paschen, Heinrich Friedrich (1865–1947), 270
Payne, William Wallace (1837–1928), 249, 258
Pedro II, Emperor of Brazil (1825–91), 23
Peirce, Benjamin (1809–80), 161, 162
photography
  astronomical, 181–2
  dry plate, 182
  solar prominences, 91–2
  wet plate, 181
  women in, 182
Pickering, Edward Charles (1846–1919), 9, 203, 206–7, 340, 345
Pickering, William Henry (1858–1938), 206–7, 209–10, 211
Pigott, Edward (1753–1825), 87
Pike's Peak, 257
Playfair, Lyon (1818–98), 135–6
Pogson, Norman (1829–91), 87
Popular Astronomy, 258
Powell, Eyre Burton (1819–1904), 67
Pritchard, Charles (1808–93), 54, 71, 85, 88, 89, 90, 93, 96, 98, 120, 159, 177
  Good Words, 177–8
Proctor, Richard Anthony (1837–88), 9, 177
proper motion, 104

radium, 275–7
Ramsay, William (1852–1916), 262, 268–9, 270–1, 275, 278–9, 336
Ranyard, Arthur Cowper (1845–94), 210–11
Rayet, Georges (1839–1906), 91
Reader, 53, 69, 70, 75–6, 110–11
Riccò, Annibale (1844–1919), 257, 258
Ritchey, George Willis (1864–1945), 254
Robinson, Thomas Romney (1792–1882), 8, 54, 56, 83, 98, 126–32, 142, 172, 325
Röntgen, Wilhelm (1845–1923), 275

Roscoe, Henry Enfield (1833–1915), 8, 20–3, 48, 89, 193
  lectures on spectrum analysis, 49–50
Royal Astronomical Society
  admission of women, 291
  founding of, 13
  membership 1860–70, 134
  response to Huggins's nebular work, 77–8
Royal Society of London
  Bakerian lecture, 205
  Government Grant Committee, 129, 198–199
  Huggins-Lockyer controversy, 241–2, 243–5
  response to Huggins's nebular work, 76–7
Rücker, Arthur William (1848–1915), 241, 244
Runge, Carl David Tolmé (1856–1927), 270–1
Russell, John Scott Russell (1808–82), 108
Rutherford, Ernest (1871–1937), 276–86
Rutherfurd, Lewis Morris (1816–92), 3–4, 23, 53, 179, 315, 333

Sabine, Edward (1788–1883), 76, 84, 132, 133–4, 151
Schmidt, Johann Friedrich Julius (1824–84), 95, 338
Schröter, Johann Hieronymous (1745–1816), 95
Schuster, Arthur (1851–1934), 8, 173, 192–4, 196, 250, 279, 286, 296
Secchi, Pietro Angelo (1818–78), 23, 53, 82–3, 88, 90–1, 111, 194, 333
See, Thomas Jefferson Jackson (1866–1962) 9, 28
Sestini, Benedetto (1816–90), 106–7, 339
Sharpey, William (1802–80), 92
Sheepshanks, Anne (1789–1876), 291
Sidereal Messenger, 249
Sidgreaves, Walter (1837–1919), 255
Sketch of the Life of Sir William Huggins, 28, 302, 316
Smyth, Charles Piazzi (1819–1900), 111
Smyth, William Henry (1788–1865), 106–7, 110
Société Centrale de Produits Chimiques, 281
Society of Arts, 135, 136
Soddy, Frederick (1877–1956), 276
solar corona, 157–8, 165, 193
  photography, out of eclipse, 257–8
solar eclipse expedition, 1870, 158–65
  American expedition, 160–1
  British government support for, 158–62
  HMS *Urgent*, 162
  self-registering spectroscope, 163
solar eclipses
  1836, 32, 91
  1842, 91
  1860, 91
  1861, 92
  1865, 92
  1867, 92
  1868, 149–54, 157
  1869, 156–8
  1870, 157–65
  1882, 192–3
  1883, 197–8
  1885, 210, 212

solar prominences, 91–3, 153–6
Somerville, Mary (1780–1872), 291, 330
South, James (1785–1867), 131
spectroscope, 11–12
Spottiswoode, William (1825–83), 173, 181
stars, 82
    colours of, 106–7
    α Bootis (Arcturus), 53, 104, 113–14
    α Canis Majoris (Sirius), 14, 47, 52, 53, 58, 67, 104,
        106, 110, 115–20, 122, 180, 183, 222–3, 226,
        333, 344
    α Lyrae (Vega), 274
    α Orionis (Betelgeuse), 49, 52, 53, 54, 58, 88, 110,
        184, 333, 334, 344
    α Tauri (Aldebaran), 52, 53, 58, 104, 333, 334
    β Cygni (Albireo), 110
    β Lyrae (Sheliak), 253, 271
    β Persei (Algol), 87
    γ Cassiopeiae, 90–1, 94
    η Argus [Carinae], 67, 87
    μ Cephei (Garnet Star), 88
    o Ceti (Mira), 87
    R Coronae, 87
stellar aberration, 107
Stewart, Balfour (1828–87), 138, 155, 330
Stokes, George Gabriel (1819–1903), 8, 21, 49, 52, 57,
        84, 110, 121, 128, 130, 136, 150, 155, 159–60,
        195–6, 197–9, 205–6, 207–8, 209–10, 211,
        213–14, 222, 225, 227, 228–9, 232, 246, 250,
        259–60, 271, 324–5, 330
Stone, Edward James (1831–97), 86, 96
Strange, Alexander (1818–76), 8, 121, 135–45, 159
    'On the insufficiency of existing national
        observatories', 137–8
    testimony, Devonshire Commission, 139–40
Strutt, John William, 3rd Baron Rayleigh (1842–1919),
        235, 275, 279, 285
Strutt, Robert John (1875–1947), 285
Struve, Otto Wilhelm von (1819–1905), 39, 66, 70,
        194, 335
Suum Cuique, 279
Swan, William (1818–94), 20

Tacchini, Pietro (1838–1905), 256, 257, 258
Tait, Peter Guthrie (1831–1901), 251
Talbot, William Henry Fox (1800–77), 20, 21
Tanner, Henry Charles Baskerville (1835–98), 153
Tempel, Wilhelm (1821–89), 69
Tennant, James Francis (1829–1915), 149–53, 154
Thalén, Tobias Robert (1827–1905), 268
theodolite, 18
thermometrics, 96–8
Thompson, Sylvanus Phillips (1851–1916),
        301, 345
Thomson, Joseph John (1856–1940), 316

Thomson, William, Lord Kelvin (1824–1907), 275,
        282, 330
Todd, David Peck (1855–1939), 9, 215
Trouvelot, Léopold (1827–95), 211
Tulse Hill Observatory
    1856, 34
    1862, 58, 333
    1870, 132, 343
    1910, 298
Tulse Hill observatory notebooks, 9–10, 323–4
    1856, 34, 38
    1866, 84
    1871–74, 173
    1882–86, 221
    motion in the line of sight, 115–20
    Wellesley College, 315
Turner, Herbert Hall (1861–1930), 3–4, 291–2
Tyndall, John (1820–93), 144, 162

Victoria, Queen of the United Kingdom and Empress of
        India (1819–1901), 23, 271–2
Vogel, Hermann Carl (1841–1907), 122, 258, 261, 333,
        334, 340
Vogel, Hermann Wilhelm (1834–98), 151, 153

Webb, Thomas William (1806–85), 36, 72
    on nebulae, 69–70
Weiss, Edmund (1837–1917), 151, 157
Wellesley College, 314–15
Wesley, William Henry (1841–1933), 9, 197–199, 207,
        211, 298, 301, 302–3, 316
Wheatstone, Charles (1802–75), 20, 21
Whewell, William (1794–1866), 13
Whitin Observatory, 314
Whiting, Sarah Frances (1846–1927), 64, 176, 298,
        314–15
willow leaves controversy, 84–6
Wolf, Charles (1827–1919), 91
Wollaston, William Hyde (1766–1828), 16–18, 49,
        110, 330
Woods, Charles Ray (1859–1920), 197–8, 213, 259
    Cape Observatory, 203–4
    Cape photographs, 208–9, 210, 211, 212–13, 214
    Riffel expedition, 199–203
    Riffel photographs, 202–3, 207–8

X-Strahlen, 275

Yerkes Observatory, 254, 292–4, 331, 345–6
Yerkes telescope, 260
Yerkes, Charles Tyson (1837–1905), 240, 256,
        272, 345
Young, Charles Augustus (1834–1908), 9, 157, 163,
        164, 165, 192–3, 206–7, 214, 224–5, 231–2,
        243–4, 246, 261, 340